TRENDS IN THEORETICAL PHYSICS II

TRENDS IN THEORETICAL PHYSICS II

Buenos Aires, Argentina November-December 1998

EDITORS
Horacio Falomir
Ricardo E. Gamboa Saraví
Fidel A. Schaposnik
Universidad Nacional de La Plata, Argentina

AIP

AIP CONFERENCE PROCEEDINGS 484

American Institute of Physics

Woodbury, New York

Editors:

Horacio Falomir

Ricardo E. Gamboa Saraví

Fidel A. Schaposnik
Departamento de Fisica
Facultad de Ciencias Exactas
Universidad Nacional de La Plata
C.C. 67, 1900 La Plata
ARGENTINA

E-mail: falomir@obelix.fisica.unlp.edu.ar
quique@dartagnan.fisica.unlp.edu.ar
fidel@athos.fisica.unlp.edu.ar

The paper "The Large N Limit of Superconformal Field Theories and Supergravity" by J. Maldacena is reprinted with permission from Adv. Theor. Math. Phys. **2** (1998) 231–252, published by International Press, P.O. Box 38-2872, Cambridge, MA 02238, U.S.A.

L.C. Catalog Card No. 99-63747
ISBN 1-56396-894-0
ISSN 0094-243X
DOE CONF- 981170

Printed in the United States of America

CONTENTS

PREFACE

The Second Meeting on Trends in Theoretical Physics organized in the framework of a European Community agreement between Santiago de Compostela University, CERN, and La Plata University took place in Buenos Aires from November 29 to December 4, 1998. During that week, physicists from Argentina, Brazil, Chile, and Uruguay met colleagues from Europe and the USA and discussed the latest developments in a wide area of theoretical physics, going from string theory to wimpzillas, from glassy systems to the CFT/AdS correspondence.

The meeting and this volume of lecture notes would not have been possible without the support of the Commission of the European Communities through the scientific cooperation agreement (C11*-CT93-0315) between our group at La Plata University, the Theory Division at CERN, and the Departamento de Física de Partículas de Santiago de Compostela.

We also wish to thank the University of La Plata, who made it possible to invite other lecturers from the USA, Europe, and Latin America, and the International Center of Theoretical Physics, the CONICET, and the SECyT, who allowed the participation of colleagues and students from different centers in Argentina and Latin America. A special thanks to the National Academy of Sciences of Argentina and its president, Dr. Eduardo Gros, for their hospitality and for providing such magnificent premises.

Moreover, we wish to thank Alberto Konig from Affari Viajes for his very efficient work regarding lodging and accommodation of lecturers and participants.

About seventy participants have been able to attend the lectures which are now compiled in this book. The reader will appreciate what we discovered during the beautiful week we spent together in Buenos Aires, namely the effort lecturers made in the quest of clarity. This, together with the benefits derived from discussions throughout the meeting, encouraged us in the plan, at the closing of the conference, of organizing a third meeting to be held in the Far South of Argentina in Tierra del Fuego in April 2000.

Horacio Falomir
Ricardo E. Gamboa Saraví
Fidel A. Schaposnik

PARTICIPANTS

Gerardo Aldazabal, CAB Bariloche.
Carlos Aragao de Carvalho, UFRJ Rio de Janeiro.
Rodrigo Aros, Universidad de Chile.
Máximo Bañados,U. de Zaragoza.
C. Gabriela Beneventano, UNLP.
Daniel Bes, TANDAR Buenos Aires.
Marcelo Ciappina, UNS Bahía Blanca.
Leticia Cugliandolo, École Normale Supérieure.
Hugo Christiansen, CBPF Rio de Janiero.
Adilson da Silva, IFT Sao Paulo.
Marcelo De Francia, UNLP.
José Edelstein, Sgo. de Compostela.
Luis Epele, UNLP.
Horacio Falomir, UNLP.
Huner Fanchiotti, UNLP.
Victoria Fernández, UNLP.
M. Fleck, IF-UFRGS Porto Alegre.
César Fosco, CAB Bariloche.
Jorge Gamboa, Univ. de Santiago de Chile.
Ricardo E. Gamboa Saraví , UNLP.
Carlos García Canal, UNLP.
Paola Giacconi, Univ. di Bologna.
Horacio Oscar Girotti, IF-UFRGS Porto Alegre.
Sonia Gonorazky, UNLP.
Nicolás Grandi, UNLP.
Jorge Griego, Univ. Montevideo.
Gabriela Grunfeld, UNLP.
Alex Kehagias, CERN.
Klaus Kirsten, Leipzig Univ.
José Labastida, Sgo. de Compostela.
Kang Li, Hangzhou Univ.
Ana López, CAB Bariloche.
Gustavo Lozano, UNLP.
Adrián Lugo, UNLP.
Juan Martín Maldacena, Harvard Univ.
Virginia Manías, UNLP.
Javier Mas, Sgo. de Compostela.
Francisco Diego Mazzitelli, Univ. de Buenos Aires.
Fernando Montani, UNLP.
Hugo Montani, CAB Bariloche.
Carlos Naón, UNLP.
Peter van Nieuwenhuizen, SUNY.
Carlos Núñez, UNLP.
Carmen Núñez, IAFE Buenos Aires.
Leonardo Packman, UNLP.
Cecilia von Reichenbach, UNLP.
Antonio Riotto, CERN.
Victor Rivelles, IFT Sao Paulo.
Gerardo Rossini, UNLP.
Joaquín Sánchez Guillén, Sgo. de Compostela.
Mariel Santangelo, UNLP.
Fidel Schaposnik, UNLP.
David Sherrington, University of Oxford.
Guillermo Silva, UNLP.
Martín Schvellinger, UNLP.
Roberto Soldati, Univ. di Bologna.
Jorge Solomin, UNLP.
Pablo Sisterna, Univ. Mar del Plata.
Roberto Trinchero, CAB Bariloche.
Marta Trobo, UNLP.
Ricardo Troncoso, CECS/ Univ. Santiago de Chile.
Jorge Zanelli, CECS/Univ. Santiago de Chile.

* * *

CHERN-SIMONS GAUGE THEORY: TEN YEARS AFTER

J. M. F. Labastida

Departamento de Física de Partículas
Universidade de Santiago de Compostela
E-15706 Santiago de Compostela, Spain
e-mail: labasti@fpaxp1.usc.es

Abstract.
A brief review on the progress made in the study of Chern-Simons gauge theory since its relation to knot theory was discovered ten years ago is presented. Emphasis is made on the analysis of the perturbative study of the theory and its connection to the theory of Vassiliev invariants. It is described how the study of the quantum field theory for three different gauge fixings leads to three different representations for Vassiliev invariants. Two of these gauge fixings lead to well known representations: the covariant Landau gauge corresponds to the configuration space integrals while the non-covariant light-cone gauge to the Kontsevich integral. The progress made in the analysis of the third gauge fixing, the non-covariant temporal gauge, is described in detail. In this case one obtains combinatorial expressions, instead of integral ones, for Vassiliev invariants. The approach based on this last gauge fixing seems very promising to obtain a full combinatorial formula. We collect the combinatorial expressions for all the Vassiliev invariants up to order four which have been obtained in this approach.

I INTRODUCTION

The connection between Chern-Simons gauge theory and the theory of knot and link invariants was established by Edward Witten ten years ago [1]. Since then the theory has been studied from a variety of points of view. Many of the standard methods in field theory have been applied, generating results which became important in the development of knot theory. The interplay between quantum field theory and knot theory has been very rich in both directions. Though the results have been more spectacular in the knot theory direction, one must not forget that the developments in Chern-Simons gauge theory have constituted a constant test of our knowledge in quantum field theory. In fact, it has been found that not always the quantum field theory methods have been able to provide the right answer. As it will be described in detail, the work of the last few years reveals that there are some issues which are not yet understood when dealing with non-covariant gauges.

The term Chern-Simons theory appears in different contexts of quantum field theory. In this conference these words have been heard at least in half of the talks. It is therefore convenient to specify which type of Chern-Simons gauge theory I will be dealing with in this paper. I will refer by the term Chern-Simons gauge theory to a quantum field theory in three-dimensions whose action is the integral on a smooth compact boundaryless three-manifold of a Chern-Simons form associated to a semi-simple non-abelian gauge group. This theory was originally considered by several authors [2], but only after the work by Witten its connection with the theory of knot and link invariants was discovered. This occurred in the summer of 1988. Some of the other theories also named by the term Chern-Simons have been recently reviewed in [3].

The presentation contained in this paper will not follow a chronological order. The development of the theory of knot and link invariants from a mathematical point of view in the last fifteen years have been very impressive and at some stages it has developed parallel to Chern-Simons gauge theory. I will try to make a correspondence from each side for each of the topics treated. Though I have witnessed the development of Chern-Simons gauge theory in detail during these ten years, I might have missed some of the corresponding achievements from the mathematical side. I apologize in advance if some omission in this respect takes place. The order in the

CP484, *Trends in Theoretical Physics II*, edited by H. Falomir, R. E. Gamboa Saraví, and F. A. Schaposnik

presentation is chosen in such a way that the progress made in these ten years can be understood starting from the basics of both, knot and Chern-Simons gauge theory. At each stage the results obtained in the context of Chern-Simons gauge theory are interpreted in the context of knot theory from a mathematical point of view.

The study of Chern-Simons gauge theory is an unusual one because it was first analyzed from a non-perturbative point of view. The original paper by Witten presents a series of non-perturbative methods which led him to show that the vevs (vevs) of the relevant operators of the theory are polynomial invariants like the Jones polynomial [4] and its generalization. Other non-perturbative analysis were made a year later and soon some of the first perturbative studies started to appear. However, only some years later, after the advent of Vassiliev invariants [5,6], the importance of the study of the perturbative series expansion was recognized. From a field theory point of view this lack of interest was understandable. Usually, field theorist can grasp only some of the perturbative aspects of their theories. Since in Chern-Simons gauge theory we had a good handle on its exact solution, why should one care about its perturbative series? It turned out that the coefficients of the perturbative series are important invariants. Their study applying perturbation theory led to interesting expressions for them. The invariance of these coefficients was clear from the beginning. What was not obvious is that they were invariants with a very special feature, they were Vassiliev invariant or invariants of finite type. Though this was known since the work by Bar-Natan [7] and Birman and Lin [8] in 1993, its proof in a quantum field theory context had to wait until 1997 [9].

Once the importance of the perturbative series expansion was realized several works addressed its study. Again the richness inherent to quantum field theory become a powerful tool and the form of this series expansion was studied for different gauge-fixings. The pioneer perturbative calculations in the covariant Landau gauge [10,11] were later extended and analyzed from a general point of view [12,13]. All these works constituted part of the inspiration for the formulation by Bott and Taubes of their configuration space integral [14]. Their integral corresponds precisely to the perturbative series expansion of the vev of a Wilson loop in Chern-Simons gauge theory in the Landau gauge. Before the work by Bott and Taubes, Kontsevich presented a different integral [15] for Vassiliev invariants. This integral turned out to correspond to the perturbative series expansion of the vev of a Wilson loop in the non-covariant light-cone gauge [16–18]. The interplay between physics and mathematics was very fruitful in these developments. This is rather clear in the case of the covariant Landau gauge. In the case of the light-cone gauge, the Chern-Simons counterpart took place much later than the formulation of the Kontsevich integral. However, as stated in the paper by Kontsevich, some of his insight to write down his integral originated from Chern-Simons gauge theory. A very recent work [19] shows from a mathematical point of view that both, the Kontsevich integral and the Bott and Taubes configuration space integral lead to the same invariants. From a field theory point of view this is just a consequence of the gauge invariance of the theory.

Gauge invariance is a powerful tool: it allows to study the theory for different gauge fixings. In the last few years a new gauge fixing has been considered: the non-covariant temporal gauge [16,20]. This gauge has the important feature that the integrals which are present in the expressions for the coefficients of the perturbative series expansion can be carried out, leading to combinatorial expressions [20]. This has been shown to be the case up to order four and it seems likely that the approach can be generalized. In this analysis a crucial role is played by the factorization theorem for Chern-Simons gauge theory proved in [21]. The resulting expressions are better presented when written in terms of Gauss diagrams for knots [22]. Some recent results from the mathematical side seem to indicate the existence of a combinatorial formula of this type [23]. At present Chern-Simons gauge theory is the only approach which have provided combinatorial expressions for all the Vassiliev invariants up to order four. Work is in progress from both sides to obtain a general combinatorial expression. Hopefully, a coherent interplay between them will provide the widely searched general combinatorial expression for Vassiliev invariants.

The paper is organized using the following table as a guide. I have listed on the left hand side the mathematical counterparts of the topics of Chern-Simons gauge theory listed on the right hand side. The sections of the paper will deal with the development of these topics from the point of view of Chern-Simons gauge theory, indicating the basic details from its knot theory counterpart. Notice that the entry in the knot-theory column corresponding to the temporal gauge has not been filled in yet.

Knot Theory	Chern-Simons Gauge Theory
Knots and links	Wilson loops
Knot and link polynomial invariants	Vevs of products of Wilson loops
Singular knots	Operators for singular knots
Invariants for singular knots	Vevs of the new operators
Finite type or Vassiliev invariants	Coeffs. of the perturbative series
Chord diagram	First coeff. of the perturbative series
{1T,4T} and {1T,AS,IHX,STU}	Lie-algebra structure of group factors
Configuration space integral	Landau gauge
Kontsevich integral	Light-cone gauge
??	Temporal gauge

The paper is organized as follows. In sect. 2, Chern-Simons gauge theory is introduced, as well as the basics of the theory of knots and links. A brief summary of some of the results obtained in the context of non-perturbative Chern-Simons gauge theory is also included in this section. In sect. 3, singular knots are introduced and their corresponding operators are defined. These operators are studied and their connection to chord diagrams is discussed. In sect. 4, Chern-Simons gauge theory is studied from a perturbative point of view. Using the covariant Landau gauge the coefficients of the perturbative series expansion are analyzed, obtaining configuration space integrals. In sect. 5, the perturbative series expansion is reobtained in the non-covariant light-cone gauge. The resulting series is the same as the Kontsevich integral for Vassiliev invariants. In sect. 6, the perturbative series expansion is analyzed in a different non-covariant gauge, the temporal gauge. The general procedure to obtain combinatorial expressions for Vassiliev invariants is presented and their explicit form is given for all the primitive invariants up to order four. Finally, in sect. 7, some concluding remarks are presented. Two tables list all the primitive Vassiliev invariants up to order four for all prime knots up to nine crossings.

II KNOTS, LINKS AND WILSON LOOPS

Let us begin recalling the basic elements of Chern-Simons gauge theory. This theory is a quantum field theory whose action is based on the Chern-Simons form associated to a non-abelian gauge group. The fundamental data in Chern-Simons gauge theory are the following: a smooth three-manifold M which will be taken to be compact, a gauge group G, which will be taken semi-simple and compact, and an integer parameter k. The action of the theory is the integral of the Chern-Simons form associated to a gauge connection A corresponding to a gauge group G:

$$S_{\rm CS}(A) = \frac{k}{4\pi}\int_M {\rm Tr}(A\wedge dA + \frac{2}{3}A\wedge A\wedge A). \tag{II.1}$$

In this expression the trace is taken in the fundamental representation of the gauge group. The action possesses the following behavior under gauge transformations,

$$A \to A + g^{-1}dg, \tag{II.2}$$

$$S_{\rm CS}(A) \to S_{\rm CS}(A) - 4\pi y k \omega(g), \tag{II.3}$$

where g is a map $g: M \to G$, $\omega(g)$ is its winding number,

$$\omega(g) = \frac{1}{48\pi^2 y}\int_M {\rm Tr}(g^{-1}dg\wedge g^{-1}dg\wedge g^{-1}dg), \tag{II.4}$$

and y the Dynkin index of the fundamental representation of G. Since the winding number (II.4) is an integer and y is a half-integer, it follows that, for integer k, the exponential,

$$\exp(iS_{\rm CS}(A)), \tag{II.5}$$

which is the quantity that enters in the computation of vevs, is invariant under gauge transformations.

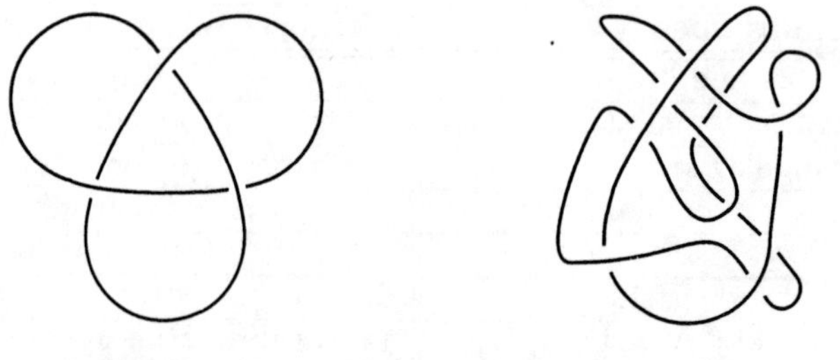

FIGURE 1. Two regular projections of the trefoil knot.

A theory characterized by an action like (II.1), which is independent of the metric on the three-manifold M, is a topological quantum field theory. In this theory, appropriate observables lead to vevs which correspond to topological invariants. Candidates to be observables of this type have to satisfy two properties. On the one hand they must be metric independent, on the other hand they must be gauge invariant. Wilson loops verify these two properties and they are therefore the paramount observables to be considered in Chern-Simons gauge theory.

Wilson loop operators correspond to the holonomy of the gauge connection A along a loop. Given a representation R of the gauge group G and a 1-cycle γ on M, it is defined as:

$$W_\gamma^R(A) = \mathrm{Tr}_R(\mathrm{Hol}_\gamma(A)) = \mathrm{Tr}_R \mathrm{P} \exp \int_\gamma A. \qquad \text{(II.6)}$$

Products of these operators are the natural candidates to obtain topological invariants after computing their vev. These vevs are formally written as:

$$\langle W_{\gamma_1}^{R_1} W_{\gamma_2}^{R_2} \cdots W_{\gamma_n}^{R_n} \rangle = \int [DA] W_{\gamma_1}^{R_1}(A) W_{\gamma_2}^{R_2}(A) \cdots W_{\gamma_n}^{R_n}(A) e^{iS_{CS}(A)}, \qquad \text{(II.7)}$$

where γ_1, γ_2,…,γ_n are 1-cycles on M and R_1, R_2 and R_n are representations of G. In (II.7), the quantity $[DA]$ denotes the functional integral measure and it is assumed that an integration over connections modulo gauge transformations is carried out. As usual in quantum field theory this integration is not well defined. Field theorists have elaborated a variety of methods to go around this problem and provide some meaning to the right hand side of (II.7). These methods fall into two categories, perturbative and non-perturbative ones, and their degree of success mostly depends on the quantum field theory under consideration. Fortunately, in Chern-Simons gauge theory these methods have been very fruitful. Indeed, in the pioneer work by Witten in 1988 he showed, using non-perturbative methods, that when one considers non-intersecting cycles γ_1, γ_2,…,γ_n without self-intersections, the vevs (II.7) lead to the polynomial invariants discovered a few years before starting with the work by V. F. Jones [4]. But before making a precise statement on what was achieved by Witten in [1] let us go through our first mathematical detour and collect some basic facts on knot theory.

Knot theory studies embeddings $\gamma : S^1 \to M$. Two of these embeddings are considered equivalent if the image of one of them can be deformed into the image of the other by an homeomorphism on M. The main goal of the theory is to classify the resulting equivalence classes. Each of these classes is a knot. Most of the study in knot theory has been carried out for the simple case $M = S^3$. This is the situation which has been widely studied from a Chern-Simons theory point of view. Chern-Simons gauge theory, however, being a formulation intrinsically three-dimensional, provides a framework to study the case of more general three-manifolds M. In this respect, Chern-Simons gauge theory seems more promising than other approaches whose formulation possesses a two-dimensional flavor.

A powerful approach to classify knots is based on the construction of knot invariants. These are quantities which can be computed taking a representative of a class and are invariant within the class, *i.e.*, are invariant under continuous deformations of the representative chosen. At present, it is not known if there exist enough knot invariants to classify knots. Vassiliev invariants are the most promising candidates but is already known that if they do classify, infinitely many of them are needed.

The problem of the classification of knots in S^3 can be reformulated in a two-dimensional framework using regular knot projections. Given a representative of a knot in S^3, deform it continuously in such a way that the projection on a plane has simple crossings. Draw the projection on the plane and at each crossing use the convention that the line that goes under the crossing is erased in a neighborhood of the crossing. The resulting diagram is a set of segments on the plane, containing the relevant information at the crossings. Two diagrams corresponding to two regular knot projections of the trefoil knot are shown in fig. 1. A given knot might have many regular projections. The first question to ask is if one can define an equivalence relation among regular projections whose equivalence classes coincide with the equivalence classes of knots. Reidemeister answered this question in the affirmative many years ago. He proved that the problem of classifying knots was equivalent to the problem of classifying knot projections modulo a series of relations among them. These relations, known as Reidemeister moves, are the ones shown in fig. 2. Invariance of a quantity under the three Reidemeister moves is called invariance under ambient isotopy. If a quantity is invariant under all but the first is said to possess invariance under regular isotopy. It is instructive to find out the way that the Reidemeister moves can be applied to the two diagrams shown in fig. 1 to show that, indeed, they are equivalent.

The formalism described for knots generalizes to the case of links. For a link of n components one considers n embeddings, $\gamma_i : S^1 \to M$, $i = 1, \ldots, n$, with no intersections among them. Again, the main problem that link theory faces is the problem of their classification modulo homeomorphisms on M. In this case one can also define regular projections and reformulate the problem in terms of their classification modulo the Reidemeister moves shown in fig. 2. Notice that $n = 1$ is just the case which corresponds to knots.

The study of knot and link invariants experimented important progress in the eighties. In 1984 V. F. Jones [4] discovered a new invariant which strongly influenced the field. He formulated the celebrated Jones polynomial, an invariant which was able to distinguish many more knots and links that previous knot invariants. To have a flavor of the type of quantities one is dealing with, let us describe it in some detail.

The Jones polynomial can be defined very simply in terms of skein relations. These are a set of rules that can be applied to the diagram of a regular knot projection to construct the polynomial invariant. They establish a relation between the invariants associated to three links which only differ in a region as shown in fig. 3. Notice that arrows have been provided to each of the segments entering fig. 3. Indeed, the Jones polynomial as well as, in general, the rest of polynomial invariants which will be discussed below are defined for oriented links. Thus, an arrow must be introduced for each of the components of a link. Though polynomial knot invariants are invariant under a reversal of its orientation, link polynomial invariants are not.

If one denotes by $V_L(t)$ the Jones polynomial corresponding to a link L, being t the argument of the polynomial, it must satisfy the skein relation:

$$\frac{1}{t} V_{L_+} - t V_{L_-} = (\sqrt{t} - \frac{1}{\sqrt{t}}) V_{L_0}, \tag{II.8}$$

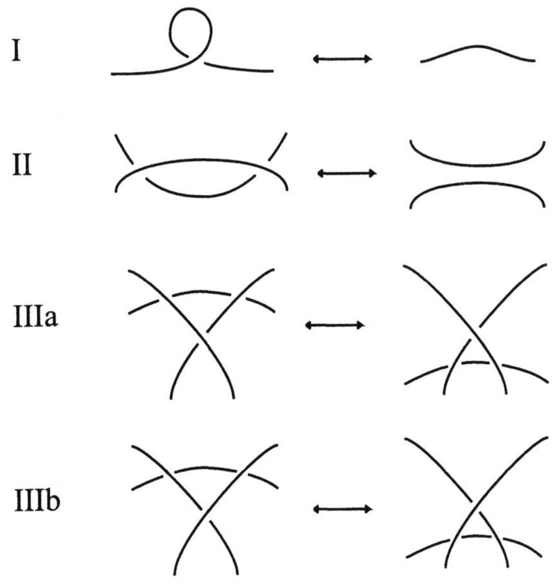

FIGURE 2. Reidemeister moves.

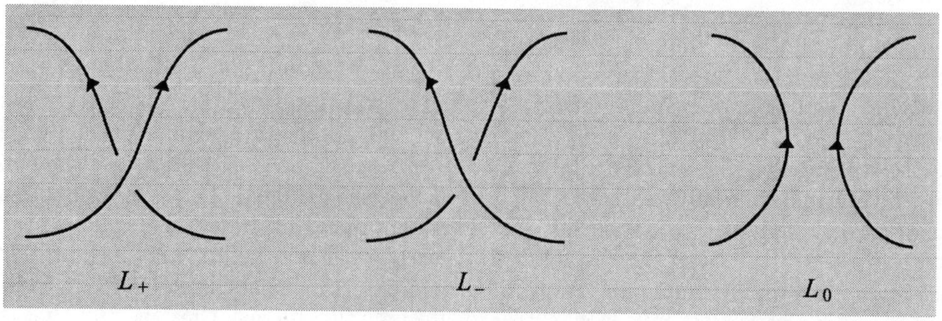

FIGURE 3. Links which enter in the skein rule of the Jones polynomial.

where L_+, L_- and L_0 are the links pictured in fig. 3. This relation plus a choice of normalization for the unknot (U) are enough to compute the Jones polynomial for any link. The standard choice for the unknot is:

$$V_U = 1, \tag{II.9}$$

though it is not the most natural one from the point of view of Chern-Simons gauge theory. For the trefoil knot shown in fig. 1 this polynomial is: $V_T = t + t^3 - t^4$. Notice that actually $V_L(t)$, in general, is not a polynomial. First, it might contain negative powers of t. Indeed, for the mirror image of the trefoil knot shown in fig. 1, $\tilde{T}$, one easily finds that $V_{\tilde{T}} = t^{-1} + t^{-3} - t^{-4}$. Actually this is just an example of a general feature of the Jones polynomial that under a reversal of the orientation of the ambient space S^3 it behaves as $V_{\tilde{L}}(t) = V_L(t^{-1})$, a property which makes it stronger than the Alexander polynomial which was not able two discriminate among links L and $\tilde{L}$ which are mirror images of each other. Second, $V_L(t)$ might contain a factor of the form $\sqrt{t}$. This is always the case if L has an even number of components. For example, for the simplest two-component link, the Hopf link, it takes the form: $V_H = \sqrt{t}(1 + t^2)$.

After Jones work in 1984, many other polynomial invariants where discovered. Two of the most celebrated ones are the HOMFLY [24] and the Kauffman [25] polynomial invariants. The first one possesses a skein rule with three entries similar to the one in fig. 3 and can be computed in the same way. The novelty is that it is a polynomial in two variables. The second one is also a polynomial in two variables but its corresponding skein rule is not as simple as in the Jones or HOMFLY polynomials. In general, the generalizations involve skein rules containing more than three terms and the computation of invariants becomes more complicated. Often other methods have to be used. Before entering in the discussion of the general framework which accounts for all these developments let us turn back to Chern-Simons gauge theory.

The pioneer work by Witten in 1988 showed that the vevs of products of Wilson loops (II.7) correspond to the Jones polynomial when one considers $SU(2)$ as gauge group and all the Wilson loops entering in the vev are taken in the fundamental representation F. For example, if one considers a knot K, Witten showed that,

$$V_K(t) = \langle W_K^F \rangle, \tag{II.10}$$

provided that one performs the identification:

$$t = \exp(\frac{2\pi i}{k+h}), \tag{II.11}$$

where $h = 2$ is the dual Coxeter number of the gauge group $SU(2)$. Witten also showed that if instead of $SU(2)$ one considers $SU(N)$ and the Wilson loop carries the fundamental representation, the resulting invariant is the HOMFLY polynomial. The second variable of this polynomial originates in this context from the N dependence. But these cases are just a sample of the general framework which Chern-Simons gauge theory offers. Taking other groups and other representations one possesses an enormous set of knot and link invariants. For some other special cases the resulting invariants correspond to specific polynomial invariants. For example, in one considers $SO(N)$ as gauge group and Wilson loops carrying the fundamental representation one is led to the Kauffman polynomial. If instead one considers $SU(2)$ as gauge group and the Wilson loops carry a representation of spin $j/2$ one rediscovers the Akutsu-Wadati [26] polynomials. Notice that the formalism allows to consider different representations for each of the components of a link, leading to the so-called colored polynomial invariants.

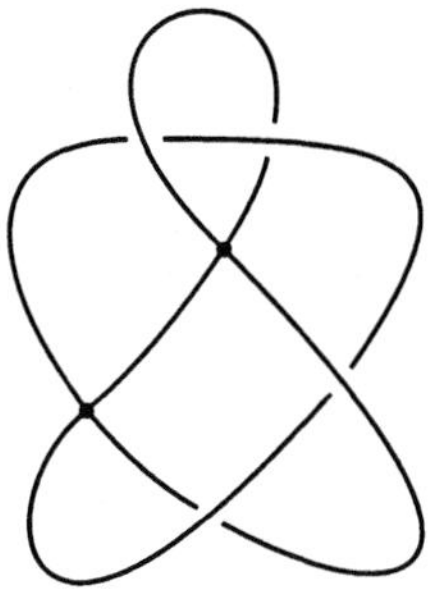

FIGURE 4. Regular projection of a singular knot with two double points.

Chern-Simons gauge theory constitutes a framework where an enormous variety of knot and link invariants can be considered. The theory can be studied for arbitrary groups and arbitrary representations. But on top of this it also provides the basis to study these invariants in more general three-manifolds. In addition, Chern-Simons gauge theory also allows to consider a more general set of observables called graphs [27] that certainly constitutes an important generalization which has not been much exploited from the field theory point of view.

The framework established from Chern-Simons gauge theory has been also formulated from a mathematical point of view. In general, the invariants inherent to Chern-Simons gauge theory have been reobtained and properly defined using contexts different than quantum field theory. This has been a very fruitful arena in the field of algebraic topology in the last ten years. There exist now a quantum group approach to polynomial invariants [28], and the general theory, including knots, links and graphs has been formulated at a categorical level [29].

A particular invariant of three-manifolds which deserves special attention is the partition function of Chern-Simons gauge theory. This quantity is hard to obtain from a field theory point of view. However, it has been properly defined from a mathematical point of view using triangulations of the three-manifold. The resulting invariant is known as the Witten-Reshetekhin-Turaev invariant [30], and it corresponds to the mentioned partition function. The partition function has been studied for some three-manifolds in [31]. The Witten-Reshetekhin-Turaev invariant can be obtained from Chern-Simons gauge theory using lattice gauge theory methods and placing the quantum field theory on a triangulated three-manifold [32]. Recently, new developments based in Chern-Simons gauge theory has led to new formulae for the partition function [33]. The approach seems very promising and it might lead to simpler computational methods for these invariants.

Many non-perturbative studies of Chern-Simons gauge theory were performed in the years following Witten's seminal work. The quantization of the theory was studied from the point of view of the operator formalism [34–36] and from more geometrical methods [37]. Also, its connection to two-dimensional conformal field theory was further elucidated [38] .A powerful method for the general computation of knot and link invariants was constructed by Kaul and collaborators [39]. Methods to compute graph invariants were also built [40]. All these works provided good setups for calculation purposes that in some situations were able to provide the answer to some open questions in knot theory. For example, general expressions for torus knots and torus links were obtained for a variety of situations using the operator formalism [41]. The problem of finding a polynomial invariant which discriminates between the two chiralities of the knots 9_{42} and 10_{71} was solved [42] using the methods developed by Kaul and collaborators. This approach was also used to show that polynomial invariants do not distinguish isotopically inequivalent mutant knots and links [43]. The connection between Chern-Simons gauge theory and rational conformal field theory was used to build knot and link invariants from any conformal field theory [44,45]. Chern-Simons gauge theory has had also important applications in the loop-representation approach to canonical quantum gravity [46,47]. Recently, graphs have also become very important in this context and it turns out that their associated Vassiliev invariants are related to physical states in the framework of canonical quantum gravity [48].

III SINGULAR KNOTS AND THEIR OPERATORS

In this section the mathematical point of view will be discussed first. This will motivate the need to consider vevs of new operators associated to loops with self-intersections. We will describe how Chern-Simons gauge theory leads, using quantum field theory techniques, to the same results as its mathematical counterpart.

In 1990, V. A. Vassiliev [5] introduced a new point of view to study the problem of classification of knots. Based on Arnold's work on singularity theory he studied the space of all smooth maps of S^1 into S^3. This includes maps with various types of singularities which divide the space into chambers, each corresponding to a knot type. Using methods of spectral sequences to obtain combinatorial conditions, Vassiliev constructed families of new invariants to characterize these chambers.

Vassiliev approach was later reformulated by Birman and Lin [8] from an axiomatic point of view. As in the original formulation by Vassiliev [5], the starting point is based on the consideration of singular knots with j double points. A representative of a singular knot with j double points consists of the image of a map from S^1 into S^3 with j simple self-intersections. Under homeomorphisms on S^3 these images form a class which constitute the singular knot itself. These singular knots can also be regularly projected. Fig. 4 shows one of these knots with two double points. The key ingredient in the construction by Birman and Lin is the observation that any knot invariant extends to generic singular knots by the Vassiliev resolution:

$$\nu(K^{j+1}) = \nu(K^j_+) - \nu(K^j_-), \tag{III.1}$$

where K^{j+1} is a singular knot with $j+1$ double points which differs from the knots K^j_+ and K^j_- only in the region shown in fig. 5. Using this extension Birman and Lin characterized the invariants of finite type or Vassiliev invariants introducing the following definition:

A Vassiliev or finite type invariant of order m is a knot invariant which is zero on the unknot and that, after extending it to singular knots, it is zero on singular knots K^j with $j > m$ singular points, and different from zero on some K^m.

Vassiliev invariants form a vector space $\mathcal{V}$. Linear combinations of Vassiliev invariants are also Vassiliev invariants. Actually, to a Vassiliev invariant of order m one can always add Vassiliev invariants of lower order obtaining additional Vassiliev invariants of order m. It is convenient to consider a filtration in the vector space of Vassiliev invariants using the order as grading. If one denotes by $\mathcal{V}_m$ the space of Vassiliev invariants of order m one has the filtration,

$$\mathcal{V}_0 \subseteq \mathcal{V}_1 \subseteq \cdots \subseteq \mathcal{V}_m \subseteq \cdots \subset \mathcal{V}, \tag{III.2}$$

which leads to the graded vector space:

$$\mathrm{gr}(\mathcal{V}) = \mathcal{V}_0 \oplus \mathcal{V}_1/\mathcal{V}_0 \oplus \mathcal{V}_2/\mathcal{V}_1 \oplus \cdots \oplus \mathcal{V}_{m+1}/\mathcal{V}_m \oplus \cdots \tag{III.3}$$

Besides introducing an axiomatic approach to Vassiliev invariants, Birman and Lin proved an important theorem in 1993 [8]. Let us consider any polynomial invariant $P_K(t)$ for a knot K. This polynomial could be any of the ones obtained from Chern-Simons gauge theory considering the vev of the corresponding Wilson loop for some group and some representation. Consider now the power series expansion:

$$Q_K(x) = P_K(\mathrm{e}^x) = \sum_{m=0}^{\infty} \nu_m(K) x^m. \tag{III.4}$$

Birman and Lin proved that if one extends the quantities $\nu_m(K)$ to Vassiliev invariants for singular knots using Vassiliev resolution (III.1), then $\nu_m(K)$ are Vassiliev invariants of order m. An immediate consequence

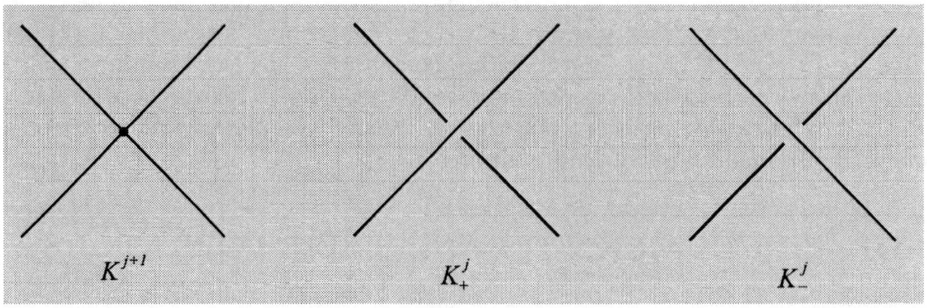

FIGURE 5. Regular projections which enter in the Vassiliev resolution of a double point.

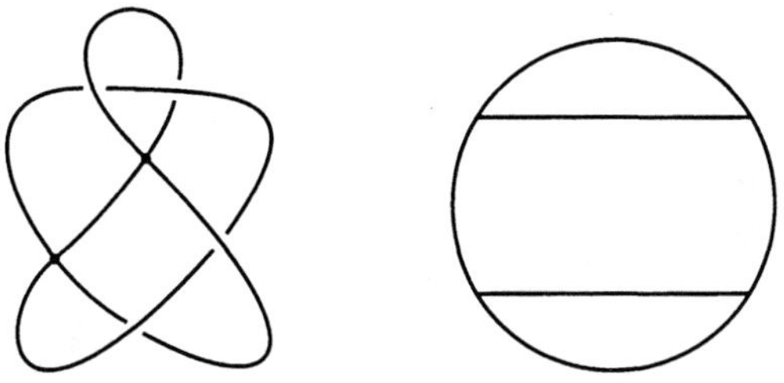

FIGURE 6. A singular knot and its corresponding chord diagram.

of this theorem is that the coefficients of the perturbative expansion associated to the vev of a Wilson loop in Chern-Simons gauge theory are Vassiliev invariants. This property of the coefficients of the perturbative series expansion has been proved using standard quantum field theory methods [9]. Before describing how this has been achieved let us make some remarks on Vassiliev invariants and introduce one of them which is particularly important.

Once Vassiliev or finite type invariants have been introduced, there are two basic questions to ask. The first one is whether or not all the known numerical knot invariants are of finite type. The answer to this question is no. There are classical numerical knot invariants like the unknotting number, the genus or the crossing number which are not Vassiliev invariants. The second question is whether or not, in the case that Vassiliev invariants classify knots, one needs an infinite sequence of Vassiliev invariants to separate knots. This question has been answer in the affirmative.

From a singular knot with m double points one can construct a particular object which determines Vassiliev invariants of order m: its chord diagram [7]. Given a singular knot K^m, its chord diagram, $CD(K^m)$, is built in the following way. Take a base point and draw the preimages of the map associated to a given representative of K^m on a circle. Then join by straight lines the pairs of preimages which correspond to each singular point. In fig. 6 the chord diagram corresponding to the singular knot of fig. 4 has been pictured. It is rather simple to observe that if $\nu(K^m)$ is a Vassiliev invariant of order m then it is completely determined by $CD(K^m)$. Indeed, if one considers two singular knots with m double points, K_1^m and K_2^m, which differ in one crossing change then $CD(K_1^m) = CD(K_2^m)$. On the other hand, by Vassiliev resolution $\nu(K_1^m) - \nu(K_2^m) = \nu(K^{m+1})$. But $\nu(K^{m+1}) = 0$ and therefore all singular knots with m double points leading to the same chord diagram have the same Vassiliev invariant of order m.

Chord diagrams play an important role in the theory of Vassiliev invariants. Since Vassiliev invariants of order m for singular knots with m double points are codified by chord diagrams one could ask if there are as many independent invariants of this kind as chord diagrams. The answer to this question is no. Chord diagrams are associated to knot diagrams and these diagrams must be considered modulo the equivalence relation dictated by the Reidemeister moves in fig. 2. These relations indeed impose some relations among chord diagrams, the so-called 1T and 4T relations [7]. They have been depicted in fig. 7.

Of particular importance is the space of chord diagrams modulo the 1T and 4T relations. This space is a graded space, graded by the number of chords:

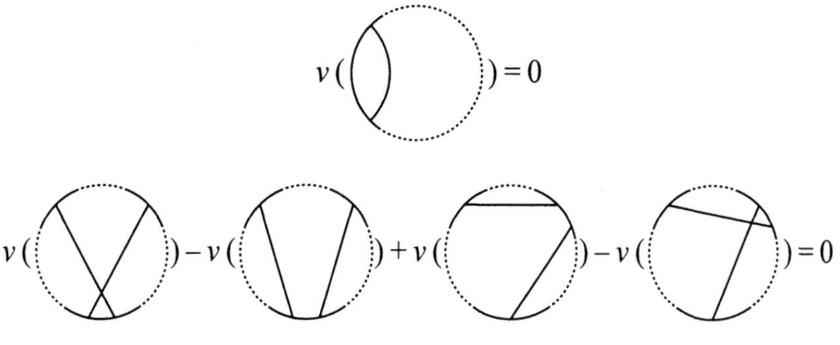

FIGURE 7. 1T and 4T relations among chord diagrams.

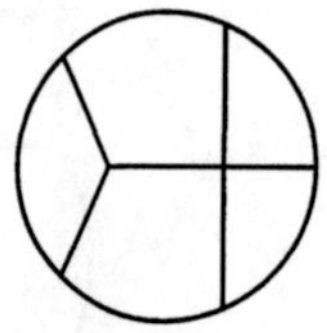

FIGURE 8. Diagram with one trivalent vertex.

$$\begin{aligned} \mathcal{A} &= \Big\{\text{chord diagrams}\Big\} \Big/ (1\text{T},\ 4\text{T}) \\ &= \mathcal{A}_0 \oplus \mathcal{A}_1 \oplus \cdots \oplus \mathcal{A}_m \oplus \cdots \end{aligned} \tag{III.5}$$

Notice that $\mathcal{A}_m$ labels all the Vassiliev invariants of order m for all knots with m double points. The dimensions of the vector spaces $\mathcal{A}_m$ are not known in general. The space $\mathcal{A}$ possesses an algebraic structure that reduces the study of the spaces $\mathcal{A}_m$ to its connected part $\hat{\mathcal{A}}_m$. If $\hat{d}_m$ denotes the dimension of the connected part of $\mathcal{A}_m$, it is known that they take the following values up to order 12 [49]:

m	1	2	3	4	5	6	7	8	9	10	11	12
$\hat{d}_m$	0	1	1	2	3	5	8	12	18	27	39	55

The general expression for the dimensions of the spaces of chord diagrams $\hat{\mathcal{A}}_m$ is an open problem which has challenged many people. As it will be discussed below, these dimensions correspond in fact to the dimensions of the spaces of primitive Vassiliev invariants.

The vector space of chord diagrams, $\mathcal{A}$, can be characterized in an equivalent way using trivalent diagrams an introducing a series of new relations. This characterization is very important because, as it will be shown below, it corresponds to the one that naturally arises from the point of view of Chern-Simons gauge theory. Let us expand the set of chord diagrams to a new set in which trivalent vertices are allowed. This means that now the lines in the interior of the circle can join a point on the circle to a point on one of the internal lines. Fig. 8 shows one of the new allowed diagrams. Notice that the point that looks like a four-valent vertex does not have any particular meaning. The new set of diagrams form a graded vector space whose grading is half the total number of vertices (internal trivalent vertices plus the previous vertices at the attachments of internal lines to the circle). Bar-Natan showed [7] that the previous space $\mathcal{A}$ in (III.5) is equivalent to the new one after modding out by the so-called 1T, AS, IHX and STU relations. The relation 1T is the previous one shown in fig. 7. The relation AS is the statement that the internal trivalent vertices are totally antisymmetric. Finally, the relations STU and IHX are the ones shown in fig. 9. Notice that in the relation STU the curved line corresponds to a piece of the circle of the diagram. The result proved by Bar-Natan in [7] is simply:

$$\mathcal{A} = \Big\{\text{trivalent diagrams}\Big\} \Big/ (1\text{T},\ \text{AS},\ \text{STU},\ \text{IHX}) \tag{III.6}$$

where $\mathcal{A}$ is the space (III.5).

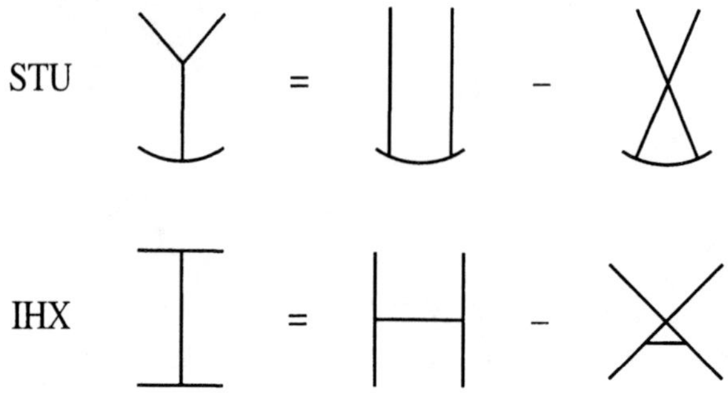

FIGURE 9. STU and IHX relations among trivalent diagrams.

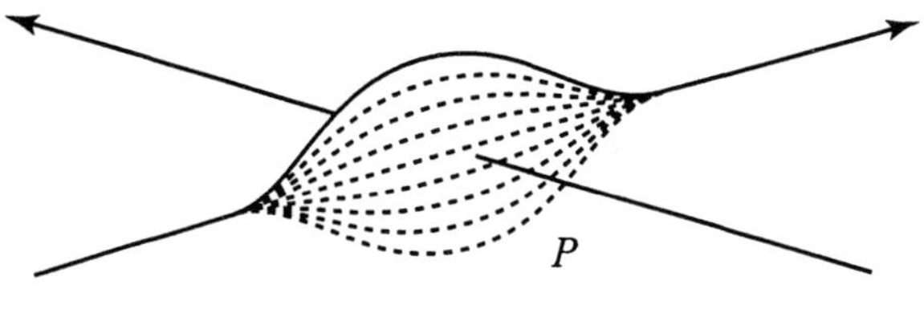

FIGURE 10. Family of paths with a intersection at the point P.

The relations AS, STU and IHX are reminiscent of a Lie-algebra structure. If one assigns totally antisymmetric structure constants f_{abc} to the internal trivalent vertices, and group generators T_a to the vertices on the circle, the STU relation is just the defining Lie-algebra relation,

$$f_{abc}T_c = T_aT_b - T_bT_a, \tag{III.7}$$

while the IHX relation corresponds to the Jacobi identity,

$$f_{abd}f_{dce} + f_{cad}f_{dbe} + f_{bcd}f_{dae} = 0. \tag{III.8}$$

We will find out below that the group factors associated to the perturbative series expansion of the vev of a Wilson loop in Chern-Simons gauge theory correspond precisely to the space $\mathcal{A}$ in (III.5) or (III.6) in its representation in terms of trivalent graphs. In that context, the 1T relation is related to framing. In order to identify this group factors with the space $\mathcal{A}$ one must consider the perturbative series in a formal sense in which group factors are not particularized to a definite group. The group factors must be regarded as objects which simply satisfy relations (III.7) and (III.8) being f_{abc} totally antisymmetric.

As follows from our discussion, singular knots play a central role in the theory of Vassiliev invariants. We must now ask about their role in Chern-Simons gauge theory. The first question to be addressed is about the natural operators that should be associated to them. Wilson loops are related to ordinary knots. What should be replacing the Wilson loop in the case of singular knots while maintaining Vassiliev resolution (III.1)? This question was answered in 1997 [9]. We will review here how this operators for singular knots are obtained using quantum field theory technics. The result will lead us to a proof of the theorem by Birman and Lin discussed above, and to make direct contact with chord diagrams.

We will begin studying the behavior of a difference of Wilson loops as the one that enters in Vassiliev resolution (III.1). Let us consider a family of smooth paths γ_u parametrized by the continuous parameter u, such that for $u = 0$ the path γ_0 possesses a self-intersection at some point P, *i.e.*, for $u = 0$ it has a double point. For $u > 0$ ($u < 0$) the path presents an overcrossing (undercrossing) near the point P. A family of paths with these features has been pictured in fig. 10. The path $\gamma_0(v)$ has a double point at $v = s_1$ and $v = t_1$ with $s_1 < t_1$. Paths $\gamma_u(v)$ with $u \neq 0$ are different from $\gamma_0(v)$ only in the region in parameter space around $v = s_1$. The derivative of $\gamma_u(v)$ with respect to u is only non-zero in that region. It vanishes away from $v = s_1$, in particular at $v = t_1$. In the two resolutions of a double point an overcrossing (undercrossing) corresponds to the one which leads to a crossing with positive (negative) sign, as depicted in fig. 11. Our goal is to study the first derivative of the vev of the Wilson loop $W^R_{\gamma_u}$ with respect to the parameter u. Due to the topological character of Chern-Simons gauge theory, one expects a step-function behaviour for $\langle W^R_{\gamma_u}\rangle$ as a function of u in the neighborhood of $u = 0$. This implies the presence of a delta function in its derivative. As shown below, this is in fact what one finds. Our goal is therefore to express

$$\langle W^R_{\gamma_+}\rangle - \langle W^R_{\gamma_-}\rangle = \int_{-\eta}^{\eta} du \frac{d}{du}\langle W^R_{\gamma_u}\rangle, \tag{III.9}$$

where η is some positive small real number, as the vev of some operator.

To compute the integrand on the right hand side of (III.9) we will use a series of well known properties of the Wilson loop and perform some formal manipulations in the integral functional inherent to vevs. Under a deformation of its path, the Wilson loop behaves as:

$$\frac{d}{du}W^R_{\gamma_u} = \oint dv\, \gamma'^{\mu}_u(v)\dot{\gamma}^{\nu}_u(v) U^R_{\gamma_u}(s,v) F_{\mu\nu}(\gamma(v)) U^R_{\gamma_u}(v,s), \tag{III.10}$$

where $U^R_{\gamma_u}(s,v)$ denotes a Wilson line which starts at s and ends at v, and $F_{\mu\nu}$ is the curvature of the connection A_μ. In (III.10) we have denoted derivatives with respect to the path parameter by a dot and derivatives with respect to u by a prime. Recall that $\gamma'_u(v)$ is only different from zero in the region in parameter space around $v = s_1$. Another important property of the Wilson line is its behavior under a functional derivation with respect to the gauge connection:

$$\frac{\delta}{\delta A^a_\mu(x)} U^R_{\gamma_u}(s,t) = \int_s^t dw\, \dot\gamma^\mu_u(w)\delta^{(3)}(x,\gamma_u(w))U^R_{\gamma_u}(s,w)T^a U^R_{\gamma_u}(w,t). \tag{III.11}$$

Relations (III.10) and (III.11) are common to any gauge theory. What makes Chern-Simons gauge theory special is that, after performing a variation of the action respect to the gauge connection one finds:

$$\frac{\delta}{\delta A^a_\mu(x)} S(A) = \frac{k}{8\pi}\epsilon^{\mu\nu\rho}F^a_{\nu\rho}(x), \tag{III.12}$$

i.e., the field equation involves just the field strength $F_{\mu\nu}$ and not its derivative like in Yang-Mills theory.

Taking into account (III.10) and (III.12), and integrating by parts in connection space, one can write the integrand on the right hand side of (III.9) as:

$$\frac{d}{du}\langle W^R_{\gamma_u}\rangle = \frac{4\pi i}{k}\int [DA] \mathrm{e}^{iS(A)} \oint dv\, \epsilon_{\mu\nu\rho}\gamma'^{\mu}_u(v)\dot\gamma^\nu_u(v)\frac{\delta}{\delta A^a_\rho(\gamma_u(v))}U^R_{\gamma_u}(s,v)T^a U^R_{\gamma_u}(v,t). \tag{III.13}$$

Using (III.11) and disregarding some subtle contributions related to framing [9] one finds:

$$\frac{d}{du}\langle W^R_{\gamma_u}\rangle = \frac{4\pi i}{k}\delta(u)\int [DA] \mathrm{e}^{iS(A)} \mathrm{Tr}\Big[T^a U^R_{\gamma_u}(s_1,t_1)T^a U_{\gamma_u}(t_1,s_1)\Big]. \tag{III.14}$$

As expected this expression involves a delta function in u. This proves in turn that for continuous deformations of the path which do not involve self-crossings the vev of a Wilson loop is invariant.

Using (III.9) and the result (III.14) one can read the operator which one must associate to a singular knot while maintaining Vassiliev resolution (III.1). Indeed, from (III.9) and (III.14) follows that

$$\frac{4\pi i}{k}\Big\langle \mathrm{Tr}\Big[T^a U^R_\gamma(s_1,t_1)T^a U^R_\gamma(t_1,s_1)\Big]\Big\rangle = \langle W^R_{\gamma_+}\rangle - \langle W^R_{\gamma_-}\rangle, \tag{III.15}$$

and therefore $(4\pi i/k)\mathrm{Tr}\Big[T^a U^R_\gamma(s_1,t_1)T^a U^R_\gamma(t_1,s_1)\Big]$ is the sought operator. Notice that in (III.15) the subindex u in the path has been suppressed. One can easily show that this operator is gauge invariant [9]. Its form is rather simple: split at the double point the Wilson loop into two Wilson lines and insert two group generators.

The result obtained for a single double point generalizes easily to deal with the general situation of n double points. Let us consider a singular knot K^n with n double points, and let us assign to each double point i a triple $\tau_i = \{s_i, t_i, T^{a_i}\}$ where s_i and t_i, $s_i < t_i$, are the values of the K^n-parameter at the double point, and T^{a_i} is a group generator. The gauge-invariant operator associated to the singular knot K^n is:

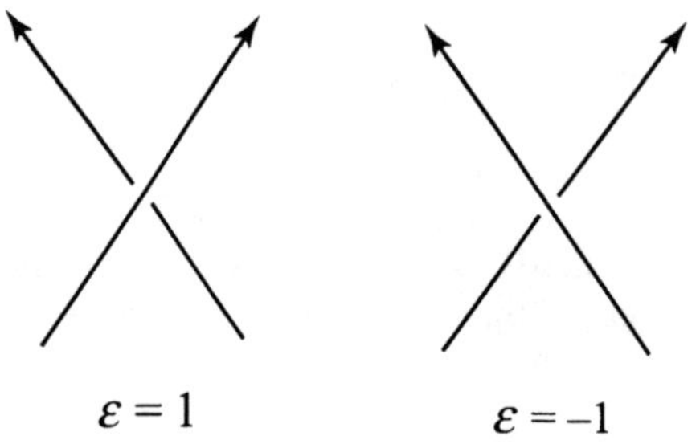

FIGURE 11. Signatures of an overcrossing and an undercrossing.

$$(\frac{4\pi i}{k})^n \mathrm{Tr}\Big[T^{\phi(w_1)}U(w_1,w_2)T^{\phi(w_2)}U(w_2,w_3)T^{\phi(w_3)}\dots$$
$$\dots U(w_{2n-1},w_{2n})T^{\phi(w_{2n})}U(w_{2n},w_1)\Big], \qquad \text{(III.16)}$$

where $\{w_i; i=1,\dots,2n\}$, $w_i < w_{i+1}$, is the set that results from ordering the values s_i and t_i, for $i=1,\dots,n$, and ϕ is a map that assigns to each w_i the group generator in the triple to which it belongs.

The singular operators (III.16) lead to some immediate applications. First, it is easily shown that they constitute a proof of the theorem by Birman and Lin discussed above. This theorem was discussed at the beginning of this section. It states that if in any polynomial knot invariant with variable t one substitutes $t \to \mathrm{e}^x$, and expand in powers of x, the coefficient of the power x^n is a Vassiliev invariant of order n (see eq. (III.4)). A Vassiliev invariant of order n vanishes for all singular knots with $n+1$ crossings. In the language of Chern-Simons gauge theory the theorem by Birman and Lin can be rephrased, simply stating that the nth-order term of the perturbative series expansion of the vev of the Wilson loop associated to a given knot is a Vassiliev invariant of order n. We will now prove that this is so with the help of the operators (III.16).

The vevs of the operators (III.16) provide an invariant for a singular knot with n double points. This singular-knot invariant can be expressed as a signed sum of 2^n invariants for non-singular knots. These 2^n invariants are the perturbative series expansion of the vev of the corresponding Wilson loops. To show that the coefficient of order n of these vevs is a Vassiliev invariant or order n is equivalent to proving that all the terms of order less than n vanish in the signed sum. But the signed sum is precisely the vev of the operator (III.16). Since the perturbative series expansion of these operators starts at order n, the proof of the theorem by Birman and Lin follows.

With the help of the operators (III.16) one makes direct contact with chord diagrams. Let us consider the vevs of the operators (III.16) at order zero. Their expressions are simply obtained by setting $U=1$ in all the Wilson lines. Let us consider a singular knot K^n with n double points, s_i, t_i, with $s_i < t_i$, for $i=1,\dots,n$, in parameter space. As usual, with these data one constructs the triples: $\tau_i = \{s_i, t_i, T^{a_i}\}$. At lowest order in perturbation theory the operators (III.16) become the group factors,

$$v_n(K^n) = (\frac{4\pi i}{k})^n \mathrm{Tr}\Big[T^{\phi(w_1)}T^{\phi(w_2)}T^{\phi(w_3)}\dots T^{\phi(w_{2n})}\Big], \qquad \text{(III.17)}$$

where the set $\{w_1, w_2, \dots, w_n\}$, with $w_j < w_{j+1}$, is obtained by ordering the set $\{s_i, t_i; i=1,\dots,n\}$, and ϕ is the induced map that assigns to each w_j the index of the group generator in the triple to which w_j belongs. The indices entering (III.17) are paired. This allows the association, to each operator (III.17), of a diagram in which the $2n$ points are distributed on a circle and the ones that possess the same value of ϕ are joined by a line. The resulting diagrams are the chord diagrams for singular knots (as the one shown in fig. 6).

The quantities which result after the assignment of Lie-algebra data to chord diagrams are called weight systems [7]. For each system one chooses a group and a representation. As it will become clear in the next section, they correspond to the group theory factors in the context of Chern-Simons perturbation theory.

IV VASSILIEV INVARIANTS AND CHERN-SIMONS PERTURBATION THEORY

Vassiliev invariants of order m for singular knots with m double points have been studied in the previous section. From a mathematical point of view these are chord diagrams and from a Chern-Simons gauge theory point of view they are group factors or weight systems associated to chord diagrams. An important question that was addressed from the mathematical side some years ago was whether or not a weight system can be integrated to construct Vassiliev invariants for non-singular knots. The answer to this question turned to be affirmative. From the side of Chern-Simons gauge theory the positive answer to the question follows simply from the existence of the perturbative series expansion. In this section we will describe some of the basics of Chern-Simons perturbation theory. In the next section we will discuss the appearance of Vassiliev invariants from the integration of weight systems.

The perturbative study of Chern-Simons gauge theory started with the works by Guadagnini, Martellini and Mintchev [10] and by Bar-Natan [11]. These works dealt basically with the lowest non-trivial order in perturbation theory. To construct the perturbative series expansion of the vev of an operator when dealing with a gauge theory one is forced to make a gauge fixing. The first analysis of the Chern-Simons perturbation theory were made in the covariant Landau gauge. Subsequent studies [12,13] in this gauge led to a framework linked to the theory of Vassiliev invariants, which culminated with the configuration space integral approach [14,50]. In the rest of this section we will describe the structure of the perturbative series expansion of the vev of a Wilson loop in the covariant Landau gauge. Chern-Simons perturbation theory in this gauge for the case of more general three-manifolds has been considered in [51].

For a perturbative analysis it is more convenient to rescale the field entering the Chern-Simons action (II.1) in such a way that the coupling constant appears in the three-vertex of the theory. Rescaling the gauge field by

$$A_\mu \to g A_\mu, \tag{IV.1}$$

the Chern-Simons action (II.1) becomes, up to a global numerical factor:

$$S'(A_\mu) = \int \mathrm{Tr}(A \wedge dA + \frac{2}{3} g A \wedge A \wedge A). \tag{IV.2}$$

This form of the action has the standard $\frac{1}{2}$ for the kinetic part since the trace of the fundamental representation is normalized as: $\mathrm{Tr}(T^a T^b) = \frac{1}{2}\delta^{ab}$. Notice that after the rescaling (IV.1) covariant derivatives contain the coupling constant g.

As stated above, to construct the perturbative series a gauge fixing must be performed. We begin choosing a Lorentz gauge in which

$$\partial^\mu A_\mu = 0. \tag{IV.3}$$

The standard Faddeev-Popov construction leads us to consider the following gauge-fixed functional integral:

$$\int [DA_\mu Dc D\bar{c} D\phi] \mathrm{e}^{iS'(A_\mu) + iS_{\mathrm{gf}}(A_\mu, c, \bar{c}, \phi)}, \tag{IV.4}$$

where

$$S_{\mathrm{gf}}(B_\mu, c, \bar{c}, \phi) = \int \mathrm{d}^3x \mathrm{Tr}(2\bar{c}\partial_\mu D^\mu c - 2\phi\partial_\mu A^\mu - \lambda\phi^2). \tag{IV.5}$$

In this action ϕ is a Lagrange multiplier which imposes the gauge condition, c and $\bar{c}$ are anticommuting Faddeev-Popov ghosts, and λ is a gauge-fixing parameter. The derivative in $D_\mu c$ is a covariant derivative. The functional integral in (IV.4) must be done over all A_μ configurations and not only over gauge orbits. As a result of the gauge fixing, the exponent in (IV.4) is invariant under the corresponding BRST transformations.

The field ϕ can be integrated out by performing a Gaussian integration in (IV.4). One finds, up to an irrelevant multiplicative factor, that the functional integral (IV.4) becomes:

$$\int [DA_\mu Dc D\bar{c}] \mathrm{e}^{iI(A_\mu, c, \bar{c})}, \tag{IV.6}$$

where

$$I(A_\mu, c, \bar{c}) = \int \mathrm{Tr}\big[\varepsilon^{\mu\nu\rho}(A_\mu \partial_\nu A_\rho + \frac{2}{3} g A_\mu A_\nu A_\rho) - \frac{1}{\lambda} A_\mu \partial^\mu \partial^\nu A_\nu + 2\bar{c}\partial_\mu D^\mu c\big]. \tag{IV.7}$$

In order to compute the vevs of operators involving Wilson loops like in (II.7) one must integrate these operators using (IV.6). After expanding the path-ordered products in the Wilson loops one ends computing the functional integral over products of gauge fields. These are basically the correlation functions of the theory and are computed using standard quantum field theory methods, which lead to a set of computational rules called Feynman rules. These rules provide terms which are organized in even powers of g. They dictate that at order g^{2m} two kinds of diagrams must be taken into account. The first group involves oriented trivalent graphs of the kind introduced in the previous section with $2m$ vertices and with the following assignments: for each internal line (gauge propagator),

$$\frac{i}{4\pi} \delta_{ab} \epsilon^{\mu\nu\rho} \frac{(x-y)_\rho}{|x-y|^3}, \tag{IV.8}$$

for each internal vertex,

$$-igf_{abc}\epsilon_{\mu\nu\rho}\int d^3x, \tag{IV.9}$$

and for each vertex on the circle,

$$g(T^a)_i^j. \tag{IV.10}$$

The second set of diagrams involves propagators and vertices related to ghost fields. Contrary to gauge field propagators, dashed lines are used for ghost field propagators. The diagrams of the second set are also trivalent diagrams but they must contain at least one dashed line. Dashed lines can not be attached to the circle. They are always attached to internal lines via the ghost-gauge field-ghost trivalent vertex. We will not reproduce here the form of the ghost-related ingredients of the Feynman rules. We will describe, however, what is their effect.

In writing the quantities associated to the Feynman rules (IV.8), (IV.9) and (IV.10) the limit $\lambda \to 0$ has been taken. This is known as the Landau gauge. It simplifies the calculations since in this case there are not infrared divergences.

After applying the Feynman rules one finds that the perturbative series contains divergent terms. Since the coupling constant g is dimensionless, the theory is, however, renormalizable. Power counting analysis [53] shows that actually the theory is superrenormalible. To deal with the divergent terms, the theory must be regularized. One can use a variety of regularizations. For most of the regularizations which have been studied it turns out that the theory is in fact finite, *i.e.*, after removing the cutoff introduced in the regularization the resulting expressions are finite. Of course, one is free to choose an arbitrary scheme using a finite renormalization. The value of k would be different in each scheme if its fixed from some "standard data". In other words, the value of k in a given scheme should be determined stating, for example, that the value of the vev of the Wilson loop for the trefoil knot divided by the partition function is a fixed quantity. Then, though working in different schemes, the computations of vevs for any other knot would agree.

Many regularizations have ben studied in the last ten years. There is a particular subset of them [52–55] that naturally leads to a scheme in which a shift in the coupling constant k occurs. In these regularizations the higher-loop contributions to the two- and three-point functions add up to a shift in k which is precisely the dual of the Coxeter number of the gauge group. This is rather remarkable because it is the same shift that appears in non-perturbative studies of Chern-Simons gauge theory in connection with its associated two-dimensional conformal field theory [1]. Why this is so is not still well understood.

In our perturbative analysis we are going to assume that higher-loop corrections to two- and three-point functions just account for the shift in k and therefore we will not consider them in the expansion. We will deal with the rest of the diagrams. The scheme is chosen so that we do not make any finite renormalization. This scheme is as good as any other but is the best to compare perturbative results to non-perturbative ones. As we will discuss in the next sections, in non-covariant gauges we will make a choice in which there is no shift and then, to compare to non-perturbative calculations, we must make a finite renormalization.

Before writing the form of the full perturbative series we must deal with another important subtlety. If one computes the first order contribution to the perturbative series expansion of the vev of a Wilson loop one finds that the resulting quantity is not a topological invariant. In the gauge fixing of the theory we have introduced a metric dependence that could lead to quantities which are not topological. This first order contribution is just a manifestation of it. Fortunately, only in this term, and in its propagation in higher order contributions, topological invariance is lost. The rest of the perturbative series expansion is truly topological. Thus, although vevs are not topological invariant quantities they fail to be so in a controllable way. The non-topological terms factorize and multiply a term which is topological. In non-perturbative studies one finds a related problem which is the need of the introduction of a framing for the knot under consideration [1]. In other words, instead of knots one must consider framed knots, or knots with a normal vector field assigned which defines a framing characterized by an integer. This integer is the number of times that the normal vector field winds around the knot. A framing can also be introduced in the perturbative approach so that the perturbative result coincides with the non-perturbative one. Let us describe how this is done.

The first order contribution to the vev of a Wilson loop has the form:

$$\begin{aligned} & g^2\mathrm{Tr}(T^aT^b)\oint dx^\mu \int^x dy^\nu \frac{i}{4\pi}\delta_{ab}\epsilon^{\mu\nu\rho}\frac{(x-y)_\rho}{|x-y|^3} \\ &= g^2\frac{1}{2}\mathrm{Tr}(T^aT^b)\oint dx^\mu \oint dy^\nu \frac{i}{4\pi}\delta_{ab}\epsilon^{\mu\nu\rho}\frac{(x-y)_\rho}{|x-y|^3}. \end{aligned} \tag{IV.11}$$

This expression is finite but depends on the shape of the knot. Let us introduce a framing and, together with the original knot, a companion knot located at the end point of the normal vector which defines the framing. If one replaces one of the original paths entering (IV.11) by the path associated to the companion knot (which will be denoted by a prime), one finds Gauss formula for the linking number:

$$g^2 \frac{1}{2} \mathrm{Tr}(T^a T^b) \oint dx^\mu \oint' dy^\nu \frac{i}{4\pi} \delta_{ab} \epsilon^{\mu\nu\rho} \frac{(x-y)_\rho}{|x-y|^3} = l. \tag{IV.12}$$

In this expression l is an integer that counts the number of times that the vector associated to the framing winds the knot. The kind of point splitting associated to the framing leads to a perturbative result that agrees with the non-perturbative analysis. Either if we introduce the framing or we leave a non-topological term we obtain a good perturbative series expansion. The corresponding term factorizes. Thus, to deal with one or the other is a matter of taste but, as in the case of the shift, agreement with the non-perturbative analysis induces the approach based on the framing. We will follow this choice. The term that factorizes has the form:

$$\exp(2\pi i l h_R), \quad h_R = \frac{1}{k+g^\vee} \mathrm{Tr}_R(T^a T^a). \tag{IV.13}$$

The quantity h_R can be identified with the conformal weight for the representation R of the associated conformal field theory. The Feynman diagrams that lead to the framing factor are those diagrams which contain isolated chords once a canonical basis from group factors has been chosen. Canonical basis play a prominent role in our discussion but before defining them let us first take a look at the general form of the perturbative series.

Once we have dealt with the issues regarding the shift and the framing we are in the position to analyze the perturbative series expansion of the vev of a Wilson loop. The Feynman rules (IV.8), (IV.9) and (IV.10) allow to split the contributions to each order in two factors: a geometrical factor which includes all the space dependence, and a group factor which includes all the group theoretical dependence. The general form is:

$$\langle W_K^R \rangle = \dim R \sum_{i=0}^{\infty} \sum_{j=1}^{d_i} \alpha_{ij}(K) r_{ij}(R) x^i, \tag{IV.14}$$

where,

$$x = \frac{2\pi i}{k} = i g^2/2 \tag{IV.15}$$

is the expansion parameter, $\dim R$ is the dimension of the representation R, $\alpha_{0,1} = r_{0,1} = 1$, and $d_0 = 1$. Notice that higher-loop corrections to the two- and three-point functions must not be included in (IV.14) so this series should correspond to the non-perturbative result without the shift. The factors $\alpha_{ij}(K)$ and $r_{ij}(R)$ appearing at each order i incorporate all the dependence dictated from the Feynman rules apart from the dependence on the coupling constant, which is contained in x. Of these two factors, in the $r_{ij}(R)$ all the group-theoretical dependence is collected. These are the group factors mentioned above. The rest is contained in the $\alpha_{ij}(K)$ or geometrical factors. They have the form of integrals over the Wilson loop corresponding to the knot K of products of propagators, as dictated by the Feynman rules. The first index in $\alpha_{ij}(K)$ denotes the order in the expansion and the second index labels the different geometrical factors which can contribute at the given order. Similarly, $r_{ij}(R)$ stands for the independent group structures which appear at order i, which are also dictated by the Feynman rules. The object d_i in (IV.14) is the dimension of the space of invariants at a given order. In our approach denotes the number of independent group structures which appear at that order. Notice that while the geometrical factors $\alpha_{ij}(K)$ are knot dependent but group and representation independent, the group factors are group and representation dependent but knot independent.

Among the basis of group factors which can be chosen in the perturbative series (IV.14) there is a special class which turns out to be very useful. We will call it the class of canonical basis. Notice that to each group factor one can associate a trivalent diagram of the kind entering (III.6) if one considers the space of all semi-simple Lie groups (as that is the most general case for which the structure constants can be chosen totally antisymmetric). We will restrict to this case in what follows.

In order to introduce the concept of canonical basis we must first deal with a sort of classification of trivalent diagrams. A trivalent diagram will be called connected diagram if it is possible to go from one propagator (or internal line) to another without ever having to go through the external circle. The diagram will be called

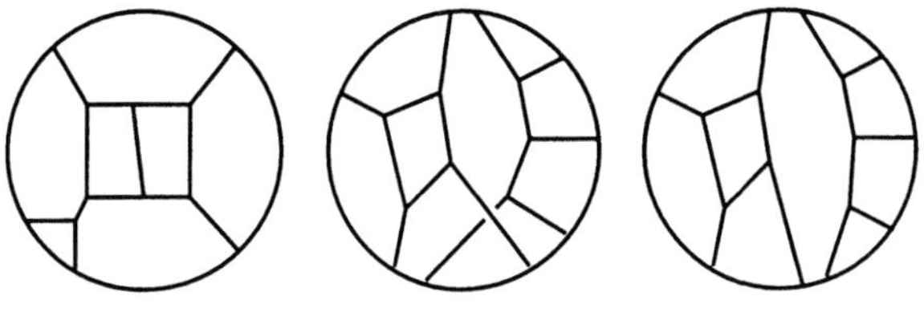

FIGURE 12. Examples of trivalent diagrams: a is connected while b and c are disconnected.

disconnected diagram if that is not possible. In this second case we say that the diagram has subdiagrams, which are the connected components of the whole diagram. Two subdiagrams are non-overlapping if we can move along the external circle meeting all the legs of one subdiagram first, and all the legs of the other second. By legs we mean internal lines attached to the external circle. If it is impossible to do that, the subdiagrams are overlapping. In fig. 12 the diagram a is connected while the diagrams b and c are disconnected. Of these last two, diagram b contains subdiagrams which are overlapping while c does not.

In the expansion (IV.14) there are many possible choices of independent groups factors $r_{ij}(R)$. Given all trivalent diagrams contributing to a given order in perturbation theory some of the resulting group factors might be related due to the Lie-algebra relations AS, STU and IHX as in (III.6). The group factors entering (IV.14) are chosen to be associated to diagrams that constitute a basis. Its elements are the $r_{ij}(R)$ in (IV.14). Of course, many choices are possible. In order to introduce our preferred choice let us first notice that due the STU relations one can always choose a basis such that the $r_{ij}(R)$ come from connected diagrams, or products of connected diagrams. That is, if there are subdiagrams, they can be chosen so that they do not overlap. The value of such an $r_{ij}(R)$ is the product of the values of its subdiagrams. This last statement follows from the fact that the part of the group factor associated to one of the subdiagrams is a diagonal matrix. Thus one can choose a basis built of connected trivalent diagrams and products of connected non-overlapping subdiagrams. A basis with this feature will be called a canonical basis. In fig. 13 a choice of canonical basis up to order six is shown.

We will denote by $r^c_{ij}(R)$ the group factors associated to a canonical basis, and $\alpha^c_{ij}(K)$ the corresponding geometrical factors. It was shown in [21] that then the perturbative series expansion (IV.14) can be written as:

$$\langle W^R_K \rangle = \dim R \exp\left\{ \sum_{i=1}^{\infty} \sum_{j=1}^{\hat{d}_i} \alpha^c_{ij}(K)\, r^c_{ij}(R)\, x^i \right\}, \tag{IV.16}$$

where $\hat{d}_i$ stands for the number of connected elements in the canonical basis at order i. Notice that $\alpha^c_{ij}(K)$ do not correspond uniquely to connected Feynman diagrams. The result (IV.16) is known as the factorization theorem. Actually, it holds for arbitrary gauges, not only in the covariant Landau gauge as one could conclude from our discussion. The dimensions $\hat{d}_i$ are precisely the ones introduced for the connected part of the space of chord diagrams (III.5). Their values up to order 12 are the ones contained in the table shown after (III.5). The geometrical factors $\alpha^c_{ij}(K)$ are a selected set of Vassiliev invariants. They are called primitive Vassiliev invariants. If they are known, the values of the whole set of Vassiliev invariants follow. These primitive Vassiliev invariants have been computed for general classes of knots as torus knots [41,56] up to order six.

The contribution at first order in (IV.16) is precisely the framing factor (IV.13). Thus, the factorization theorem (IV.16) shows also its factorization. The rest of the terms in the exponent of (IV.16) are knot invariants. The series contained in that exponent was analyzed by Bott and Taubes [14] in their work on the configuration space for Vassiliev invariants. They showed that the integral expression entering the geometrical factors $\alpha^c_{ij}(K)$ are convergent [14,50]. Further work on the subject has led to a proof of their invariance [13,57].

The explicit expression for the integrals entering in the second order contribution was first presented in [10]. It was later analyzed in detail by Bar-Natan [11]. It takes the form:

$$\alpha^c_{21}(K) = \frac{1}{4} \oint dx^\mu \int^x dy^\nu \int^y dz^\rho \int^z dw^\sigma \, \Delta_{\mu\rho}(x-z)\Delta_{\nu\sigma}(y-w)$$

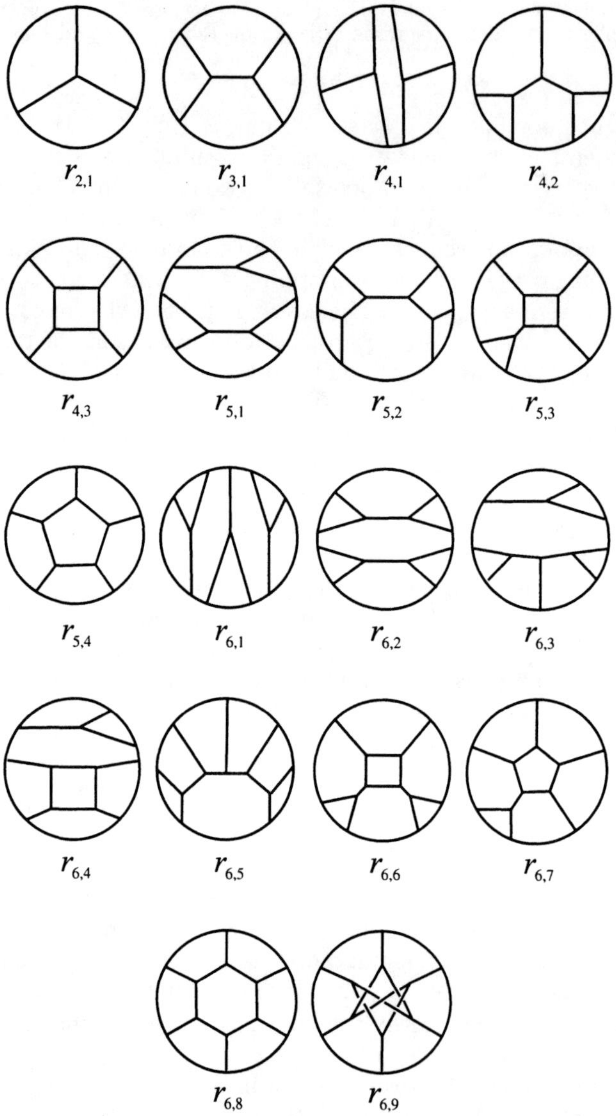

FIGURE 13. Choice of a canonical basis up to order 6.

$$-\frac{1}{16}\oint dx^\mu \int^x dy^\nu \int^y dz^\rho \int d^3\omega\, v_{\mu\nu\rho}(x,y,z;\omega), \tag{IV.17}$$

where,

$$\Delta_{\mu\nu}(x-y) = \frac{1}{\pi}\epsilon_{\mu\rho\nu}\frac{(x-y)^\rho}{|x-y|^3}, \tag{IV.18}$$

and,

$$v_{\mu\nu\rho}(x,y,z;\omega) = \Delta_{\mu\sigma_1}(x-\omega)\Delta_{\nu\sigma_2}(y-\omega)\Delta_{\rho\sigma_3}(z-\omega)\epsilon^{\sigma_1\sigma_2\sigma_3}. \tag{IV.19}$$

This invariant turns out to be the total twist of the knot and coincides mod 2 with the Arf invariant. The integral expression for the order three invariant, $\alpha^c_{31}(K)$, associated to the group factor r_{31} shown in fig. 13 was first presented in [12]. We reproduce it here for completeness:

$$\begin{aligned}
\alpha^c_{31}(K) = {} & \frac{1}{8}\oint dx_1^{\mu_1}\int^{x_1} dx_2^{\mu_2}\int^{x_2} dx_3^{\mu_3}\int^{x_3} dx_4^{\mu_4}\int^{x_4} dx_5^{\mu_5}\int^{x_5} dx_6^{\mu_6} \\
& \big[\Delta_{\mu_1\mu_4}(x_1-x_4)\Delta_{\mu_2\mu_6}(x_2-x_6)\Delta_{\mu_3\mu_5}(x_3-x_5) \\
& +\Delta_{\mu_1\mu_3}(x_1-x_3)\Delta_{\mu_2\mu_5}(x_2-x_5)\Delta_{\mu_4\mu_6}(x_4-x_6) \\
& +\Delta_{\mu_1\mu_5}(x_1-x_5)\Delta_{\mu_2\mu_4}(x_2-x_4)\Delta_{\mu_3\mu_6}(x_3-x_6)\big] \\
& +\frac{1}{4}\oint dx_1^{\mu_1}\int^{x_1} dx_2^{\mu_2}\int^{x_2} dx_3^{\mu_3}\int^{x_3} dx_4^{\mu_4}\int^{x_4} dx_5^{\mu_5}\int^{x_5} dx_6^{\mu_6} \\
& \big[\Delta_{\mu_1\mu_4}(x_1-x_4)\Delta_{\mu_2\mu_5}(x_2-x_5)\Delta_{\mu_3\mu_6}(x_3-x_6)\big] \\
& -\frac{1}{32}\oint dx_1^{\mu_1}\int^{x_1} dx_2^{\mu_2}\int^{x_2} dx_3^{\mu_3}\int^{x_3} dx_4^{\mu_4}\int^{x_4} dx_5^{\mu_5}\int d^3y \\
& \big[\Delta_{\mu_1\nu_1}(x_1-y)\Delta_{\mu_2\mu_5}(x_2-x_5)\Delta_{\mu_3\nu_3}(x_3-y)\Delta_{\mu_4\nu_4}(x_4-y)\epsilon^{\nu_1\nu_3\nu_4} \\
& +\Delta_{\mu_1\mu_3}(x_1-x_3)\Delta_{\mu_2\nu_2}(x_2-y)\Delta_{\mu_4\nu_4}(x_4-y)\Delta_{\mu_5\nu_5}(x_5-y)\epsilon^{\nu_2\nu_4\nu_5} \\
& +\Delta_{\mu_1\nu_1}(x_1-y)\Delta_{\mu_2\mu_4}(x_2-x_4)\Delta_{\mu_3\nu_3}(x_3-y)\Delta_{\mu_5\nu_5}(x_5-y)\epsilon^{\nu_1\nu_3\nu_5} \\
& +\Delta_{\mu_1\nu_1}(x_1-y)\Delta_{\mu_2\nu_2}(x_2-y)\Delta_{\mu_3\mu_5}(x_3-x_5)\Delta_{\mu_4\nu_5}(x_4-y)\epsilon^{\nu_1\nu_2\nu_4} \\
& +\Delta_{\mu_1\mu_4}(x_1-x_4)\Delta_{\mu_2\nu_2}(x_2-y)\Delta_{\mu_3\nu_3}(x_3-y)\Delta_{\mu_5\nu_5}(x_5-y)\epsilon^{\nu_2\nu_3\nu_5}\big] \\
& +\frac{1}{128}\oint dx_1^{\mu_1}\int^{x_1} dx_2^{\mu_2}\int^{x_2} dx_3^{\mu_3}\int^{x_3} dx_4^{\mu_4}\int d^3y\int d^3z \\
& \big[\Delta_{\mu_1\nu_1}(x_1-y)\Delta_{\mu_2\sigma_2}(x_2-z)\Delta_{\mu_3\sigma_3}(x_3-z)\Delta_{\mu_4\nu_4}(x_4-y)\Delta_{\alpha\beta}(y-z)\epsilon^{\nu_1\alpha\nu_4}\epsilon^{\sigma_3\beta\sigma_2} \\
& +\Delta_{\mu_1\nu_1}(x_1-y)\Delta_{\mu_2\nu_2}(x_2-y)\Delta_{\mu_3\sigma_3}(x_3-z)\Delta_{\mu_4\sigma_4}(x_4-z)\Delta_{\alpha\beta}(y-z)\epsilon^{\nu_2\alpha\nu_1}\epsilon^{\sigma_4\beta\sigma_3}\big]
\end{aligned} \tag{IV.20}$$

Properties of the primitive Vassiliev invariants $\alpha^c_{21}(K)$ and $\alpha^c_{31}(K)$ have been studied in [58]. In these works the integral expressions for $\alpha^c_{21}(K)$ and $\alpha^c_{31}(K)$ were studied in the flat-knot limit and combinatorial expressions of the kind that will be presented below from the study in the temporal gauge were obtained.

V LIGHT-CONE GAUGE AND THE KONTSEVICH INTEGRAL

In the previous section, the perturbative series expansion of the vev of a Wilson loop in the covariant Landau gauge has been analyzed. A similar analysis can be carried out for the operators associated to singular knots which were introduced in sect. 3. The lowest order contributions in such an expansion are the group factors corresponding to a chord diagrams. These group factors form a weight system. Before the development of operators for singular knots the following question was raised: could weight systems on chord diagrams be integrated to obtain invariants for non-singular knots? Kontsevich answered this question affirmatively in 1993. He showed that a weight system on chord diagrams of order m determines a unique Vassiliev invariant on non-singular knots.

Kontsevich theorem is constructive in the sense that it provides an explicit expression for the Vassiliev invariant for non-singular knots. This expression is known as the Kontsevich integral. We will provide its explicit form below, after deriving it from the quantum field theory side. In the context of Chern-Simons gauge theory Kontsevich theorem is contained in our construction of operators for singular knots. One should think of the first order contribution of the vev of a singular knot as the weight system, and of the full perturbative series for vevs of Wilson loops as the integration of the weight system. Thus, the perturbative series expansion of the vev of a Wilson loop should correspond to the Kontsevich integral. This does not seem to be the case as the configuration space integrals found in the previous section have a different form than the integrals contained in Kontsevich theorem. This fact poses a puzzle which, however, is solved very simply.

Gauge theories can be studied in different gauges. The analysis carried out in the previous sections dealt with the covariant Landau gauge but many others could have been used. It turns out that the perturbative series which results in the non-covariant light-cone gauge leads to the Kontsevich integral. To show this is the main goal of this section. Vevs of gauge invariant operators are independent of the gauge chosen and therefore the expressions obtained in the light-cone gauge should be equivalent to the ones obtained in any other gauge. In this sense the analysis of operators for singular knots in Chern-simons gauge theory constituted a general proof, or gauge independent proof, of Kontsevich theorem. The configuration space integrals of the previous section can equally be regarded as an integration of weight systems on chord diagrams. Actually, very recently, it has been shown from the mathematical side [19] that both series of integrals are equivalent. A ratification of the gauge independence of gauge invariant operators mentioned above.

In order to deal with non-covariant gauges let us introduce a unit constant vector n. The gauge-fixing conditions of axial type which we will discuss in this and in the next section correspond to the choice:

$$n^{\mu}A_{\mu} = 0. \tag{V.1}$$

In the case of the light-cone gauge the unit vector n satisfies the condition $n^2 = 0$. The gauge fixed action of the theory is built following the standard Faddeev-Popov construction which leads to a functional integral as in (IV.4) where now:

$$S_{\rm gf}(B_{\mu}, c, \bar{c}, \phi) = \int {\rm d}^3x {\rm Tr}(\phi n^{\mu}A_{\mu} + bn^{\mu}D_{\mu}c + \lambda\phi^2). \tag{V.2}$$

Again, ϕ is a Lagrange multiplier which imposes the gauge condition, c and $\bar{c}$ are anticommuting Faddeev-Popov ghosts, and λ is a gauge-fixing parameter.

The higher-loop analysis in axial-type gauges is a very delicate issue. It requires to take into consideration some specific prescription to regulate unphysical poles. Fortunately, this analysis has been done for the case of Chern-Simons gauge theory in [59]. In these works it is shown that a regulator can be chosen so that the effect of higher-loop contributions for two- and three-point functions is a shift of the parameter k as the one discussed in the previous section in the covariant Landau gauge. Though strictly speaking this has been proved at one loop, it is believed that, as in the case of covariant gauges, it holds at any order in higher loops. We will assume that higher-loop contributions to the two- and the three-point functions account for the shift of the parameter k. Notice that the gauge condition (V.1) does not fix the gauge completely. One still has gauge invariance under gauge transformations in the direction normal to n. The presence of this residual gauge invariance is a source of problems which, as it will be discussed below, are not yet solved. Another study of light-cone perturbation theory can be found in [60].

We will work in the limit $\lambda \to 0$. The theory greatly simplifies in this case. The propagator for the gauge field in momentum space acquires the simple form:

$$\delta_{ab}\frac{\lambda}{(np)^2}(p_{\mu}p_{\nu} - \frac{i}{\lambda}(np)\epsilon_{\mu\nu\rho}n^{\rho}) \to -i\epsilon_{\mu\nu\rho}\frac{n^{\rho}}{np}\delta_{ab}. \tag{V.3}$$

This propagator is orthogonal to the n-direction. This implies that there is no coupling to the ghosts fields and, furthermore, that there is no gauge field self-coupling. Thus there is only a Feynman rule to be taken into account to compute the vevs of operators: the one associated to the gauge field propagator. The group factors that remain in this case correspond just to chord diagrams. The fact that in this gauge no group factors with trivalent vertices have to be taken into account is a quantum field theory ratification of Bar-Natan theorem among the equivalence of the two representations, (III.5) and (III.6), of the space $\mathcal{A}$.

Non-covariant gauges share the problem of the presence of unphysical poles in their propagators [61]. This is manifest in (V.3). Several prescriptions have been proposed to avoid these unphysical poles. Usually, a prescription is chosen so that some particular properties of the correlation functions are fulfilled. In the light-cone gauge there is a natural prescription which is motivated by the simple form that (V.3) takes in coordinate

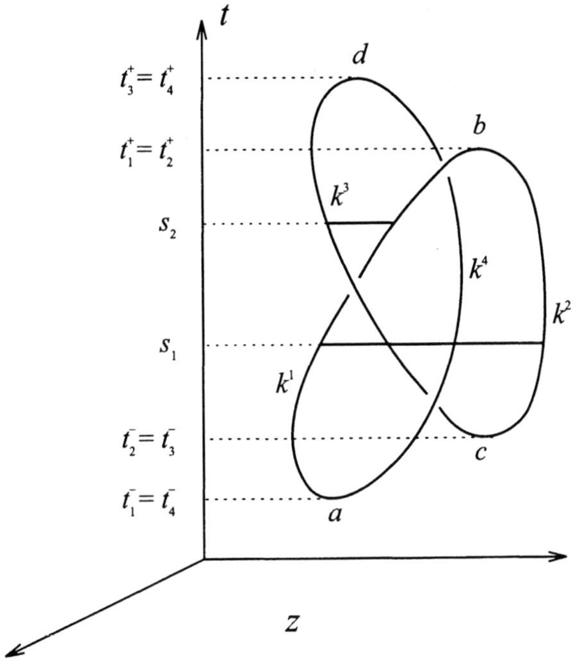

FIGURE 14. Example of a Morse knot.

space after performing a Wick rotation. As it will be shown below, this prescription leads to Kontsevich integral.

In the light-cone gauge it is convenient to introduce light-cone coordinates. Let us choose the unit vector n as $(0, 1, -1)$. The new coordinates are:

$$x^+ = x^1 + x^2, \quad x^- = x^1 - x^2, \tag{V.4}$$

and the new light-cone components for the gauge connection become:

$$A_+ = A_1 + A_2, \quad A_- = A_1 - A_2. \tag{V.5}$$

The form of the propagator (V.3) in coordinate space is more conveniently expressed after performing a Wick rotation to Euclidean space. A point in Euclidean space will be denoted as (t, z), where $z = x^1 + ix^2$. After introducing $A_z = A_1 + iA_2$ and $A_{\bar{z}} = A_1 - iA_2$ one finds that there exist a prescription to deal with the unphysical pole in (V.3) which leads to:

$$\begin{aligned} &\langle A^a_{\bar{z}}(x) A^b_m(x') \rangle = 0, \\ &\langle A^a_m(x) A^b_n(x') \rangle = \delta^{ab} \epsilon_{mn} \frac{1}{2\pi i} \frac{\delta(t-t')}{z-z'}, \end{aligned} \tag{V.6}$$

with $m, n = \{0, z\}$, and ϵ_{mn} is antisymmetric with $\epsilon_{0z} = 1$. This is the gauge-field propagator that we will use in our analysis of the perturbative expansion of the vev of a Wilson loop.

Before entering into the analysis of the perturbative series expansion corresponding to the vev a Wilson loop, we must discuss the potential problems that might be encountered because of the particular form of the gauge-field propagator (V.6). This propagator is singular when its two end-points coincide. Actually, it is particularly singular in this situation because both the numerator and the denominator lead to divergences. In the light-cone gauge we have two special kinds of singularities, there may be situations in which only one, the numerator or the denominator, leads to a divergence. In order to avoid singularities from the numerator one is forced to avoid paths with sections in which t, the first component of a generic point (t, z) is constant. This constraint implies that one must consider Morse knots. A Morse knot is a knot in which t is a Morse function on it. A Morse knot is characterized by $2n$ extrema, half of them maxima, and the other half minima. An example of a Morse knot is depicted in fig. 14.

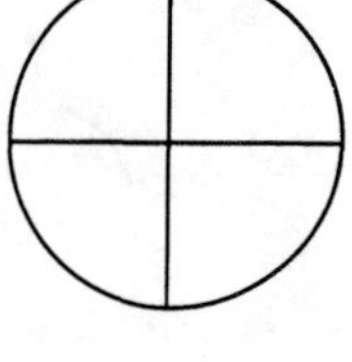

FIGURE 15. Group factor at second order.

Another potential problem due to the structure of (V.6) comes from situations in which the two end-points of the propagator are close to one of the extrema of a Morse knot since the denominator then vanishes. To solve this problem one introduces a regularization procedure based on the introduction of a framing. The resulting invariants will correspond to invariants of framed knots. As in the case of the Landau gauge one finds [17] that a term like (IV.13) factorizes. One is left then with a perturbative series expansion with group factors which correspond to chord diagrams without isolated chords. The ones with isolated chords just contribute to the framing factor.

In order to obtain the form of the perturbative series of the vev of a Wilson loop we begin by writing all the contributions at a given order m. To carry this out we must consider all possible ways of connecting $2m$ points on the Morse knot by m propagators, taking into account the regularization described above, *i.e.*, with one point of the propagator attached to K and the other to its companion knot K_ε, and then path-order integrating. The path-order integration can be split into a sum such that in each term enters a path-ordered integration along $2m$ curves among the set k^i, k^i_ε, $i = 1, \ldots, 2n$. This set of curves builds the Morse knot under consideration. A given term in this sum might contain propagators joining k^i and k^i_ε. In this case one must introduce a factor $1/2$. The contributions coming from propagators joining k^i and k^j_ε, with $i \neq j$, have also a factor $1/2$ since double counting occurs. Accordingly, propagators joining different curves must be replaced by $1/2$ the sum of their two possible choices of attaching their end-points.

To each rearrangement of the m propagators corresponds a group factor. These are easily obtained just using the rule (IV.10). To fix ideas we will present in detail the second-order contribution, $m = 2$, for a particular group factor, the one shown in fig. 15. This contribution is of the form

$$(ig)^2 \int_0^1 ds_1 \int_{s_1}^1 ds_2 \int_{s_2}^1 ds_3 \int_{s_3}^1 ds_4 \dot{x}^{\mu_1}(s_1)\dot{x}^{\mu_2}(s_2)\dot{x}^{\mu_3}(s_3)\dot{x}^{\mu_4}(s_4)$$
$$\langle A^{a_1}_{\mu_1}(x(s_1)) A^{a_3}_{\mu_3}(x(s_3))\rangle \langle A^{a_2}_{\mu_2}(x(s_2)) A^{a_4}_{\mu_4}(x(s_4))\rangle \mathrm{Tr}(T^{a_1}T^{a_2}T^{a_3}T^{a_4}). \qquad (V.7)$$

We will now write more explicitly one of the contributions to this multiple integral taking into account the form of the propagator (V.6). The delta function in this propagator imposes very strong restrictions on the possible contributions. Its presence implies that the only non-vanishing configurations are those in which the two end-points of each propagator are at the same height. To be more concrete let us consider the computation of (V.7) for the trefoil knot shown in fig. 14. This knot is made out of four curves, k^1, k^2, k^3 and k^4, whose end-points are the four critical points a, b, c and d. The heights of these critical points are:

$$\begin{aligned} a &\to t_1^- = t_4^-, \\ b &\to t_1^+ = t_2^+, \\ c &\to t_2^- = t_3^-, \\ d &\to t_3^+ = t_4^+. \end{aligned} \qquad (V.8)$$

They are depicted in fig. 14.

To obtain the contributions it is convenient to divide in four parts the circle that represents the knot in fig. 14 and join these parts by lines representing the propagators, taking into account the ordering of the four points to which they are attached. This ordering and the delta function in the height imply that no line can have its two end-points attached to the same part. They also imply that there are no contributions in which two end-points of different lines are attached to one part and the other two to another part. The contributions are easily depicted on the knot itself. One has been pictured in fig. 14. For each contribution, one must compute a sign, which takes into account the direction in which one travels along the knot in the new parametrization. To be more explicit, let us write, for example, the integral associated to the contribution shown in fig. 14. It takes the form:

$$(ig)^2 \frac{1}{(2\pi i)^2} \frac{1}{2^2} \int_{t_2^- < s_1 < s_2 < t_1^+} ds_1 ds_2$$

$$\left(\frac{\dot{z}_1(s_1) - \dot{z}_2'(s_1)}{z_1(s_1) - z_2'(s_1)} + \frac{\dot{z}_1'(s_1) - \dot{z}_2(s_1)}{z_1'(s_1) - z_2(s_1)} \right) \left(\frac{\dot{z}_3(s_2) - \dot{z}_1'(s_2)}{z_3(s_2) - z_1'(s_2)} + \frac{\dot{z}_3'(s_2) - \dot{z}_1(s_2)}{z_3'(s_2) - z_1(s_2)} \right), \tag{V.9}$$

where the primes denote the companion knot. The data entering this integral (V.9) are shown in fig. 14. Notice that this integral is not divergent if we take the limit $\varepsilon \to 0$ before performing the integration. This feature is common to all the contributions corresponding to the group factor under consideration. Only the contributions related to framing are potentially divergent. One could therefore remove in (V.9) the terms with primes and the factor $1/2^2$. The integral to be computed takes the form:

$$(ig)^2 \frac{1}{(2\pi i)^2} \int_{t_2^- < s_1 < s_2 < t_1^+} ds_1 ds_2 \frac{\dot{z}_1(s_1) - \dot{z}_2(s_1)}{z_1(s_1) - z_2(s_1)} \frac{\dot{z}_3(s_2) - \dot{z}_1(s_2)}{z_3(s_2) - z_1(s_2)}. \tag{V.10}$$

One of the two integrations can easily be performed, leading to:

$$(ig)^2 \frac{1}{(2\pi i)^2} \int_{t_2^- < s_2 < t_1^+} ds_2 \log \left(\frac{z_1(s_2) - z_2(s_2)}{z_1(t_2^-) - z_2(t_2^-)} \right) \frac{\dot{z}_3(s_2) - \dot{z}_1(s_2)}{z_3(s_2) - z_1(s_2)}. \tag{V.11}$$

Notice that, as argued before, this integral is finite. Although z_1 and z_2 get close to each other when $s_2 \to t_1^+$, the singularity in the integrand, being logarithmic, is too mild to lead to a divergent result.

We are now in a position to write the form of the general contribution. Notice that the most significant fact of our previous discussion is the presence of the delta function in the height of the propagator (V.6). It implies that the only non-vanishing configurations of the propagators are those in which their two end-points have the same height; in other words, only contributions in which the line representing the propagator is horizontal do not vanish. This observation allows us to rearrange the contributions to the perturbative series expansion in the following way. Consider all possible pairings $\{z_i(s), z_j'(s)\}$ of curves k^i and k_ε^j, $i, j = 1, \ldots, 2n$, where $2n$ is the number of extrema of the Morse knot under consideration. A contribution at order m in perturbation theory will involve a path-ordered integral in the heights $s_1 < \ldots < s_r < \ldots < s_m$ of a product of m propagators:

$$\prod_{r=1}^{m} \frac{dz_{i_r}(s_r) - dz_{j_r}'(s_r)}{z_{i_r}(s_r) - z_{j_r}'(s_r)}. \tag{V.12}$$

This product is characterized by a set of m ordered pairings, each one labelled by a pair of numbers (i_r, j_r) with $r = 1, \ldots, m$. We will denote an ordered pairing of m propagators generically by P_m. One must take into account all possible ordered pairings, *i.e.*, one must sum over all the possible P_m. The group factor that corresponds to each ordered pairing P_m is simply obtained by placing the group generators at the end-points of the propagators and taking the trace of the product, which results after traveling along the knot. The resulting group factor will be denoted by $R(P_m)$.

Another important fact that must be taken into account is the presence of a product of signs. For each pairing $P_m = \{(i_r, j_r), r = 1, \ldots, m\}$ there will be a contribution from their product. Certainly, the result will be a sign that will depend on the ordered pairing P_m. We will denote such a product by $s(P_m)$:

We are now in a position to write the full expression for the contribution to the perturbative series expansion at order m. It takes the form:

$$(ig^2)^m \left(\frac{1}{2\pi i}\right)^m \frac{1}{2^m} \sum_{P_m} \int_{t_{P_m}^- < t_1 < \ldots < t_r < \ldots < t_m < t_{P_m}^+} s(P_m) \prod_{r=1}^{m} \frac{dz_{i_r}(t_r) - dz_{j_r}'(t_r)}{z_{i_r}(t_r) - z_{j_r}'(t_r)} R(P_m), \tag{V.13}$$

where $t_{P_m}^+$ and $t_{P_m}^-$ are highest and lowest heights, which can be reached by the last and first propagators of a given ordered pairing P_m. This expression corresponds to the Kontsevich integral for framed knots as presented in [62]. If one sets to zero all the group factors associated to diagrams with isolated chords one can disregard the primes in (V.13). We define $Z_m(K)$ as the resulting contribution at order m divided by the dimension of the representation:

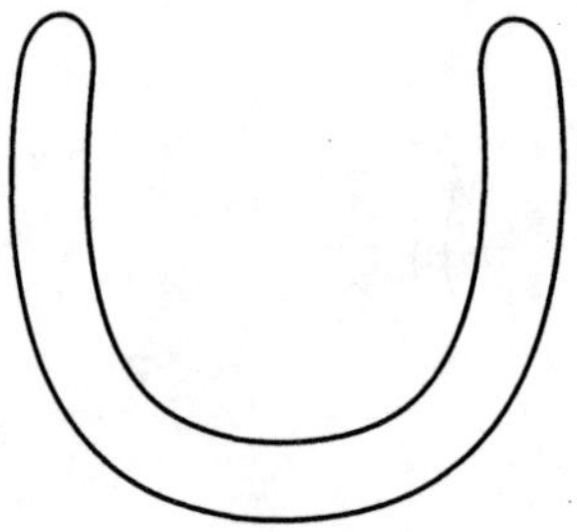

FIGURE 16. Unknot with four extrema.

$$Z_m(K) = \frac{(ig^2)^m}{\dim R}\left(\frac{1}{2\pi i}\right)^m \frac{1}{2^m} \sum_{P_m} \int_{t_{P_m}^- < t_1 < \ldots < t_r < \ldots < t_m < t_{P_m}^+} s(P_m) \prod_{r=1}^{m} \frac{dz_{i_r}(t_r) - dz_{j_r}(t_r)}{z_{i_r}(t_r) - z_{j_r}(t_r)} R(P_m), \tag{V.14}$$

This is the expression originally obtained by Kontsevich [15].

It is well known that the Kontsevich integral is not a knot invariant. As first pointed out by Kontsevich himself [15], it has to be corrected by a subtle factor to have full invariance. Indeed, if the shape of a Morse knot is modified in such a way that the number of extrema changes, the Kontsevich integral (V.14) is not invariant. Kontsevich proposed the solution to this lack of invariance adding a factor now known as the Kontsevich factor. If we denote by U the unknot with the shape shown in fig. 16, it turns out that the coefficients of the power expansion of:

$$\frac{1 + \sum_{m=1}^{\infty} Z_m(K)}{(1 + \sum_{m=1}^{\infty} Z_m(U))^{\frac{n}{2}}}, \tag{V.15}$$

where n denotes the number of extrema of the knot K, are truly Vassiliev invariants. The denominator of this expression is the so-called Kontsevich factor. It is not known how to obtain this factor using quantum field theory arguments. This is one of the open problems from the physical side. The study of gauge theories in non-covariant gauges presents many problems and, as it is the case here, in many occasions they lead to results which do not agree with their covariant-gauge counterparts. Often the cause for the discrepancy is linked to the ambiguity in the choice of a prescription to avoid non-physical poles. Most likely this is not the case at least for Chern-Simons gauge theory. As it is described in the next section, one finds that the Kontsevich factor is needed even if no specific prescription is taken. It is likely that the problem is related to the fact that there is a residual gauge invariance. As we discussed above, if one does not consider Morse knots, due to the residual gauge invariance, one gets divergent contributions. One needs to reduce to a single point all horizontal directions to avoid those divergences. Precisely when the number of points on the horizontal direction changes, *i.e.*, when the knot is deformed so that the number of extrema gets modified one encounters the non-invariance of the Kontsevich integral (V.14). It is believed that a proper treatment of the residual gauge invariance will lead to understand the origin of the Kontsevich factor from a field theoretical point of view.

VI TEMPORAL GAUGE AND COMBINATORIAL FORMULAE

The perturbative studies of Chern-Simons gauge theory that we have presented have led to two types of integral expressions for Vassiliev invariants. These expressions are not very useful from a computational point of view. Formulae of combinatorial type should be much preferred. There are some indications that a general combinatorial formula for Vassiliev invariants exists. On the one hand, the work on the study of the configuration space integrals in the limit of flat knots [58] originated a combinatorial expression for the Vassiliev invariant of order three. On the other hand, recent work from the mathematical side [63,23] supports this conjecture. The search for the combinatorial formula led to the study of Chern-Simons gauge theory in the non-covariant temporal gauge [20]. This turns out to be the more suitable gauge to carry out all the intermediate integrals and obtain combinatorial expressions. No other approach has provided a combinatorial expression for the two primitive Vassiliev invariants at order four. The temporal gauge has been also treated in [16,64].

The starting point of the analysis in the temporal gauge is the same as in the light-cone gauge. The gauge-fixing condition (V.1) is the same but now n is a unit vector of the form $n^\mu = (1,0,0)$. The gauge-fixed action and the gauge-field propagator have also the same general form (V.2) and (V.3). As before, the propagator presents a pole at $np = 0$, and a prescription to regulate it is needed. As observed in the previous sections, to construct the perturbative series expansion of the vev of a Wilson loop, we need the Fourier transform of (V.3). We will work in the limit $\lambda \to 0$. In the temporal gauge the momentum-space integral that has to be carried out has the form:

$$\Delta(x_0, x_1, x_2) = \int_M \frac{\mathrm{d}^3 p}{(2\pi)^3} \frac{\mathrm{e}^{i(p^0 x_0 + p^1 x_1 + p^2 x_2)}}{p^0}. \tag{VI.1}$$

This integral is ill-defined due to the pole at $p^0 = 0$. To make sense of it a prescription has to be given to circumvent the pole. We will not take a precise prescription. Instead, we will work in a rather general framework. Let us first analyze the dependence of $\Delta(x_0, x_1, x_2)$ in (VI.1) on x_0. The pole in p^0 is avoided if, instead of (VI.1), one analyses the derivative of $\Delta(x_0, x_1, x_2)$ with respect to x_0. Considering $\Delta(x_0, x_1, x_2)$ as a distribution one obtains:

$$\frac{\partial \Delta}{\partial x_0} = i\delta(x_0)\delta(x_1)\delta(x_2). \tag{VI.2}$$

Integrating this expression with respect to x_0, one finds that any prescription would lead to a result of the following form:

$$\Delta(x_0, x_1, x_2) = \frac{i}{2}\mathrm{sign}(x_0)\delta(x_1)\delta(x_2) + f(x_1, x_2), \tag{VI.3}$$

where $f(x_1, x_2)$ is a prescription-dependent distribution. The important consequence of the result (VI.3) is that the dependence of $\Delta(x_0, x_1, x_2)$ on x_0 has to be in the form $\mathrm{sign}(x_0)\delta(x_1)\delta(x_2)$. This observation will be crucial in our analysis. We will actually work with the rather general formula (VI.3) for $\Delta(x_0, x_1, x_2)$. This form of the propagator will allow us to introduce the notion of kernels of a Vassiliev invariants and to design a procedure to compute combinatorial expressions for these invariants.

As for the issue of the framing and the shift, a similar analysis as the one in the light-cone gauge holds in this case. In the situation under consideration, however, it is more useful to consider a specific framing, the one in which the twist is zero or vertical framing. The linking number of a framed knot can be expressed as the sum of the writhe plus the twist associated to its regular projection. The writhe is the sum of the signatures of the crossings and the twist is the number of times that the framed knot, seen as a band, twists around its middle axis. Choosing a framing so that the twist is zero implies that the linking number coincides with the writhe. Thus a dependence on the writhe factorizes. This information turns out to be very useful in the analysis of the perturbative series.

We will review the salient facts of the analysis of the perturbative series expansion of the vev of a Wilson loop in the temporal gauge. the complete analysis can be found in [20]. Given a knot K and one of its regular knot projections, $\mathcal{K}$, on the x_1, x_2-plane which is a Morse knot in the x_1 and x_2 directions, the perturbative series expansion for the vev of the corresponding Wilson loop has the form:

$$\langle W_K^R \rangle = \langle W_{\mathcal{K}}^R \rangle_{\text{temp}} \, \langle W_U^R \rangle^{b(\mathcal{K})}, \tag{VI.4}$$

being,

$$\frac{1}{\dim R} \langle W_K^R \rangle = 1 + \sum_{i=1}^{\infty} v_i(K) x^i, \tag{VI.5}$$

and,

$$\frac{1}{\dim R} \langle W_{\mathcal{K}}^R \rangle_{\text{temp}} = 1 + \sum_{i=1}^{\infty} \hat{v}_i(\mathcal{K}) x^i. \tag{VI.6}$$

In these expressions x denotes the coupling constant (IV.15). The function $b(\mathcal{K})$ is the exponent of the Kontsevich factor, which has been conjectured to be [20],

$$b(\mathcal{K}) = \frac{1}{12}(n_{x_1} + n_{x_2}), \tag{VI.7}$$

where n_{x_1} and n_{x_2} are the critical points of the regular projection $\mathcal{K}$ in both, the x_1 and the x_2 directions. In (VI.4) U denotes the unknot and $\langle W_{\mathcal{K}}^R\rangle_{\rm temp}$ is the vev of the Wilson loop corresponding to the regular projection $\mathcal{K}$ as computed perturbatively in the temporal gauge with the standard Feynman rules of the theory. Notice that though each of the factors on the right hand side of (VI.4) depends on the regular projection chosen, the left hand side does not. While the coefficients $v_i(K)$ of the series (VI.5) are Vassiliev invariants the coefficients $\hat{v}_i(\mathcal{K})$ of (VI.6) are not. The latter depend on the regular projection chosen.

An explicit combinatorial form (no integrals left) of the coefficients $\hat{v}_i(\mathcal{K})$ in (VI.6) would lead to a general combinatorial formula for Vassiliev invariants. Unfortunately, this has not been obtained yet at all orders. Only part of the contributions entering $\hat{v}_i(\mathcal{K})$ have been explicitly written at all orders. These are the *kernels* introduced in [20]. The kernels are quantities which depend on the knot projection chosen and therefore are not knot invariants. However, at a given order i a kernel differs from an invariant of type i by terms that vanish in signed sums of order i. The kernel contains the part of a Vassiliev invariant which is the last in becoming zero when performing signed sums, in other words, a kernel vanishes in signed sums of order $i+1$ but does not in signed sums of order i. In some sense the kernel represents the most fundamental part of a Vassiliev invariant, *i.e.*, the part that survives a maximum number of signed sums. Kernels plus the structure of the perturbative series expansion seem to contain enough information to reconstruct the full Vassiliev invariants. This was shown in [20] up to order four. The results obtained there will be presented below and rewritten in a more compact form. A summary of this approach has been presented in [22].

The expression for the kernels results after considering only the part of the propagator (VI.3) which contains the sign function. This part involves a double delta function and therefore all the integrals can be performed. The result is a combinatorial expression in terms of crossing signatures after distributing propagators among all the crossings. Of course, the contribution from this part does not depend on the function $f(x_1, x_2)$ entering (VI.3). The general expression can be written in a universal form much in the spirit of the universal form of the Kontsevich integral [15]. Let us consider a knot K with a regular knot projection $\mathcal{K}$ containing n crossings. Let us choose a base point on $\mathcal{K}$ and let us label the n crossings by $1, 2, \ldots, n$ as we pass for first time through each of them when traveling along $\mathcal{K}$ starting at the base point. The universal expression for the kernel associated to $\mathcal{K}$ has the form:

$$\mathcal{N}(\mathcal{K}) = \sum_{k=0}^{\infty}\left(\sum_{m=1}^{k}\sum_{\substack{p_1,\ldots,p_m=1\\ p_1+\cdots+p_m=k}}^{k}\sum_{\substack{i_1,\ldots,i_m=1\\ i_1<\cdots<i_m}}^{n}\frac{\varepsilon_{i_1}^{p_1}\cdots\varepsilon_{i_m}^{p_m}}{(p_1!\cdots p_m!)^2}\sum_{\substack{\sigma_1,\ldots,\sigma_m\\ \sigma_1\in P_1,\ldots,\sigma_m\in P_m}}\mathcal{T}(i_1,\sigma_1;\ldots;i_m,\sigma_m)\right). \tag{VI.8}$$

In this equation P_m denotes the permutation group of p_m elements. The factors in the inner sum, $\mathcal{T}(i_1,\sigma_1;\ldots;i_m,\sigma_m)$, are group factors which are computed in the following way: given a set of crossings, $i_1,\ldots,i_m$, and a set of permutations, $\sigma_1,\ldots,\sigma_m$, with $\sigma_1 \in P_1,\ldots,\sigma_m \in P_m$, the corresponding group factor $\mathcal{T}(i_1,\sigma_1;\ldots;i_m,\sigma_m)$ is the result of taking a trace over the product of group generators which is obtained after assigning $p_1,\ldots,p_m$ group generators to the crossings $i_1,\ldots,i_m$ respectively, and placing each set of group generators in the order which results after traveling along the knot starting from the base point. The first time that one encounters a crossing i_j a product of p_j group generators is introduced; the second time the product is similar, but with the indices rearranged according to the permutation $\sigma_j \in P_j$.

The universal formula (VI.8) for the kernels can be written in a more useful way collecting all the coefficients multiplying a given group factor. Recall that the group factors can be labeled by chord diagrams. At order k one has a term for each of the inequivalent chord diagrams with k chords. Denoting chord diagrams by D, equation (VI.8) can be written as:

$$\mathcal{N}(\mathcal{K}) = \sum_{D} N_D(\mathcal{K})D, \tag{VI.9}$$

where the sum extends to all inequivalent chord diagrams. Our next task is to derive from (VI.8) the general form of the kernels $N_D(\mathcal{K})$. The concept of kernel can be extended to include singular knots by considering signed sums of (VI.9), or, following [9], introducing vevs of the operators for singular knots. If $\mathcal{K}^j$ denotes a regular projection of a knot K^j with j simple singular crossings or double points, the corresponding universal form for the kernel possesses an expansion similar to (VI.9):

$$\mathcal{N}(\mathcal{K}^j) = \sum_{D} N_D(\mathcal{K}^j)D. \tag{VI.10}$$

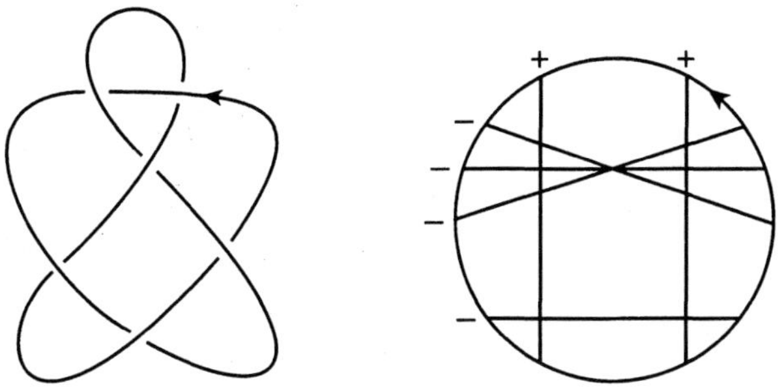

FIGURE 17. A regular knot projection and its corresponding Gauss diagram.

The general results about singular knots proved in [9] lead to two important features for (VI.10). On the one hand, finite type implies that $N_D(\mathcal{K}^j) = 0$ for chord diagrams D with more than j chords. On the other hand, $N_D(\mathcal{K}^j) = 2^j \delta_{D,D(\mathcal{K}^j)}$, where $D(\mathcal{K}^j)$ is the chord diagram corresponding to the singular knot projection $\mathcal{K}^j$. As observed above, kernels constitute the part of a Vassiliev invariant which survives a maximum number of signed sums.

To compute $N_D(\mathcal{K})$ we will introduce first the notion of the set of labeled chord subdiagrams of a given chord diagram. We will denote this set by S_D. This set is made out of a selected set of labeled chord diagrams that we now define. A *labeled chord diagram* of order p is a chord diagram with p chords and a set of positive integers $k_1, k_2, \ldots, k_p$, which will be called labels, such that each chord has one of these integers attached. The set S_D is made out of labeled chord diagrams which satisfy two conditions. These conditions are fixed by the form of the series entering the kernels (VI.8). We will call the elements of S_D labeled chord subdiagrams of the chord diagram D. They are defined as follows. A *labeled chord subdiagram* of a chord diagram D with k chords is a labeled chord diagram of order p with labels $k_1, k_2, \ldots, k_p$, $p \leq k$, such that the following two conditions are satisfied: *a)* $k_1 + k_2 + \cdots + k_p = k$; *b)* there exist elements $\sigma_1 \in P_{k_1}, \sigma_2 \in P_{k_2}, \ldots, \sigma_p \in P_{k_p}$ of the permutation groups $P_{k_1}, P_{k_2}, \ldots, P_{k_p}$ such that, after replacing the j-th chord diagram by k_j chords arranged according to the permutation σ_j, for $j = 1, \ldots, p$, the resulting chord diagram is homeomorphic to D. The number of ways that permutations $\sigma_1 \in P_{k_1}, \sigma_2 \in P_{k_2}, \ldots, \sigma_p \in P_{k_p}$ can be chosen is called the multiplicity of the labeled chord subdiagram. We will denote the multiplicity of a given labeled chord subdiagram, $s \in S_D$, by $m_D(s)$.

The chord diagram D itself can be regarded as a labeled chord subdiagram such that its labels, or positive integers attached to its chords, are 1. It has multiplicity 1. All the elements of S_D except D have a number of chords smaller than the number of chords of D. Not all labeled chord diagrams are subdiagrams of D. However, given a labeled chord diagram with labels $k_1, k_2, \ldots, k_p$ there can be different sets of permutations leading to D. The number of these different sets is the multiplicity introduced above. The elements of the sets S_D for all chord diagrams D up to order four which do not have disconnected subdiagrams are the following:

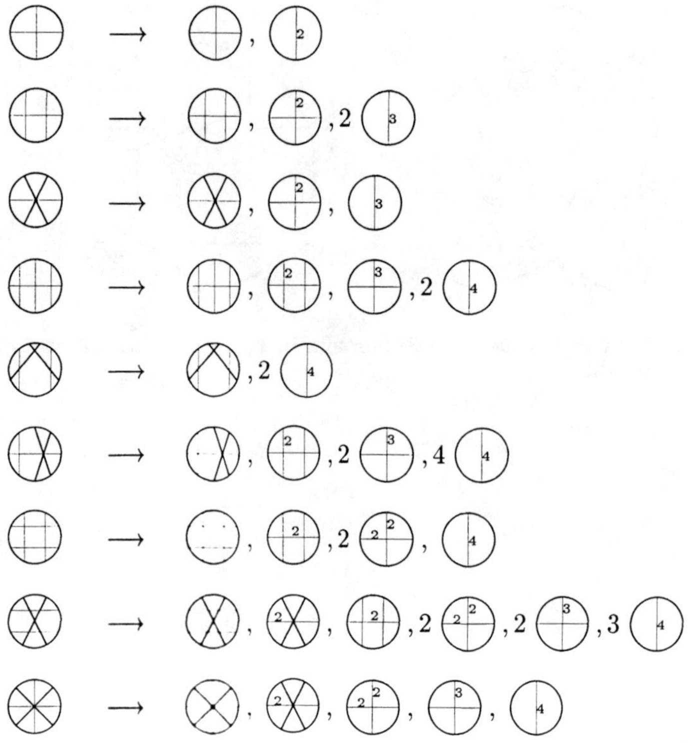

(VI.11)

The numbers accompanying each labeled chord subdiagram denote their multiplicity. When no number is attached to a chord of a labeled chord diagram it should be understood that the corresponding label is 1.

In order to write our final expression for the kernels we need to recall the notion of Gauss diagram. Given a regular projection $\mathcal{K}$ of a knot K we can associate to it its Gauss diagram $G(\mathcal{K})$. The regular projection $\mathcal{K}$ can be regarded as a generic immersion of a circle into the plane enhanced by information on the crossings. The Gauss diagram $G(\mathcal{K})$ consists of a circle together with the preimages of each crossing of the immersion connected by a chord. Each chord is equipped with the sign of the signature of the corresponding crossing. An example of Gauss diagram has been pictured in fig. 17. Gauss diagrams are useful because they allow to keep track of the sums involving the crossings which enter in (VI.8) in a very simple form. Let us consider a chord diagram D and one of its labeled chord subdiagrams $s \in S_D$. Let us assume that s has p chords and labels $k_1, k_2, \cdots, k_p$. We define the product,

$$\langle s, G(\mathcal{K})\rangle, \tag{VI.12}$$

as the sum over all the embeddings of s into $G(\mathcal{K})$, each one weighted by a factor,

$$\frac{\varepsilon_1^{k_1}\varepsilon_2^{k_2}\cdots\varepsilon_p^{k_p}}{(k_1!k_2!\cdots k_p!)^2}, \tag{VI.13}$$

where $\varepsilon_1, \varepsilon_2, \ldots, \varepsilon_p$ are the signatures of the chords of $G(\mathcal{K})$ involved in the embedding. Using (VI.12) the kernels $N_D(\mathcal{K})$ entering (VI.9) can be written as,

$$N_D(\mathcal{K}) = \sum_{s\in S_D} m_D(s)\langle s, G(\mathcal{K})\rangle, \tag{VI.14}$$

where $m_D(s)$ denotes the multiplicity of the labeled subdiagram $s \in S_D$ relative to the chord diagram D.

The product (VI.12) possesses important properties. First, it is independent of the base point chosen for the regular projection $\mathcal{K}$ and, correspondingly, for the Gauss diagram $G(\mathcal{K})$. Second, it is of finite type. This means that if s has j chords, the result of computing a signed sum of order higher than j is zero. Recall that signed sums of order k are used to define quantities associated to singular knot projections with k double points, as the ones entering (VI.9). A signed sum of order k contains 2^k terms which correspond to the possible ways of resolving k double points into overcrossings and undercrossings. Each one has a sign which corresponds to the product of the signatures of the crossings involved in the k double points. If s is a labeled chord diagram with j chords and all its labels take value one, the order-j signed sum is 2^j if the configuration of the singular projection with j double points associated to such a sum corresponds to the chord diagram s; otherwise its value is zero. This fact leads to the result mentioned above stating that:

$$N_D(\mathcal{K}^j) = 2^j \delta_{D,D(\mathcal{K}^j)}, \tag{VI.15}$$

where $D(\mathcal{K}^j)$ is the chord diagram corresponding to the singular knot projection associated to the signed sum. Of course, the product (VI.12) vanishes if the number of chords of s is bigger than the number of chords of the Gauss diagram $G(\mathcal{K})$.

The products (VI.12) can be regarded as quantities of finite type associated to Gauss diagrams G whether or not these correspond to a regular projection of a knot. Gauss diagrams can be studied as abstract objects characterized by chord diagrams with signs assigned to their chords. It is clear that in such a general context the quantities $\langle s, G\rangle$, as defined in (VI.12), are of finite type. In other words, if s has j chords and G is an abstract Gauss diagram, the product $\langle s, G\rangle$ vanishes under signed sums of order higher than j. This observation leads to conjecture that the product (VI.12) might play an interesting role in the theory of virtual knots [65,23].

The terms $\langle s, G(\mathcal{K})\rangle$ entering (VI.14) are related to the quantities $\chi(\mathcal{K})$ defined in [20]. It is straightforward to obtain the following relations:

$\langle$ 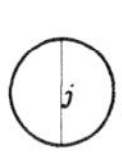 $, G(\mathcal{K})\rangle = \frac{1}{(j!)^2}\chi_1(\mathcal{K}),$ $\langle$  $, G(\mathcal{K})\rangle = \frac{1}{(j!)^2} n(\mathcal{K}),$ j even,

$\langle$ 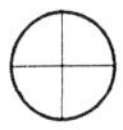 $, G(\mathcal{K})\rangle = \chi_2^A(\mathcal{K}),$ $\langle$ 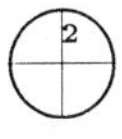 $, G(\mathcal{K})\rangle = \frac{1}{4}\chi_2^B(\mathcal{K}),$

$\langle$ 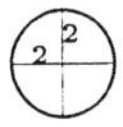 $, G(\mathcal{K})\rangle = \frac{1}{16}\chi_2^C(\mathcal{K}),$ $\langle$  $, G(\mathcal{K})\rangle = \chi_3^A(\mathcal{K}),$

$\langle$ 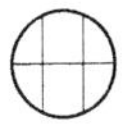 $, G(\mathcal{K})\rangle = \chi_3^B(\mathcal{K}),$ $\langle$  $, G(\mathcal{K})\rangle = \frac{1}{4}\chi_3^C(\mathcal{K}),$

$\langle$ 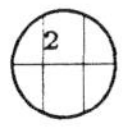 $, G(\mathcal{K})\rangle = \frac{1}{4}\chi_3^D(\mathcal{K}),$ $\langle$ 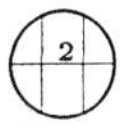 $, G(\mathcal{K})\rangle = \frac{1}{4}\chi_3^E(\mathcal{K}),$

$\langle$  $, G(\mathcal{K})\rangle = \chi_4^A(\mathcal{K}),$ $\langle$  $, G(\mathcal{K})\rangle = \chi_4^B(\mathcal{K}),$

$\langle$ 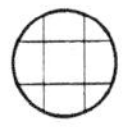 $, G(\mathcal{K})\rangle = \chi_4^C(\mathcal{K}),$ $\langle$ 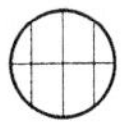 $, G(\mathcal{K})\rangle = \chi_4^D(\mathcal{K}),$

$\langle$  $, G(\mathcal{K})\rangle = \chi_4^E(\mathcal{K}),$ $\langle$  $, G(\mathcal{K})\rangle = \chi_4^F(\mathcal{K}).$

(VI.16)

Notice that in the second relation $n(\mathcal{K})$ denotes the number of crossings of the regular projection $\mathcal{K}$. The rest of the quantities on the right hand side of (VI.16) were defined in [20].

In [20], Vassiliev invariants up to order four were expressed in terms of these quantities and the crossing signatures. The strategy to obtain them was to start with the kernels (VI.14) and exploit the properties of the perturbative series expansion of Chern-Simons gauge theory. A special role in the construction was played by the factorization theorem proved in [21]. In order to discuss some of the steps followed in [20] we will describe in detail the computation of the combinatorial expression for the Vassiliev invariant of order two. Let us begin considering a canonical basis for the group factors where diagrams with isolated chords are included. In fig. 18 this basis has been depicted up to order four. Tildes have been used to denote the diagrams with

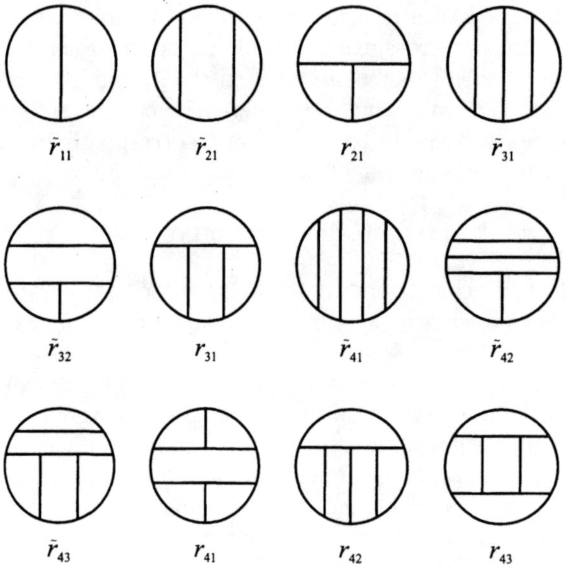

FIGURE 18. Choice of canonical basis up to order four which includes diagrams with isolated chords.

isolated chords. These diagrams are included because they provide useful information when working in the vertical framing. Instead of factorizing them out as in the previous sections, we will keep them in our analysis. In the analysis of the perturbative series it is important to know the expressions of all the chord diagrams in terms of the elements of the canonical basis in fig. 18. These expressions has been collected in figs. 19 and 20.

The perturbative series expansions entering (VI.5) and (VI.6) get some modifications relative to their form in (IV.14). We will write them in the form:

$$\frac{1}{\dim R}\langle W_K^R\rangle = 1+\sum_{i=1}^{\infty}\sum_{j=1}^{d_i}\alpha_{ij}(K)r_{ij}(R)x^i+\sum_{i=1}^{\infty}\sum_{j=1}^{\tilde{d}_i}\gamma_{ij}(K)\tilde{r}_{ij}(R)x^i,$$
$$\frac{1}{\dim R}\langle W_K^R\rangle_{\text{temp}} = 1+\sum_{i=1}^{\infty}\sum_{j=1}^{d_i}\hat{\alpha}_{ij}(\mathcal{K})r_{ij}(R)x^i+\sum_{i=1}^{\infty}\sum_{j=1}^{\tilde{d}_i}\hat{\gamma}_{ij}(\mathcal{K})\tilde{r}_{ij}(R)x^i. \tag{VI.17}$$

Notice that we have split the perturbative series into two sums. In the first sum the group factors, and their corresponding coefficients, are those appearing in (IV.14), while in the second sum they are all the non-primitive elements coming from diagrams with isolated chords. The quantities $r_{ij}(R)$ and $\tilde{r}_{ij}(R)$ denote the respective group factors (whose corresponding chord diagrams up to order four are depicted in fig. 18), while d_i and $\tilde{d}_i$ are the dimension of their basis at order i. As for the geometrical factors, $\alpha_{ij}(K)$ and $\gamma_{ij}(K)$ denote the Vassiliev invariants, primitive or not, we are looking for, while $\hat{\alpha}_{ij}(\mathcal{K})$ and $\hat{\gamma}_{ij}(\mathcal{K})$ are just the geometrical coefficients in the canonical basis of the perturbative Chern-Simons theory in the temporal gauge.

The strategy is the following. First, the behavior of the unknown integrals entering $\hat{\alpha}_{ij}(\mathcal{K})$ and $\hat{\gamma}_{ij}(\mathcal{K})$ is analyzed; then the whole invariant is built, taking into account the corresponding global term as dictated by (VI.4). Since, as shown in [21], the perturbative series expansion of the vev of the Wilson loop exponentiates in terms of the primitive basis elements, we have the following simple relation among primitives:

$$\alpha_{ij}(K) = \hat{\alpha}_{ij}(\mathcal{K}) + b(\mathcal{K})\,\alpha_{ij}(U). \tag{VI.18}$$

Let us begin with the analysis of $\hat{v}_i(\mathcal{K})$ in (VI.6). At first order we have no correction term (recall we are using the vertical framing), and the temporal gauge series provides the full regular invariant:

$$v_1(K) = \hat{v}_1(\mathcal{K}) = \left(\hat{\gamma}_{11}^E(\mathcal{K}) + \hat{\gamma}_{11}^D(\mathcal{K})\right)\tilde{r}_{11}(R) \tag{VI.19}$$

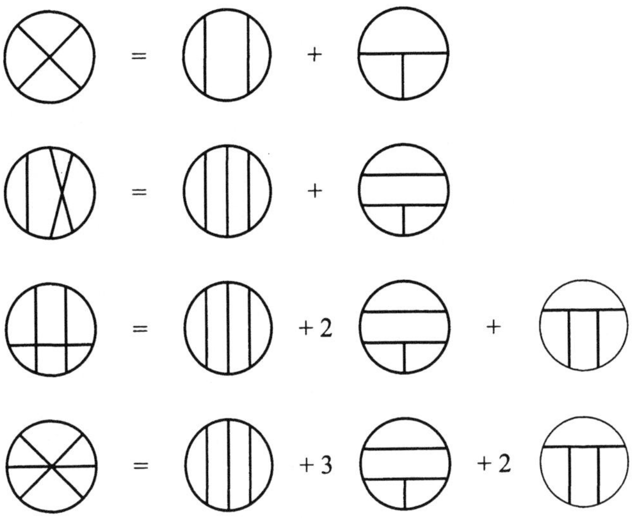

FIGURE 19. Expansion of chord diagrams in the canonical basis: orders two and three.

In this expression we have written the geometrical factor $\hat{\gamma}_{11}(\mathcal{K})$ as a sum of two parts. The first one $\hat{\gamma}_{11}^E(\mathcal{K})$ is built from the part of the propagator (VI.3) not containing the unknown distribution f. The second one, $\hat{\gamma}_{11}^D(\mathcal{K})$, depends on f. This type of decomposition can be done in general though at higher orders is more complicated. The integrals made out of the f-dependent part of the propagator (VI.3) will be denoted by a superindex D and a subindex which will label the chord diagram it comes from.

The general calculation requires a more subtle labelling. Given a chord diagram, each chord in it represents the propagator of the theory. Our propagator (VI.3) contains two pieces: the explicit one, which leads to the signatures of the crossings, and the f-dependent one. The integrals arising from perturbation theory are a sum over all the possible ways of identifying the chords with each of them. So for a given diagram we will end up with different types of D integrals, depending on how many f-terms they contain. When all the propagators in the integral are of this kind, it will be denoted simply by D_{ij}. If only one chord stands for the signature-dependent part, its evaluation will simply result in a crossing sign, ε_m, plus a restriction of the original integration domain. The chord standing for this factor is attached to the mth crossing, which means that the ordered integration domain of the other chords of the D integral is now limited by the position of that crossing. The resulting integral is written as:

$$\varepsilon_m \hat{\alpha}_{ij}^{D_m}\,, \tag{VI.20}$$

with the subindex of D denoting that one of the chords in the diagram is attached to the mth crossing.

More involved cases arise when the integrand contains two signature-dependent terms of the propagator (VI.3). In this case one must distinguish three subcases: both are attached to the same crossing, both are attached to different crossings and they have the pattern of the second case in fig. 21, and finally, both are attached to different crossings but this time they have the pattern of the third case in fig. 21. The set of pairs of crossings corresponding to the second case will be denoted by $\mathcal{C}_a$. The one corresponding to the third by $\mathcal{C}_b$. As only invariants up to order four will be considered, there is no need to handle the case where three or more signature-dependent terms of the propagator (but not all) are fixed to crossings. When the contribution does not contain f-dependent terms, the integral may be read from the kernels (VI.8). It will be denoted by $\hat{\alpha}_{ij}^E$.

From the expression (VI.8) for the kernels one easily finds, extracting the $k = 1$ contribution:

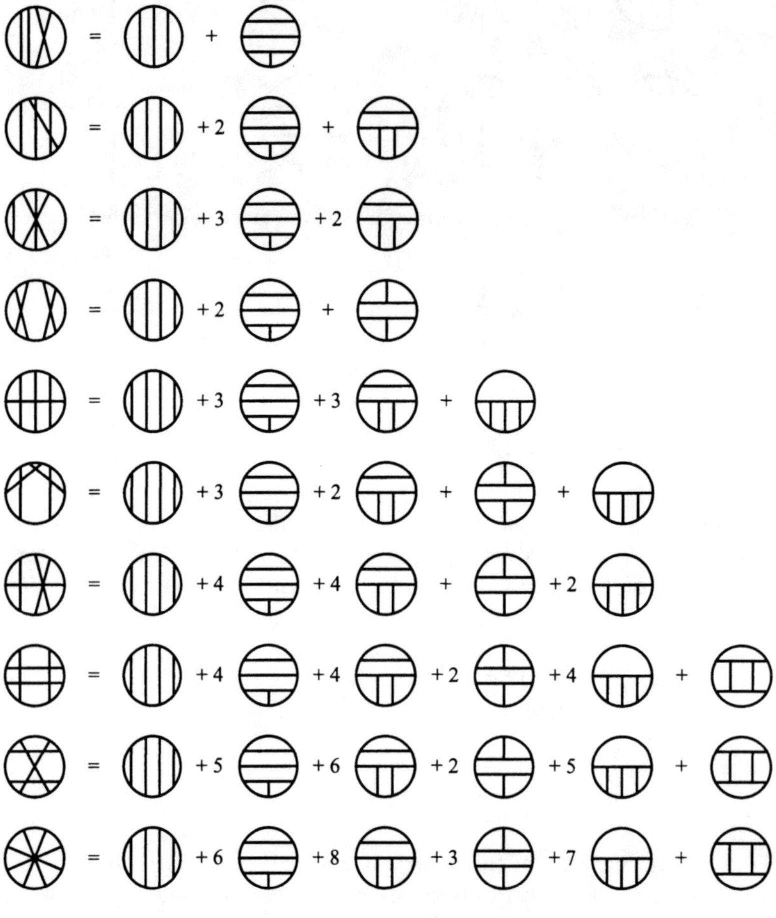

FIGURE 20. Expansion of chord diagrams in the canonical basis: order four.

$$\hat{\gamma}^E_{11}(\mathcal{K}) = \sum_{i=1}^{n} \varepsilon_i, \tag{VI.21}$$

where n is the number of crossings in $\mathcal{K}$. This corresponds to the writhe, or linking number in the vertical framing, which is known to be the correct answer for $v_1(K)$. Thus we must have

$$\hat{\gamma}^D_{11}(\mathcal{K}) = 0, \tag{VI.22}$$

which agrees with our general arguments, showing that contributions with an odd number of f-dependent terms vanish.

At order two, the series expansion of (VI.6) can be expressed as:

$$\hat{v}_2(\mathcal{K}) = \left(\hat{\gamma}^E_{21}(\mathcal{K}) + \hat{\gamma}^D_{21}(\mathcal{K})\right)\tilde{r}_{21}(R) + \left(\hat{\beta}^E_{21}(\mathcal{K}) + \hat{\beta}^D_{21}(\mathcal{K})\right)s_{21}(R), \tag{VI.23}$$

where $s_{21}(R)$ is the group factor corresponding to the diagram on the left-hand side of the first row in fig. 19. We will denote by s_{ij} group diagrams which appear at intermediate steps but do not belong to the chosen canonical basis of fig. 18. Their associated geometric factors will be denoted by β_{ij}. Notice that in (VI.23) we have not included terms of the form $\sum_{m=1}^{n} \varepsilon_m \alpha_{ij}^{D_m}$, since they have an odd number of f-dependent terms, and should not contribute. Terms of even order in the perturbative series expansion are invariant under a space reflection, while terms of odd order change sign. This implies that at even (odd) orders there are only contributions with an even (odd) number of f-dependent terms. In terms of the group factors of the chosen canonical basis (see fig. 18), the last expression, after using the first equation in fig. 19, takes the form:

$$\begin{aligned}\hat{v}_2(\mathcal{K}) &= \hat{\gamma}_{21}(\mathcal{K})\tilde{r}_{21}(R) + \hat{\alpha}_{21}(\mathcal{K})r_{21}(R) \\ &= \left(\hat{\gamma}^E_{21}(\mathcal{K}) + \hat{\beta}^E_{21}(\mathcal{K}) + \hat{\gamma}^D_{21}(\mathcal{K}) + \hat{\beta}^D_{21}(\mathcal{K})\right)\tilde{r}_{21}(R) + \left(\hat{\beta}^E_{21}(\mathcal{K}) + \hat{\beta}^D_{21}(\mathcal{K})\right)r_{21}(R)\end{aligned} \tag{VI.24}$$

The computation of the two signature-dependent terms in (VI.24), $\hat{\gamma}^E_{21}$ and $\hat{\beta}^E_{21}$ is easily obtained from the kernels (VI.8). One finds:

$\hat{\gamma}^E_{21} = \langle$ 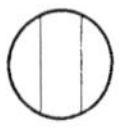 $+$ 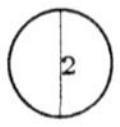 $, G(\mathcal{K})\rangle$ (VI.25)

$\hat{\beta}^E_{21} = \langle$ 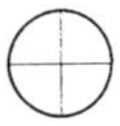 $+$ 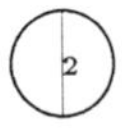 $, G(\mathcal{K})\rangle$ (VI.26)

Adding them up one obtains:

$$\hat{\gamma}^E_{21} + \hat{\beta}^E_{21} = \frac{1}{2}\left(\sum_{i=1}^{n} \varepsilon_i\right)^2. \tag{VI.27}$$

According to the factorization theorem [21] this is the whole non-primitive regular invariant of order two, $\gamma_{21} = \frac{1}{2}(\sum \varepsilon_i)^2$. Thus, we conclude that the order-two D-type terms must satisfy:

$$\hat{\gamma}^D_{21} + \hat{\beta}^D_{21} = 0. \tag{VI.28}$$

One more relation is needed to get rid of the two known quantities $\hat{\gamma}^D_{21}$ and $\hat{\beta}^D_{21}$. A new relation is obtained taking into account (VI.18). One easily finds:

$\alpha_{21}(K) = \langle$ 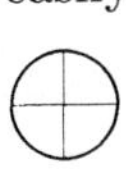 $+$ 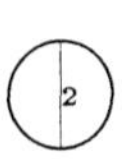 $, G(\mathcal{K})\rangle + \hat{\beta}^D_{21}(\mathcal{K}) + b(\mathcal{K})\,\alpha_{21}(U),$ (VI.29)

where $\alpha_{21}(U)$ stands for the value of this invariant for the unknot. The function $b(\mathcal{K})$ is the unknown exponent in the global factor in (VI.4). Using the fact that $\hat{\beta}^D_{21}(\mathcal{K})$ and $b(\mathcal{K})$ are equal in $\mathcal{K}$ and $\alpha(\mathcal{K})$, being $\alpha(\mathcal{K})$ the ascending diagram of $\mathcal{K}$, and that the latter is equivalent under ambient isotopy to the unknot, one finds:

$\hat{\beta}^D_{21}(\mathcal{K}) = \alpha_{21}(U)\,[1 - b(\mathcal{K})] - \langle$ 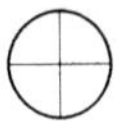 $+$ 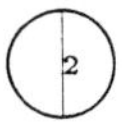 $, G(\alpha(\mathcal{K}))\rangle.$ (VI.30)

The final expression for the invariant is:

$\alpha_{21}(K) = \alpha_{21}(U) + \langle$  $, G(\mathcal{K})\rangle - \langle$  $, G(\alpha(\mathcal{K}))\rangle,$ (VI.31)

where $\alpha_{21}(U)$ stands for the value of α_{21} for the unknot. Recall that the ascending diagram $\alpha(\mathcal{K})$ of a knot projection $\mathcal{K}$ is defined as the diagram obtained by switching, when traveling along the knot from a base point, all the undercrossings to overcrossings. Ascending diagrams enter often in the combinatorial expressions and it is convenient introduce the following notation. A bar over a quantity $L(\mathcal{K})$ indicates that the same quantity for the ascending diagram has to be subtracted, *i.e.*:

$$\bar{L}(\mathcal{K}) = L(\mathcal{K}) - L(\alpha(\mathcal{K})) \qquad \text{(VI.32)}$$

where $\alpha(\mathcal{K})$ denotes the standard ascending diagram of $\mathcal{K}$. Using this notation, the final form for the only primitive Vassiliev invariant at order two is:

$\alpha_{21}(K) = \alpha_{21}(U) + \langle$  $, \bar{G}(\mathcal{K})\rangle.$ (VI.33)

The combinatorial expression (VI.33) agrees with the formulae given in [58] and [66]. Notice that its dependence on $b(\mathcal{K})$ has disappeared, so up to this order we do not get any condition on this function. The analysis at higher orders, however, imposes relations fro the function $b(\mathcal{K})$. All the resulting relations are consistent with the ansatz (VI.7). It is important to remark that the derivation of (VI.33) that we have presented is much simpler than the one in the covariant gauge obtained in [58]. This simplicity is rooted in the special features of the temporal gauge that permits to have the compact expression (VI.8) for the kernels, which are the essential building blocks of the combinatorial expressions for Vassiliev invariants.

The procedure followed at second order has been implemented in [20] for orders three and four. We will reproduce here the resulting combinatorial expressions. At order three there is only one primitive invariant. It takes the form:

$\alpha_{31}(K) = \langle$  $+$  $+ 2$ 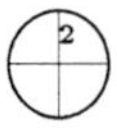 $, G(\mathcal{K})\rangle - \sum_{i=1}^{n} \varepsilon_i(\mathcal{K}) \Big[\langle$ 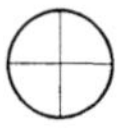 $, G(\alpha(\mathcal{K}))\rangle\Big]_i.$ (VI.34)

Several comments are in order to explain the quantities entering this expression. The sum is over all crossings i, $i = 1, \ldots, n$, and $\varepsilon_i(\mathcal{K})$ denotes the corresponding signature. The square brackets $[\ ]_i$ enclosing a quantity $L(\mathcal{K})$ denote:

$$\Big[L(\mathcal{K})\Big]_i = L(\mathcal{K}) - L(\mathcal{K}_{i_+}) - L(\mathcal{K}_{i_-}), \qquad \text{(VI.35)}$$

where the regular projection diagrams $\mathcal{K}_{i_+}$ and $\mathcal{K}_{i_-}$ are the ones which result after the splitting of $\mathcal{K}$ at the crossing point i as shown in the first row of fig. 21. It is clear from the list (VI.16) that these two invariants can be written in terms of the products (VI.12) and the crossing signatures.

Combinatorial expressions for the two primitive invariants at order four have been presented in [20]. Their construction is based on the use of the kernels (VI.14) and the properties of the perturbative series expansion. As in the case of previous orders, these invariants are expressed in terms of the products (VI.12) and the crossing signatures. Their form is more complicated than the ones at lower orders. At order four there are two primitive Vassiliev invariants. We will make the same choice of basis as in [20]. The diagrams associated to them are: $r_{4,2}$ and $r_{4,3}$ in fig. 13. They turn out to be:

$\alpha_{42}(K) = \alpha_{42}(U) + \langle 7$  $+ 5$  $+ 4$ 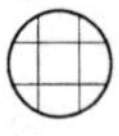 $+ 2$  $+$ 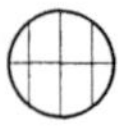 $+$ 

$+ 8$  $+ 2$ 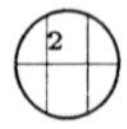 $+ 8$ 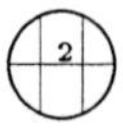 $+ \frac{1}{6}$  $, \bar{G}(\mathcal{K})\rangle$

$+ \sum_{\substack{i,j\in\mathcal{C}_a \\ i>j}} \bar{\varepsilon}_{ij}(\mathcal{K}) \Big(\Big[\langle$ 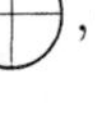 $, G(\alpha(\mathcal{K}))\rangle\Big]_{ij}^{a} - 2\Big[\langle$ 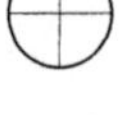 $, G(\alpha(\mathcal{K}))\rangle\Big]_i - 2\Big[\langle$  $, G(\alpha(\mathcal{K}))\rangle\Big]_j \Big)$

$+ \sum_{\substack{i,j\in\mathcal{C}_b \\ i>j}} \bar{\varepsilon}_{ij}(\mathcal{K}) \Big(\Big[\langle$  $, G(\alpha(\mathcal{K}))\rangle\Big]_{ij}^{b} - \Big[\langle$ 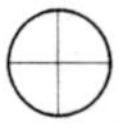 $, G(\alpha(\mathcal{K}))\rangle\Big]_i - \Big[\langle$ 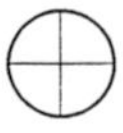 $, G(\alpha(\mathcal{K}))\rangle\Big]_j \Big),$

(VI.36)

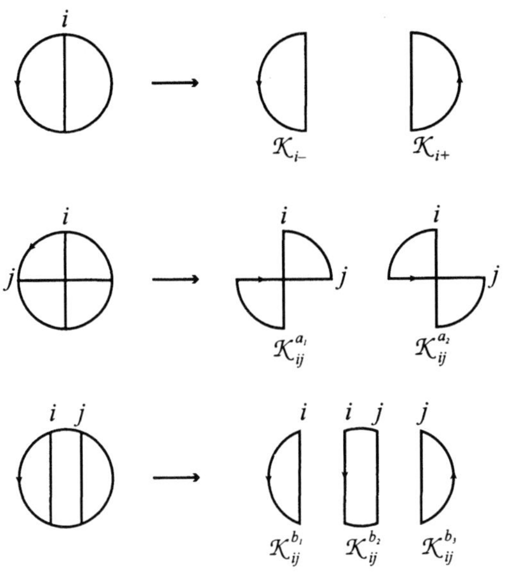

FIGURE 21. Splitting a knot into other knots.

and,

$$\alpha_{43}(K) = \alpha_{43}(U) + \langle \circledast + \otimes + \oplus + 2\,\oplus_{2} - \frac{1}{6}\oplus, \bar{G}(\mathcal{K})\rangle$$
$$+ \sum_{\substack{i,j \in \mathcal{C}_a \\ i>j}} \bar{\varepsilon}_{ij}(\mathcal{K}) \Big(\big[\langle \oplus, G(\alpha(\mathcal{K}))\rangle\big]^{a}_{ij} - \big[\langle \oplus, G(\alpha(\mathcal{K}))\rangle\big]_{i} - \big[\langle \oplus, G(\alpha(\mathcal{K}))\rangle\big]_{j} \Big). \tag{VI.37}$$

In these expressions the explicit dependence on the signatures appears in the quantities $\bar{\varepsilon}_{ij}(\mathcal{K})$ which are:

$$\bar{\varepsilon}_{ij}(\mathcal{K}) = \varepsilon_{ij}(\mathcal{K}) - \varepsilon_{ij}(\alpha(\mathcal{K})) = \varepsilon_i(\mathcal{K})\varepsilon_j(\mathcal{K}) - \varepsilon_i(\alpha(\mathcal{K}))\varepsilon_j(\alpha(\mathcal{K})). \tag{VI.38}$$

The sums in which these products are involved are over double splittings of the knot projection $\mathcal{K}$ at the crossings i and j. There are two ways of carrying out these double splittings, depending on the configuration associated to the crossings i and j. These are shown in the second and third rows of fig. 21. In the first one the regular projection $\mathcal{K}$ is split into two while in the second one it is split into three. Splittings of the first type build the set $\mathcal{C}_a$. The ones of the second type build $\mathcal{C}_b$. While only the first one is involved in the invariant α_{43}, both appear in α_{42}. The new quantities entering the sums are:

$$\begin{aligned} \Big[L(\mathcal{K})\Big]^{a}_{ij} &= L(\mathcal{K}) - L(\mathcal{K}^{a_1}_{ij}) - L(\mathcal{K}^{a_2}_{ij}), \\ \Big[L(\mathcal{K})\Big]^{b}_{ij} &= L(\mathcal{K}) - L(\mathcal{K}^{b_1}_{ij}) - L(\mathcal{K}^{b_2}_{ij}) - L(\mathcal{K}^{b_3}_{ij}), \end{aligned} \tag{VI.39}$$

where $\mathcal{K}^{a_1}_{ij}, \mathcal{K}^{a_2}_{ij}, \mathcal{K}^{b_1}_{ij}, \mathcal{K}^{b_2}_{ij}$ and $\mathcal{K}^{b_3}_{ij}$ are the knot projections which originate after a double splitting of $\mathcal{K}$, as denoted in fig. 21. As in previous orders, in the expressions (VI.36) and (VI.37), the quantities $\alpha_{42}(U)$ and $\alpha_{43}(U)$ correspond to the value of these invariants for the unknot. It has been proved in [20] that the combinatorial expressions for $\alpha_{42}(K)$ and $\alpha_{43}(K)$ in (VI.36) and (VI.37) are invariant under Reidemeister moves of fig. 2.

Vassiliev invariants constitute vector spaces and their normalization can be chosen in such a way that they are integer-valued. Once their value for the unknot has been subtracted off they can be presented in many basis in which they are integers. We will chose here a particular basis in which the numerical values for the invariants up to order four are rather simple:

$$
\begin{aligned}
\nu_2(K) &= \frac{1}{4}\tilde{\alpha}_{21}(K),\\
\nu_3(K) &= \frac{1}{8}\tilde{\alpha}_{31}(K),\\
\nu_4^1(K) &= \frac{1}{8}(\tilde{\alpha}_{42}(K) + \tilde{\alpha}_{43}(K)),\\
\nu_4^2(K) &= \frac{1}{4}(\tilde{\alpha}_{42}(K) - 5\tilde{\alpha}_{43}(K)).
\end{aligned}
\tag{VI.40}
$$

In these equations the tilde indicates that the value for the unknot has been subtracted, *i.e.*, $\tilde{\alpha}_{ij}(K) = \alpha_{ij}(K) - \alpha_{ij}(U)$. In Tables 1 and 2 we have collected the value of the Vassiliev invariants (VI.41) for all prime knots up to nine crossings. Notice that we could have chosen a basis where all the values for the trefoil knot are 1 just redefining $\nu_4^1(K)$ into $\nu_4^1(K) - 2\nu_4^2(K)$. We have no done so because $\nu_4^1(K)$, as defined in (VI.41), has a simple shape when plotted versus $\nu_2(K)$. Actually, the resulting shape has features similar to the shape which results after plotting $\nu_3(K)$ versus $\nu_2(K)$ (see [22] for more comments in this respect). The similar behavior observed for $|\nu_3(K)|$ and $\nu_4^2(K)$ is expected from their general form for torus knots. As it was shown in [67] and [56], for a torus knot characterized by two coprime integers p and q these invariants are the following polynomials in p and q:

$$
\begin{aligned}
\nu_2(p,q) &= \frac{1}{24}(p^2-1)(q^2-1),\\
\nu_3(p,q) &= \frac{1}{144}(p^2-1)(q^2-1)pq,\\
\nu_4^1(p,q) &= \frac{1}{288}(p^2-1)(q^2-1)p^2q^2,\\
\nu_4^2(p,q) &= \frac{1}{720}(p^2-1)(q^2-1)(2p^2q^2-3p^2-3q^2-3).
\end{aligned}
\tag{VI.41}
$$

The explicit expression of Vassiliev invariants as polynomials in p and q is known up to order six [67]. Of course, up to order four their value agree with the ones computed explicitly from equations (VI.36) and (VI.37), as it can be checked explicitly from the tables collected below. The only torus knots up to nine crossings are 3_1, 5_1, 7_1, 8_{19} and 9_1, whose associated coprime integers are (3,2), (5,2), (7,2), (4,3) and (9,2), respectively.

It would be desirable to write the invariants in such a way that signatures and split sums do not appear. Even better would be to possess expressions where terms involving ascending diagrams are not present. It is

TABLE 1. Primitive Vassiliev invariants up to order four for all prime knots up to eight crossings.

Knot	ν_2	ν_3	ν_4^1	ν_4^2	Knot	ν_2	ν_3	ν_4^1	ν_4^2
3_1	1	1	3	1	8_5	−1	−3	1	−41
4_1	−1	0	2	-3	8_6	−2	−3	7	-36
5_1	3	5	25	11	8_7	2	−2	4	22
5_2	2	3	13	4	8_8	2	−1	5	12
6_1	−2	−1	7	−12	8_9	−2	0	14	-34
6_2	−1	−1	3	−13	8_{10}	3	−3	15	15
6_3	1	0	0	7	8_{11}	−1	−2	2	-27
7_1	6	14	98	46	8_{12}	−3	0	14	−17
7_2	3	6	32	13	8_{13}	1	−1	−1	17
7_3	5	11	73	25	8_{14}	0	0	4	−16
7_4	4	8	50	8	8_{15}	4	7	37	18
7_5	4	8	46	24	8_{16}	1	−1	−1	17
7_6	1	2	8	-1	8_{17}	−1	0	6	−19
7_7	−1	1	−1	3	8_{18}	1	0	4	−9
8_1	−3	−3	13	−31	8_{19}	5	10	60	35
8_2	0	−1	3	30	8_{20}	2	2	8	6
8_3	−4	0	30	−40	8_{21}	0	−1	−1	−14
8_4	−3	1	21	-39					

TABLE 2. Primitive Vassiliev invariants up to order four for all prime knots with nine crossings.

Knot	ν_2	ν_3	ν_4^1	ν_4^2
9_1	10	30	270	130
9_2	4	10	62	32
9_3	9	26	228	87
9_4	7	19	151	51
9_5	6	15	115	20
9_6	7	18	134	77
9_7	5	12	78	47
9_8	0	2	8	-8
9_9	8	22	180	80
9_{10}	8	22	188	48
9_{11}	4	-9	57	10
9_{12}	1	3	15	1
9_{13}	7	18	142	45
9_{14}	-1	2	-6	5
9_{15}	2	-5	25	4
9_{16}	6	14	94	62
9_{17}	-2	0	6	-2
9_{18}	6	15	107	52
9_{19}	-2	1	3	4
9_{20}	2	4	20	6
9_{21}	3	-6	36	-3
9_{22}	-1	1	1	7
9_{23}	5	11	69	41
9_{24}	1	2	6	-5
9_{25}	0	1	11	-14

Knot	ν_2	ν_3	ν_4^1	ν_4^2
9_{26}	0	1	-5	2
9_{27}	0	1	3	-6
9_{28}	1	0	-2	3
9_{29}	1	-2	2	11
9_{30}	-1	-1	5	-9
9_{31}	2	2	8	6
9_{32}	-1	2	-2	-11
9_{33}	1	-1	3	1
9_{34}	-1	0	2	-3
9_{35}	7	18	150	13
9_{36}	3	-7	39	15
9_{37}	-3	1	13	-7
9_{38}	6	14	98	46
9_{39}	2	-4	24	-10
9_{40}	-1	-1	3	-13
9_{41}	0	1	-9	18
9_{42}	-2	0	10	-18
9_{43}	1	2	14	-13
9_{44}	0	1	-1	10
9_{45}	2	-4	20	6
9_{46}	-2	-3	3	-20
9_{47}	-1	-2	-6	5
9_{48}	3	-5	29	-5
9_{49}	6	14	102	30

not known if this is possible even for the few orders in which combinatorial expressions for the invariants exist. There are indications however that in order to achieve such a goal arrow diagrams as the ones used in [63] have to be introduced. The effect of the introduction of these diagrams is to reduce the amount of embeddings entering the product (VI.12) to a selected set. Both, the expressions and the amount of calculation could notably simplify if this is possible. This issue is under investigation.

VII CONCLUDING REMARKS

In this paper I have presented a brief review of the developments in the context of Chern-Simons gauge theory since the connection between this theory and the theory of knots and links was discovered in 1988. My presentation has started from the basics of both, the physical and mathematical theories, away from the chronological order. I hope to have convinced the reader that the interplay between the physical and mathematical approaches has been very fruitful.

Chern-Simons gauge theory has received the attention of many theoretical physicist and it has been studied from many different points of view. The non-perturbative study led to discover a close connection with the theory of knots and links, which was later further analyzed in detail. Perturbative studies led to new insights in the theory of Vassiliev invariants. These invariant appear in the context of Chern-Simons gauge theory as the coefficients of the perturbative series expansion. Their structure have been analyzed working out the perturbative series expansions for different gauge fixings. Gauge theories are very rich in this respect since they can be studied in different gauges, each providing a particular structure for the coefficients of the perturbative series expansion.

The perturbative analysis of the theory in the covariant Landau gauge led to covariant integral expressions for Vassiliev invariants. These expressions are also known as configuration space integrals. Though important, their from is rather complicated for explicit computations. Simpler expressions for the coefficients appear in non-covariant gauges. In the light-cone gauge one recovers the Kontsevich integral. In the temporal gauge one finds combinatorial expressions. In non-covariant gauges there is an important issue that has not yet been solved. In both, the light-cone and the temporal gauges, one must introduce a factor to find agreement with results in the covariant Landau gauge or with results from a non-perturbative point of view. The origin of this factor, called Kontsevich factor, is not understood. There should exist a field theory argument to justify its presence. The analysis in the temporal gauge indicates that the origin of the factor is not due to a bad choice of the prescription to avoid unphysical poles, a standard problem when dealing with non-covariant gauges. It seems it is related to the presence of a residual gauge invariance. Further work in this respect should be done. Its solution might shed some light on general problems related to non-covariant gauges.

The perturbative analysis in the temporal gauge leads to combinatorial expressions for Vassiliev invariants. We have constructed an approach that avoids integral expressions making use of the factorization theorem. It works very successfully up to order four and, indeed, up to now, this is the only framework which has provided combinatorial expressions for the two primitive Vassiliev invariants at order four. The approach opens a variety of investigations. Certainly, a generalization of the reconstruction procedure from the kernels up to order four should be constructed. This could lead to a general combinatorial formula for Vassiliev invariants. The approach is also well suited to obtain combinatorial expressions for Vassiliev invariants for links, a field which has not been much explored up to now [68]. Another context in which our approach could be also very powerful is in the study of vevs of graphs, quantities that plays an important role in recent developments in the canonical approach to quantum gravity [48]. Vassiliev invariants for graphs constitute a rather unexplored field which could lead to new sets of important invariants.

It is known that polynomial invariants, *i.e.*, vevs of Wilson loops, do not classify knots. On the one hand, polynomial invariants do not distinguish knots that are not invertible (knots which are not equivalent to the ones obtained after reversing their orientation). On the other hand, polynomial invariants do not detect mutant knots which are not equivalent [43]. The question that immediately arises is whether or not Vassiliev invariants, being the coefficients of the power series expansion of Chern-Simons gauge theory, have a chance to classify knots. Fortunately, the answer to this question is affirmative. The polynomial invariants, or vevs of Wilson loops, are associated to a group and a representation (weight system). On the other hand, Vassiliev invariants are the coefficients of the perturbative power series expansion of Chern-Simons gauge theory. Only after making a particular choice of group and representation one makes contact between this series and a polynomial invariant. But one could consider the series as a formal one whose group factors are just the diagrams of the space $\mathcal{A}$ in (III.5) or (III.6). This space might be bigger than the space of all weight systems and therefore the possibility for Vassiliev invariants to classify knots could still be open. This seems to be the case as shown in [69]. Thus, the set of coefficients of the formal perturbative power series expansion (Vassiliev invariants) is bigger than the set of vevs for any representation and any semi-simple gauge group (polynomial invariants). This is a promising indication that Vassiliev invariants might classify knots. However, simple questions as whether Vassiliev invariants ever detect nonivertibility remains open.

Let me just finish bringing to the attention of the reader the problem of the dimensions of each of the elements of the graded vector space of Vassiliev invariants (III.5) or (III.6). This problem is a very important counting problem which can be addressed from both, a diagrammatic and a group-theoretical point of view. The values of the dimensions are known only up to order 12 [49]. No insight on the problem has been obtained from Chern-Simons gauge theory. Could some field-theoretical method be used to obtain the general solution to this challenging counting problem? This is another important open question which certainly deserves further investigation.

ACKNOWLEDGMENTS

I would like to thank my collaborators in Chern-Simons gauge theory all along these ten years, M. Alvarez, L. Alvarez-Gaumé, J. M. Isidro, P. M. Llatas, M. Mariño, E. Pérez and A. V. Ramallo. Their insights have led to understand many of the issues discussed in this paper. I would also like to thank the organizers of the workshop "Trends in Theoretical Physics II", for inviting me to deliver a lecture and for their warm hospitality. I acknowledge funds provided by the European Commission, which supports the collaboration network 'CERN-Santiago de Compostela-La Plata' under contract C11*-CT93-0315, for making possible the organization of the workshop. This work is supported in part by DGICYT under grant PB96-0960.

REFERENCES

1. E. Witten, *Commun. Math. Phys.* **121** (1989) 351.
2. J. Schonfeld, *Nucl. Phys.* **B185** (1981) 157; R. Jackiw ans S. Templeton, *Phys. Rev.* **D23** (1981) 2291; S. Deser, R. Jackiw ans S. Templeton, *Phys. Rev. Lett.* **48** 975 and *Ann. Phys.* **140** (1984) 372.
3. G.V. Dunne, "Aspects of Chern-Simons Theory", Les Houches Lectures 1998, hep-th/9902115.
4. V. F. R. Jones, *Bull. AMS* **12** (1985) 103; *Ann. of Math.* **126** (1987) 335.
5. V. A. Vassiliev, "Cohomology of knot spaces", *Theory of singularities and its applications*, Advances in Soviet *Mathematics*, vol. 1, *Americam Math. Soc.*, Providence, RI, 1990, 23-69.
6. J.S. Birman, *Bull. AMS* **28** (1993) 253.
7. D. Bar-Natan, *Topology* **34** (1995) 423.
8. J.S. Birman and X.S. Lin, *Invent. Math.* **111** (1993) 225.
9. J.M.F. Labastida and E. Pérez, *Nucl. Phys.* **B527** (1998) 499, hep-th/9712139.
10. E. Guadagnini, M. Martellini and M. Mintchev, *Phys. Lett.* **B227** (1989) 111 and **B228** (1989) 489; *Nucl. Phys.* **B330** (1990) 575.
11. D. Bar-Natan, "Perturbative aspects of Chern-Simons topological quantum field theory", Ph.D. Thesis, Princeton University, 1991.
12. M. Alvarez and J.M.F. Labastida, *Nucl. Phys.* **B395** (1993) 198, hep-th/9110069, and **B433** (1995) 555, hep-th/9407076; Erratum, ibid. **B441** (1995) 403.
13. D. Altschuler and L. Friedel, *Commun. Math. Phys.* **187** (1997) 261 and **170** (1995) 41.
14. R. Bott and C. Taubes, *Jour. Math. Phys.* **35** (1994) 5247.
15. M. Kontsevich, *Advances in Soviet Math.* **16**, Part 2 (1993) 137.
16. A.S. Cattaneo, P. Cotta-Ramusino, J. Frohlich and M. Martellini, *J. Math. Phys.* **36** (1995) 6137.
17. J.M.F. Labastida and E. Pérez, *J. Math. Phys.* **39** (1998) 5183; hep-th/9710176.
18. L. Kauffman, "Witten's Integral and Kontsevich Integral", preprint.
19. S. Poirier, "Rationality Results for the Configuration Space Integral of Knots", preprint, math.GT/9901028 and "The Limit Configuration Space Integral for Tangles and the Kontsevich Integral", preprint, math.GT/9902058.
20. J.M.F. Labastida and E. Pérez, "Combinatorial Formulae for Vassiliev Invariants from Chern-Simons Gauge Theory", CERN and Santiago de Compostela preprint, CERN-TH/98-193, US-FT-11/98; hep-th/9807155.
21. M. Alvarez and J.M.F. Labastida, *Commun. Math. Phys.* **189** (1997) 641, q-alg/9604010.
22. J.M.F. Labastida and E. Pérez, "Vassiliev Invariants in the Context of Chern-Simons Gauge Theory", Santiago de Compostela preprint, US-FT-18/98; hep-th/9812105.
23. M. Goussarov, M. Polyak and O. Viro, "Finite Type Invariants of Classical and Virtual Knots", preprint, 1998, math.GT/9810073.
24. P. Freyd, D. Yetter, J. Hoste, W.B.R. Lickorish, K. Millet and A. Ocneanu, *Bull. AMS* **12** (1985) 239.
25. L.H. Kauffman, *Trans. Am. Math. Soc.* **318** (1990) 417.
26. Y. Akutsu and M. Wadati, *J. Phys. Soc. Jap.* **56** (1987) 839 and 3039.
27. E. Witten, *Nucl. Phys.* **B322** (1989) 629 and **B330** (1990) 285.
28. C. Kassel, M. Rosso and V. Turaev, "Quantum groups and knot invariants", Panoramas et syntheses 5, Societe Mathematique de France, 1997.
29. C. Kassel and V. Turaev, "Chord diagram invariants of tangles and graphs", University of Strasbourg preprint, 1995.
30. N.Y. Reshetikhin and V. Turaev, *Invent. Math.* **103** (1991) 547.
31. D.S. Freed and R.E. Gompf , *Commun. Math. Phys.* **141** (1991) 79.
32. R. Gambini, private communication.
33. R.K. Kaul, "Chern-Simons Theory, Knot Invariants, Vertex Models and Three-Manifold Invariants, hep-th/9804122.
34. M. Bos and V.P. Nair, *Phys. Lett.* **B223** (1989) 61 and *Int. J. Mod. Phys.* **A5** (1990) 959.
35. J.M.F. Labastida and A.V. Ramallo, *Phys. Lett.* **B227** (1989) 92 and **B228** (1989) 214.
36. S. Elitzur, G. Moore, A. Schwimmer and N. Seiberg, *Nucl. Phys.* **B326** (1989) 108.
37. S. Axelrod, S. Della Pietra and E. Witten, *j. Diff. Geom.* **33** (1991) 787.
38. J. Frohlich and C. King, *Commun. Math. Phys.* **126** (1989) 167.
39. R.K. Kaul and T.R. Govindarajan, *Nucl. Phys.* **B380** (1992) 293 and **B393** (1993) 392; P. Ramadevi, T.R. Govindarajan and R.K. Kaul, *Nucl. Phys.* **B402** (1993) 548; R.K. Kaul, *Commun. Math. Phys.* **162** (1994) 289.
40. S. Martin, *Nucl. Phys.* **B338** (1990) 244.
41. J.M.F. Labastida, P.M. Llatas and A.V. Ramallo, *Nucl. Phys.* **B348** (1991) 651; J.M.F. Labastida and M. Mariño, *Int. J. Mod. Phys.* **A10** (1995) 1045; J.M.F. Labastida and E. Pérez, *J. Math. Phys.* **37** (1996) 2013.
42. P. Ramadevi, T.R. Govindarajan and R.K. Kaul, *Mod. Phys. Lett.* **A9** (1994) 3205.

43. P. Ramadevi, T.R. Govindarajan and R.K. Kaul, *Mod. Phys. Lett.* **A10** (1995) 1635.
44. J. M. Isidro, J. M. F. Labastida and A. V. Ramallo, *Phys. Lett.* **B282** (1992) 63; *Nucl. Phys.* **B398** (1993) 187; I. P. Ennes, P. Ramadevi, A. V. Ramallo and J. M. Sanchez de Santos, *Int. J. Mod. Phys.* **A13** (1998) 2931.
45. R.K. Kaul, *Nucl. Phys.* **B417** (1994) 267; P. Ramadevi, T.R. Govindarajan and R.K. Kaul, *Nucl. Phys.* **B422** (1994) 291.
46. Carlo Rovelli and Lee Smolin, *Phys. Rev. Lett.* **68** (1992) 431; *Nucl. Phys.* **B385** (1992) 587.
47. B. Bruegmann, R. Gambini and J. Pullin, *Phys. Rev. Lett.* **61** (1988) 1155; *Nucl. Phys.* **B331** (1990) 80.
48. R. Gambini, J. Griego and J. Pullin, *Phys. Lett.* **B425** (1998) 41 and *Nucl. Phys.* **B534** (1998) 675.
49. J. A. Kneissler, "The number of primitive Vassiliev invariants up to degree 12", Math. Inst. Bonn, 1997, q-alg/9706022.
50. D. Thurston, "Integral expressions for the Vassiliev knot Invariants", Harvard University senior thesis, April 1995; math/9901110.
51. S. Axelrod and I.M. Singer, "Chern-Simons Perturbation Theory", 1991, hep-th/9110056 and *J. Diff. Geom.* **39** (1994) 173.
52. W. Chen, G.W. Semenoff, and Y.-S. Wu, *Mod. Phys. Lett.* **A5** (1990) 1833.
53. L. Alvarez-Gaumé, J.M.F. Labastida and A. V. Ramallo, *Nucl. Phys.* **B334** (1990) 103; G. P. Korchemsky, *Mod. Phys. Lett.* **A6** (1991) 727.
54. C.P. Martin, *Phys. Lett.* **B241** (1990) 513; G. Giavarini, C.P. Martin, F. Ruiz Ruiz, *Nucl. Phys.* **B381** (1992) 222, *Phys. Lett.* **B314** (1993) 328, *Phys. Rev.* **D47** (1993) 5536 and *Phys. Lett.* **B332** (1994) 345.
55. M. Asorey and F. Falceto, *Phys. Lett.* **B241** (1990) 31; M. Asorey, F. Falceto, J.L. Lopez and G. Luzon, *Phys. Rev.* **D49** (1994) 5377 and *Nucl. Phys.* **B429** (1994) 344.
56. S. Willerton, "On Universal Vassiliev Invariants, Cabling, and Torus Knots", University of Melbourne preprint (1998).
57. S.-W. Yang, "Feynman integral, knot invariant and degree theory of maps", National Taiwan University preprint, September 1997; q-alg/9709029.
58. A.C. Hirshfeld and U. Sassenberg, *Journal of Knot Theory and its Ramifications* **5** (1996) 489 and **5** (1996) 805.
59. G. Leibbrandt and C.P. Martin, *Nucl. Phys.* **B377** (1992) 593 and **B416** (1994) 351.
60. A. Brandhuber, S. Emery, M. Langer, O. Piguet, M. Schweda and S.P. Sorella, *Helv. Phys. Acta* **66** (1993) 551.
61. G. Leibbrandt, *Rev. Mod. Phys.* **59** (1987) 1067.
62. T.Q.T. Le and J. Murakami, "The universal Vassiliev-Kontsevich invariant for framed oriented links", Max-Plank Institut fur Mathematik preprint, 1994.
63. M. Goussarov, M. Polyak and O. Viro, *Int. Math. Res. Notices* **11** (1994) 445.
64. J.F.W.H. van de Wetering, *Nucl. Phys.* **B379** (1992) 172.
65. L. H. Kauffman, "Virtual Knot Theory", preprint, 1998.
66. P. Lannes, *L'Enseignement Math.* **39** (1993) 295.
67. M. Alvarez and J.M.F. Labastida, *Journal of Knot Theory and its Ramifications* **5** (1996) 779; q-alg/9506009.
68. M. Alvarez, J.M.F. Labastida and E. Pérez, *Nucl. Phys.* **B488** (1997) 677.
69. Pierre Vogel, *Algebraic structures on modules of diagrams,* Université Paris VII preprint, July 1995.

The quantum mass of the susy kink and the BPS bound

Peter van Nieuwenhuizen[1]

Institute for Theoretical Physics
State University of New York at Stony Brook
Stony Brook, New York, 11794-3840, USA

Abstract. Recently, the long-standing problem of calculating the quantum mass of supersymmetric **and nonsupersymmetric** solitons $1+1$ dimensions in an unambiguous way, and the issue of whether or not it satisfies the BPS bound at the quantum level, was solved [1-4]. We discuss here the kink. One basic idea is to compute $\frac{\partial}{\partial m}\sum\omega$ (which converges) instead of $\sum\omega$, and to fix the integration constant by imposing the renormalization condition that the vacuum energies of the topological sector and the nontopological sector become equal when m and other dimensional constants (if present) tend to zero. Another new idea is to derive the boundary conditions from the symmetries of the action and the topology of the classical solution. For the bosonic fluctuations about the kink we then find **antiperiodic** boundary conditions, while for the fermionic fluctuations we find **twisted** boundary conditions. The almost universally used periodic or antiperiodic boundary conditions are shown to be inconsistent for fermions (!). The central charge Z in the supersymmetry algebra is classically given by the usual total derivative, but at the quantum level an "anomaly" (an extra term of order $\hbar$) is found in Z. This term is calculated by using a higher-derivative regularization scheme which preserves supersymmetry. Regulating the supersymmetry generators in this way, the BPS bound remains saturated for the supersymmetric kink, but both the mass and the central charge receive (equal) quantum corrections.

I INTRODUCTION

More than two decades ago the quantization of (supersymmetric) solitons was studied. The more recent duality between extended objects and pointlike objects in string theory and quantum field theory has brought interest in the quantum properties of solitons back. In this contribution we report on a recent solution to the 20-year old problem of how to compute the quantum mass of a (supersymmetric) soliton, and whether the BPS bound remains saturated at the quantum level.

Classical solitons (by which we mean static nonsingular solutions of nonlinear field equations with finite energy) become at the quantum level a background field about which quantum fluctuations occur. The original field $\varphi(x,t)$ in the action is then considered to be a sum of the soliton field $\varphi_{sol}(x)$ and a quantum field, $\varphi(x,t)=\varphi_{sol}(x)+\eta(x,t)$. We consider $1+1$ dimensional relativistic field theories, and assume that the soliton has a translational (and only a translational) zero mode. In that case one writes

$$\varphi(x,t)=\varphi_{sol}(x-X(t))+\sum{}'q^m(t)\eta_m(x-X(t)) \tag{I.1}$$

where $\eta_m(x)$ are the nonzero modes (solutions of the linearized field equations in the presence of the soliton with nonzero eigenvalues). Since $\frac{d}{dx}\varphi_{sol}(x)$ is the (normalizable) zero mode for translations, one has not lost any degrees of freedom. Denoting the set of variables $\{X(t),q^m(t)\}$ by $u^I(t)$, the Lagrangian can be written as $L=\frac{1}{2}\dot{u}^I g_{IJ}\dot{u}^J - V(u)$ where $g_{IJ}(u)=\int \partial\varphi(x,t)/\partial u^I \partial\varphi(x,t)/\partial u^J dx$. The Hamiltonian is then simply given by $H=\frac{1}{2}\pi_I g^{IJ}\pi_J + V(u)$ where $\pi_I(t)=\{P(t),\pi_m(t)\}$ are the canonically conjugate momenta. To cast this "quantum mechanical" Hamiltonian into a form which resembles more the Hamiltonian of a $1+1$ dimensional field theory, one introduces fields

---

[1] NSF Grant PHY 972-2101. E-mail: vannieu@insti.physics.sunysb.edu

CP484, *Trends in Theoretical Physics II*, edited by H. Falomir, R. E. Gamboa Saraví, and F. A. Schaposnik

$$\eta(x,t) \equiv \sum{}' q^m(t)\eta_m(x - X(t))$$
$$\pi(x,t) \equiv \sum{}' \pi_m(t)\eta_m(x - X(t)) \tag{I.2}$$

One finds then for the classical Hamiltonian the following expression

$$H_{sol} = M_{cl} + \int_{-\infty}^{\infty} \left[\frac{1}{2}\pi(x,t)^2 + \frac{1}{2}\eta'(x,t)^2 + \frac{1}{2}\eta^2 V''(\varphi_{sol})\right] dx$$
$$+ \frac{[P + (\pi, \eta')]^2}{2M_{cl}} \frac{1}{1 + (\eta', \varphi'_{sol})/M_{cl}} + \int \sum_{n=3}^{\infty} \frac{1}{n!}\eta^n V^{(n)}(\varphi_{sol}) dx \tag{I.3}$$

(The notation (f, g) is used for the inner products $\int f^*(x)g(x)dx$. Of course $(\varphi'_{sol}, \eta_m) = 0$ because φ'_{sol} and η_m have different eigenvalues of the kinetic operator).

At the quantum level there are two sources of modifications of the Hamiltonian
(i) Operator orderings. These are fixed by requiring that the generators $H = \int T_{00}dx, P = \int T_{01}dx$ and $L = \int xT_{00}dx$ satisfy the Poincaré algebra.[5] The solution is that $H = \frac{1}{2}\pi_I g^{IJ}\pi_J$ should be replaced by

$$\hat{H} = \frac{1}{2} g(u)^{-\frac{1}{4}} \pi_I g^{IJ}(u) \pi_J g(u)^{-\frac{1}{4}} \tag{I.4}$$

where $[Q, P] = i\hbar$ and $[q^m, \pi_n] = i\hbar\delta^m_n$.
(ii) Renormalization. In the $1 + 1$ dimensional field theories we consider, only mass renormalization is needed to remove infinities

$$m_0^2 = m^2 + \delta m^2 \; ;$$ 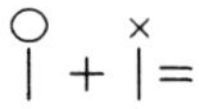 $$= 0 \tag{I.5}$$

and δm^2 is fixed by a suitable renormalization condition. For the kink, we require that tadpoles vanish in the vacuum without soliton. All terms with one or more factors δm^2 are denoted by ΔH.[1]

¿From this expression of the quantum Hamiltonian we read off that at the one-loop level the quantum corrections $M^{(1)}$ to the classical mass M_{cl} of a soliton at rest ($\langle P \rangle = 0$) are given by $\langle H \rangle_{sol} = \sum \frac{1}{2}\hbar\omega + \langle \Delta H \rangle$, where ω are the solutions of

$$\left[-\frac{d^2}{dx^2} + V''(\varphi_{sol}(x))\right] \eta(x) = \omega^2 \eta(x) \tag{I.6}$$

To regulate the sum over frequencies we put the system in a box. To determine these solutions, we need to specify the boundary conditions. The sum diverges (quadratically), hence it has to be regularized. One considers the difference of the energy of the topological vacuum and the energy of the trivial vacuum: $\sum \frac{1}{2}\hbar\omega - \sum \frac{1}{2}\hbar\omega^{(0)}$, but even this expression diverges (logarithmically). Adding the mass counter term $\delta M = \langle \Delta H \rangle$, one finds a finite expression. Adding the sum over fermionic (F) frequencies to the bosonic (B) frequencies of supersymmetric models one finds

$$M^{(1)} = \sum \frac{1}{2}\hbar\omega_B - \frac{1}{2}\sum \omega_B^{(0)} - \sum \frac{1}{2}\hbar\omega_F + \sum \frac{1}{2}\hbar\omega_F^{(0)} + \delta M \tag{I.7}$$

The problem is then to give meaning to these infinite sums, in particular, to determine the finite parts in an unambiguous way. Two decades of conflicting results can be summarized as follows: the expression for $M^{(1)}$ is finite but
(i) as observed by Schonfeld [7], different boundary conditions given different $M^{(1)}$.
(ii) different regularization schemes give different $M^{(1)}$. In particular, susy practitioners prefer to cut both sums off at the same energy. We shall call this procedure "energy cut-off". On the other hand, lattice gauge field theorists prefer to cut both series off at the same value $n = N$. We shall call this "mode cut-off". Note

---

[1)] For applications to path integrals it is useful to **rewrite** $\hat{H}$ in Weyl-ordered form, because this is the form of the Hamiltonian which is used for path integrals. Weyl-reordering of $\hat{H}$ in (1.4) leads to an extra term ΔV of order $\hbar^2$ (terms linear in $\hbar$ cancel).[6] The complete quantum Hamiltonian then reads $\hat{H}_{\text{ren}} = (\frac{1}{2}\pi_I g^{IJ}\pi_J)_W + V(u) + \Delta V + \Delta H$.

that there are various mode cut-offs, depending on whether one includes the bound states and/or zero modes in the sum over n. As announced in the abstract, we shall find unique topological boundary conditions, and compute $\frac{d}{dm}\sum\omega$ instead of $\sum\omega$. Imposing the physical renormalization condition that if the soliton disappears ($m \to 0$), also $M^{(1)}$ vanishes, we find a unique answer.

The results on the quantum mass of supersymmetric and non-supersymmetric solitons are due to Nastase, Rebhan, Stephanov and van Nieuwenhuizen.[1,2] The next issue concerns the BPS bound, and here we discuss results for supersymmetric solitons by Shifman, Vainshtein and Voloshin.[3] Naive canonical quantization leads to $\{Q^+, Q^-\} = 2\int_{-\infty}^{\infty} \frac{d}{dx}(\int U(\varphi)d\varphi)dx$, where for the kink $\int U(\varphi)d\varphi \equiv W(\varphi) = \sqrt{\frac{\lambda}{2}}\left(\frac{1}{3}\varphi^3 - \frac{\mu^2}{\lambda}\varphi - \frac{\delta\mu^2}{\lambda}\varphi\right)$. Taking the vacuum expectation value of this expression, using $\langle\eta^2(x)\rangle = \frac{1}{2\pi}\int_0^{\Lambda}\frac{dk}{\sqrt{k^2+m^2}}$ and $\delta m^2 = \frac{\lambda\hbar}{\pi}\int_0^{\Lambda}\frac{dk}{(k^2+m^2)^{1/2}}$, one finds that all quantum corrections to Z cancel. As a result, the BPS bound seems violated.[1]

However, as well-known, anomalies seem absent in naive canonical quantization, but if one regulates the composite operators, extra terms of the form $\hbar M^2/M^2$ are found which give a nonvanishing contribution at the quantum level when $M^2 \to \infty$. The problem is then to find a suitable regularization scheme. What does suitable mean? For us it means maintaining supersymmetry. As such one may use old ideas of B. Lee, Slavnov and others, and add terms with higher derivatives to the action. For gauge theories this only regulates from 2 loops on, but for the models we consider also the 1-loop graphs are regularized. To maintain the possibility of using standard canonical formalism, we only add higher space derivatives. This violates relativistic invariance, but since already the classical soliton breaks Lorentz invariance, this is not a problem.

Constructing the Noether current for supersymmetry transformations, one can then compute $\{Q^+, Q^-\} = 2Z$ and find an extra term in Z. This term yields a finite one-loop correction to Z [3], and with this result for Z, and our earlier result for $M^{(1)}$ [2], the BPS bound is saturated at the one-loop level.

We believe that our result for the one-loop corrections to the mass of the kink is correct because it saturates the BPS bound, and because it agrees with results obtained from the Yang-Baxter equation.[8] The latter method uses, of course, quite different physical ideas. Our result disagrees with the many articles in the literature that found no correction to the mass. However, most of these articles do not use regularization. In 4 dimensions, the phenomenon of "multiplet shortening" allowed Olive and Witten to conclude that the BPS bound must be satisfied at the quantum level. In our 2-dimensional case, there is no multiplet shortening for the $N = 1$ theory we consider, hence saturation of the BPS bound is not necessary. Yet, it occurs.

II THE QUANTUM MASS

The action for the kink is

$$\mathcal{L} = -\frac{1}{2}(\partial_\mu\varphi)^2 - \frac{1}{2}U^2(\varphi);\ U(\varphi) = \sqrt{\frac{\lambda}{2}}\left(\varphi^2 - \frac{\mu_0^2}{\lambda}\right) \tag{II.1}$$

In the trivial vacuum, setting $\mu_0^2 = \mu^2 + \delta\mu^2$, we have $\langle\varphi\rangle = \pm\frac{\mu}{\sqrt{\lambda}}$, and decomposing $\varphi = \mu/\sqrt{\lambda} + \eta$ we obtain

$$\mathcal{L} = -\frac{1}{2}(\partial_\mu\eta)^2 - \mu^2\eta^2 - \mu\sqrt{\lambda}\eta^3 - \frac{1}{4}\lambda\eta^4 + \frac{1}{2}\delta\mu^2\left(\eta^2 + \frac{2\mu}{\sqrt{\lambda}}\eta\right) \tag{II.2}$$

We fix $\delta\mu^2$ by requiring that there are no tadpoles

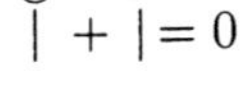

$$= 0 \Rightarrow \delta\mu^2 = \frac{3\lambda\hbar}{2\pi}\int_0^{\Lambda}\frac{dk}{\sqrt{k^2+m^2}} \quad \textit{by a finite correction} \tag{II.3}$$

The mass $m^2 \equiv 2\mu^2$ is the renormalized mass of the meson in the action. The physical mass m_{ph}^2 is obtained from the pole in the propagator, and differs from m^2

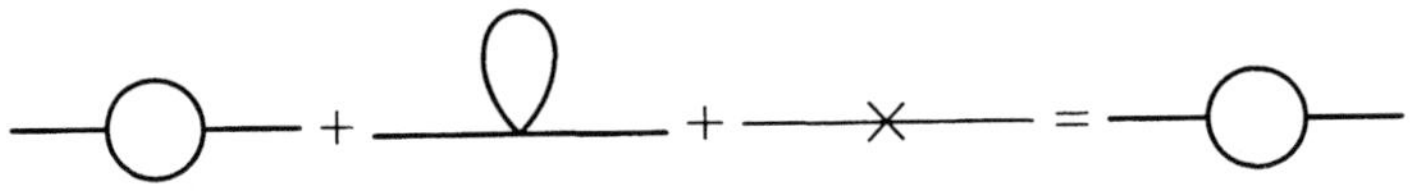

$$= m_{ph}^2 = m^2 - \frac{\sqrt{3}}{2}\hbar\lambda \tag{II.4}$$

but we will not use it.

In the topological sector there is a kink with a "winding" number

$$\varphi_{sol}(x) = \frac{\mu}{\sqrt{\lambda}} \tanh \mu \left(\frac{x - x_0}{\sqrt{2}} \right) \; ; \; M_{cl} = \frac{2\sqrt{2}\mu^3}{3\lambda} \tag{II.5}$$

This is a solution of the classical field equations (so with μ but without $\delta\mu^2$). Decomposing $\varphi = \varphi_{sol}(x) + \eta(x,t)$, we find to first order in $\delta\mu^2$ a correction to the mass

$$\begin{aligned}\delta M = \langle \Delta H \rangle &= \int_{-\infty}^{\infty} \left[\frac{\lambda}{4} \left(\varphi_{sol}^2 - \frac{\mu^2 + \delta\mu^2}{\lambda} \right)^2 - \frac{\lambda}{4} \left(\varphi_{sol}^2 - \frac{\mu^2}{\lambda} \right)^2 \right] dx \\ &= -\frac{1}{2}\delta\mu^2 \int_{-\infty}^{\infty} \left(\varphi_{sol}^2 - \frac{\mu^2}{\lambda} \right) dx = \frac{m}{\lambda}\delta\mu^2 \end{aligned} \tag{II.6}$$

In fact, $M_{cl}(\mu_0) = M_{cl}(\mu) + \delta M$. The quantum fluctuations $\eta(x,t) = \eta(x)e^{-i\omega t}$ are expanded into solutions of

$$\left(-\frac{d}{dx^2} - \mu^2 + 3\lambda\varphi_{sol}^2(x) \right) \eta = \omega^2 \eta \tag{II.7}$$

There are two bound states (one of which is the translational zero mode) and a continuum

$$\omega_0 = 0, \omega_B = \frac{1}{2}\sqrt{3}m, \omega = \sqrt{k^2 + m^2} \tag{II.8}$$

For large $x, \eta(x) \longrightarrow \exp i \left(kx \pm \frac{1}{2}\delta(k) \right)$ where the phase shift $\delta(k)$ is given by

$$\delta(k) = -2 \text{ arctg } (3mk/(m^2 - 2k^2)) \tag{II.9}$$

a. Bosonic boundary conditions

To discretize the sum over ω we enclose the system in a "box" of length L and impose suitable boundary conditions. Then first L is sent to infinity, and next Λ is removed by $\Lambda \to \infty$. The simplest (but incorrect) choice would be periodic boundary conditions (PBC)

$$\begin{aligned} k_n \frac{L}{2} + \frac{1}{2}\delta(k_n) &= -k_n \frac{L}{2} - \frac{1}{2}\delta(k_n) + 2\pi n \\ dkL + \delta'(k)dk &= 2\pi dn = dk^{(0)}L \end{aligned} \tag{II.10}$$

The 1-loop correction to the mass is given by $M^{(1)} = \delta M + \sum \left(\frac{1}{2}\hbar\omega_n - \frac{1}{2}\hbar\omega_n^{(0)} \right)$. With an energy cut-off (e.c.) one finds then[2]

$$M_{e.c.}^{(1)} = \delta M + \frac{1}{2}\hbar\omega_B + \frac{1}{2}\hbar \int_{-\Lambda}^{\Lambda} \sqrt{k^2 + m^2}\, \delta'(k) \frac{dk}{2\pi} = \frac{\hbar m}{4\sqrt{3}} \tag{II.11}$$

We used the same cut-off Λ in δM as in the last term, the justification being that at high energies one does not feel the presence of the soliton. With mode cut-off (m.c.) one finds instead[3]

$$\begin{aligned} M_{m.c.}^{(1)} &= \delta M + \frac{1}{2}\hbar\omega_B + \frac{\hbar}{2} \sum_{-N+1}^{N-1} \sqrt{(k_n)^2 + m^2} - \frac{\hbar}{2} \sum_{-N}^{N} \sqrt{(k_n^{(0)})^2 + m^2} \\ &= \frac{\hbar\, m}{4\sqrt{3}} - \frac{3\hbar m}{2\pi} \end{aligned} \tag{II.12}$$

---

2) Use $\int_{-\infty}^{\infty} dx(1 + x^2)^{-1/2}(1 + 4x^2)^{-1} = 2\sqrt{3}\pi/9$.

3) Write the two sums as $S = 2 \cdot \hbar/2 \sum_{n=1}^{N} (\sqrt{k_{n-1}^2 + m^2} - \sqrt{k_n^{(0)2} + m^2}) - \hbar m$, and use $k_{n-1}L + \delta(k_{n-1}) = 2\pi(n-1)$ and $k_n^{(0)}L = 2\pi n$. Then with the $\delta(k)$ in (2.15) one finds $S = -\frac{\hbar}{2\pi} \int_0^{\Lambda} \frac{d}{dk}\sqrt{k^2 + m^2}\delta(k)dk - \hbar m$.

Note that we took 2 continuum modes less in the topological sector because there are 2 bound states. More generally, for an arbitrary continuous function $f(k^2+m^2)$ the difference between energy cut-off and mode cut-off is

$$\frac{1}{\pi}f(k^2+m^2)[\delta(k)]_0^\infty + 2f(m^2) = \lim_{k\to\infty}\frac{1}{\pi}f(k^2+m^2)\delta(k) \tag{II.13}$$

For $f=\frac{\hbar}{2}\sqrt{k^2+m^2}$ this yields $3\hbar m/(2\pi)$. This difference is positive because the kink is attractive: at fixed energy more modes fit in the topological sector than in the trivial sector. Phrased differently, the lattice for energy cut-off is finer then for mode cut-off.

Clearly, energy cut-off and mode cut-off give different results for PBC. There are many other choices of boundary conditions, and in general e.c. and m.c. give different results.

The main idea of our new approach to boundary conditions is to view the fields $\varphi(x,t)$ as basic, and the decomposition into a classical background and quantum deviations as only useful for perturbative calculations. With this essentially nonperturbative point of view we read off from the background solution what the boundary conditions should be, but then we forget about the background solutions, and only consider $\varphi(x,t)$. There are two sectors in the kink: a trivial sector which has PBC (periodic boundary conditions), and a topological sector with APB.

We now present what we believe to be the correct boundary conditions.[2] We note that the action has the Z_2 symmetry $\varphi\to-\varphi$. Since the kink is antiperiodic, $\varphi_{sol}(x)=-\varphi_{sol}(-x)$, we require that the whole field $\varphi(x,t)$ should satisfy antiperiodic boundary conditions (APB), but then also the quantum fields $\eta(x,t)$ should satisfy APB.

$$\eta(-L/2)=-\eta(L/2)\ ;\ \eta'(-L/2)=-\eta'(L/2) \tag{II.14}$$

(The kink in an infinite volume satisfies actually $\eta'(-\infty)=\eta'(\infty)=0$, but in a finite box we can find a solution of the classical field equations with APB. Use the analogy with a particle rolling down a hill). The APB are similar to a Moebius strip: there is no real boundary.

These boundary conditions give the quantization rule

$$k_nL+\delta(k_n)=\pi+2\pi n \tag{II.15}$$

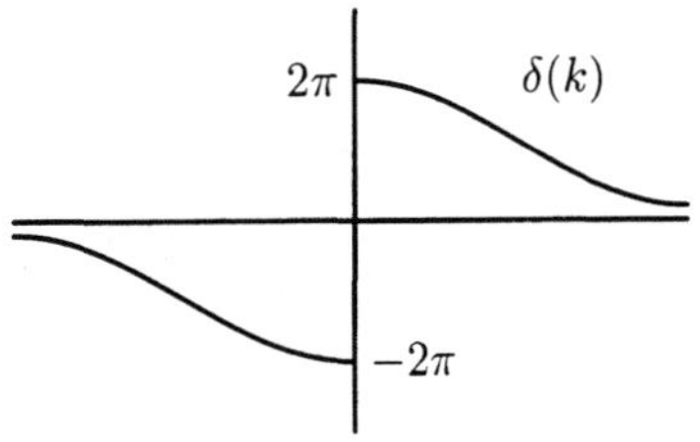

So $k_n=\ldots 3\pi/L,\pi/L,-\pi/L,-3\pi/L\ldots$, as $n=\ldots 2,1,-2,-3\ldots$. We can assign the "unclaimed" n's, namely $n=0$ and $n=-1$, to the two bound states. (This may only seem amusing, but has a deeper justification due to Levinson's theorem [4]).

b. Fermionic boundary conditions

The supersymmetric extension of the action for the kink reads

$$\mathcal{L}=-\frac{1}{2}(\partial_\mu\varphi)^2-\frac{1}{2}U^2-\frac{1}{2}\bar{\psi}\partial\!\!\!/\psi-\frac{1}{2}U'\bar{\psi}\psi \tag{II.16}$$

Because there are now also fermionic tadpoles, the value of $\delta\mu^2$ changes, namely $\delta\mu^2$ is a factor 1/3 smaller than (2.3) (but not zero!). The action has the Z_2 symmetry $\varphi\to-\varphi,\psi\to\gamma_1\gamma_0\psi$ (the mass term violates chiral symmetry but $U'\bar{\psi}\psi$ is invariant). The fermionic fluctuations $\psi=\psi(x)e^{-i\omega t}$ satisfy $[\partial\!\!\!/+U'(\varphi_{sol})]\psi(x)e^{-i\omega t}=0$, and after multiplication by $(-\partial\!\!\!/+U')$ we obtain

$$\left[-\partial_x^2-\omega^2+(U')^2-\gamma^1U''\varphi'_{sol}\right]\psi=0 \tag{II.17}$$

Note that $\varphi'_{sol}=-U(\varphi_{sol})$, and all U in this equation depend on φ_{sol}. Define $\psi_\pm=P_\pm\psi$ with $P_\pm=\frac{1}{2}(1\pm\gamma^1)$. Then the equation for ψ_+ is the same as for η in (2.7), while ψ_- follows from

$$\partial_0 \gamma^0 \psi_- + (\partial_x + U')\psi_+ = 0 \tag{II.18}$$

We use the representation $\gamma^1 = \begin{pmatrix} 1 & 0 \\ 0 & -1 \end{pmatrix}, \gamma^0 = \begin{pmatrix} 0 & -1 \\ 1 & 0 \end{pmatrix}$ and find then for $x \to \pm\infty$

$$\psi = \begin{pmatrix} \psi_+ \\ \psi_- \end{pmatrix} = Re \begin{pmatrix} 1 \\ -e^{\pm i\theta/2} \end{pmatrix} e^{-i\omega t + i(kx \pm \frac{1}{2}\delta(k))} \tag{II.19}$$

where $e^{i\theta/2} = \frac{k-im}{\omega}$. (We used that $U' \to \pm m$ as $x \to \pm\infty$). So both ψ_+ and ψ_- are given by cosines. Clearly, (2.19) satisfies (2.18) because $P_\pm \partial_0 \gamma^0 \psi = -i\omega\gamma^0\psi_\mp = \pm i\omega\psi_\pm = -P_\pm(\partial_x + U')\psi = -(\partial_x + U')\psi_\pm$, since $+i\omega(-e^{\pm i\theta/2}) = -ik \mp m$. The new (topological) boundary conditions for the fermion become now

$$\psi_+(-L/2) = -\psi_-(L/2);\ \psi_-(L/2) = -\psi_+(L/2) \tag{II.20}$$

We call them twisted boundary conditions because they are antiperiodic (as for the bosons) and at the same time change the spinor indices.[4] Imposing these boundary conditions on (2.19) leads to the following quantization condition

$$kL + \delta(k) + \frac{1}{2}\theta(k) = 2\pi n + \pi \tag{II.21}$$

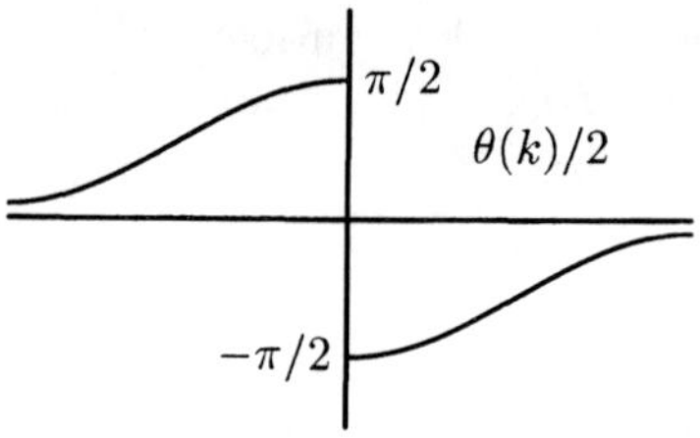

This condition holds both for ψ_+ and ψ_-. Note PBC or ABC for ψ_+ and ψ_- are inconsistent!!

c. Renormalization condition

Having determined the frequencies ω_n for the continuum modes in the box, we must now evaluate the sums in $M^{(1)}$. Instead we first sum over $\frac{\partial}{\partial m}\omega$, and afterwards integrate over m. Since $\delta(k)$ only depends on k/m and $k_n L + \delta(k_n) = 2\pi n + \pi$, we find for the bosonic case

$$\frac{dk}{dm} = -\frac{1}{L}\frac{d}{dm}\delta(k) = \frac{1}{L}\frac{k}{m}\frac{d}{dk}\delta(k);\ \frac{d}{dm}\sqrt{(k^2+m^2)} = \frac{1}{\sqrt{k^2+m^2}}\left(m + k\frac{dk}{dm}\right) \tag{II.22}$$

Then

$$\begin{aligned} \frac{d}{dm}M^{(1)} &= \frac{d}{dm}\delta M + \frac{d}{dm}\frac{1}{2}\hbar\omega_B + \frac{1}{2}\hbar\int_{-\infty}^{\infty}\frac{1}{\sqrt{k^2+m^2}} \\ &\left[\left(m + \frac{1}{L}\frac{k^2}{m}\delta'(k)\right)\left(\frac{Ldk}{2\pi}\right)\left(1 + \frac{1}{L}\delta'(k)\right) - mL\frac{dk}{2\pi}\right] \\ &= \left(\frac{1}{4\sqrt{3}} - \frac{3}{2\pi}\right)\hbar m \end{aligned} \tag{II.23}$$

Hence we get the same answer as from mode cut-off. This result also agrees with the result obtained by Dashen, Hasslacher and Neveu from a semiclassical approach [9].

---

[4] The helicity eigenstates are $\psi_+ + \psi_-$ and $\psi_+ - \psi_-$ and one is periodic while the other is antiperiodic. In string terminology one is in the R sector while the other is in the NS sector.

TABLE 1. Summary of one-loop corrections to the mass of the kink or the supersymmetric kink. We consider three regularization schemes: energy cut-off, mode cut-off, and differentiation w.r.t. $\frac{\partial}{\partial m}$. We also consider two boundary conditions in the kink sector for the boson: periodic or antiperiodic (=topological). For the fermion both PBC and ABC are inconsistent, as explained in the text, so we only consider topological boundary conditions. In the trivial sector (the sector without soliton) we use periodic boundary conditions both for the fermions and the bosons. The correct answer for the susy soliton is $-m/2\pi$ (times $\hbar$).

kink/superkink	energy cut-off	mode cut-off	$\frac{\partial}{\partial m}$ regularization
bosons PBC, fermions top	$\frac{m}{4\sqrt{3}} \Big/ 0$	$\frac{m}{4\sqrt{3}} - \frac{3m}{2\pi} \Big/$ diverges	$\frac{m}{4\sqrt{3}} - \frac{3m}{2\pi} \Big/ -\frac{m}{2\pi}$
bosons ABC, fermions top	$\frac{m}{4\sqrt{3}} \Big/ 0$	$\frac{m}{4\sqrt{3}} - \frac{3m}{2\pi} \Big/ \frac{\pi-2}{4\pi}m$	$\frac{m}{4\sqrt{3}} - \frac{3\pi}{2\pi} \Big/ -\frac{m}{2\pi}$

d. The quantum mass of the susy kink

Our general analysis of boundary conditions instructs us to use PBC for bosons and fermions in the trivial vacuum. Then for the susy system, the bosonic and fermionic vacuum fluctuations ω_n cancel pairwise. Also the bound states (and zero modes) cancel in pairs.

In the topological sector we find the same result for the bosons as before, with $\delta'(k)$ in (2.23). For the fermionic fluctuations we now find the same results, but with $\delta'(k) + \frac{1}{2}\theta'(k)$ and with an overall minus sign. Thus the terms with $\delta'(k)$ cancel, and one finds[5]

$$\frac{d}{dm}\delta M^{(1)} = \frac{d}{dm}\delta M - \frac{1}{2}\hbar \int_{-\infty}^{\infty} \frac{1}{m}\sqrt{k^2+m^2}\,\frac{1}{2}\theta'(k)\frac{dk}{2\pi} = -\hbar/2\pi \tag{II.24}$$

Hence, the one loop correction to the mass of the susy kink is

$$M^{(1)} = -(\hbar/2\pi)m \tag{II.25}$$

This result disagrees with the results from mode cut-off or energy cut-off with PBC or ABC, as Table 1 shows.

III THE QUANTUM CENTRAL CHARGE Z

The central charge Z appears in the susy algebra as follows

$$\{Q^\alpha, \bar{Q}_\beta\} = (\gamma^\mu)^\alpha{}_\beta P_\mu + (\tau^3)^\alpha{}_\beta Z \tag{III.1}$$

All fields are Heisenberg operators, and this relation is exact to all orders in $\hbar$. Without regularization, using equal-time canonical commutation relations for the Heisenberg fields, one finds that Z is the x-integral of a total x-derivative, and one obtains

$$Z = W(x=+\infty) - W(x=-\infty)\ ;\ W = \int U(\varphi)d\varphi \tag{III.2}$$

If one then substitutes $\varphi = \varphi_{sol}(x) + \eta(x,t)$ and takes the vacuum expectation value of Z, one finds two results
(i) at the classical level (with φ replaced by φ_{sol}) $Z = -M_{cl}$ so that the BPS bound is satisfied
(ii) at the one-loop level, the contribution from the counter term cancels the one-loop correction. More in detail

5) Use $\int_{-\infty}^{\infty} dx(1+x^2)^{-3/2} = 2$.

$$W = \sqrt{\frac{\lambda}{2}}\left(\frac{1}{3}\varphi^3 - \frac{\mu^2}{\lambda}\varphi - \frac{\delta\mu^2}{\lambda}\varphi\right) \tag{III.3}$$

and taking for $\langle\eta^2\rangle$ the propagator at large x (the propagator of the nontopological sector), the contributions with $\delta\mu^2$ and $\langle\eta^2\rangle$ sum up to zero. Thus it seems that there are no quantum corrections to the central charge.

However, it is well-known that one must regulate composite operators in canonical quantization. We now discuss an approach based on higher-derivative regularization [3]. It preserves rigid supersymmetry and does not introduce further fields (contrary to the Pauli-Villars method). To stay in the canonical framework with only first-order time derivatives of fields, we only add further space derivatives. This breaks, of course, Lorentz invariance but since the classical soliton solution already breaks Lorentz invariance, this is no major drawback. We consider the following action

$$\mathcal{L} = \left[-\frac{1}{2}(\partial_\mu\varphi)^2 - \frac{1}{2}\bar\psi\partial\!\!\!/\psi + \frac{1}{2}F^2\right] + \left[-\frac{1}{2}(\partial_\mu\partial_x\varphi)^2/M^2\right.$$
$$\left. -\frac{1}{2}\partial_x\bar\psi)\partial\!\!\!/(\partial_x\psi/M^2 + \frac{1}{2}(\partial_x F)^2/M^2\right] + \left[-\frac{1}{2}U'\bar\psi\psi + UF\right] \tag{III.4}$$

Each of the three expressions in square brackets is separately invariant under $\delta\varphi = \bar\epsilon\psi, \delta\psi = (\partial\!\!\!/\varphi + F)\epsilon$ and $\delta F = \bar\epsilon\partial\!\!\!/\psi$. One can rewrite this action in superspace but there is no advantage in doing so. We now construct the Noether currents.

Parametrizing the variation of $\mathcal{L}$ for local ϵ in an unambiguous way as

$$\delta\mathcal{L} = (\partial_\mu\bar\epsilon)j^\mu + \partial_\mu[\bar\epsilon k_0^\mu + (\partial_x\bar\epsilon)k_1^\mu] \tag{III.5}$$

where j^μ, k_0^μ and k_1^μ may contain terms with δ_x^μ, we equate this expression for $\delta\mathcal{L}$ to that expression for $\delta\mathcal{L}$ which is obtained when $\mathcal{L}$ is viewed as a functional of $\varphi, \partial_\mu\varphi$ and $\partial_\mu\partial_x\varphi$. By equating the variations with $\bar\epsilon, \partial_\mu\bar\epsilon$ and $\partial_\mu\partial_x\bar\epsilon$ one finds a hierarchy of relations between j^μ, k_0^μ and k_1^μ. One finds then that j^μ is conserved on-shell, where

$$j^\mu = -(\partial\!\!\!/\varphi + U)\gamma^\mu\psi + (\partial\!\!\!/\varphi)\gamma^\mu\frac{\partial_x^2}{M^2}\psi$$
$$+\delta_x^\mu\left[\left(\frac{\partial_x F}{M^2}\right)\partial\!\!\!/\psi - F\frac{\partial\!\!\!/\partial_x\psi}{M^2} + \left(\frac{\Box\partial_x\varphi}{M^2}\right)\psi - (\Box\varphi)\frac{\partial_x\psi}{M^2}\right] \tag{III.6}$$

The noncanonical terms with $\partial_x\ddot\varphi$ and $\ddot\varphi$ in j^1 can be eliminated by adding an identically conserved term $\epsilon^{\mu\nu}\partial_\nu A$ to j^μ with suitable $A(A = -\partial\!\!\!/\varphi\gamma^0 \overleftrightarrow{\partial_x}\psi)$. This yields[6]

$$j^0 = -\left[\partial\!\!\!/(1-\partial_x^2/M^2)\varphi\right]\gamma^\mu\psi - U\gamma^\mu\psi \tag{III.7}$$

The canonical commutation relations yield the following Dirac brackets

$$\{\psi^\alpha(x,t), \psi^\beta(y,t)\} = \hbar\delta^{\alpha\beta}\left(1 - \frac{\partial_x^2}{M^2}\right)^{-1}\delta(x-y)$$
$$[\varphi(x,t), \dot\varphi(y,t)] = i\hbar\left(1 - \frac{\partial_x^2}{M^2}\right)^{-1}\delta(x-y) \tag{III.8}$$

By evaluating the following anticommutator

$$2\zeta^0(x) = \frac{1}{2}tr\left(\{j^0(x), \bar Q\}\tau_3\right) = \int\left[(1-\partial_x^2/M^2)\varphi'(x)\right]\{\psi(x), \psi^T(y)\}U(y)dy$$
$$+\int U(x)\{\psi(x), \psi^T(y)\}\left(1 - \frac{\partial_y^2}{M^2}\right)\varphi'(y)dy = \left[\left(1 - \frac{\partial_x^2}{M^2}\right)\varphi'\right]\left[(1 - \frac{\partial_x^2}{M^2})^{-1}U\right] + U\varphi'$$
$$\tag{III.9}$$

---

6) The terms with $\dot\psi$ can be removed by using the field equations. This is under study.

we find the integrand of Z. Note that only j^0 is needed for this calculation since $Q = \int j^0 dx$, and j^0 has a rather simple form. By subtracting $\varphi' U$ from the first term (and adding it to the second term) and writing $\varphi' U$ as $-\varphi' \left(1 - \frac{\partial_x^2}{M^2}\right) F$, one finds, as expected, that ζ^0 is again a total ∂_x derivative, but there is an extra term in the integrand of order M^{-2}

$$\zeta^0 = \partial_x \left[W - \frac{1}{2M^2} \left(\varphi' \partial_x F - \varphi'' F \right) \right], \quad W \equiv \int U(\varphi) d\varphi, \; F = -\left(1 - \partial_x^2 / M^2\right)^{-1} U \qquad \text{(III.10)}$$

As we now explain, taking the vacuum expectation value, the terms with two ∂_x derivatives yield a result proportional to M^2 which cancels the overall factor M^{-2} and leads to a nonvanishing quantum correction to Z.

$$\begin{aligned} \langle \zeta^0 \rangle &= \partial_x \left[\langle W \rangle + \frac{1}{2M^2} \langle \eta'(x) \frac{1}{1 - \partial_x^2/M^2} \eta'(x) \rangle U' - \frac{1}{2M^2} \langle \eta''(x) \frac{1}{1 - \partial_x^2/M^2} \eta(x) \rangle U' \right] \\ &= \partial_x \left[\langle W \rangle + \frac{1}{M^2} \langle \eta'(x) \frac{1}{1 - \partial_x^2/M^2} \eta(x) \rangle U' \right] \end{aligned} \qquad \text{(III.11)}$$

This is the same mechanism that yields anomalies although one might call it an "anti-anomaly" because with this extra term a "symmetry" (saturation of the BPS bound) is restored. Already in [2] a conjecture was made that a topological anomaly would be present in the ultraviolet divergences, but in [3] it was clearly and conclusively shown that the "anomaly" resides in the ultraviolet sector of the central charge.

The kinetic terms from which the φ propagator is obtained, read

$$\mathcal{L} = \frac{1}{2} \varphi \left[1 - \frac{\partial_x^2}{M^2} \right] \Box \varphi - \frac{1}{2} U \left(1 - \frac{\partial_x^2}{M^2} \right)^{-1} U \qquad \text{(III.12)}$$

Expanding φ into $\varphi_0 + \eta$ where φ_0 is the kink solution of the higher-derivative theory, we find far away from the kink that φ_0 becomes constant, and the propagator is given by the inverse of

$$\left(1 - \frac{\partial_x^2}{M^2} \right) \Box - U' \left(1 - \frac{\partial_x^2}{M^2} \right)^{-1} U' - U'' \left(1 - \frac{\partial_x^2}{M^2} \right)^{-1} U \qquad \text{(III.13)}$$

Substituting $\varphi = \varphi_0$ into U', the propagator becomes in momentum space

$$\frac{1}{1 + p_x^2/M^2} \frac{1}{p^2 - (U')^2 (1 + p_x^2/M^2)^{-2} - U''(1 + p_x^2/M^2)^{-2} U} \qquad \text{(III.14)}$$

It is then clear that in $\langle \partial_x \eta \left(1 - \frac{\partial_x^2}{M^2} \right)^{-1} \partial_x \eta \rangle$ the extra factor p_x^2 in the numerator leads after integration over the momenta to a term proportional to M^2. Introducing $p/M = q$ we find

$$\langle \partial_x \eta \left(1 - \frac{\partial_x^2}{M} \right)^{-1} \partial_x \eta \rangle = M^2 \int \frac{d^2 q}{(2\pi)^2} \frac{q_x^2}{(1 + q_x^2)^2} \frac{-i}{q^2 + \mathcal{O}(M^{-2})} = \frac{M^2}{4\pi} \qquad \text{(III.15)}$$

First of all, it is clear that the integral has been properly regulated since it is finite. One finds then that the result for the central charge is equal to (minus) the mass correction in (2.25). Hence **the BPS bound is saturated at the one-loop level.**

IV DISCUSSION

In general, boundary conditions on the quantum fluctuations induce surface effects. We imposed boundary conditions which follow from the symmetry of the action (a Z_2 symmetry) and the boundary properties of the classical bosonic vacuum solution (periodic in the trivial sector, antiperiodic in the topological sector). This led to unique topological boundary conditions[7] which close the system on itself, like a Moebius strip. For the

[7] The other Z_2 symmetry, $\psi \to -\psi$, leads to $\psi(-L/2) = -\psi(L/2)$ in the trivial sector, but redefining ψ by a parity transformation $\psi(x,t) \to \gamma^0 \psi(-x,t)$, one recovers PBC. The same holds in the topological section.

bosonic fluctuations around the kink we get then ABC, but for the fermions we **derive** (not an assumption!) twisted boundary conditions. This removes one source of ambiguity.

Another source of ambiguity are the ultraviolet divergences. (The infrared divergences are regulated by putting the system in a box). We regulated the infinite sum over modes by first evaluating $\frac{d}{dm}(\sum \omega)$, and then integrating over m, setting the integration constant to zero. This is a renormalization condition, based on the physical condition that as $m \to 0$, the soliton disappears, and also $M^{(1)} \to 0$. (One could also have used the higher-derivative model to regulate $M^{(1)}$, but then one would have to compute the kink solution for the higher-derivative theory, and the fluctuation spectrum. Our method is for simpler).

The results of [2] concern the computations of the mass of soliton, and are also valid for nonsupersymmetric systems with a Z_2 symmetry. On the other hand, in ref. [3] the central charge is computed for supersymmetric systems and it is there shown that the BPS bound is saturated. The results of [2] for supersymmetric systems confirm the results of [3].

We refer to reference [2] for an analysis of the 2-loop corrections in the $N = 1$ sine-Gordon model. They are finite, nonzero, and unambiguous. In the $N = 2$ susy kink system, we find $Z = M^{(1)}$ with $M^{(1)} = 0$ because here bosonic and fermionic contributions cancel mode-by-mode.[2] Since for $N = 2$ multiplet shortening occurs, the result $M^{(1)} = Z$ is in agreement with the general arguments of Olive and Witten.

A problem under study is the following. Can one introduce, quite generally, a new variational principle in field theory, according to which one considers the set of all (consistent) boundary conditions, and selects that boundary condition which yields the lowest energy? One might view a quantum system with different boundary conditions as being in different "states", and due to quantum fluctuations, the system would cascade down to the boundary condition where the energy is minimal. For theories with symmetries (as in our case), this lowest-energy boundary condition would then have to correspond to a symmetry of the action, thus giving a dynamical derivation of our topological boundary conditions.

ACKNOWLEDGMENTS

I thank Horatiu Nastase, Misha Stephanov and Tony Rebhan for the collaboration in which the results of the quantum mass of solitons were found, and André Litvintsev for discussions. To future sponsors, I would like to write that relatively small conferences of this kind have far more impact than mega meetings since they allow intense discussions during the talks. To the organizers and participants, thanks for a wonderful conference and friendly atmosphere, and to Hugo Montani and his charming family thanks for helping me to climb "El Catedral" in Bariloche.

REFERENCES

1. A. Rebhan and P. van Nieuwenhuizen, *Nucl. Phys. B* **508** (1997) 449. (Correct bosonic mass but [2] supercedes this article).
2. H. Nastase, M. Stephanov, P. van Nieuwenhuizen and A. Rebhan, *Nucl. Phys. B* **542** (1999) 471. (Correct bosonic and susy mass, anomaly in Z not included).
3. M. Shifman, A. Vainshtein and M. Voloshin, *Phys. Rev. D* **59** (1999) 045016. (Correct Z, no independent calculation of the mass).
4. N. Graham and A. Jaffe, *Nucl. Phys. B* **544** (1999) 432. (Claims agreement with result for mass in [2] and central charge in [3]. Their work is based on a phase shift analysis, but see the criticism in [3].)
5. E. Tomboulis, *Phys. Rev. D* **12** (1975) 1678.
6. J.L. Gervais and A. Jevicki, *Nucl. Phys. B* **110** (1976) 93 and 113.
7. J.F. Schonfeld, *Nucl. Phys. B* **161** (1979) 125.
8. C. Ahn, D. Bernard and A. Le Clair, *Nucl. Phys. B* **246** (1990) 409; A.C. Ahn, *Nucl. Phys. B* **354** (1991) 57.
9. R. Dashen, B. Hasslacher and A. Neveu, *Phys. Rev. D* **10** (1974) 4114 and 4130, *Phys. Rev. D* **12** (1975) 2443.

The Large N Limit of Superconformal field theories and supergravity [1]

Juan Maldacena[2]

Lyman Laboratory of Physics, Harvard University, Cambridge, MA 02138, USA

Abstract. We show that the large N limit of certain conformal field theories in various dimensions include in their Hilbert space a sector describing supergravity on the product of Anti-deSitter spacetimes, spheres and other compact manifolds. This is shown by taking some branes in the full M/string theory and then taking a low energy limit where the field theory on the brane decouples from the bulk. We observe that, in this limit, we can still trust the near horizon geometry for large N. The enhanced supersymmetries of the near horizon geometry correspond to the extra supersymmetry generators present in the superconformal group (as opposed to just the super-Poincare group). The 't Hooft limit of 3+1 $\mathcal{N} = 4$ super-Yang-Mills at the conformal point is shown to contain strings: they are IIB strings. We conjecture that compactifications of M/string theory on various Anti-deSitter spacetimes is dual to various conformal field theories. This leads to a new proposal for a definition of M-theory which could be extended to include five non-compact dimensions.

I GENERAL IDEA

In the last few years it has been extremely fruitful to derive quantum field theories by taking various limits of string or M-theory. In some cases this is done by considering the theory at geometric singularities and in others by considering a configuration containing branes and then taking a limit where the dynamics on the brane decouples from the bulk. In this paper we consider theories that are obtained by decoupling theories on branes from gravity. We focus on conformal invariant field theories but a similar analysis could be done for non-conformal field theories. The cases considered include N parallel D3 branes in IIB string theory and various others. We take the limit where the field theory on the brane decouples from the bulk. At the same time we look at the near horizon geometry and we argue that the supergravity solution can be trusted as long as N is large. N is kept fixed as we take the limit. The approach is similar to that used in [1] to study the NS fivebrane theory [2] at finite temperature. The supergravity solution typically reduces to $p+2$ dimensional Anti-deSitter space (AdS_{p+2}) times spheres (for D3 branes we have $AdS_5 \times S^5$). The curvature of the sphere and the AdS space in Planck units is a (positive) power of $1/N$. Therefore the solutions can be trusted as long as N is large. Finite temperature configurations in the decoupled field theory correspond to black hole configurations in AdS spacetimes. These black holes will Hawking radiate into the AdS spacetime. We conclude that excitations of the AdS spacetime are included in the Hilbert space of the corresponding conformal field theories. A theory in AdS spacetime is not completely well defined since there is a horizon and it is also necessary to give some boundary conditions at infinity. However, local properties and local processes can be calculated in supergravity when N is large if the proper energies involved are much bigger than the energy scale set by the cosmological constant (and smaller than the Planck scale). We will conjecture that the full quantum M/string-theory on AdS space, plus suitable boundary conditions is dual to the corresponding brane theory. We are not going to specify the boundary conditions in AdS, we leave this interesting problem for the future. The $AdS\times$(spheres) description will become useful for large N, where we can isolate some local processes from the question of boundary conditions. The supersymmetries of both theories agree, both are given by the superconformal group. The superconformal group has twice the amount of supersymmetries of the

---

[1] From Adv. Theor. Math. Phys. **2** (1998) 231-252. Reprinted with permission of International Press.

[2] malda@pauli.harvard.edu

CP484, *Trends in Theoretical Physics II*, edited by H. Falomir, R. E. Gamboa Saraví, and F. A. Schaposnik
1999 American Institute of Physics 1-56396-894-0

corresponding super-Poincare group [3,4]. This enhancement of supersymmetry near the horizon of extremal black holes was observed in [5,6] precisely by showing that the near throat geometry reduces to $AdS\times$(spheres). AdS spaces (and branes in them) were extensively considered in the literature [7–13], includding the connection with the superconformal group.

In section 2 we study $\mathcal{N}=4$ d=4 $U(N)$ super-Yang-Mills as a first example, we discuss several issues which are present in all other cases. In section 3 we analyze the theories describing M-theory five-branes and M-theory two-branes. In section 4 we consider theories with lower supersymmetry which are related to a black string in six dimensions made with D1 and D5 branes. In section 5 we study theories with even less supersymmetry involving black strings in five dimensions and finally we mention the theories related to extremal Reissner-Nordström black holes in four spacetime dimensions (these last cases will be more speculative and contain some unresolved puzzles). Finally in section 6 we make some comments on the relation to matrix theory.

II D3 BRANES OR $\mathcal{N}=4$ $U(N)$ SUPER-YANG-MILLS IN D=3+1

We start with type IIB string theory with string coupling g, which will remain fixed. Consider N parallel D3 branes separated by some distances which we denote by r. For low energies the theory on the D3 brane decouples from the bulk. It is more convenient to take the energies fixed and take

$$\alpha' \to 0 \ , \qquad U \equiv \frac{r}{\alpha'} = \text{fixed} \ . \tag{II.1}$$

The second condition is saying that we keep the mass of the stretched strings fixed. As we take the decoupling limit we bring the branes together but the the Higgs expectation values corresponding to this separation remains fixed. The resulting theory on the brane is four dimensional $\mathcal{N}=4$ $U(N)$ SYM. Let us consider the theory at the superconformal point, where $r=0$. The conformal group is SO(2,4). We also have an $SO(6)\sim SU(4)$ R-symmetry that rotates the six scalar fields into each other[3]. The superconformal group includes twice the number of supersymmetries of the super-Poincare group: the commutator of special conformal transformations with Poincare supersymmetry generators gives the new supersymmetry generators. The precise superconformal algebra was computed in [3] . All this is valid for any N.

Now we consider the supergravity solution carrying D3 brane charge [14]

$$\begin{aligned} ds^2 &= f^{-1/2} dx_{||}^2 + f^{1/2}(dr^2 + r^2 d\Omega_5^2) \ , \\ f &= 1 + \frac{4\pi g N \alpha'^2}{r^4} \ , \end{aligned} \tag{II.2}$$

where $x_{||}$ denotes the four coordinates along the worldvolume of the three-brane and $d\Omega_5^2$ is the metric on the unit five-sphere[4]. The self dual five-form field strength is nonzero and has a flux on the five-sphere. Now we define the new variable $U \equiv \frac{r}{\alpha'}$ and we rewrite the metric in terms of U. Then we take the $\alpha' \to 0$ limit. Notice that U remains fixed. In this limit we can neglect the 1 in the harmonic function (II.2) . The metric becomes

$$ds^2 = \alpha' \left[\frac{U^2}{\sqrt{4\pi g N}} dx_{||}^2 + \sqrt{4\pi g N} \frac{dU^2}{U^2} + \sqrt{4\pi g N} d\Omega_5^2 \right] \ . \tag{II.3}$$

This metric describes five dimensional Anti-deSitter (AdS_5) times a five-sphere[5]. We see that there is an overall α' factor. The metric remains constant in α' units. The radius of the five-sphere is $R_{sph}^2/\alpha' = \sqrt{4\pi g N}$, and is the same as the "radius" of AdS_5 (as defined in the appendix). In ten dimensional Planck units they are both proportional to $N^{1/4}$. The radius is quantized because the flux of the 5-form field strength on the 5 sphere is quantized. We can trust the supergravity solution when

$$gN \gg 1 \ . \tag{II.4}$$

---

3) The representation includes objects in the spinor representations, so we should be talking about SU(4), we will not make this, or similar distinctions in what follows.

4) We choose conventions where $g \to 1/g$ under S-duality.

5) See the appendix for a brief description of AdS spacetimes.

When N is large we have approximately ten dimensional flat space in the neighborhood of any point[6]. Note that in the large N limit the flux of the 5 form field strength per unit Planck (or string) 5-volume becomes small.

Now consider a near extremal black D3 brane solution in the decoupling limit (II.1). We keep the energy density on the brane worldvolume theory (μ) fixed. We find the metric

$$ds^2 = \alpha' \left\{ \frac{U^2}{\sqrt{4\pi g N}} \left[-(1 - U_0^4/U^4)dt^2 + dx_i^2 \right] + \sqrt{4\pi g N} \frac{dU^2}{U^2(1 - U_0^4/U^4)} + \sqrt{4\pi g N} d\Omega_5^2 \right\}, \tag{II.5}$$

$$U_0^4 = \frac{2^7}{3} \pi^4 g^2 \mu.$$

We see that U_0 remains finite when we take the $\alpha' \to 0$ limit. The situation is similar to that encountered in [1] . Naively the whole metric is becoming of zero size since we have a power of α' in front of the metric, and we might incorrectly conclude that we should only consider the zero modes of all fields. However, energies that are finite from the point of view of the gauge theory, lead to proper energies (measured with respect to proper time) that remain finite is in α' units (or Planck units, since g is fixed). More concretely, an excitation that has energy ω (fixed in the limit) from the point of view of the gauge theory, will have proper energy $E_{proper} = \frac{1}{\sqrt{\alpha'}} \frac{\omega (gN4\pi)^{1/4}}{U}$. This also means that the corresponding proper wavelengths remain fixed. In other words, the spacetime action on this background has the form $S \sim \frac{1}{\alpha'^4} \int d^{10}x \sqrt{G} R + \cdots$, so we can cancel the factor of α' in the metric and the Newton constant, leaving a theory with a finite Planck length in the limit. Therefore we should consider fields that propagate on the AdS background. Since the Hawking temperature is finite, there is a flux of energy from the black hole to the AdS spacetime. Since $\mathcal{N} = 4$ d=4 $U(N)$ SYM is a unitary theory we conclude that, for large N, *it includes in its Hilbert space the states of type IIB supergravity on* $(AdS_5 \times S_5)_N$, where subscript indicates the fact that the "radii" in Planck units are proportional to $N^{1/4}$. In particular the theory contains gravitons propagating on $(AdS_5 \times S_5)_N$. When we consider supergravity on $AdS_5 \times S_5$, we are faced with global issues like the presence of a horizon and the boundary conditions at infinity. It is interesting to note that the solution is nonsingular [15]. The gauge theory should provide us with a specific choice of boundary conditions. It would be interesting to determine them.

We have started with a quantum theory and we have seen that it includes gravity so it is natural to think that this correspondence goes beyond the supergravity approximation. We are led to the conjecture that *Type IIB string theory on* $(AdS_5 \times S^5)_N$ *plus some appropriate boundary conditions (and possibly also some boundary degrees of freedom) is dual to* $\mathcal{N}$ *=4 d=3+1* $U(N)$ *super-Yang-Mills.* The SYM coupling is given by the (complex) IIB string coupling, more precisely $\frac{1}{g_{YM}^2} + i\frac{\theta}{8\pi^2} = \frac{1}{2\pi}(\frac{1}{g} + i\frac{\chi}{2\pi})$ where χ is the value of the RR scalar.

The supersymmetry group of $AdS_5 \times S^5$, is known to be the same as the superconformal group in 3+1 spacetime dimensions [3], so the supersymmetries of both theories are the same. This is a new form of "duality": a large N field theory is related to a string theory on some background, notice that the correspondence is non-perturbative in g and the $SL(2,Z)$ symmetry of type IIB would follow as a consequence of the $SL(2,Z)$ symmetry of SYM[7]. It is also a strong-weak coupling correspondence in the following sense. When the effective coupling gN becomes large we cannot trust perturbative calculations in the Yang-Mills theory but we can trust calculations in supergravity on $(AdS_5 \times S^5)_N$. This is suggesting that the $\mathcal{N} = 4$ Yang-Mills master field is the anti-deSitter supergravity solution (similar ideas were suggested in [17]). Since N measures the size of the geometry in Planck units, we see that quantum effects in $AdS_5 \times S^5$ have the interpretation of $1/N$ effects in the gauge theory. So Hawking radiation is a $1/N$ effect. It would be interesting to understand more precisely what the horizon means from the gauge theory point of view. IIB supergravity on $AdS_5 \times S^5$ was studied in [7,9].

The above conjecture becomes nontrivial for large N and gives a way to answer some large N questions in the SYM theory. For example, suppose that we break $U(N) \to U(N-1) \times U(1)$ by Higgsing. This corresponds to putting a three brane at some point on the 5-sphere and some value of U, with world volume directions along the original four dimensions ($x_{||}$). We could now ask what the low energy effective action for the light

---

6) In writing (II.4) we assumed that $g \leq 1$, if $g > 1$ then the condition is $N/g \gg 1$. In other words we need large N , not large g.

7) This is similar in spirit to [16] but here N is not interpreted as momentum.

U(1) fields is. For large N (II.4) it is the action of a D3 brane in $AdS_5 \times S^5$. More concretely, the bosonic part of the action becomes the Born-Infeld action on the AdS background

$$\begin{aligned} S &= -\frac{1}{(2\pi)^3 g}\int d^4x h^{-1}\left[\sqrt{-Det(\eta_{\alpha\beta} + h\partial_\alpha U\partial_\beta U + U^2 h g_{ij}\partial_\alpha\theta^i\partial_\beta\theta^j + 2\pi\sqrt{h}F_{\alpha\beta})} - 1\right], \\ h &= \frac{4\pi g N}{U^4}, \end{aligned} \tag{II.6}$$

with $\alpha, \beta = 0, 1, 2, 3$, $i, j = 1, .., 5$; and g_{ij} is the metric of the unit five-sphere. As any low energy action, (II.6) is valid when the energies are low compared to the mass of the massive states that we are integrating out. In this case the mass of the massive states is proportional to U (with no factors of N). The low energy condition translates into $\partial U/U \ll U$ and $\partial\theta^i << U$, etc.. So the nonlinear terms in the action (II.6) will be important only when gN is large. It seems that the form of this action is completely determined by superconformal invariance, by using the broken and unbroken supersymmetries, in the same sense that the Born Infeld action in flat space is given by the full Poincare supersymmetry [18]. It would be very interesting to check this explicitly. We will show this for a particular term in the action. We set $\theta^i = const$ and $F = 0$, so that we only have U left. Then we will show that the action is completely determined by broken conformal invariance. This can be seen as follows. Using Lorentz invariance and scaling symmetry (dimensional analysis) one can show that the action must have the form

$$S = \int d^{p+1}x U^{p+1} f(\partial_\alpha U \partial^\alpha U/U^4), \tag{II.7}$$

where f is an arbitrary function. Now we consider infinitesimal special conformal transformations

$$\begin{aligned} \delta x^\alpha &= \epsilon^\beta x_\beta x^\alpha - \epsilon^\alpha(x^2 + \tfrac{\tilde{R}^4}{U^2})/2, \\ \delta U &\equiv U'(x') - U(x) = -\epsilon^\alpha x_\alpha U, \end{aligned} \tag{II.8}$$

where ϵ^α is an infinitesimal parameter. For the moment $\tilde{R}$ is an arbitrary constant. We will later identify it with the "radius" of AdS, it will turn out that $\tilde{R}^4 \sim gN$. In the limit of small $\tilde{R}$ we recover the more familiar form of the conformal transformations (U is a weight one field). Usually conformal transformations do not involve the variable U in the transformations of x. For constant U the extra term in (II.8) is a translation in x, but we will take U to be a slowly varying function of x and we will determine $\tilde{R}$ from other facts that we know. Demanding that (II.7) is invariant under (II.8) we find that the function f in (II.7) obeys the equation

$$f(z) + const = 2\left(z + \frac{1}{\tilde{R}^4}\right) f'(z) \tag{II.9}$$

which is solved by $f = b[\sqrt{1 + \tilde{R}^4 z} - a]$. Now we can determine the constants $a, b, \tilde{R}$ from supersymmetry. We need to use three facts. The first is that there is no force (no vacuum energy) for a constant U. This implies $a = 1$. The second is that the ∂U^2 term (F^2 term) in the $U(1)$ action is not renormalized. The third is that the only contribution to the $(\partial U)^4$ term (an F^4 term) comes from a one loop diagram [19]. This determines all the coefficients to be those expected from (II.6) including the fact that $\tilde{R}^4 = 4\pi g N$. It seems very plausible that using all 32 supersymmetries we could fix the action (II.6) completely. This would be saying that (II.6) is a consequence of the symmetries and thus not a prediction[8]. However we can make very nontrivial predictions (though we were not able to check them). For example, if we take g to be small (but N large) we can predict that the Yang-Mills theory contains strings. More precisely, in the limit $g \to 0$, gN = fixed $\gg 1$ ('t Hooft limit) we find free strings in the spectrum, they are IIB strings moving in $(AdS_5 \times S^5)_{gN}$.[9] The sense in which

---

8) Notice that the action (II.6) includes a term proportional to v^6 similar to that calculated in [20] . Conformal symmetry explains the agreement that they would have found if they had done the calculation for 3+1 SYM as opposed to 0+1.

9) In fact, Polyakov [21] recently proposed that the string theory describing bosonic Yang-Mills has a new dimension corresponding to the Liouville mode φ, and that the metric at $\varphi = 0$ is zero due to a "zig-zag" symmetry. In our case we see that the physical distances along the directions of the brane contract to zero as $U \to 0$. The details are different, since we are considering the $\mathcal{N} = 4$ theory.

these strings are present is rather subtle since there is no energy scale in the Yang-Mills to set their tension. In fact one should translate the mass of a string state from the AdS description to the Yang-Mills description. This translation will involve the position U at which the string is sitting. This sets the scale for its mass. As an example, consider again the D-brane probe (Higgsed configuration) which we described above. From the type IIB description we expect open strings ending on the D3 brane probe. From the point of view of the gauge theory these open strings have energies $E = \frac{U}{(4\pi g N)^{1/4}}\sqrt{N_{open}}$ where N_{open} is the integer charaterizing the massive open string level. In this example we see that α' disappears when we translate the energies and is replaced by U, which is the energy scale of the Higgs field that is breaking the symmetry.

Now we turn to the question of the physical interpretation of U. U has dimensions of mass. It seems natural to interpret motion in U as moving in energy scales, going to the IR for small U and to the UV for large U. For example, consider a D3 brane sitting at some position U. Due to the conformal symmetry, all physics at energy scales ω in this theory is the same as physics at energies $\omega' = \lambda\omega$, with the brane sitting at $U' = \lambda U$.

Now let us turn to another question. We could separate a group of D3 branes from the point were they were all sitting originally. Fortunately, for the extremal case we can find a supergravity solution describing this system. All we have to do is the replacement

$$\frac{N}{U^4} \to \frac{N-M}{U^4} + \frac{M}{|\vec{U}-\vec{W}|^4}, \tag{II.10}$$

where $\vec{W} = \vec{r}/\alpha'$ is the separation. It is a vector because we have to specify a point on S^5 also. The resulting metric is

$$ds^2 = \alpha'\left[U^2 \frac{1}{\sqrt{4\pi g}\left(N-M+\frac{MU^4}{|\vec{U}-W|^4}\right)^{1/2}}dx_{\|}^2 + \sqrt{4\pi g}\frac{1}{U^2}\left(N-M+\frac{MU^4}{|\vec{U}-W|^4}\right)^{1/2} d\vec{U}^2\right]. \tag{II.11}$$

For large $U \gg |W|$ we find basically the solution for $(AdS_5 \times S^5)_N$ which is interpreted as saying that for large energies we do not see the fact that the conformal symmetry was broken, while for small $U \ll |W|$ we find just $(AdS_5 \times S^5)_{N-M}$, which is associated to the CFT of the unbroken $U(N-M)$ piece. Furthermore, if we consider the region $|\vec{U}-\vec{W}| \ll |\vec{W}|$ we find $(AdS_5 \times S^5)_M$, which is described by the CFT of the $U(M)$ piece.

We could in principle separate all the branes from each other. For large values of U we would still have $(AdS_5 \times S^5)_N$, but for small values of U we would not be able to trust the supergravity solution, but we naively get N copies of $(AdS_5 \times S^5)_1$ which should correspond to the $U(1)^N$.

Now we discuss the issue of compactification. We want to consider the YM theory compactified on a torus of radii R_i, $x_i \sim x_i + 2\pi R_i$, which stay fixed as we take the $\alpha' \to 0$ limit. Compactifying the theory breaks conformal invariance and leaves only the Poincare supersymmetries. However one can still find the supergravity solutions and follow the above procedure, going near the horizon, etc. The AdS piece will contain some identifications. So we will be able to trust the supergravity solution as long as the physical length of these compact circles stays big in α' units. This implies that we can trust the supergravity solution as long as we stay far from the horizon (at $U = 0$)

$$U \gg \frac{(gN)^{1/4}}{R_i}, \tag{II.12}$$

for all i. This is a larger bound than the naive expectation $(1/R_i)$. If we were considering near extremal black holes we would require that U_0 in (II.5) satisfies (II.12), which is, of course, the same condition on the temperature gotten in [22].

The relation of the three-brane supergravity solution and the Yang-Mills theory has been studied in [23–26] . All the calculations have been done for near extremal D3 branes fall into the category described above. In particular the absorption cross section of the dilaton and the graviton have been shown to agree with what one would calculate in the YM theory [24,25]. It has been shown in [26] that some of these agreements are due to non-renormalization theorems for $\mathcal{N} = 4$ YM. The black hole entropy was compared to the *perturbative* YM calculation and it agrees up to a numerical factor [23]. This is not in disagreement with the correspondence we were suggesting, It is expected that large gN effects change this numerical factor, this problem remains unsolved.

Finally notice that the group $SO(2,4) \times SO(6)$ suggests a twelve dimensional realization in a theory with two times [27].

III OTHER CASES WITH 16 → 32 SUPERSYMMETRIES, M5 AND M2 BRANE THEORIES

Basically all that we have said for the D3 brane carries over for the other conformal field theories describing coincident M-theory fivebranes and M-theory twobranes. We describe below the limits that should be taken in each of the two cases. Similar remarks can be made about the entropies [28], and the determination of the probe actions using superconformal invariance. Eleven dimensional supergravity on the corresponding AdS spaces was studied in [8,10,11,15].

A M5 brane

The decoupling limit is obtained by taking the 11 dimensional Planck length to zero, $l_p \to 0$, keeping the worldvolume energies fixed and taking the separations $U^2 \equiv r/l_p^3$ = fixed [29]. This last condition ensures that the membranes stretching between fivebranes give rise to strings with finite tension.

The metric is[10]

$$\begin{aligned} ds^2 &= f^{-1/3} dx_{||}^2 + f^{2/3}(dr^2 + r^2 d\Omega_4^2), \\ f &= 1 + \frac{\pi N l_p^3}{r^3}, \end{aligned} \tag{III.1}$$

We also have a flux of the four-form field strength on the four-sphere (which is quantized). Again, in the limit we obtain

$$ds^2 = l_p^2 [\frac{U^2}{(\pi N)^{1/3}} dx_{||}^2 + 4(\pi N)^{2/3} \frac{dU^2}{U^2} + (\pi N)^{2/3} d\Omega_4^2] , \tag{III.2}$$

where now the "radii" of the sphere and the AdS_7 space are $R_{sph} = R_{AdS}/2 = l_p(\pi N)^{1/3}$. Again, the "radii" are fixed in Planck units as we take $l_p \to 0$, and supergravity can be applied if $N \gg 1$.

Reasoning as above we conclude that this theory contains seven dimensional Anti-deSitter times a four-sphere, which for large N looks locally like eleven dimensional Minkowski space.

This gives us a method to calculate properties of the large N limit of the six dimensional (0,2) conformal field theory [30]. The superconformal group again coincides with the algebra of the supersymmetries preserved by $AdS_7 \times S^4$. The bosonic symmetries are $SO(2,6) \times SO(5)$ [4]. We can do brane probe calculations, thermodynamic calculations [28], etc.

The conjecture is now that *the (0,2) conformal field theory is dual to M-theory on* $(AdS_7 \times S^4)_N$, the subindex indicates the dependence of the "radius" with N.

B M2 brane

We now take the limit $l_p \to 0$ keeping $U^{1/2} \equiv r/l_p^{3/2}$ = fixed. This combination has to remain fixed because the scalar field describing the motion of the twobrane has scaling dimension 1/2. Alternatively we could have derived this conformal field theory by taking first the field theory limit of D2 branes in string theory as in [31–33], and then taking the strong coupling limit of that theory to get to the conformal point as in [34–36]. The fact that the theories obtained in this fashion are the same can be seen as follows. The D2 brane gauge theory can be obtained as the limit $\alpha' \to 0$, keeping $g_{YM}^2 \sim g/\alpha'$ = fixed. This is the same as the limit of M-theory two branes in the limit $l_p \to 0$ with $R_{11}/l_p^{3/2} \sim g_{YM}$ = fixed. This is a theory where one of the Higgs fields is compact. Taking $R^{11} \to \infty$ we see that we get the theory of coincident M2 branes, in which the SO(8) R-symmetry has an obvious origin.

The metric is

---

10) In our conventions the relation of the Planck length to the 11 dimensional Newton constant is $G_N^{11} = 16\pi^7 l_p^9$.

$$ds^2 = f^{-2/3}dx_{||}^2 + f^{1/3}(dr^2 + r^2 d\Omega_7^2) ,$$

$$f = 1 + \frac{2^5\pi^2 N l_p^6}{r^6} , \qquad \text{(III.3)}$$

and there is a nonzero flux of the dual of the four-form field strength on the seven-sphere. In the decoupling limit we obtain $AdS_4 \times S^7$, and the supersymmetries work out correctly. The bosonic generators are given by $SO(2,3) \times SO(8)$. In this case the "radii" of the sphere and AdS_4 are $R_{sph} = 2R_{AdS} = l_p(2^5\pi^2 N)^{1/6}$.

The entropy of the near extremal solution agrees with the expectation from dimensional analysis for a conformal theory in 2+1 dimensions [28], but the N dependence or the numerical coefficients are not understood.

Actually for the case of the two brane the conformal symmetry was used to determine the v^4 term in the probe action [37], we are further saying that conformal invariance determines it to all orders in the velocity of the probe. Furthermore the duality we have proposed with M-theory on $AdS_4 \times S^7$ determines the precise numerical coefficient.

When M-theory is involved the dimensionalities of the groups are suggestive of a thirteen dimensional realization [38].

IV THEORIES WITH 8 → 16 SUPERSYMMETRIES, THE D1+D5 SYSTEM

Now we consider IIB string theory compactified on M^4 (where $M^4 = T^4$ or $K3$) to six spacetime dimensions. As a first example let us start with a D-fivebrane with four dimensions wrapping on M^4 giving a string in six dimensions. Consider a system with Q_5 fivebranes and Q_1 D-strings, where the D-string is parallel to the string in six dimensions arising from the fivebrane. This system is described at low energies by a 1+1 dimensional (4,4) superconformal field theory. So we take the limit

$$\alpha' \to 0 , \qquad \frac{r}{\alpha'} = \text{fixed} , \qquad v \equiv \frac{V_4}{(2\pi)^4\alpha'^2} = \text{fixed} , \qquad g_6 = \frac{g}{\sqrt{v}} = \text{fixed} \qquad \text{(IV.1)}$$

where V_4 is the volume of M^4. All other moduli of M^4 remain fixed. This is just a low energy limit, we keep all dimensionless moduli fixed. As a six dimensional theory, IIB on M^4 contains strings. They transform under the U-duality group and they carry charges given by a vector q^I. In general we can consider a configuration where $q^2 = \eta_{IJ}q^I q^J \neq 0$ (the metric is the U-duality group invariant), and then take the limit (IV.1).

This theory has a branch which we will call the Higgs branch and one which we call the Coulomb branch. On the Higgs branch the 1+1 dimensional vector multiplets have zero expectation value and the Coulomb branch is the other one. Notice that the expectation values of the vector multiplets in the Coulomb branch remain fixed as we take the limit $\alpha' \to 0$.

The Higgs branch is a SCFT with (4,4) supersymmetry. This is the theory considered in [39]. The above limit includes also a piece of the Coulomb branch, since we can separate the branes by a distance such that the mass of stretched strings remains finite.

Now we consider the supergravity solution corresponding to D1+D5 branes [40]

$$ds^2 = f_1^{-1/2} f_5^{-1/2} dx_{||}^2 + f_1^{1/2} f_5^{1/2}(dr^2 + r^2 d\Omega_3^2) ,$$

$$f_1 = \left(1 + \tfrac{g\alpha' Q_1}{vr^2}\right) , \qquad f_5 = \left(1 + \tfrac{g\alpha' Q_5}{r^2}\right) , \qquad \text{(IV.2)}$$

where $dx_{||}^2 = -dt^2 + dx^2$ and x is the coordinate along the D-string. Some of the moduli of M^4 vary over the solution and attain a fixed value at the horizon which depends only on the charges and some others are constant throughout the solution. The three-form RR-field strength is also nonzero.

In the decoupling limit (IV.1) we can neglect the 1's in f_i in (IV.2) and the metric becomes

$$ds^2 = \alpha' \left[\frac{U^2}{g_6\sqrt{Q_1 Q_5}} dx_{||}^2 + g_6\sqrt{Q_1 Q_5}\frac{dU^2}{U^2} + g_6\sqrt{Q_1 Q_5} d\Omega_3^2 \right] . \qquad \text{(IV.3)}$$

The compact manifold $M^4(Q)$ that results in the limit is determined as follows. Some of its moduli are at their fixed point value which depends only on the charges and not on the asymptotic value of those moduli at infinity (the notation $M^4(Q)$ indicates the charge dependence of the moduli) [41] [11]. The other moduli, that were constant in the black hole solution, have their original values. For example, the volume of M^4 has the fixed point value $v_{fixed} = Q_1/Q_5$, while the six dimensional string coupling g_6 has the original value. Notice that there is an overall factor of α' in (IV.3) which can be removed by canceling it with the factor of α' in the Newton constant as explained above. We can trust the supergravity solution if Q_1, Q_5 are large, $g_6 Q_i \gg 1$. Notice that we are talking about a six dimensional supergravity solution since the volume of M^4 is a constant in Planck units (we keep the Q_1/Q_5 ratio fixed). The metric (IV.2) describes three dimensional AdS_3 times a 3-sphere. The supersymmetries work out correctly, starting from the 8 Poincare supersymmetries we enhance then to 16 supersymmetries. The bosonic component is $SO(2,2) \times SO(4)$. In conformal field theory language $SO(2,2)$ is just the $SL(2,R) \times SL(2,R)$ part of the conformal group and SO(4) $\sim SU(2)_L \times SU(2)_R$ are the R-symmetries of the CFT [43].

So the conjecture is that *the 1+1 dimensional CFT describing the Higgs branch of the D1+D5 system on* M^4 *is dual to type IIB string theory on* $(AdS_3 \times S^3)_{Q_1 Q_5} \times M^4(Q)$. The subscript indicates that the radius of the three sphere is $R^2_{sph} = \alpha' g_6 \sqrt{Q_1 Q_5}$. The compact fourmanifold $M^4(Q)$ is at some particular point in moduli space determined as follows. The various moduli of M^4 are divided as tensors and hypers according to the (4,4) supersymmetry on the brane. Each hypermultiplet contains four moduli and each tensor contains a modulus and an anti-self-dual B-field. (There are five tensors of this type for T^4 and 21 for $K3$). The scalars in the tensors have fixed point values at the horizon of the black hole, and those values are the ones entering in the definition of $M^4(Q)$ (Q indicates the dependence on the charges). The hypers will have the same expectation value everywhere. It is necessary for this conjecture to work that the 1+1 dimensional (4,4) theory is indendent of the tensor moduli appearing in its original definition as a limit of the brane configurations, since $M^4(Q)$ does not depend on those moduli. A non renomalization theorem like [44,45] would explain this. We also need that the Higgs branch decouples from the Coulomb branch as in [46,47].

Finite temperature configurations in the 1+1 conformal field theory can be considered. They correspond to near extremal black holes in AdS_3. The metric is the same as that of the BTZ 2+1 dimensional black hole [48], except that the angle of the BTZ description is not periodic. This angle corresponds to the spatial direction x of the 1+1 dimensional CFT and it becomes periodic if we compactify the theory[12] [49–51] [13]. All calculations done for the 1D+5D system [39,53,54] are evidence for this conjecture. In all these cases [54] the nontrivial part of the greybody factors comes from the AdS part of the spacetime. Indeed, it was noticed in [55] that the greybody factors for the BTZ black hole were the same as the ones for the five-dimensional black hole in the dilute gas approximation. The dilute gas condition $r_0, r_n \ll r_1 r_5$ [53] is automatically satisfied in the limit (IV.1) for finite temperature configurations (and finite chemical potential for the momentum along $\hat{x}$). It was also noticed that the equations have an $SL(2,R) \times SL(2,R)$ symmetry [56], these are the isometries of AdS_3, and part of the conformal symmetry of the 1+1 dimensional field theory. It would be interesting to understand what is the gravitational counterpart of the full conformal symmetry group in 1+1 dimensions.

V THEORIES WITH $4 \to 8$ SUPERSYMMETRIES

The theories of this type will be related to black strings in five dimensions and Reissner-Nordström black holes in four dimensions. This part will be more sketchy, since there are several details of the conformal field theories involved which I do not completely understand, most notably the dependence on the various moduli of the compactification.

11) The fixed values of the moduli are determined by the condition that they minimize the tension of the corresponding string (carrying charges q^I) in six dimensions [41]. This is parallel to the discussion in four dimensions [42]

12) I thank G. Horowitz for many discussions on this correspondence and for pointing out ref. [49] to me. Some of the remarks the remarks below arose in conversations with him.

13) The ideas in [49–51] could be used to show the relation between the AdS region and black holes in M-theory on a light like circle. However the statement in [49–51] that the $AdS_3 \times S^3$ spacetime is U-dual to the full black hole solution (which is asymptotic to Minkowski space) should be taken with caution because in those cases the spacetime has identifications on circles that are becoming null. This changes dramatically the physics. For examples of these changes see [32,52].

A Black string in five dimensions

One can think about this case as arising from M-theory on M^6 where M^6 is a CY manifold, $K3 \times T^2$ or T^6. We wrap fivebranes on a four-cycle $P_4 = p^A \alpha_A$ in M^6 with nonzero triple self intersection number, see [57]. We are left with a one dimensional object in five spacetime dimensions. Now we take the following limit

$$l_p \to 0 \qquad (2\pi)^6 v \equiv V_6/l_p^6 = \text{fixed} \qquad U^2 \equiv r/l_p^3 = \text{fixed} \ , \tag{V.1}$$

where l_p is the eleven dimensional Planck length. In this limit the theory will reduce to a conformal field theory in two dimensions. It is a (0,4) CFT and it was discussed in some detail in a region of the moduli space in [57]. More generally we should think that the five dimensional theory has some strings characterized by charges p^A, forming a multiplet of the U-duality group and we are taking a configuration where the triple self intersection number p^3 is nonzero (in the case $M^6 = T^6$, $p^3 \equiv D \equiv D_{ABC}p^A p^B p^C$ is the cubic E_6 invariant).

We now take the corresponding limit of the black hole solution. We will just present the near horizon geometry, obtained after taking the limit. Near the horizon all the vector moduli are at their fixed point values [58]. So the near horizon geometry can be calculated by considering the solution with constant moduli. We get

$$ds^2 = l_p^2 \left[\frac{U^2 v^{1/3}}{D^{1/3}}(-dt^2 + dx^2) + \frac{D^{2/3}}{v^{2/3}} \left(4\frac{dU^2}{U^2} + d\Omega_2^2 \right) \right] \ . \tag{V.2}$$

In this limit M^6 has its vector moduli equal to their fixed point values which depend only on the charge while its hyper moduli are what they were at infinity. The overall size of M^6 in Planck units is a hypermultiplet, so it remains constant as we take the limit (V.1. We get a product of three dimensional AdS_3 spacetime with a two-sphere, $AdS_3 \times S^2$. Defining the five dimensional Planck length by $l_{5p}^3 = l_p^3/v$ we find that the "radii" of the two sphere and the AdS_3 are $R_{sph} = R_{AdS}/2 = l_{5p}D^{1/3}$. In this case the superconformal group contains as a bosonic subgroup $SO(2,2) \times SO(3)$. So the R-symmetries are just $SU(2)_R$, associated to the 4 rightmoving supersymmetries.

In this case we conjecture that this (0,4) conformal field theory is dual, for large p^A, to M-theory on $AdS_3 \times S^2 \times M_p^6$. The hypermultiplet moduli of M_p^6 are the same as the ones entering the definition of the (0,4) theory. The vector moduli depend only on the charges and their values are those that the black string has at the horizon. A necessary condition for this conjecture to work is that the (0,4) theory should be independent of the original values of the vector moduli (at least for large p). It is not clear to me whether this is true.

Using this conjecture we would get for large N a compactification of M theory which has five extended dimensions.

B Extremal 3+1 dimensional Reissner-Nordström

This section is more sketchy and contains an unresolved puzzle, so the reader will not miss much if he skips it.

We start with IIB string theory compactified on M^6, where M^6 is a Calabi-Yau manifold or $K3 \times T^2$ or T^6. We consider a configuration of $D3$ branes that leads to a black hole with nonzero horizon area. Consider the limit

$$\alpha' \to 0 \qquad (2\pi)^6 v \equiv \frac{V_6}{\alpha'^3} = \text{fixed} \qquad U \equiv \frac{r}{\alpha'} = \text{fixed} \ . \tag{V.3}$$

The string coupling is arbitrary. In this limit the system reduces to quantum mechanics on the moduli space of the three-brane configuration.

Taking the limit (V.3) of the supergravity solution we find

$$ds^2 = \alpha' \left[\frac{U^2}{g_4^2 N^2} dt^2 + g_4^2 N^2 \frac{dU^2}{U^2} + g_4^2 N^2 d\Omega_2^2 \right] \tag{V.4}$$

where N is proportional to the number of D3 branes. We find a two dimensional AdS_2 space times a two-sphere, both with the same radius $R = l_{4p}N$, where $l_{4p}^2 = g^2\alpha'/v$. The bosonic symmetries of $AdS_2 \times S^2$ are $SO(2,1) \times SO(3)$. This superconformal symmetry seems related to the symmetries of the *chiral* conformal field theory that was proposed in [59] to describe the Reissner-Nordström black holes. Here we find a puzzle, since in the limit (V.3) we got a quantum mechanical system and not a 1+1 dimensional conformal field theory. In the limit (V.3) the energy gap (mentioned in [59,60]) becomes very large[14]. So it looks like taking a large N limit at the same time will be crucial in this case. These problems might be related to the large ground state entropy of the system.

If this is understood it might lead to a proposal for a non perturbative definition of M/string theory (as a large N limit) when there are four non-compact dimensions.

It is interesting to consider the motion of probes on the AdS_2 background. This corresponds to going into the "Coulomb" branch of the quantum mechanics. Dimensional analysis says that the action has the form (II.7) with $p = 0$. Expanding f to first order we find $S \sim \int dt \frac{\dot{U}^2}{U^3} \sim \int dt v^2/r^3$, which is the dependence on r that we expect from supergravity when we are close to the horizon. A similar analysis for Reissner-Nordström black holes in five dimensions would give a term proportional to $1/r^4$ [17]. It will be interesting to check the coefficient (note that this is the *only* term allowed by the symmetries, as opposed to [17]).

VI DISCUSSION, RELATION TO MATRIX THEORY

By deriving various field theories from string theory and considering their large N limit we have shown that they contain in their Hilbert space excitations describing supergravity on various spacetimes. We further conjectured that the field theories are dual to the full quantum M/string theory on various spacetimes. In principle, we can use this duality to give a definition of M/string theory on flat spacetime as (a region of) the large N limit of the field theories. Notice that this is a non-perturbative proposal for defining such theories, since the corresponding field theories can, *in principle*, be defined non-perturbatively. We are only scratching the surface and there are many things to be worked out. In [61] it has been proposed that the large N limit of D0 brane quantum mechanics would describe eleven dimensional M-theory. The large N limits discussed above, also provide a definition of M-theory. An obvious difference with the matrix model of [61] is that here N is not interpreted as the momentum along a compact direction. In our case, N is related to the curvature and the size of the space where the theory is defined. In both cases, in the large N limit we expect to get flat, non-compact spaces. The matrix model [61] gives us a prescription to build asymptotic states, we have not shown here how to construct graviton states, and this is a very interesting problem. On the other hand, with the present proposal it is more clear that we recover supergravity in the large N limit.

This approach leads to proposals involving five (and maybe in some future four) non-compact dimensions. The five dimensional proposal involves considering the 1+1 dimensional field theory associated to a black string in five dimensions. These theories need to be studied in much more detail than we have done here.

It seems that this correspondence between the large N limit of field theories and supergravity can be extended to non-conformal field theories. An example was considered in [1], where the theory of NS fivebranes was studied in the $g \to 0$ limit. A natural interpretation for the throat region is that it is a region in the Hilbert space of a six dimensional "string" theory[15]. And the fact that contains gravity in the large N limit is just a common feature of the large N limit of various field theories. The large N master field seems to be the anti-deSitter supergravity solutions [17].

When we study non extremal black holes in AdS spacetimes we are no longer restricted to low energies, as we were in the discussion in higher dimensions [44,54]. The restriction came from matching the AdS region to the Minkowski region. So the five dimensional results [53,54] can be used to describe arbitrary non-extremal black holes in three dimensional Anti-deSitter spacetimes. This might lead us to understand better where the degrees of freedom of black holes really are, as well as the meaning of the region behind the horizon. The question of the boundary conditions is very interesting and the conformal field theories should provide us with some definite boundary conditions and will probably explain us how to interpret physically spacetimes with horizons. It would be interesting to find the connection with the description of 2+1 dimensional black holes proposed by Carlip [63].

In [8,13] super-singleton representations of AdS were studied and it was proposed that they would describe the dynamics of a brane "at the end of the world". It was also found that in maximally supersymmetric cases it reduces to a free field [8]. It is tempting therefore to identify the singleton with the center of mass degree of

---

14) I thank A. Strominger for pointing this out to me.

15) This possibility was also raised by [62], though it is a bit disturbing to find a constant energy flux to the UV (that is how we are interpreting the radial dimension).

freedom of the branes [6,13]. A recent paper suggested that super-singletons would describe all the dynamics of *AdS* [51]. The claim of the present paper is that all the dynamics of *AdS* reduces to previously known conformal field theories.

It seems natural to study conformal field theories in Euclidean space and relate them to deSitter spacetimes.

Also it would be nice if these results could be extended to four-dimensional gauge theories with less supersymmetry.

ACKNOWLEDGMENTS

I thank specially G. Horowitz and A. Strominger for many discussions. I also thank R. Gopakumar, R. Kallosh, A. Polyakov, C. Vafa and E. Witten for discussions at various stages in this project. My apologies to everybody I did not cite in the previous version of this paper. I thank the authors of [64] for pointing out a sign error.

This work was supported in part by DOE grant DE-FG02-96ER40559.

APPENDIX

$D = p + 2$-dimensional anti-deSitter spacetimes can be obtained by taking the hyperboloid

$$-X_{-1}^2 - X_0^2 + X_1^2 + \cdots + X_p^2 + X_{p+1}^2 = -R^2 , \qquad (A.1)$$

embedded in a flat D+1 dimensional spacetime with the metric $\eta = Diag(-1,-1,1,\cdots,1)$. We will call R the "radius" of *AdS* spacetime. The symmetry group $SO(2,D-1) = SO(2,p+1)$ is obvious in this description. In order to make contact with the previously presented form of the metric let us define the coordinates

$$\begin{aligned} U &= (X_{-1} + X_{p+1}) \\ x_\alpha &= \frac{X_\alpha R}{U} \qquad\qquad \alpha = 0,1,\cdots,p \\ V &= (X_{-1} - X_{p+1}) = \frac{x^2 U}{R^2} + \frac{R^2}{U} . \end{aligned} \qquad (A.2)$$

The induced metric on the hyperboloid (A.1) becomes

$$ds^2 = \frac{U^2}{R^2} dx^2 + R^2 \frac{dU^2}{U^2} . \qquad (A.3)$$

This is the form of the metric used in the text. We could also define $\tilde{U} = U/R^2$ so that metric (A.3) has an overall factor of R^2, making it clear that R is the overall scale of the metric. The region outside the horizon corresponds to $U > 0$, which is only a part of (A.1). It would be interesting to understand what the other regions in the *AdS* spacetime correspond to. For further discussion see [65].

REFERENCES

1. J. Maldacena and A. Strominger, hep-th/9710014.
2. N. Seiberg, Phys. Lett. **B408** (1997) 98, hep-th/9705221
3. R. Haag, J. Lopuszanski and M. Sohnius, Nucl. Phys. **B88** (1975) 257.
4. W. Nahm, Nucl. Phys. **B135** (1978) 149.
5. G. Gibbons, Nucl. Phys. **B207**, (1982) 337; R. Kallosh and A. Peet, Phys. Rev. **D46** (1992) 5223, hep-th/9209116; S. Ferrara, G. Gibbons, R. Kallosh, Nucl. Phys. **B500** (1997) 75, hep-th/9702103.
6. G. Gibbons and P. Townsend, Phys. Rev. Lett. **71** (1993) 5223, hep-th/9307049.
7. Look for "gauged" supergravities in *Supergravities in Diverse Dimensions* , Vol. 1 and 2, A. Salam and E. Sezgin, (1989), North-Holland.

8. C. Frondal, Phys. Rev. **D26** (82) 1988; D. Freedman and H. Nicolai, Nucl. Phys. **B237** (84) 342; K. Pilch, P. van Nieuwenhuizen and P. Townsend, Nucl. Phys. **B242** (84) 377; M. Günaydin, P. van Nieuwenhuizen and N. Warner, Nucl. Phys. **B255** (85) 63; M. Günaydin and N. Warner, Nucl. Phys. **B272** (86) 99; M. Günaydin, N. Nilsson, G. Sierra and P. Townsend, Phys. Lett. **B176** (86) 45; E. Bergshoeff, A. Salam, E. Sezgin and Y. Tanii, Phys. Lett. **205B** (1988) 237; Nucl. Phys. **D305** (1988) 496; E. Bergshoeff, M. Duff, C. Pope and E. Sezgin, Phys. Lett. **B224** (1989) 71;
9. M. Günaydin and N. Marcus, Class. Quant. Grav. **2** (1985) L11; Class. Quant. Grav. **2** (1985) L19; H. Kim, L. Romans and P. van Nieuwenhuizen, Phys. Lett. **143B** (1984) 103; M. Günaydin, L. Romans and N. Warner, Phys. Lett. 154B (1985) 268; Phys. Lett. 164B (1985) 309; Nucl. Phys. **B272** (1986) 598
10. M. Blencowe and M. Duff, Phys. Lett. **203B** (1988) 229; Nucl. Phys. **B310** (1988) 387.
11. H. Nicolai, E. Sezgin and Y. Tanii, Nucl. Phys. **B305** (1988) 483.
12. M. Duff, G. Gibbons and P. Townsend, hep-th/9405124.
13. P. Claus, R. Kallosh and A. van Proeyen, hep-th/9711161.
14. G. Horowitz and A. Strominger, Nucl. Phys. **B360** (1991) 197.
15. G. Gibbons, G. Horowitz and P. Townsend, hep-th/9410073.
16. L. Susskind, hep-th/9611164; O. Ganor, S. Ramgoolam and W. Taylor IV, Nucl. Phys. **B492** (1997) 191, hep-th/9611202.
17. M. Douglas, J. Polchinski and A. Strominger, hep-th/9703031.
18. M. Aganagic, C. Popescu and J. Schwarz, Nucl. Phys. **B495** (1997) 99, hep-th/9612080.
19. M. Dine and N. Seiberg, Phys. Lett. **B409** (1997) 239, hep-th/9705057.
20. K. Becker, M. Becker, J. Polchinski and A. Tseytlin, Phys. Rev. **D56** (1997) 3174, hep-th/9706072.
21. A. Polyakov, hep-th/9711002.
22. T. Banks, W. Fishler, I. Klebanov and L. Susskind, hep-th/9709091.
23. S. Gubser, I. Klebanov and A. Peet, Phys. Rev. **D54** (1996) 3915, hep-th/9602135; A. Strominger, unpublised.
24. I. Klebanov, Nucl. Phys. **B496** (1997) 231, hep-th/9702076.
25. S. Gubser, I. Klebanov and A. Tseytlin, Nucl. Phys. **B499** (1997) 217, hep-th/9703040.
26. S. Gubser and I. Klebanov, hep-th/9708005.
27. C. Vafa, Nucl. Phys. **B469** (1996) 403, hep-th/9602022.
28. I. Klebanov and A. Tseytlin, Nucl. Phys. **B475** (1996) 164, hep-th/9604089.
29. A. Srominger, Phys. Lett. **B383** (1996) 44, hep-th/9512059.
30. E. Witten, hep-th/9507121, N. Seiberg and E. Witten, Nucl. Phys. **B 471** (1996) 121, hep-th/9603003
31. J. Maldacena, Proceedings of Strings'97, hep-th/9709099.
32. N. Seiberg, Phys. Rev. Lett. **79** (1997) 3577, hep-th/9710009.
33. A. Sen, hep-th/9709220.
34. N. Seiberg, hep-th/9705117.
35. S. Sethi and L. Susskind, Phys. Lett. **B400** (1997) 265, hep-th/9702101.
36. T. Banks and N. Seiberg, Nucl. Phys. **B497** (1997) 41, hep-th/9702187.
37. T. Banks, W. Fishler, N. Seiberg and L. Susskind, Phys. Lett. **B408** (1997) 111, hep-th/9705190.
38. I. Bars, Phys. Rev. **D55** (1997) 2373, hep-th/9607112.
39. A. Strominger and C. Vafa, Phys. Lett. **B379** (1996) 99, hep-th/9601029.
40. G. Horowitz, J. Maldacena and A. Strominger, Phys. Lett. **B383** (1996) 151, hep-th/9603109.
41. L. Andrianopoli, R. D'Auria and S. Ferrara, Int. J. Mod .Phys **A12** (1997) 3759, hep-th/9612105.
42. S. Ferrara, R. Kallosh and A. Strominger, Phys. Rev. **D52** (1995) 5412, hep-th/9508072; S. Ferrara and R. Kallosh, Phys. Rev. **D54** (1996) 1514, hep-th/9602136; Phys.Rev. **D54** (1996) 1525, hep-th/960309.
43. J.C. Breckenridge, R.C. Myers, A.W. Peet and C. Vafa, Phys. Lett. **B391** (1997) 93, hep-th/9602065.
44. J. Maldacena, Phys. Rev. **D55** (1997) 7645, hep-th/9611125.
45. D. Diaconescu and N. Seiberg, JHEP07(1997)001, hep-th/9707158.
46. O. Aharony, M. Berkooz, S. Kachru, N. Seiberg and E. Silverstein, hep-th/9707079.
47. E. Witten, hep-th/9707093.
48. Bañados, Teitelboim and Zanelli, Phys. Rev. Lett. **69** (1992) 1849, hep-th/9204099.
49. S. Hyun, hep-th/9704005.
50. H. Boonstra, B. Peeters and K. Skenderis, hep-th/9706192
51. K. Sfetsos and K. Skenderis, hep-th/9711138.
52. S. Hellerman and J. Polchinski, hep-th/9711037.
53. G. Horowitz and A. Strominger, Phys. Rev. Lett. **77** (1996) 2368, hep-th/9602051.
54. A. Dhar, G. Mandal and S. Wadia Phys. Lett. **B388** (1996) 51, hep-th/9605234; S. Das and S. Mathur, Nucl. Phys. **B478** (1996) 561, hep-th/9606185; Nucl.Phys. **B482** (1996) 153, hep-th/9607149; J. Maldacena and A. Strominger, Phys. Rev. **D55** (1997) 861, hep-th/9609026; S. Gubser, I. Klebanov, Nucl. Phys. **B482** (1996) 173, hep-th/9608108; C. Callan, Jr., S. Gubser, I. Klebanov and A. Tseytlin, Nucl. Phys. **B489** (1997) 65, hep-th/9610172; I. Klebanov

and M. Krasnitz Phys. Rev. **D55** (1997) 3250, hep-th/9612051; I. Klebanov, A. Rajaraman and A. Tseytlin Nucl. Phys. **B503** (1997) 157, hep-th/9704112; S. Gubser, hep-th/9706100; H. Lee, Y. Myung and J. Kim, hep-th/9708099; K. Hosomichi, hep-th/9711072.

55. D. Birmingham, I. Sachs and S. Sen, hep-th/9707188.
56. M. Cvetic and F. Larsen, Phys. Rev. **D56** (1997) 4994, hep-th/9705192; hep-th/9706071.
57. J. Maldacena, A. Strominger and E. Witten, hep-th/9711053.
58. A. Chamseddine, S. Ferrara, G. Gibbons and R Kallosh, Phys. Rev. **D55** (1997) 3647, hep-th/9610155.
59. J. Maldacena and A. Strominger, Phys. Rev. **D56** (1997) 4975, hep-th/9702015.
60. J. Maldacena and L. Susskind, Nucl .Phys. **B475** (1996) 679, hep-th/9604042.
61. T. Banks, W. Fischler, S. Shenker and L. Susskind, hep-th/9610043.
62. O. Aharony, S. Kachru, N. Seiberg and E. Silverstein, private communication.
63. S. Carlip, Phys. Rev. **D51** (1995) 632, gr-qc/9409052.
64. R. Kallosh, J. Kumar and A. Rajaraman, hep-th/9712073.
65. S. Hawking and J. Ellis, *The large scale structure of spacetime*, Cambrige Univ. Press (1973), and references therein.

On supergravity duals of four-dimensional $\mathcal{N} = 1$ SCFTs

A. Kehagias

Theory Division,
CERN, 1211 Geneva 23,
Switzerland

Abstract. We present D3-brane solutions in type IIB string theory which preserve 1/4 of supersymmetry. Their near-horizon geometry is $AdS^5 \times X^5$ where X^5 is an appropriate Einstein space and they are candidates for supergravity duals of certain four-dimensional superconformal field theories.

One of the well-known vacua of type IIB supergravity is the $AdS_5 \times S^5$ one, first described in [1]. The non-vanishing fields here are the metric and a Freund-Rubin type anti-self-dual five form. This background has received a lot of attention recently due to its conjectural connection to $\mathcal{N} = 4$ $SU(N)$ supersymmetric YM theory at large N [2–4]. According to this conjecture [2], the large N limit of certain superconformal field theories (SCFT) can be described in terms of Anti de-Sitter (AdS) supergravity and correlation functions of the SCFT theory which lives in the boundary of AdS can be expressed in terms of the bulk theory [3], [4]. In particular, the four-dimensional $\mathcal{N} = 4$ supersymmetric $SU(N)$ Yang-Mills theory is described by the type IIB string theory on $AdS_5 \times S^5$ where the radius of both the AdS_5 and S^5 are proportional to N. Then, the large N limit in field theory corresponds to the type IIB supergravity vacuum. The symmetry of the latter is $SO(4,2) \times SU(4)$ which is just the even subgroup of the $SU(2,2|4)$ superalgebra.

The anti-de Sitter group $SO(4,2)$ has supersymmetric extensions, in addition to the $\mathcal{N} = 4$ algebra $SU(2,2|4)$ corresponding to the $AdS_5 \times S^5$ vacuum, the superalgebras $SU(2,2|2)$ and $SU(2,2|1)$ as well. It is natural to expect that there exist type IIB vacua of the form $AdS^5 \times X^5$ which realize these superalgebras for appropriate five-dimensional manifolds X^5 . The even subgroups of $SU(2,2|2)$ and $SU(2,2|1)$ are $SO(4,2) \times U(2)$ and $SO(4,2) \times U(1)$, respectively, and they are realized by conformal field theories with less supersymmetries, namely, $\mathcal{N} = 2$ and $\mathcal{N} = 1$ superconformal Yang-Mills theories. In fact, in [5] it has been proposed that the supergravity duals of these theories are type IIB supergravity on $AdS_5 \times S^5/\Gamma$ where Γ is appropriate subgroup of $SU(2)$ or $SU(3)$. However, this interpretation has a drawback. Namely, as it has been shown both in string-theory [6] and in field theory context [7], [8], the correlation functions of the $\mathcal{N} = 0,1,2$ theories constructed by orbifolds are the same as those of the $\mathcal{N} = 4$ theory in the large N limit. Thus, one may ask if there exist type IIB vacua, other than the orbifold ones which realize the $\mathcal{N} = 2$ and $\mathcal{N} = 1$ superalgebra $SU(2,2|2)$ and $SU(2,2|1)$ and corresponds to genuine SCFT. This question has been worked out in [9]. The result is that the supergravity dual of the $\mathcal{N} = 1$ SCFT is $AdS_5 \times X^5$. The manifold X^5 is found to be Einstein which admits a $U(1)$ fibration over a a complex surface S with positive first Chern class. The later turns out to be $\mathbf{P}^1 \times \mathbf{P}^1$, $\mathbf{P}^2$ or $\mathbf{P}_{n_1,\ldots,n_k}$, a blow up of $\mathbf{P}^2$ at k points with $3 \leq k \leq 8$. The $\mathbf{P}^2$ corresponds to the maximally supersymmetric $AdS^5 \times S^5$ vacuum [1] and the $\mathbf{P}^1 \times \mathbf{P}^1$ to the $AdS^5 \times T^{1,1}$ [10]. As we will describe below, these vacua are realized as the near horizon geometry of $D3$-branes with Ricci-flat but not flat (except for the maximal $\mathcal{N} = 4$ SCFT) transverse space. In particular, the transverse space turns out to be a cone over X^5. It has then been realised in [11] that this corresponds to putting the $D3$-branes at conifold points of Calabi-Yau manifolds [12]. The $\mathcal{N} = 1$ SCFTs corresponding to $D3$-branes at conifolds has been studied in [11], [13]- [19].

The same analysis can also be applied to M2 and M5 branes. In the maximal supersymmetric case, their near horizon geometry is $AdS_{4,7} \times S^{7,4}$. The M5 brane for example breaks half of the supersymmetries and its near horizon geometry is $AdS_7 \times S^4$ supporting a $\mathcal{N} = 2$ SCFT. The supergravity dual of the $\mathcal{N} = 1$ SCFT is now $AdS_7 \times S^4/\Gamma$ where $\Gamma \in SU(2)$ and this has been worked out in [20]. Similarly, the M2-branes breaks half of the supersymmetries and their near horizon geometry is $AdS_4 \times S^7$. The dual field theory is three-dimensional

CP484, *Trends in Theoretical Physics II*, edited by H. Falomir, R. E. Gamboa Saraví, and F. A. Schaposnik

$\mathcal{N}=8$ SCFT. Theories with less supersymmetries can be obtained by orbifolds [20] or by M2 branes with transverse space a cone over an appropriate seven-dimensional Einstein manifold X^7 [11], [21]- [23].

Here, we will review the $D3$-brane solutions sited at conifold points which support $\mathcal{N}=1$ SCFT. We will follow basically [9]. For a more complete discussion the reader may consult [16].

The massless bosonic fields of the type IIB superstring theory are the graviton g_{MN}, the dilaton ϕ and the antisymmetric tensor B^1_{MN} in the NS-NS sector and the axion χ, the two-form B^2_{MN} and the self-dual four-form field A_{MNPQ} in the R-R sector. The fermionic superpartners are a complex Weyl gravitino ψ_M $(\gamma^{11}\psi_M=\psi_M)$ and a complex Weyl dilatino λ $(\gamma^{11}\lambda=-\lambda)$. The theory has two supersymmetries generated by two supercharges of the same chirality. In addition there exist a conserved $U(1)$ charge, which generates rotations of the two supersymmetries. This $U(1)$ is just the R-symmetry group of the type IIB theory. The graviton and the four-form field are neutral under $U(1)$, the antisymmetric tensors have charge $q=1$, the scalars have $q=2$, whereas the gravitino and the dilatino have charges $q=1/2$ and $q=3/2$, respectively.

The two scalars of the theory can be combined into a complex one, $\tau=\tau_1+i\tau_2$, defined by $\tau=\chi+ie^{-\phi}$, which parametrizes an $SL(2,\mathbf{R})/U(1)$ coset space. The theory has a global $SL(2,\mathbf{R})$ and a local $U(1)$ symmetry. One may define the quantities

$$P_M=-\epsilon_{\alpha\beta}V^\alpha_+\partial_M V^\beta_+=ie^{2i\theta}\frac{\partial_M\tau}{2\tau_2}\,,\quad Q_M=-i\epsilon_{\alpha\beta}V^\alpha_+\partial_M V^\beta_-=\partial_M\theta-\frac{\partial_M\tau_1}{2\tau_2}\,, \tag{1}$$

where Q_M is a composite $U(1)$ gauge connection and P_M has charge $U(1)$ charge $q=2$.. We also define the complex three-form

$$G_{KMN}=-\sqrt{2}i\delta_{\alpha\beta}V^\alpha_+H^\beta_{KMN}=-i\frac{e^{i\theta}}{\sqrt{\tau_2}}(\tau H^1_{KMN}+H^2_{KMN})\,,$$

with charge $q=1$ as well as the five-form field strength F_{MNPQR}

$$F_{MNPQR}=5\partial_{[M}A_{NPQR]}-\frac{5}{4}Im\left((B^1+iB^2)_{MN}(H^1-iH^2)_{PQR}\right)\,,$$

with charge $q=0$.

The supersymmetry transformations of the dilatino and gravitino in a pure bosonic background are [1], [24]

$$\delta\lambda=i\gamma^M P_M\epsilon^*-\frac{i}{24}\gamma^{MNK}G_{MNK}\epsilon\,, \tag{2}$$

$$\delta\psi_M=D_M\epsilon+\frac{i}{480}\gamma^{M_1\cdots M_5}F_{M_1\cdots M_5}\gamma_M\epsilon-\frac{1}{96}\left(\gamma_M{}^{NKL}G_{NKL}-9\gamma^{NL}G_{MNL}\right)\epsilon^*\,, \tag{3}$$

where

$$D_M\epsilon=(\partial_M+\frac{1}{4}\omega_M{}^{AB}\gamma_{AB}-\frac{1}{2}iQ_M)\epsilon,$$

and ϵ is a complex Weyl spinor $(\gamma^{11}\epsilon=\epsilon)$. The five-form F_{MNPQR} satisfies

$$F_{MNPQR}=-\frac{1}{5!}\varepsilon_{MNPQRSTUVW}F^{STUVW}\,, \tag{4}$$

i.e., it is anti-self dual. This can be seen from the supersymmetry transformation of the gravitino eq.(3) since the factor $\gamma^{M_1\cdots M_5}\gamma_M\epsilon$ and the chirality of ϵ projects out the self-dual part of F_{MNPQR}.

The on-shell closure of the supersymmetry algebra specifies the field equations [1], [24] which for the bosonic fields of type IIB supergravity turn out to be

$$D^M P_M=\frac{1}{24}G_{MNK}G^{MNK}\,,\quad D^M G_{MNK}=P^M G^*_{MNK}-\frac{2}{3}iF_{MNKPQ}G^{MPQ}\,,$$

$$R_{MN}=P_M P^*_N+P^*_M P_N+\frac{1}{6}F_{MKLPQ}F_N{}^{KLPQ}$$

$$+\frac{1}{8}\left(G_M{}^{PQ}G^*_{NPQ}+G_N{}^{PQ}G^*_{MPQ}-\frac{1}{6}g_{MN}G^*_{KPQ}G^{KPQ}\right)\,. \tag{5}$$

We will now present supersymmetric D7- and D3-branes backgrounds which preserve some supersymmetry. The non-vanishing bosonic fields are then the graviton g_{MN}, the complex scalar τ and the four-form potential A_{MNKL}. The conditions for unbroken supersymmetry turn out to be

$$\gamma^M P_M \epsilon = 0\,, \tag{6}$$

$$D_M \epsilon + \frac{i}{480}\gamma^{M_1\cdots M_5} F_{M_1\cdots M_5}\gamma_M \epsilon = 0\,. \tag{7}$$

We will assume that the ten-dimensional space-time is of the form $M^4 \times B^6$. We split the coordinates x^M as $x^M = (x^\mu, x^m)$ where $(\mu = 0,\cdots,3\ m = 1,\cdots,6)$. The γ-matrices split accordingly as

$$\gamma^\mu = \Gamma^\mu \otimes 1\,, \qquad \gamma^m = \Gamma^5 \otimes \Gamma^m\,, \tag{8}$$

where Γ^μ, Γ^m are $SO(1,3)$ and $SO(6)$ Γ-matrices respectively. We also define four- and six-dimensional chirality matrices Γ^5, Γ^7 as $\Gamma^5 = i\Gamma^0\cdots\Gamma^3$, $\Gamma^7 = -i\Gamma^4\cdots\Gamma^9$, so that $\gamma^{11} = \Gamma^5\Gamma^7$ and $(\Gamma^5)^2 = (\Gamma^7)^2 = 1$. In the representation (8), Γ^μ are real and hermitian apart from Γ^0 which is anti-hermitian while Γ^a as well as Γ^5 and Γ^7 are imaginary and hermitian. The topology of space-time allows a non-zero five-form field of the form

$$F_{\mu\nu\rho\kappa m} = \epsilon_{\mu\nu\rho\kappa} F_m\,, \qquad F_{mn\ell pq} = \epsilon_{mn\ell pqr}\tilde{F}^r\,, \tag{9}$$

where $F_m, \tilde{F}_m$ are vectors in B^6 which depend on x^m only. They are not independed since the (anti) self-duality condition eq.(4) gives $F_m = \tilde{F}_m$. Moreover, we assume that the complex scalar τ depends only on the B^6 coordinates so that $P_M = (0, P_m)$. One may easily verify that

$$i\gamma^{M_1\cdots M_5} F_{M_1\cdots M_5} = 5!\left(\Gamma^5 \otimes F^m\Gamma_m\Gamma^7 - 1 \otimes F^m\Gamma_m\right)\,, \tag{10}$$

so that eqs.(6,7) turns out to be

$$\Gamma^5 \otimes \Gamma^m P_m \epsilon = 0\,, \tag{11}$$

$$D_\mu \epsilon + \frac{1}{2}(\Gamma^\mu \otimes F^m\Gamma_m)\epsilon = 0\,, \tag{12}$$

$$D_m \epsilon - \frac{s}{2}F_m \epsilon + \frac{s}{2}(1 \otimes F^n\Gamma_{mn})\epsilon = 0\,, \tag{13}$$

where s is the four-dimensional chirality of ϵ $\left(\Gamma^5 \otimes 1\epsilon = s\epsilon\right)$. To solve the above equations, we assume that the metric is of the form

$$ds^2 = e^{2A(x^m)}\eta_{\mu\nu}dx^\mu dx^n + e^{2B(x^m)}h_{mn}dx^m dx^n\,, \tag{14}$$

where h_{mn} is the metric on B^6. Then, eqs.(12,13) are written as

$$\partial_\mu \epsilon + \frac{1}{2}\left(s\partial_n A - F_n\right)\Gamma_\mu \otimes \Gamma^n \epsilon = 0\,, \tag{15}$$

$$\nabla_m \epsilon - \frac{s}{2}F_m \epsilon + \frac{1}{2}\left(\partial_n B + sF_n\right) 1 \otimes \Gamma_m{}^n \epsilon = 0\,, \tag{16}$$

where $\nabla_m = \partial_m + \frac{1}{4}\omega_{mab}(h)\Gamma^{ab} - \frac{i}{2}Q_m$ is the gauge spin-covariant derivative with respect to the metric h_{mn}. By splitting the spinor ϵ as $\epsilon = e^{A/2}\theta \otimes \eta$, with $\Gamma^5\theta = \theta$, $\Gamma^7\eta = \eta$, eqs.(15, 16) give

$$F_m = \partial_m A\,, \qquad B = -A\,, \tag{17}$$

and eq.(16) is then reduced to

$$\nabla_m \eta = 0\,. \tag{18}$$

Thus, the number of unbroken supersymmetries is determined by the number of gauge-covariantly constant spinors η. The integrability condition of eq.(18) is

$$R_{mn}(h) = P_m^* P_n + P_m P_n^*\,, \tag{19}$$

which combined with the field equations

$$R_{MN} = P_M P_N^* + P_N P_M^* + \frac{1}{96} F_M{}^{KLPQ} F_{NKLPQ} \, ,$$

gives that $A(x^m)$ is harmonic [25],

$$\frac{1}{\sqrt{h}} \partial_m \left(\sqrt{h} h^{mn} \partial_n e^{-4A} \right) = 0 \, . \tag{20}$$

Thus, finally, what remain to be solved are eqs.(18,20) and the supersymmetric condition

$$\Gamma^m P_m \eta = 0 \, . \tag{21}$$

If the transverse space B^6 is compact, the solution to eq.(20) turns out to be $A = const.$ In that case, we get just the standard F-theory compactification on a elliptically-fibered CY_4. In this case, the existence of gauge-covariantly constant spinors eq.(18) specifies the base of the fibration to be Kähler. If we assume that $Q_m = 0$ is constant, then η is covariant constant and we have compactification in type IIB theory on $K3 \times T^2$ or on CY_3 manifolds. $D3$-brane solutions are obtained only for non-compact transverse space and this is the case we will consider now. In particular, we will assume that the metric h_{mn} on B^6 takes the form

$$h_{mn} dx^m dx^n = dr^2 + r^2 g_{ij} dx^i dx^j \, , \quad (i,j = 1, ..., 5) \, , \tag{22}$$

where g_{ij} is the metric of a five-dimensional compact space X^5 so that B^6 is a cone over X^5. In this case, eq.(20) gives

$$e^{-4A} = 1 + \frac{Q}{r^4} \, ,$$

where $Q = 4\pi g_s N \alpha'$ and the ten-dimensional metric takes the form

$$ds^2 = \left(1 + \frac{Q}{r^4}\right)^{-1/2} (-dt^2 + dx_1^2 + dx_2^2 + dx_3^2) + \left(1 + \frac{Q}{r^4}\right)^{1/2} \left(dr^2 + r^2 g_{ij} dx^i dx^j\right) \, . \tag{23}$$

This is reduced to the standard D3-brane solution [26] if X^5 is S^5, i.e., when the metric (22) is flat. The near-horizon geometry at $r \to 0$ turns out to be

$$ds_h^2 = \frac{r^2}{\sqrt{Q}} \left(-dt^2 + dx_1^2 + dx_2^2 + dx_3^2\right) + \frac{\sqrt{Q}}{r^2} dr^2 + \sqrt{Q} g_{ij} dx^i dx^j \, , \tag{24}$$

and, thus, it is $AdS^5 \times X^5$. It is now clear why we have assume a cone-like metric for the transverse space B^6. It is the form which leads to a near horizon geometry of the form $AdS_5 \times X^5$. Having assumed a different metric we would have obtained a completely different geometry.

Let us now return to the gauge-covariant Killing spinor equation (18) which will determine the possible spaces X^5. With the metric (22), eq.(18) turns out to be

$$\left(\partial_r - \frac{i}{2} Q_r\right) \eta = 0 \, , \quad \overset{\circ}{\nabla}_i \eta + \frac{1}{4} \Gamma_i{}^r \eta = 0 \, , \tag{25}$$

where $\overset{\circ}{\nabla}_i = \partial_i + \frac{1}{4}\omega_{ijk}(h)\Gamma^{jk} - \frac{i}{2} Q_i$ is the gauge-covariant derivative on X^5. For $Q_m = 0$, these equations have been studied in [27], [28]. Since the metric (22) satisfies

$$R_{rr}(h) = R_{ri} = 0 \, , \quad R_{ij}(h) = R_{ij}(g) - 4 g_{ij} \, , \tag{26}$$

we get the integrability condition of eq.(25)

$$\begin{aligned} 0 = P_r^* P_r &= P_r^* P_i + P_r P_i^* \, , \\ R_{ij}(g) &= P_i^* P_j + P_i P_j^* + 4 g_{ij} \, . \end{aligned} \tag{27}$$

Thus, $P_r = 0$ and the complex scalar τ is independent of r. We may split now the γ-matrices Γ^i as

$$\Gamma^i = \Sigma^i \otimes \sigma^1 \,, \quad \Gamma^r = 1 \otimes \sigma^3 \,,$$

where Σ^i are $SO(5)$ γ matrices. The spinor η split accordingly as $\eta = \eta_0 \otimes \xi$. Then, eq.(25) turns out to be

$$\overset{\circ}{\nabla}_i \eta_0 = \mp \frac{1}{2} \Sigma_i \eta_0 \,, \tag{28}$$

so that η_0 is a Killing spinor on X^5. The number of independent solutions to eq.(28) specifies the number of unbroken symmetries.

The Killing spinor equation (28) has extensively been studied in the Kaluza-Klein supergravity context. There, the isometries of X^5 appear as the gauge group in four dimensions. In our case, the isometries of X^5 will lead to gauged supergravities in AdS^5 and in view of the AdS/CFT correspondence will appear as global symmetries in the boundary CFT. Since, the supersymmetric boundary CFT will have at least a global $U(1)$ corresponding to the R-symmetry of the minimal $\mathcal{N} = 1$ case, X^5 will necessarily have a $U(1)$ isometry. It is then natural to assume that X^5 is a $U(1)$ bundle over a four-dimensional space with metric of the form

$$g_{ij} dx^i dx^j = g_{ab} dx^a dx^b + 4(d\psi + A_a dx^a)^2 \,, \quad (a, b = 1, ..., 4) \,. \tag{29}$$

If the four-dimensional space is a complex Kähler surface $\mathcal{S}$ with metric g_{ab}, then $A_a dx^a$ is the $U(1)$ connection with field strength proportional to the Kähler two-form of the base $\mathcal{S}$, J_{ab}, i.e.,

$$F_{ab} = i J_{ab} \,. \tag{30}$$

In this case, the Killing spinor equation (28) can be solved as in the seven-dimensional case considered in [29]. In addition, the condition eq.(21) gives that

$$\Gamma^a P_a \eta + \Gamma^\psi P_\psi \eta = 0 \,, \tag{31}$$

where $x^i = (x^a, \psi), a = 1, ..., 4$. By using an explicit representation for the γ-matrices (Γ^a, Γ^ψ), one may verify that eq.(31) is satisfied if τ is a holomorphic function of the complex coordinates $z^1 = x^1 + ix^2$, $z^2 = x^3 + ix^4$ and independent of ψ. With the metric (29), the integrability conditions eq.(27) are then reduced to

$$R_{ab} = P_a^* P_b + P_a P_b^* + 6 g_{ab} \,. \tag{32}$$

In the absence of D7-branes, ($P_a = 0$) we have that

$$R_{ab} = 6 g_{ab} \,, \tag{33}$$

and the obvious solution is then $\mathbf{P}^2$, the complex projective space. In that case, X^5 is a $U(1)$ bundle over $\mathbf{P}^2$, which is just the five-sphere S^5 and we recover the standard D3-brane solution. However, there are other solutions as well. Namely, every complex compact surface $\mathcal{S}$ which satisfies eq.(33) provides a solutions as well. Such surfaces have positive first Cern-class $c_1 > 0$ as opposed to the CY's which have vanishing c_1. Surfaces with $c_1 > 0$ are known as del Pezzo surfaces. These include $\mathbf{P}^1 \times \mathbf{P}^1$, or $\mathbf{P}^2_{n_1 \ldots n_k}$, the surface which is obtained by blowing up $\mathbf{P}^2$ at k generic points (no three-points are collinear and no six points are in one quadratic curve in $\mathbf{P}^2$) with $0 \leq k \leq 8$. Thus, the solutions to eq.(33) is reduced to find all del Pezzo surfaces which admit Kähler-Einstein metrics [30]- [34]. A complex surface now admits a Kähler-Einstein metric if and only if its group of automorphism is reductive [30], [31]. Surfaces which admit Kähler-Einstein metrics are for example Fermat cubics in $\mathbf{P}^3$ [32]. As have been proven in [33], del Pezzo surfaces which admit Kähler-Einstein metrics are $\mathbf{P}^2$, $\mathbf{P}^1 \times \mathbf{P}^1$ and $\mathbf{P}^2_{n_1 \ldots n_k}$ with $3 \leq k \leq 8$. In the case of a $\mathbf{P}^2$ base X^5 is just S^5 and we have a $\mathcal{N} = 4$ SCFT while for the other case we will get a $\mathcal{N} = 1$ SCFT. In particular, the SCFT corresponding to the $\mathbf{P}^1 \times \mathbf{P}^1$ base have been found in [11].

REFERENCES

1. J.H. Schwarz, Nucl. Phys. B 226 (1983)269.
2. J. Maldacena, hep-th/9711200.
3. S.S. Gubser, I.R. Klebanov and A.M. Polyakov, hep-th/9802109.
4. E. Witten, hep-th/9802150; hep-th/9803131.
5. S. Kachru and E. Silverstein, hep-th/9802183
6. A. Lawrence, N. Nekrasov and C. Vafa, hep-th/9803015.
7. M. Bershadsky, Z. Kakushadze and C. Vafa, hep-th/9803076; M. Bershadsky, A. Johansen, hep-th/9803249.
8. Z. Kakushadze, hep-th/9803214, hep-th/9804184.
9. A. Kehagias, Phys. Lett. B 435 (1998)337, hep-th/9805131.
10. L. Romans, Phys. Lett. B 153 (1985)392.
11. I. Klebanov and E. Witten, hep-th/9807080.
12. P. Candelas and X. de la Ossa, Nucl. Phys. B 342 (1990)246.
13. C. Ahn, K. Oh and R. Tatar, hep-th/9808143, hep-th/9806041.
14. O. Aharony, A. Fayyazuddin and J. Maldacena, J. High Energy Phys. 07 (1998)013, hep-th/9806159.
15. S.S. Gubser and I.R. Klebanov, hep-th/9808075.
16. D.R. Morrison and M.R. Plesser, hep-th/9810201.
17. K. Dasgupta, S. Mukhi, hep-th/9811139.
18. S. Gubser, N. Nekrasov and S. Shatashvili, hep-th/9811230.
19. E. Lopez, hep-th/9812025.
20. S. Ferrara, A. Kehagias, H. Partouche and A. Zaffaroni, Phys. Lett. B 431 (1998)42, hep-th/9803109.
21. J.M. Figueroa-O'Farrill, hep-th/9807149.
22. B.S. Acharya, J.M. Figueroa-O'Farrill, C.M. Hull and B. Spence, hep-th/9808014.
23. K. Oh and R. Tatar, hep-th/9810244.
24. J.H. Schwarz and P. West, Phys. Lett. B 126 (1983)301; P.S. Howe and P. West, Nucl. Phys. B 238 (1984)181.
25. M.J. Duff and J.X. Lu, Phys. Lett. B 273 (1991)409.
26. G.T. Horowitz and A. Strominger, Nucl. Phys. B 360 (1991)197.
27. Th. Friedrich and I. Kath, J. Differential Geom. 29 (1989)263; Commun. Math. Phys. 133 (1989) 543.
28. C. Bär, Commun. Math. Phys. 154 (1993) 509.
29. C.N. Pope and N.P. Warner, Class. Quantum Grav. 2 (1985)L1.
30. A.L. Besse, *"Einstein Manifolds"*, Springer-Verlag, 1987.
31. A. Futaki, *"Kähler-Einstein Metrics and Integral Invariants"*, Springer-Verlag, 1988.
32. Y.T. Siu, Ann. Math. 127 (1988)585.
33. G. Tian and S.T. Yau, Commun. Math. Phys. 112 (1987)175.
34. G. Tian, Invent. Math. 89 (1987)225; 101 (1990)101; 130 (1997)1.

DUALITY SYMMETRY IN THE SCHWARZ-SEN MODEL

H. O. Girotti

Instituto de Física, Universidade Federal do Rio Grande do Sul
Caixa Postal 15051, 91501-970 - Porto Alegre, RS, Brazil.

Abstract. The duality symmetric but non manifestly covariant action proposed by Schwarz and Sen is canonically quantized in the Coulomb gauge. The resulting theory turns out to be, nevertheless, relativistically invariant. It is shown, afterwards, that the Schwarz-Sen model naturally emerges when duality is implemented as a local symmetry of sourceless electrodynamics. This implies in the equivalence of these theories at the quantum level.

I INTRODUCTION

The equations of motion of the four-dimensional low energy effective field theory for the bosonic sector of the heterotic string, are invariant under SL(2,R) duality transformations of the massless fields involved. This, so called S-duality symmetry, is a symmetry of the equations of motion but not of the corresponding action. On the other hand, the low energy effective field theory action retains the target space duality symmetry (T-duality) of string theory. It would be desirable to achieve S-duality at the level of the action in the hope of attaining results similar to those given by the T-duality symmetry. However, it is difficult to conciliate duality symmetry and manifest Lorentz covariance.

The difficulty of writing a non trivial action involving only a finite-component self-dual covariant form is well known. In fact, to construct a duality symmetric action being manifestly covariant one must introduce auxiliary fields. The first attempts in this direction involved an infinite number of auxiliary fields [1].

Duality symmetric actions being manifestly covariant and containing a finite number of auxiliary fields have also been constructed but they are not polynomials. These actions were found by Pasti, Sorokin and Tonin (PST) [2, 3]. For the case of electrodynamics, the PST action covariantizes the action found earlier by Deser and Teitelboim [4] and rediscovered by Schwarz and Sen [5], which is duality symmetric but not manifestly covariant. The introduction of sources into the manifestly covariant and non-manifestly covariant versions of these duality symmetric actions has also been achieved [6].

The present work is essentially dedicated to establish the quantum mechanical equivalence of the Schwarz-Sen and Maxwell theories in the case free of sources. We start by quantizing the Schwarz-Sen model in the Coulomb gauge. The resulting quantum theory will be shown to be, nevertheless, relativistically invariant. This is our Section II. In Section III we begin by recalling that the Maxwell action, when formulated in the Coulomb gauge, remains invariant under a set of non-local duality transformations derived by Deser and Teitelboim [4]. Additional fields are, afterwards, brought into the theory in order to make these transformations local. Correspondingly, a new expression for the generating functional of Green functions is derived. In Section IV we demonstrate that the Green functions generating functionals for the Maxwell and Schwarz-Sen theories are rigorously identical. Section V contains the conclusions.

II QUANTIZATION OF THE SCHWARZ-SEN MODEL

The duality symmetric action proposed by Schwarz-Sen [5] involves two gauge potentials $A^{\mu,a} (0 \leq \mu \leq 3,\ 1 \leq a \leq 2)$ and reads[1]

CP484, *Trends in Theoretical Physics II*, edited by H. Falomir, R. E. Gamboa Saraví, and F. A. Schaposnik

$$S = -\frac{1}{2}\int d^4x \left(B^{a,i}\epsilon_{ab}E^{b,i} + B^{a,i}B^{a,i}\right) \ , \tag{1}$$

where

$$E^{a,i} = -F^{a,0i} = -\left(\partial^0 A^{a,i} - \partial^i A^{a,0}\right) \ , \tag{2.a}$$

$$B^{a,i} = -\frac{1}{2}\epsilon^{ijk}F^a_{jk} = -\epsilon^{ijk}\partial_j A^a_k \ , \tag{2.b}$$

and $1 \leq i,j,k \leq 3$. S is separately invariant under the local gauge transformations

$$A^{a,0} \to A^{a,0} + \Psi^a \ , \tag{3.a}$$

$$A^{a,i} \to A^{a,i} - \partial^i \Lambda^a \ , \tag{3.b}$$

and under the global $SO(2)$ rotations

$$A^{\mu a} \to A'^{\mu a} = A^{\mu a}\cos\theta + \epsilon^{ab}A^{\mu b}\sin\theta \ . \tag{4}$$

Of course, (4) reduces to the usual discrete duality transformation for $\theta = \pi/2$. However, the Lagrangian density in (1),

$$\mathcal{L} = \frac{1}{2}\epsilon^{jki}(\partial_j A^a_k)\epsilon_{ab}(\partial_0 A^b_i) - \frac{1}{2}\epsilon^{jki}(\partial_j A^a_k)\epsilon_{ab}(\partial_i A^b_0) - \frac{1}{4}F^{a,jk}F^a_{jk} \ , \tag{5}$$

is not a Lorentz scalar. The use of the equations of motion deriving from (5),

$$\epsilon^{ijk}\epsilon_{ab}\partial_0\partial_j A^b_k + \partial_j\left(\partial^j A^{a,i} - \partial^i A^{a,j}\right) = 0 \ , \tag{6}$$

allows for the elimination from S of one of the gauge fields, the action for the remaining one being the conventional Maxwell action.

Within the Hamiltonian framework, the Schwarz-Sen model is characterized by the canonical Hamiltonian (H_c)

$$H_c = \int d^3x \left[\frac{1}{2}\epsilon^{jki}(\partial_j A^a_k)\epsilon_{ab}(\partial_i A^b_0) + \frac{1}{4}F^{a,jk}F^a_{jk}\right] \ . \tag{7}$$

Furthermore, the system possesses the primary constraints

$$\Omega^a_0 \equiv \pi^a_0 \approx 0 \ , \tag{8.a}$$

$$\Omega^a_i \equiv \pi^a_i + \frac{1}{2}\epsilon_{ab}\,\epsilon_{ijk}\,\partial^j A^{b,k} \approx 0 \ , \tag{8.b}$$

where we have designated by π^a_μ the momentum canonically conjugate to $A^{a,\mu}$. Then, the total Hamiltonian (H') is given by $H' = H_c + \int d^3x \left(u^{a,0}\Omega^a_0 + u^{a,i}\Omega^a_i\right)$, where the u's are Lagrange multipliers. Persistence in time of Ω^a_0 produces neither secondary constraints nor determines the Lagrange multipliers. On the other hand, persistence in time of the primary constraints $\{\Omega^a_i\}$ does not lead to the existence of secondary constraints but determines partially the Lagrange multipliers $\{u^a_i\}$. Indeed, since the Poisson bracket

---

1) This section is mainly based on Refs. [7,8]. Our space-time metric is $g_{00} = -g_{11} = -g_{22} = 1$ while ϵ_{ab} designates a generic element of the two-dimensional antisymmetric unit matrix ($\epsilon_{12} = +1$).

$$[\Omega^a_i(\vec{x})\,,\,\Omega^b_j(\vec{y})]_P = -\epsilon_{ab}\,\epsilon_{ijk}\,\partial^j_x \delta(\vec{x}-\vec{y}) \tag{9}$$

does not vanish, $\dot{\Omega}^a_i = [\Omega^a_i, H']_P \approx 0$ yields $u^{a,i} = \epsilon_{ab}(B^{b,i} - \partial^i \phi^b)$, where ϕ^a is an arbitrary scalar. Thus,

$$\Omega^a(\vec{x}) = \partial^i \Omega^a_i(\vec{x}) \approx 0 \tag{10}$$

and $\Omega^a_0 \approx 0$ are the first-class constraints in the theory.

To isolate the second-class constraints from (8.b), we split Ω^a_i into longitudinal (L) and transversal (T) components, namely, $\Omega^a_i = \Omega^a_{Li} + \Omega^a_{Ti}$, where $\Omega^a_{Li} = -\frac{\partial_i \partial^j}{\nabla^2}\Omega^a_j$, $\Omega^a_{Ti} = \left(g^j_i + \frac{\partial_i \partial^j}{\nabla^2}\right)\Omega^a_j$ and $\nabla^2 \equiv -\partial_j \partial^j$. The first-class constraint (10) only involves the longitudinal components Ω^a_{Li} and states that these components vanish individually. Then, the second-class constraints are

$$\Omega^a_{Ti} = \pi^a_{Ti} + \frac{1}{2}\epsilon_{ab}\,\epsilon_{ijk}\,\partial^j A^{b,k}_T \approx 0\ . \tag{11}$$

The determination of the constraint structure is over. It only remains to be mentioned that the gauge potential $A^{a,\mu}$, when acted upon by the generator of infinitesimal gauge transformations, $G = \int d^3x\,(\Psi^a \Omega^a_0 + \Lambda^a \Omega^a)$, undergoes the change $A^{a,\mu} \to A^{a,\mu} + \delta A^{a,\mu}$ with $\delta A^{a,0} = [A^{a,0}, G]_P = \Psi^a$ and $\delta A^{a,i} = [A^{a,i}, G]_P = -\partial^i \Lambda^a$, in agreement with Eqs.(3).

We shall next quantize the theory by means of the Dirac bracket quantization procedure [9–12]. To this end, we start by fixing the gauge through the subsidiary conditions

$$\chi^{a,0} \equiv A^{a,0} \approx 0\ , \tag{12.a}$$

$$\chi^a \equiv \partial_i A^{a,i} \approx 0\ . \tag{12.b}$$

The fact that the Coulomb condition and $A^{a,0} \approx 0$ are, when acting together, accessible gauge conditions is a peculiarity of the model under analysis. We now recall that, according to the quantization procedure being used, the equal-time commutation algebra is to be abstracted from the corresponding Dirac bracket algebra, the constraint and gauge conditions thereby translating into strong operator relations. After some calculations one finds that

$$\left[A^{a,i}_T(\vec{x})\,,\,A^{b,j}_T(\vec{y})\right] = -i\epsilon_{ab}\,\epsilon^{ijk}\,\frac{\partial^x_k}{\nabla^2}\delta(\vec{x}-\vec{y})\ , \tag{13.a}$$

$$\left[A^{a,i}_T(\vec{x})\,,\,\pi^b_{Tj}(\vec{y})\right] = \frac{i}{2}\delta_{ab}\left(g^i_j + \frac{\partial^i_x \partial^x_j}{\nabla^2}\right)\delta(\vec{x}-\vec{y})\ , \tag{13.b}$$

$$\left[\pi^a_{Ti}(\vec{x})\,,\,\pi^b_{Tj}(\vec{y})\right] = \frac{i}{4}\epsilon_{ab}\,\epsilon_{ijk}\partial^k_x \delta(\vec{x}-\vec{y})\ , \tag{13.c}$$

while the Hamiltonian operator reads

$$H = \frac{1}{4}\int d^3x F^{a,jk} F^a_{jk} = -\frac{1}{2}\int d^3x B^{a,j} B^a_j\ . \tag{14}$$

One may wonder on whether the right hand side of (14) is afflicted by ordering ambiguities. This not so, since

$$\left[B^{a,i}(\vec{x})\,,\,B^{b,j}(\vec{y})\right] = i\,\epsilon_{ab}\,\epsilon^{ijk}\,\partial^x_k \delta(\vec{x}-\vec{y})\ . \tag{15}$$

The main object of interest is the field commutator at different space-time points. To find it, we must first solve the Heisenberg equations of motion deriving from (13) and (14), namely,

$$\mathcal{D}^{(-)ab}_{ik} A^{b,k}_T = 0\ , \tag{16.a}$$

$$\partial_0 \pi^a_{Ti} = \frac{1}{2}\partial^j F^a_{ji}\ , \tag{16.b}$$

where

$$\mathcal{D}_{ik}^{(\pm)ab} \equiv g_{ik}\delta_{ab}\partial_0 \pm \epsilon_{ab}\epsilon_{ijk}\partial^j \ . \tag{17}$$

Notice that, in the Coulomb gauge, the Lagrange equation of motion (6) can be casted as

$$\epsilon^{jli}\partial_l \mathcal{D}_{ik}^{(-)ab} A_T^{b,k} = 0 \Longrightarrow \mathcal{D}_{ik}^{(-)ab} A_T^{b,k} = \partial_i \xi^a \ . \tag{18}$$

Since $\partial^i \mathcal{D}_{ik}^{(-)ab} A_T^{b,k} = 0$, the function ξ^a must verify $\nabla^2 \xi^a = 0$ but is otherwise arbitrary. Thus, the Lagrangian and the Hamiltonian formulations lead to equivalent equations of motions only after the introduction of a regularity requirement at spatial infinity. This situation resembles that encountered in connection with the theory of the two-dimensional $(x^0, x^1, x^\pm = 1/\sqrt{2}(x^0 \pm x^1))$ self-dual field (Φ) proposed by Floreanini and Jackiw [13,14], where the equations of motion in the Lagrangian and Hamiltonian formulations turn out to be, respectively, $\partial_1 \partial_- \Phi = 0$ and $\partial_- \Phi = 0$. We also recall that in order to solve $\partial_- \Phi = 0$ one starts by realizing that $\partial_- \Phi = 0 \Longrightarrow \partial_+ \partial_- \Phi = 0 \Longrightarrow \Box \Phi = 0$. The solutions of $\partial_- \Phi = 0$ are then contained in the field of solutions of $\Box \Phi = 0$. We shall follow here a similar approach, since

$$\mathcal{D}_{ik}^{(-)ab} A_T^{b,k} = 0 \Longrightarrow \mathcal{D}^{(+)ca,li} \mathcal{D}_{ik}^{(-)ab} A_T^{b,k} = 0 \Longrightarrow \Box A_T^{c,l} = 0 \ . \tag{19}$$

The solving of $\Box A_T^{a,i} = 0$ leads to

$$A_T^{a,i}(x) = \int d^3y D(x-y) \overleftrightarrow{\partial}_y^0 A_T^{a,i}(y) \ , \tag{20}$$

where $D(x-y)$ is the zero-mass Pauli-Jordan delta function and $(A \overleftrightarrow{\partial}^k B) \equiv A\partial^k B - B\partial^k A$. The combined use of this last equation and (13) allowed us to find the following explicit form for the field commutator at different space-time points

$$\begin{aligned} &[A_T^{a,i}(x) \, , \, A_T^{b,j}(y)] \\ &= i \left[\delta_{ab} \left(g^{ij} + \frac{\partial_x^i \partial_x^j}{\nabla_x^2} \right) - \epsilon_{ab}\epsilon^{ijk} \frac{\partial_k^x \partial_0^x}{\nabla_x^2} \right] D(x-y) \ . \end{aligned} \tag{21}$$

One can verify, by applying $\mathcal{D}_{ki}^{(-)ca}(x)$ to both sides of (21), that the field configurations entering the just mentioned commutator are in fact solutions of (16.a).

Now, the function $D(x-y)$ can be given as the sum of a positive plus a negative frequency part and we, therefore, can write

$$A_T^{a,i}(x) = A_T^{a,i(+)}(x) + A_T^{a,i(-)}(x) \ , \tag{22}$$

where

$$\begin{aligned} &A_T^{a,i(\pm)}(x) \\ &= \frac{1}{(2\pi)^{3/2}} \int \frac{d^3k}{\sqrt{2|\vec{k}|}} \exp[\pm i(|\vec{k}|x^0 - \vec{k}\cdot\vec{x})] \sum_{\lambda=1}^{2} \varepsilon_\lambda^{a,i}(\vec{k}) a_\lambda^{(\pm)}(\vec{k}) \end{aligned} \tag{23}$$

and $\varepsilon_\lambda^{a,i}(\vec{k}), \lambda = 1, 2$, are unit norm polarization vectors. By going back with (23) into (21) one obtains

$$\sum_{\lambda,\lambda'=1}^{2} \varepsilon_{\lambda}^{a,i}(\vec{k})\varepsilon_{\lambda'}^{b,j}(\vec{k'}) \left[a_{\lambda}^{(-)}(\vec{k}) \, , \, a_{\lambda'}^{(+)}(\vec{k'})\right]$$
$$= \left[-\delta_{ab}\left(g^{ij} + \frac{k^i k^j}{|\vec{k}|}\right) + \epsilon_{ab}\epsilon^{ijl}\frac{k_l}{|\vec{k}|}\right] \delta(\vec{k} - \vec{k'}) \ , \tag{24}$$

while all others commutators vanish. The polarization vectors are to be found by replacing (23) into the gauge condition (12.b) and the equation of motion (16.a). One arrives to

$$\sum_{\lambda=1}^{2} \vec{\varepsilon}_{\lambda}^{\,a}(\vec{k}) \times \vec{\varepsilon}_{\lambda}^{\,b}(\vec{k}) = -\,2\,\epsilon_{ab}\frac{\vec{k}}{|\vec{k}|} \ . \tag{25}$$

On the other hand, the Coulomb gauge polarization vectors span, by construction, the space orthogonal to $\vec{k}$, i.e.,

$$\sum_{\lambda=1}^{2} \varepsilon_{\lambda}^{a,i}(\vec{k})\,\varepsilon_{\lambda}^{a,j}(\vec{k}) = -\left(g^{ij} + \frac{k^i k^j}{|\vec{k}|^2}\right) \ . \tag{26}$$

By using (25) and (26) we solve at once for the commutator in (24),

$$\left[a_{\lambda}^{(-)}(\vec{k}) \, , \, a_{\lambda'}^{(+)}(\vec{k'})\right] = \delta_{\lambda\lambda'}\,\delta(\vec{k} - \vec{k'}) \ . \tag{27}$$

Thus the space of states is, as expected, a Fock space with positive definite metric.

Hence, the quantized Schwarz-Sen model is a physically sensible quantum field theory. Our next task is to demonstrate that this theory is also relativistically invariant. We are therefore looking for a set of composite operators $\{\Theta_{\mu\nu}\}$ which may serve as Poincaré densities. We shall build them by following the rules that are used to construct the symmetric (Belinfante) energy-momentum tensor in a manifestly Lorentz invariant theory. In this way we find

$$\Theta_{00} = -\frac{1}{2}\,B^{a,i}\,B_i^a \ , \tag{28.a}$$
$$\Theta_{0i} = \Theta_{i0} = -\frac{1}{2}\,\epsilon_{ijk}\,\epsilon_{ab}B^{a,j}\,B^{b,k} \ , \tag{28.b}$$
$$\Theta_{ij} = \Theta_{ji} = -B_i^a\,B_j^a + g_{ij}\,B^{a,l}\,B_l^a \ . \tag{28.c}$$

Thus, Θ is symmetric and free of ordering ambiguities but we can not yet decide on whether or not it is a tensor. As for the equal-time commutator algebra obeyed by the components of Θ, it is fully determined by the commutator (15). In particular, one can corroborate that

$$[\,\Theta^{00}(x^0,\vec{x})\,,\,\Theta^{00}(x^0,\vec{y})\,]$$
$$= -\,i\,\{\,\Theta^{0k}(x^0,\vec{x}) + \Theta^{0k}(x^0,\vec{y})\,\}\;\partial_k^x\delta(\vec{x}-\vec{y}) \ , \tag{29.a}$$
$$[\,\Theta^{00}(x^0,\vec{x})\,,\,\Theta^{0k}(x^0,\vec{y})\,]$$
$$= -\,i\,\{\,\Theta^{kj}(x^0,\vec{x}) - g^{kj}\,\Theta^{00}(x^0,\vec{y})\,\}\;\partial_j^x\,\delta(\vec{x}-\vec{y}) \ , \tag{29.b}$$
$$[\,\Theta^{0k}(x^0,\vec{x})\,,\,\Theta^{0j}(x^0,\vec{y})\,]$$
$$= i\,\{\,\Theta^{0k}(x^0,\vec{y})\,\partial_x^j + \Theta^{0j}(x^0,\vec{x})\,\partial_x^k\,\}\;\delta(\vec{x}-\vec{y}) \ . \tag{29.c}$$

As known [15], positivity requires that a singular Schwinger term, proportional to $(\partial^3)\delta(\vec{x}-\vec{y})$, must also be present in the right hand side of Eq.(29.b). The definition of $\Theta^{\mu\nu}$ can, in fact, be altered as to yield such term. But, since Schwinger terms do not contribute to the algebra of integrated charges, we have omitted them in Eqs.(29). From Eqs.(29) then follows that the charges

$$P^\mu \equiv \int d^3x\, \Theta^{0\mu} \;, \tag{30.a}$$

$$J^{\mu\nu} \equiv \int d^3x\, (\Theta^{0\mu}x^\nu - \Theta^{0\nu}x^\mu) \;, \tag{30.b}$$

obey the Poincaré algebra, i.e.,

$$[P^\mu, P^\nu] = 0 \;, \tag{31}$$

$$[J^{\mu\nu}, P^\sigma] = i\,(g^{\mu\sigma}P^\nu - g^{\nu\sigma}P^\mu) \;, \tag{32}$$

$$[J^{\mu\nu}, J^{\rho\sigma}] = i\,(g^{\mu\rho}J^{\nu\sigma} + g^{\nu\sigma}J^{\mu\rho} - g^{\mu\sigma}J^{\nu\rho} - g^{\nu\rho}J^{\mu\sigma}) \;. \tag{33}$$

It takes just a few more steps to demonstrate that Θ is a tensor. Indeed, the additional equal-time commutators $\left[\Theta^{ij}(x^0,\vec{x})\,,\,\Theta^{00}(x^0,\vec{y})\right]$ and $\left[\Theta^{ij}(x^0,\vec{x})\,,\,\Theta^{0k}(x^0,\vec{y})\right]$ can also be readily evaluated by using (28) and (15). These results and (29) can be collected into

$$\left[P^\mu\,,\,\Theta^{\alpha\beta}\right] = -i\,\partial^\mu\Theta^{\alpha\beta} \;, \tag{34.a}$$

$$\left[J^{\mu\nu}\,,\,\Theta^{\alpha\beta}\right] = -i\,(x^\nu\partial^\mu - x^\mu\partial^\nu)\,\Theta^{\alpha\beta}$$

$$-i\,(\Theta^{\mu\alpha}g^{\nu\beta} + \Theta^{\mu\beta}g^{\nu\alpha} - \Theta^{\nu\alpha}g^{\mu\beta} - \Theta^{\nu\beta}g^{\mu\alpha}) \;, \tag{34.b}$$

which are, respectively, the translation and rotation transformation laws to be obeyed by a second-rank tensor. The purported proof of relativistic invariance of the quantized Schwarz-Sen theory is now complete.

What remains to be done is to demonstrate that the Coulomb gauge formulation of the quantized Schwarz-Sen theory is in fact covariant. Since translations and ordinary rotations do not destroy the Coulomb gauge condition we concentrate on Lorentz boosts. By using (30), (28), (2.b) and (13.a) one arrives to

$$-i\left[J^{0k}\,,\,A_T^{a,i}\right] = (x^0\,\partial^k - x^k\,\partial^0)A_T^{a,i} - \epsilon_{ab}\epsilon^{klj}\frac{\partial^i\partial_l}{\nabla^2}A_{Tj}^b \;. \tag{35}$$

The term proportional to ϵ_{ab} signalizes that gauge potentials corresponding to different values of a get mixed by Lorentz boosts. This does not occur for ordinary rotations. Furthermore, the mixing term in (35) describes an operator gauge transformation, which, as one easily verifies, makes this commutator compatible with the transversality condition $\partial_i A_T^{a,i} = 0$. Hence, under Lorentz boosts, the field $A_T^{a,i}$ undergoes, besides the usual vector transformation, an operator gauge transformation which restores the Coulomb gauge in the new Lorentz frame.

As for the Nother's charge associated with the $SO(2)$ symmetry (4), it is straightforward to verify that it can be written as

$$Q = -\frac{1}{2}\int d^3x\epsilon^{jik}(\partial_j A_{Ti}^a)A_{Tk}^a = \frac{1}{2}\int d^3x B^{ak}A_{Tk}^a \;. \tag{36}$$

Observe that Q is a SO(2) invariant Chern-Simons term. Thus, up to surface terms, it is gauge invariant. It is also metric independent and so its algebraic form also holds for curved spaces. The use of (13.a) enables one to verify that Q indeed generates the infinitesimal $SO(2)$ rotations

$$[Q, A^b_{Tj}(y)] = -i\epsilon^{ba} A^a_{Tj}(y) \ . \tag{37}$$

Furthermore, in terms of the creation and anhilation operators of (23) the operator Q is found to read

$$Q = i \int d^3k (a_1^\dagger a_2 - a_2^\dagger a_1) \tag{38}$$

and becomes diagonal,

$$Q = \int d^3k (a_L^\dagger a_L - a_R^\dagger a_R) \ , \tag{39}$$

in the base of circularly polarized operators, defined by

$$a_R^\dagger = \frac{a_1^\dagger + i a_2^\dagger}{\sqrt{2}} \ , \tag{40.a}$$

$$a_L^\dagger = \frac{a_1^\dagger - i a_2^\dagger}{\sqrt{2}} \ . \tag{40.b}$$

From (39) one sees that, in a generic state, Q counts the number of left minus right polarized photons. It is easily checked that Q commutes with all the generators of the conformal group as should be the case for an internal symmetry generator.

The last part of this Section is dedicated to present the functional quantization of the Schwarz-Sen theory in the Coulomb gauge. Clearly, the constraints (8.a) and (12.a) can be used to eliminate the phase-space variables π^a_0, $A^{a,0}$ from the outset. On the other hand, the constraints $\Omega^a \approx 0$, $\chi^a \approx 0$ have vanishing Poisson brackets with those in the set $\{\Omega^a_{Ti} \approx 0\}$. This means that the Faddeev-Popov determinant split as follows

$$\det(\nabla^2) \ \det{}^{1/2}(\epsilon^{ab}\epsilon^{ijk}\partial_j) \ , \tag{41}$$

which after taking into account the functional relationship

$$\det{}^{1/2}(\epsilon_{ab}\epsilon_{ijk}\partial^j) = \det(\epsilon_{ijk}\partial^j) \ , \tag{42}$$

reduces to

$$\det(\nabla^2) \ \det(\epsilon^{ijk}\partial_j) \ . \tag{43}$$

Clearly, the first factor in (41) is the determinant of the matrix whose elements are $[\Omega^a(\vec{x}), \chi^b(\vec{y})]_P$, while the second is the determinant of the matrix whose elements are given at (9). Hence, for the model under analysis, the phase-space generating functional of Green functions ($\tilde{W}$) is given by

$$\begin{aligned} \tilde{W} &= \int [\mathcal{D}A^{a,i}] [\mathcal{D}\pi^a_i] \, \delta[\partial_i A^{a,i}] \, \delta[\partial^i \pi^a_i] \det(\nabla^2) \\ &\times \delta[\pi^a_i + \frac{1}{2}\epsilon^{ab}\epsilon_{ijk}\partial^j A^{b,k}] \det(\epsilon^{ijk}\partial_j) \, e^{i\tilde{S}_{eff}} \ , \end{aligned} \tag{44}$$

where

$$\tilde{S}_{eff} = \int d^4x \left[\pi^a_i \dot{A}^{ia} - \frac{1}{2}(\vec{\nabla} \times \vec{A}^a)^2 \right] \ , \tag{45}$$

is the corresponding effective action.

III LOCAL DUALITY TRANSFORMATIONS FOR MAXWELL THEORY

As is well known, for the free Maxwell field, in the Coulomb gauge, the phase-space Green functions generating functional is given by[2]

$$W = \int [\mathcal{D}A^i]\,[\mathcal{D}\pi_i]\, \det(\nabla^2)\, \delta[\partial_i A^i]\, \delta[\partial^i \pi_i]\, e^{iS_{eff}} \ , \tag{46}$$

where the effective action (S_{eff}) reads

$$S_{eff} = \int d^4x \left[\pi_i \dot{A}^i - \left(-\frac{1}{2}\pi^i \pi_i - \frac{1}{2} B^i B_i \right) \right] \ . \tag{47}$$

Here, $B^i = -\epsilon^{ijk}\partial_j A_k$ is the i-th component of the magnetic field, while π_i denotes the momentum canonically conjugate to A^i. As shown in Ref. [4], up to surface terms, S_{eff} remains invariant under the non-local duality transformations

$$A^i \to A'^i = A^i + \delta_D A^i; \;\; \delta_D A^i = \theta\, \nabla^{-2}\, \epsilon^{ijk}\, \partial_j \pi_k \ , \tag{48.a}$$

$$\pi_i \to \pi'_i = \pi_i + \delta_D \pi_i; \;\; \delta_D \pi_i = \theta\, \epsilon_{ijk}\, \partial^j A^k \ . \tag{48.b}$$

The point we would like to stress now is that these transformations can be made local by introducing the auxiliary fields C^i, and their corresponding canonical conjugate momenta P_i, defined as follows

$$\nabla^2 C^i = \epsilon^{ijk}\, \partial_j \pi_k \ , \tag{49.a}$$

$$P_i = \epsilon_{ijk}\, \partial^j A^k \ , \tag{49.b}$$

$$\partial_i C^i = \partial^i P_i = 0 \ . \tag{49.c}$$

In fact, from Eqs.(48) and (49) one verifies that

$$\delta_D A^i = \theta\, C^i, \quad \delta_D C^i = -\theta\, A^i, \tag{50.a}$$

$$\delta_D \pi_i = \theta\, P_i, \quad \delta_D P_i = -\theta\, \pi_i. \tag{50.b}$$

Our next task consists in reformulating the Maxwell theory in terms of the fields A^i, π_i, C^i and P_i. To this end, we first recall that

$$\left(\prod \delta[\ ]\right) (\vec{\nabla} \times \vec{C})^2 = \left(\prod \delta[\ ]\right) \pi_i \pi_i \ , \tag{51.a}$$

$$\left(\prod \delta[\ ]\right) \frac{1}{2} P_i \dot{C}^i = \left(\prod \delta[\ ]\right) \frac{1}{2} \pi_i \dot{A}^i \ , \tag{51.b}$$

where the following definition

$$\prod \delta[\ ] \equiv \delta[\partial_i A^i]\, \delta[\partial^i \pi_i]\, \delta[\partial_i C^i] \delta[\partial^i P_i]\, \delta[C^i - \nabla^{-2} \epsilon^{ijk} \partial_j \pi_k] \delta[P_i - \epsilon_{ijk} \partial^j A^k] \ , \tag{52}$$

---

[2] This section is mainly based on Ref. [8].

has been introduced. As consequence,

$$\left(\prod \delta[\ ]\right) S_{eff} = \left(\prod \delta[\ ]\right) \tilde{\tilde{S}}_{eff} , \tag{53}$$

where

$$\tilde{\tilde{S}}_{eff} \equiv \int d^4x \left\{\frac{1}{2}\pi_i \dot{A}^i + \frac{1}{2}P_i \dot{C}^i - \frac{1}{2}\left[(\vec{\nabla} \times \vec{A})^2 + (\vec{\nabla} \times \vec{C})^2\right]\right\}. \tag{54}$$

Correspondingly, the Coulomb gauge generating functional W of Maxwell theory can be cast as

$$\begin{aligned} W = \int & [\mathcal{D}A^i]\,[\mathcal{D}\pi_i]\,\delta[\partial_i A^i]\,\delta[\partial^i \pi_i] \\ & \times [\mathcal{D}C^i]\,[\mathcal{D}P_i]\,\delta[\partial_i C^i]\delta[\partial^i P_i]\ \det(\nabla^2) \\ & \times \delta[C^i - \nabla^{-2}\epsilon^{ijk}\partial_j \pi_k]\delta[P_i - \epsilon_{ijk}\partial^j A^k]\ e^{i\tilde{\tilde{S}}_{eff}} . \end{aligned} \tag{55}$$

It is through this form of W that we shall make contact with the Schwarz-Sen model. Needless to say, W in (55) is, by construction, invariant under the set of local duality transformations (50). We learnt in this Section that, by means of an appropriate enlargement of the phase-space, one can incorporate duality as a local symmetry of sourceless electrodynamics.

IV EQUIVALENCE OF THE MAXWELL AND SCHWARZ-SEN THEORIES

We have now at hand two duality symmetric theories. One is the Schwarz-Sen theory, whose phase space is spanned by the variables $A^{1,i}$, $A^{2,i}$, π_i^1 and π_i^2. The other one is Maxwell theory, whose phase-space variables are A^i, C^i, π_i and P_i. We shall prove, in this Section, that these theories are quantum mechanically equivalent[3].

The initial step toward this proof consists in identifying those variables whose behavior under infinitesimal duality transformations is the same. For the coordinates, the task is easy. Indeed, under infinitesimal duality transformations the Schwarz-Sen coordinates $A^{a,i}$ change as follows (see (4))

$$\delta_D A^{1,i} = \theta\, A^{2,i},\ \ \delta_D A^{2,i} = -\,\theta\, A^{1,i} . \tag{56}$$

Therefore, if one sets

$$A^{1,i} = A^i , \tag{57}$$

one obtains, from (50.a) and (56),

$$A^{2,i} = C^i . \tag{58}$$

The situation is slightly more involved for the momenta. By combining (8.b), (56) and (57) one finds

$$\delta_D \pi_i^1 = -\frac{1}{2}\epsilon_{ijk}\partial^j \delta_D A^{2,k} = \frac{\theta}{2}\epsilon_{ijk}\partial^j A^{1,k} = \frac{\theta}{2}\epsilon_{ijk}\partial^j A^k . \tag{59}$$

On the other hand, (48.b) enables one to write

$$\frac{\theta}{2}\epsilon_{ijk}\partial^j A^k = \frac{1}{2}\delta_D \pi_i . \tag{60}$$

Therefore,

3) This Section is mainly based on Ref. [8]

$$\pi_i^1 = \frac{1}{2}\pi_i \ . \tag{61}$$

Through a similar calculation, which uses (49.b) instead of (48.b), one arrives to

$$\pi_i^2 = \frac{1}{2} P_i \ . \tag{62}$$

We shall next take advantage of the identifications above to rewrite the generating functional of Maxwell theory in terms of the Schwarz-Sen variables. This change of integration variables in (55) leads to

$$\begin{aligned} W = & \int [\mathcal{D}A^{a,i}]\,[\mathcal{D}\pi_i^a]\,\delta[\partial_i A^{a,i}]\,\delta[\partial^i \pi_i^a]\,\det(\nabla^2) \\ & \times\ \delta[A^{2,i} - 2\,\nabla^{-2}\epsilon^{ijk}\partial_j \pi_k^1]\delta[2\,\pi_i^2 - \epsilon_{ijk}\partial^j A^{1,k}] \\ & \times\ e^{i\tilde{S}_{eff}} \ . \end{aligned} \tag{63}$$

We emphasize that, when written in terms of the Schwarz-Sen variables, the Maxwell action ($\tilde{\tilde{S}}_{eff}$) becomes the Schwarz-Sen action ($\tilde{S}_{eff}$). However, since the integration measure in (63) does not appear to be that in (44), a few more algebraic manipulations will be required to establish the equivalence between these theories. Let L^{ik} be the differential operator

$$L^{ik} \equiv \epsilon^{ijk}\partial_j \ . \tag{64}$$

Then,

$$\delta[\partial_i A^{2,i}]\,A^{2,i} = \delta[\partial_i A^{2,i}]\,L^{ik}\nabla^{-2}L_{km}A^{2,m} \ , \tag{65}$$

which, in particular, implies that

$$\begin{aligned} & \delta[\partial_i A^{2,i}]\,\delta[A^{2,i} - 2\,\epsilon^{ijk}\,\nabla^{-2}\,\partial_j \pi_k^1] \\ & = \delta[\partial_i A^{2,i}]\det{}^{-1}\left(-\epsilon^{ijk}\partial_j\nabla^{-2}\right)\,\delta[\epsilon_{klm}\partial^l A^{2,m} + 2\,\pi_k^1] \ . \end{aligned} \tag{66}$$

On the other hand, from the eigenvalue equation

$$\delta[\partial_j \phi^j]\,L^{ik}\nabla^{-2}L_{km}\,\phi^m = -\,\delta[\partial_j\phi^j]\,\phi^i \ , \tag{67}$$

one learns that

$$\det\left(L^{ik}\nabla^{-2}L_{km}\right) = \mathit{constant} \ , \tag{68}$$

or, equivalently,

$$\det{}^{-1}\left(-\epsilon^{ijk}\partial_j\nabla^{-2}\right) = \mathit{constant} \times \det\left(\epsilon_{klm}\partial^l\right) \ . \tag{69}$$

By going back with (66) and (69) into (63) one finds that

$$\begin{aligned} W = & \ \mathit{constant} \times \int [\mathcal{D}A^{a,i}]\,[\mathcal{D}\pi_i^a]\,\delta[\partial_i A^{a,i}]\,\delta[\partial^i \pi_i^a]\,\det(\nabla^2) \\ & \times\ \delta[\pi_i^a + \frac{1}{2}\epsilon^{ab}\epsilon_{ijk}\partial^j A^{b,k}]\det(\epsilon^{ijk}\partial_j) \\ & \times\ \exp\left\{ i \int d^4x \left[\pi_i^a A^{a,i} - \frac{1}{2}\left(\vec{\nabla}\times\vec{A^a}\right)^2\right]\right\} \\ = & \ \mathit{constant} \times \tilde{W} \ . \end{aligned} \tag{70}$$

The purported proof of equivalence between the Maxwell and the Schwarz-Sen theories is now complete.

V CONCLUSIONS

We started in this work by canonically quantizing the Schwarz-Sen theory in the Coulomb gauge. The resulting theory turned out to be physically sensible. The Lagrangian density defining the theory is not covariant but, nevertheless, we were able of constructing a set charges verifying the Poincaré algebra.

Later, we showed that the phase space associated with the Coulomb gauge formulation of Maxwell theory can be conveniently enlarged in order to accommodate duality as a local symmetry.

By analyzing the behavior under duality transformation, we were able of identifying the phase-space variables of the Maxwell theory with those of the Schwarz-Sen theory. It was then possible to demonstrate that the corresponding Green functions generating functionals were the same.

REFERENCES

1. B. McClain, Y. S. Wu, and F.Yu, *Nucl.Phys.* **B343**, 689 (1990); C. Wotzasek, *Phys. Rev. Lett.* **66**, 129 (1991); F. P. Devecchi and M. Henneaux, *Phys. Rev.* **D54**, 1606 (1996); I. Martin and A. Restuccia, Phys. Lett. **B323**, 311 (1994); N. Berkovits, **Phys. Lett. B388**, 743 (1996).
2. P. Pasti, D. Sorokin, and M. Tonin, *Phys. Lett.* **B352**, 59 (1995); ibid. *Phys. Rev.* **D52**, 4277 (1995); ibid. *Phys. Rev.* **55**, 6292 (1997).
3. A. Khoudeir and N. Pantoja, Phys. Rev. **D53**, 5974 (1996).
4. S. Deser and C. Teitelboim, Phys. Rev. **D13**, 1592 (1976).
5. J. H. Schwarz and A. Sen, Nucl. Phys. **B411**, 35 (1994).
6. R. Medina and N. Berkovits, Phys. Rev. **D56**, 6388 (1997).; S. Deser, A. Gomberoff, M. Henneaux, and C. Teitelboim, **hep-th** 9702184.
7. H. O. Girotti, *Phys. Rev.* **D55**, 5136 (1997).
8. H. O. Girotti, M. Gomes, V. Rivelles and A. J. da Silva, *Phys. Rev.* **D56**, 6615 (1997).
9. P. A. M. Dirac, *Lectures on Quantum Mechanics*, Belfer Graduate School on Science, Yeshiva University (New York, 1964).
10. E. S. Fradkin and G. A. Vilkovisky, CERN preprint TH 2332 (1977), unpublished.
11. K. Sundermeyer, *Constrained Dynamics* (Springer-Verlag, Berlin, 1982).
12. H. O. Girotti, *Classical and quantum dynamics of constrained systems*, lectures in Proc. of the Vth Jorge Andre Swieca Summer School, O. J. P. Eboli, M. Gomes and A. Santoro editors (World Scientific, Singapore, 1990) p1-77.
13. R. Floreanini and R. Jackiw, *Phys. Rev. Lett.* **59**, 1873 (1987).
14. M. E. V. Costa and H. O. Girotti, *Phys.Rev. Lett.* **60**, 1771 (1988); F. P. Devecchi and H. O. Girotti, *Phys. Rev.* **D49**, 4302 (1994).
15. David G. Boulware and S. Deser, J. Math. Phys. **8**, 1468 (1967).

Integrable theories in any dimension: a perspective

Orlando Alvarez[a], L.A. Ferreira[b] and J. Sánchez Guillén[c]

[a] *Department of Physics*
University of Miami
P.O. Box 248046
Coral Gables, FL 33124, USA

[b] *Instituto de Física Teórica - IFT/UNESP*
Rua Pamplona 145
01405-900 São Paulo-SP, BRAZIL

[c] *Departamento de Física de Partículas,*
Facultad de Física
Universidad de Santiago
E-15706 Santiago de Compostela, SPAIN

Abstract. We review the developments of a recently proposed approach to study integrable theories in any dimension. The basic idea consists in generalizing the zero curvature representation for two-dimensional integrable models to space-times of dimension $d+1$ by the introduction of a d-form connection. The method has been used to study several theories of physical interest, like self-dual Yang-Mills theories, Bogomolny equations, non-linear sigma models and Skyrme-type models. The local version of the generalized zero curvature involves a Lie algebra and a representation of it, leading to a number of conservation laws equal to the dimension of that representation. We discuss the conditions a given theory has to satisfy in order for its associated zero curvature to admit an infinite dimensional (reducible) representation. We also present the theory in the more abstract setting of the space of loops, which gives a deeper understanding and a more simple formulation of integrability in any dimension.

I INTRODUCTION

This paper addresses the long standing problem of the generalization of the zero curvature in classical integrability, from two dimensions to theories defined on space-times of any dimension. It consists of a review of the ideas proposed in [1] to implement that generalization, and also of the subsequent developments.

Integrability and, in a more general sense, construction of solutions and constants of motion, are important issues, given the relevance of nonperturbative aspects of field theories in High Energy Physics , Statistical Mechanics, Solid State Physics and Gravity.

In $2d$, an impressive understanding of integrable theories has been achieved, which can be essentially encoded in the zero curvature formulation. In fact, the equations of motion of such theories, including relativistic invariant ones, can be naturally expressed in terms of a flat connection as $[\partial_\nu - A_\nu, \partial_\mu - A_\mu] = 0$. The simplest (abelian) example is that of a free scalar field ϕ, with $A_\mu = -\epsilon_{\mu\nu}\partial^\mu\phi$. The equations of motion of the Sine-Gordon models and its generalizations, the so called Affine Toda systems, are also expressed as zero curvature with A_μ taking values in an affine algebra [2]. One of the remarkable facts is that the algebraic structure involved in the zero curvature does not describe, in general, any explicit symmetry of the original equations of motion. In the case of theories presenting soliton solutions, a great unification is achieved in the framework of affine algebras, where there exists algebraic equivalent methods, like dressing of vacuum solutions or solitonic specialization giving the interesting solitonic solutions in a systematic and simple way [3].

The zero curvature formulation also provides a way of constructing conserved charges. This is easily seen choosing a finite two dimensional space-time with periodic boundary conditions on the time component of the connection. By Stokes calculus, any power N of the $Tr(P\exp(\int_{-L}^{L} A_x(x,t)dx))^N$ is conserved in the

CP484, *Trends in Theoretical Physics II*, edited by H. Falomir, R. E. Gamboa Saraví, and F. A. Schaposnik

time evolution (the path ordering P is for the non-abelian general case). Despite the great advances in $2d$ classical integrability, and also on some relations between integrability in *parameter space* and solutions of supersymmetric gauge theories in $4d$ [4], very little was done in generalizing this beautiful picture to any dimension in *space time*, and understand *directly* those solutions in terms of generalized soliton dynamics, most interesting by itself. But expressing relativistic invariant field equations, known and possibly new ones, in such geometrical way which yields their integrals of motion, cannot be simple in higher dimensions. First of all, relativistic invariance requires higher rank tensors and connections, in principle with complicated gauge transformations. Besides, in $2d$ the Lorentz transformations are rather trivial $x_\pm \rightarrow \lambda^{\pm 1} x_\pm$ and they can be imitated by the grading operator in the algebra, just making the forward (backward) light-cone to be positive (negative), incorporating also holomorphicity directly.

The main difficulties in extending of the integrability concepts to higher dimensions are associated to non-locality issues that rise when dealing with higher rank connections. Those problems can be circunvented by the introduction of auxiliary connections that allow for parallel transport. In addtion, once the invariant equations have been formulated, to get their integrals of motion one needs a generalization of Stokes calculus, which is directly related to the first problem of the transformation of the generalized connections. In fact Stokes formulae are problematic for the relevant non-abelian case even in $2d$ [5]. Therefore, the starting point of [1] was to obtain a simple local expression for the non-abelian Stokes theorem, as we review in section II.

Our approach to generalize the zero curvature is based on the observation that the conservation laws can be expressed as the condition for the invariance of some quantity under the deformation of hypersurfaces. We are in fact constructing higher dimensional non-abelian Gauss-type laws. The main difficult is related to the implementation of the calculus of the variation of the hypersurfaces and the requirement that at the end one wants *local* conservation laws. The key observation for implementing the calculus was based on the formula satisfied by the Wilson loop W_c, calculated on a closed countour C, i.e.

$$\frac{dW_c}{dt} = W_c T(F, 2\pi, t) \tag{I.1}$$

where t parametrizes the variations of the countour that keep a given fixed point x_0 fixed, and

$$T(F, 2\pi, t) \equiv \int_0^{2\pi} d\sigma W^{-1} F_{\mu\nu} W \frac{dx^\mu}{d\sigma}\frac{dx^\nu}{dt}, \tag{I.2}$$

and $F_{\mu\nu}$ is the curvature of the connection A_μ defining the Wilson line W, i.e.

$$\frac{dW}{d\sigma} = -A_\mu \frac{dx^\mu}{d\sigma} W \tag{I.3}$$

and the integral in T is along the curve parameter σ. Eq.(I.1), which is easily obtained from Interaction Picture method in [1], expresses loop independence when the connection is flat and it is essentially non-local. This simple formulation gives the idea for the zero curvature generalization to $3d$: introduce a functional of the fields V defined by its variation under deformations of the loops through the equation

$$\frac{dV}{dt} = VT(B, 2\pi, t) \tag{I.4}$$

where $B_{\mu\nu}$ is a general antisymmetric tensor functional and

$$T(B, 2\pi, t) \equiv \int_0^{2\pi} d\sigma W^{-1} B_{\mu\nu} W \frac{dx^\mu}{d\sigma}\frac{dx^\nu}{dt}, \tag{I.5}$$

One looks then for the conditions for V, now a surface integral, to be independent of the surface scanned by the different loops. Notice that the integrand in $T(B, 2\pi, t)$ is a local function. The form of T will be essential to guarantee Lorentz and gauge invariant time evolution. Generalizing directly the previous 2d Interaction picture computation varying now with respect to a new orthogonal parameter τ, we obtain a rather complicated expression, albeit with a transparent geometrical and physical meaning. It is a generalized non-abelian Gauss law and, as shown in detail in section II A, it corresponds to parallel transport in the space of loops. This geometrical theory tell us how to proceed in any dimension and makes very natural the translation of the surface independence to different *local* equations of motion and/or integrability conditions, which are deduced directly in section III. A natural condition is the flatness of the first connection $F_A = 0$, which guarantees independence of how the surface is scanned by loops, and then a sufficient condition for the vanishing of the non-local curvature associated to V is the constant covariance of the tensor connection $D_A B = 0$. This nesting of constant covariance and vanishing of the commutator of two $d-forms$ and flatness of the lower $d-1$, $d-2$... curvatures (or possibly some of them) is generic and it is worked out in detail also for the $4d$ (see [1] for details).

The Lorentz covariant and gauge invariant formalism is very general and one has lots of algebraic and topological structures to explore which support those local formulations and which will yield new formulations of known theories or new ones, with a systematic method to get their integrals of motion and solutions. In $3d$, the most simple case is choosing $B_{\mu\nu}(0)$ on an abelian subalgebra. Notice that $B_{\mu\nu}$ can then be uniquely defined at any point x by parallel transport under A_μ. The integrability conditions in this case correspond to the BF theory, Chern-Simons and other "topological" theories in 3d. That the construction works for these topological cases is reassuring and one can define now naturally new observables like invariants for link crossings or study topological defects, but the importance of conserved quantities is, of course, with time evolution. On the other hand very little is clearly understood about non-perturbative solutions and classical integrability for the situation when there is full relativistic dynamics in dimension higher than 2. So we concentrated in [1] on the simplest genuine $2+1$ theory, the $O(3)$ nonlinear model, although most of the approach can be used for a general sigma model as shown later in [6,7]. In the four dimensional case, it was shown in [1] how to express self-dual Yang-Mills and BPS theory in such zero curvature formulation of local integrability conditions for volume independence. We obtain also a reduction of the equations which exhibit directly their conservation character.

One of the interesting aspects of [1] is that many theories presenting the local zero curvature are not integrable in the sense of possessing an infinite number of conservation laws. However, some of those theories contain integrable submodels that do present an infinite number of conserved currents. We discuss in section IV the conditions a given theory has to satisfy to present an infinity number of conserved local currents. The main point is that its zero curvature representation has to involve infinite dimensional representations of a Lie algebra, or equivalently as we explain, and infinite dimensional non-semisimple Lie algebra of the Poincaré type.

II GEOMETRICAL APPROACH TO INTEGRABILITY

The zero curvature condition in two dimensional spacetime, know as the Zakharov-Shabat equation [8] is given by

$$F_{\mu\nu} \equiv [\partial_\mu + A_\mu, \partial_\nu + A_\nu] = 0 \qquad \mu, \nu = 0, 1 \tag{II.1}$$

One of the consequences of it is that it leads to conservation laws. In order to see that, consider a quantity W defined by paralell transport with a connection A_μ,

$$\frac{dW}{d\sigma} + A_\mu \frac{dx^\mu}{d\sigma} W = 0 \tag{II.2}$$

Consider now a closed curve Γ, and let Σ be a two dimensional surface having Γ as its boundary. The nonabelian Stokes theorem states that the quantity W can be determined either by integrating (II.2) on Γ or on Σ. More precisely one has

$$W(\Gamma) = P\exp\left(\int_\Gamma d\sigma A_\mu \frac{dx^\mu}{d\sigma}\right) = \mathcal{P}\exp\left(\int_\Sigma d\tau\, d\sigma W^{-1} F_{\mu\nu} W \frac{dx^\mu}{d\sigma}\frac{dx^\nu}{d\tau}\right) \tag{II.3}$$

where P and $\mathcal{P}$ mean path and surface ordering respectively (see [1] for details).

Therefore, if the connection A_μ is flat, i.e. (II.1) holds true, then W is equal to unity for closed loops. Then, conserved quantities are constructed as follows. First we consider the case where spacetime is a cylinder. At a fixed time t_0 consider a loop γ_0 beginning and ending at x_0. At a later fixed time t_1 consider a loop γ_1 also beginning and ending at x_0. Let γ_{01} be a path connecting (t_0, x_0) with (t_1, x_0). The flat connection allows us to integrate the parallel transport equation (II.2) along two different paths obtaining $W(\gamma_0) = W(\gamma_{01})^{-1} W(\gamma_1) W(\gamma_{01})$. We first observe that $W(\gamma_0)$ transforms under a gauge transformation $g(x)$ as $W(\gamma_0) \to g(x_0) W(\gamma_0) g(x_0)^{-1}$. The conserved quantity should be gauge invariant. If χ is a character for the group G we have that $\chi(W(\gamma_0))$ will be gauge invariant. Also $\chi(W(\gamma_0)) = \chi(W(\gamma_1))$. Thus we can construct a conserved gauge invariant quantity

$$\chi(W(\gamma_0)) \tag{II.4}$$

for every independent character of the group. These are the constants of motion in the zero curvature construction. Note that the data needed to compute $\chi(W(\gamma_0))$ is all determined at time t_0.

In the case where the spacetime is two dimensional Minkowski space one has to impose physically sensible boundary conditions at spatial infinity. Note that $P\exp\left(\int_{-\infty}^{\infty} A_x dx\right)$ is not gauge invariant if one allows nontrivial gauge transformations at infinity. In setting up the problem one has to choose the correct physical boundary conditions which may for example require that A_0 vanishes at infinity. Depending on the details one can construct a conserved quantity by a slight modification of the construction above.

The conserved quantities (II.4) are nonlocal because in general the connection A_μ lies in a nonabelian algebra. However, in cases like the affine Toda models [2], it is possible to get local conservation laws by gauge transforming A_μ into an abelian subalgebra.

The basic idea in [1] to bring such concepts to higher dimensions, is to introduce quantities integrated over hypersurfaces and to find the conditions for them to be independent of deformations of the hypersurfaces which keep their boundaries fixed. Such an approach will certainly lead to conservation laws in a manner very similar to the two dimensional case. However, the main problem of that it is how to introduce non-linear zero curvatures keeping things as local as possible. The way out is to introduce auxiliary connections to allow for parallel transport. The number of possibilities of implementing those ideas increase with the dimensionality of space-time. However, the simplest scenario is that where, in a space-time of dimension $d+1$, one introduces a rank d antisymmetric tensor $B_{\mu_1\mu_2\dots\mu_d}$ and a vector A_μ. The idea can perhaps be best stated using a formulation in "loop space". On a $d+1$ dimensional space-time M one considers the space $\Omega^{d-1}(M, x_0)$ of $d-1$ dimensional closed hypersurfaces based at a fixed point $x_0 \in M$. One then introduces on such "higher loop space" a 1-form $\mathcal{A}$ which is basically the quantity $W^{-1}B_{\mu_1\mu_2\dots\mu_d}W$ integrated over the closed hypersurfaces (see [1] for details). The quantity W is defined in terms of the vector A_μ through (II.2). However, for W to be independent of the way one integrates it from x_0 to a given point on the hypersurface, one has to assume that A_μ is flat, i.e.

$$F_{\mu\nu} = [D_\mu, D_\nu] = \partial_\mu A_\nu - \partial_\nu A_\mu + [A_\mu, A_\nu] = 0\,; \qquad \mu,\nu = 0,1,2\dots d \tag{II.5}$$

with

$$D_\mu \cdot \equiv \partial_\mu \cdot + [A_\mu\,,\,\cdot\,] \tag{II.6}$$

Roughly speaking a d dimensional closed hypersurface in M, based at x_0, corresponds to a (one dimensional) loop in $\Omega^{d-1}(M, x_0)$. Therefore, the condition to have things independent of deformation of hypersurfaces translates in such "higher loop space" to the zero curvature condition for $\mathcal{A}$, namely

$$\mathcal{F} = \delta\mathcal{A} + \mathcal{A}\wedge\mathcal{A} = 0 \tag{II.7}$$

The relation (II.7) (together with (II.5)) is the generalization of the zero curvature (II.1) to higher dimensions proposed in [1].

The idea of constructing conserved quantities using (II.7) is the same as the two dimensional case decribed above. Indeed, the condition (II.7) implies that the path ordered exponential of $\mathcal{A}$ along a closed loop in the higher loop space $\Omega^{d-1}(M, x_0)$ is independent of the loop. Following arguments similar to those below (II.3), one sees that the conserved quantities are path ordered exponentials of $\mathcal{A}$ along closed loop at a given fixed time. Since loops in $\Omega^{d-1}(M, x_0)$ correspond to d-dimesnional closed surfaces in the $d+1$-dimensional spcaetime M, one gets that such quantities are indeed integrated over the physical space. Although the analogy with the two dimensional case looks straitghforward, the implementation is quite involved due to the ordering of the integration, the respect to gauge invariance and boundary conditions. We refer to [1] for those very important explanations.

In order to make the formulas more explicit we give below the discussion of the $2+1$ dimensional case.

A The zero curvature in three dimensions

Assume we have a principal bundle $P \to M$ with connection. For a fixed point $x_0 \in M$ let $\Omega(M, x_0)$ be the space of all loops based at x_0:

$$\Omega(M, x_0) = \{\gamma : S^1 \to M \mid \gamma(0) = x_0\} \, . \tag{II.8}$$

we now want to construct a principal G-bundle over $\Omega(M, x_0)$ with connection. Note that the structure group of the bundle is a finite dimensional group not a loop group. It will be the trivial bundle $\mathcal{P} = \Omega(M, x_0) \times G$. Conceptually the bundle is constructed as follows. Over $x_0 \in M$ the bundle $P \to M$ has fiber P_{x_0} which is isomorphic to G. All loops in $\Omega(M, x_0)$ have $x_0 \in M$ as a starting point so we can consider them having P_{x_0} in common. This is the common fiber in the cartesian product $\Omega(M, x_0) \times G$. Mathematically we have a natural map $\pi : \Omega(M, x_0) \to M$ given by $\pi(\gamma) = x_0$. The bundle $\mathcal{P}$ is just the pullback bundle $\pi^* P$, see [9]. Since $\mathcal{P} \to \Omega(M, x_0)$ is a trivial bundle we can put the trivial connection on it. There is a more interesting connection one can put on it which exploits the connection on the bundle $P \to M$. Consider a Lie algebra valued 2-form B on M such that under the transition function ϕ we have $B \to \phi^{-1} B \phi$. Let $W(\sigma)$ be the parallel transport operator from the point $x(0) = x_0$ to the point $x(\sigma)$ along the loop γ. We can assign the Lie algebra valued 1-form

$$\mathcal{A}[x(\sigma)] = \int_0^{2\pi} d\sigma \, W(\sigma)^{-1} B_{\mu\nu}(x(\sigma)) W(\sigma) \frac{dx^\mu}{d\sigma} \delta x^\nu(\sigma) \tag{II.9}$$

The transformation laws of the above are determined by $\phi(x_0)$ which is clearly associated with the common fiber P_{x_0}. Thus we can define a connection on $\mathcal{P}$ by

$$\varpi = -dg g^{-1} + g \left(\int_0^{2\pi} d\sigma \, W(\sigma)^{-1} B_{\mu\nu}(x(\sigma)) W(\sigma) \frac{dx^\mu}{d\sigma} \delta x^\nu(\sigma) \right) g^{-1} \, .$$

Thus we can treat $\mathcal{A}$ as the connection in a certain trivialization. The curvature is given by $\mathcal{F} = \delta \mathcal{A} + \mathcal{A} \wedge \mathcal{A}$.

What do we mean when we say that we want to have parallel transport independent of path in $\Omega(M, x_0)$? Look at the space $\Omega(M, x_0)$ and consider a curve Γ in $\Omega(M, x_0)$ parametrized by τ such that $\Gamma(0)$ is the "constant curve" x_0. Note that for fixed τ, $\Gamma(\tau)$ is a curve $x_\tau(\sigma)$ for $\sigma \in [0, 2\pi]$ in M. Thus it is convenient to "write" Γ as $x(\sigma, \tau)$. The statement that parallel transport be independent of the choice of curve $\Gamma \in \Omega(M, x_0)$ with fixed starting and ending points is the statement that the curvature vanish. If one wants the parallel transport between points in $\Omega(M, x_0)$ to be independent of path then parallel transport should be path independent in M. The reason for this is that a loop in $\Omega(M, x_0)$ beginning at the trivial loop may be viewed as a map from the square $[0, 2\pi]^2$ to M such that $\partial[0, 2\pi]^2$ gets mapped to x_0. To get the same result for two different sets of "constant" τ curves associated with the same closed 2-submanifold in M, one needs ordinary parallel transport to be path independent, i.e. $F = 0$.

In order to perform the curvature computation we need the standard result:

$$\begin{aligned} W(\sigma)^{-1} \delta W(\sigma) &= -W(\sigma)^{-1} A_\mu(x(\sigma)) W(\sigma) \delta x^\mu(\sigma) \\ &\quad + \int_0^\sigma d\sigma' \, W(\sigma')^{-1} F_{\mu\nu}(x(\sigma')) W(\sigma') \frac{dx^\mu}{d\sigma'} \delta x^\nu(\sigma') \, . \end{aligned}$$

We also need definition (II.9). Let δ be the exterior derivative on the space $\Omega(M, x_0)$ and thus $\delta^2 = 0$ and

$$\delta x^\mu(\sigma) \wedge \delta x^\nu(\sigma') = -\delta x^\nu(\sigma') \wedge \delta x^\mu(\sigma) \, .$$

Computing the curvature $\mathcal{F} = \delta \mathcal{A} + \mathcal{A} \wedge \mathcal{A}$ is tedious and we refer to [1] for the details. The result is

$$\begin{aligned} \mathcal{F} = &-\frac{1}{2} \int_0^{2\pi} d\sigma \, W(\sigma)^{-1} \left[D_\lambda B_{\mu\nu} + D_\mu B_{\nu\lambda} + D_\nu B_{\lambda\mu} \right] (x(\sigma)) W(\sigma) \\ &\quad \times \frac{dx^\lambda}{d\sigma} \delta x^\mu(\sigma) \wedge \delta x^\nu(\sigma) \\ &- \int_0^{2\pi} d\sigma \int_0^\sigma d\sigma' \, \left[F^W_{\kappa\mu}(x(\sigma')), B^W_{\lambda\nu}(x(\sigma)) \right] \frac{dx^\kappa}{d\sigma'} \frac{dx^\lambda}{d\sigma} \delta x^\mu(\sigma') \wedge \delta x^\nu(\sigma) \\ &+ \frac{1}{2} \int_0^{2\pi} d\sigma \int_0^{2\pi} d\sigma' \, \left[B^W_{\kappa\mu}(x(\sigma')), B^W_{\lambda\nu}(x(\sigma)) \right] \frac{dx^\kappa}{d\sigma'} \frac{dx^\lambda}{d\sigma} \delta x^\mu(\sigma') \wedge \delta x^\nu(\sigma) \end{aligned} \tag{II.10}$$

The vanishing of $\mathcal{F}$, i.e. relation (II.7), is our generalization of the zero curvature to higher dimensions. It implies hypersurface indepence and conservation laws as we explained above.

III LOCAL INTEGRABILITY CONDITIONS

The condition (II.7) that the loop space curvature should vanish is local in $\Omega^{d-1}(M, x_0)$, but it is highly non-local in the spacetime M. We now discuss some sufficient local conditions for the vanishing of (II.10).

Let $\mathcal{G}$ be a Lie algebra and R be a representation of it. We introduce the nonsemisimple Lie algebra $\mathcal{G}_R$ as

$$\begin{aligned} [T_a\,,\,T_b] &= f_{ab}^c T_c \\ [T_a\,,\,P_i] &= P_j R_{ji}\,(T_a) \\ [P_i\,,\,P_j] &= 0 \end{aligned} \tag{III.1}$$

where T_a constitute a basis of $\mathcal{G}$ and P_i a basis for the abelian ideal P (representation space). The fact that R is a matrix representation, i.e.

$$[R\,(T_a)\;,\,R\,(T_b)] = R\,([T_a\,,\,T_b]) \tag{III.2}$$

follows from the Jacobi identities.

We take the connection A_μ to be in $\mathcal{G}$ and the rank d antisymmetric tensor $B_{\mu_1\mu_2\ldots\mu_d}$ to be in P, i.e.

$$A_\mu = A_\mu^a T_a\,, \qquad B_{\mu_1\mu_2\ldots\mu_d} = B^i_{\mu_1\mu_2\ldots\mu_d} P_i \tag{III.3}$$

Then a set of sufficient *local* conditions for the vanishing of the curvature $\mathcal{F}$ in (II.7) is given by

$$D_\mu \tilde{B}^\mu = 0\,; \qquad F_{\mu\nu} = 0 \tag{III.4}$$

where we have introduced the dual of $B_{\mu_1\mu_2\ldots\mu_d}$ as

$$\tilde{B}^\mu \equiv \frac{1}{d!}\,\varepsilon^{\mu\mu_1\mu_2\ldots\mu_d}\,B_{\mu_1\mu_2\ldots\mu_d} \tag{III.5}$$

Indeed, in the $2+1$ dimensional case the conditions (III.1) and (III.3) imply that

$$W^{-1} B_{\mu\nu} W \in P \tag{III.6}$$

and therefore the commutator in the last term of (II.10) vanishes, since P is abelian. The first condition in (III.4) in the $2+1$ dimensional case reads

$$D_\lambda B_{\mu\nu} + D_\mu B_{\nu\lambda} + D_\nu B_{\lambda\mu} = 0 \tag{III.7}$$

Therefore, together with $F_{\mu\nu} = 0$, it implies that the remaining terms of (II.10) vanish too.

The relations (III.4) are the *local* integrability conditions which we introduce for theories defined on a spacetime of any dimension, and which constitutes a generalization of the zero curvature condition (II.1) in two dimensions. They lead to local conservation laws. Indeed, since the connection A_μ is flat it can be written as

$$A_\mu = -\partial_\mu W\, W^{-1} \tag{III.8}$$

and consequently (III.4) imply that the currents

$$J_\mu \equiv W^{-1}\,\tilde{B}_\mu\,W \tag{III.9}$$

are conserved

$$\partial_\mu\,J^\mu = 0 \tag{III.10}$$

The zero curvature conditions (III.4) are invariant under the gauge transformations

$$A_\mu \rightarrow g\, A_\mu\, g^{-1} - \partial_\mu g\, g^{-1}$$
$$\tilde{B}_\mu \rightarrow g\, \tilde{B}_\mu\, g^{-1} \qquad \text{(III.11)}$$

and

$$A_\mu \rightarrow A_\mu$$
$$\tilde{B}_\mu \rightarrow \tilde{B}_\mu + \varepsilon_{\mu\mu_1\ldots\mu_d} D^{\mu_1} \alpha^{\mu_2\ldots\mu_d} \equiv \tilde{B}_\mu + D^\nu \tilde{\alpha}_{\mu\nu} \qquad \text{(III.12)}$$

where we have introduced the dual $\tilde{\alpha}_{\mu\nu} \equiv \varepsilon_{\mu\nu\mu_2\ldots\mu_d}\alpha^{\mu_2\ldots\mu_d}$. In (III.11) g is an element of the group obtained by exponentiating the Lie algebra $\mathcal{G}$. The transformations (III.12) are symmetries of (III.4) as a consequence of the fact that the connection A_μ is flat, i.e. $[D_\mu\, ,\, D_\nu] = 0$. In addition, the parameters $\alpha^{\mu_1\ldots\mu_{d-1}}$ take values in the abelian ideal P.

The currents (III.9) are invariant under the transformations (III.11), but under (III.12) they transform as

$$J_\mu \rightarrow J_\mu + \varepsilon_{\mu\mu_1\ldots\mu_d}\partial^{\mu_1}\left(W^{-1}\,\alpha^{\mu_2\ldots\mu_d}\, W\right) = J_\mu + \partial^\nu\left(W^{-1}\,\tilde{\alpha}_{\mu\nu}\, W\right) \qquad \text{(III.13)}$$

The transformations (III.11) and (III.12) do not commute and their algebra is isomorphic to the non-semisimple algebra $\mathcal{G}_R$ introduced in (III.1). The nontrivial gauge transformations allow in principle the dressing of vacuum solutions to obtain general ones, as discussed in [1].

IV INTEGRABLE SUBMODELS

The number of conserved currents one gets from (III.9) is equal to the dimension of the representation of $\mathcal{G}$, defined by the generators P_i of $\mathcal{G}_R$. Consequently, the notion of integrability in such approach is related to infinite dimensional representations, or equivalently to infinite dimensional non-semisimple Lie algebras of the type (III.1). That is similar to the two dimensional case where the appearence of infinte number of charges is also linked to infinite dimensional Lie algebras. However, those are in general of the affine type, and it is now quite well understood the role they have in soliton theory and exact methods of construction of solutions. In our approach, the role of algebraic structures like (III.1) is not fully understood yet, but we believe it must have profound consequences in the study of higher dimensional integrable theories.

One point that became clear recently is that of integrable submodels. It is known that conditions like self-duality in Yang-Mills theories and the Bogomolny equations in gauge theories with symmetry spontaneously broken by a Higgs in the adjoint representation, play a crucial role in the construction of submodels which possess properties of solvability not present in the full theory. Those conditions lead to a saturation of a bound on the Euclidean action in the case of self-dual Yang-Mills and on the energy in the case of Bogomolny equations.

In our approach a similar thing happens, however involving quite different structures. It does lead to conditions for integrable submodels, but apparently not to saturation of bounds. In the examples studied so far what one gets is the following. Suppose that using the gauge symmetries (III.11) and (III.12) one can find a gauge where the connection A_μ can be split as [6]

$$A_\mu = A_\mu^S + A_\mu^K \qquad \text{(IV.1)}$$

where $A_\mu^{S/K}$ are the components of A_μ in the decomposition

$$\mathcal{G} = S + K \qquad \text{(IV.2)}$$

with K being a subalgebra of $\mathcal{G}$, and S its complement in $\mathcal{G}$. Suppose in addition that

$$\left[A_\mu^S\, ,\, \tilde{B}^\mu\right] = 0 \qquad \text{(IV.3)}$$

Then the first zero curvature (III.4) becomes

$$\partial^\mu \tilde{B}_\mu + \left[A^{\mu K}\, ,\, \tilde{B}_\mu\right] = \left(\partial^\mu \tilde{B}_{\mu,i} + A_a^{\mu K}\tilde{B}_{\mu,j} R\left(K_a\right)_{ij}\right) P_i \qquad \text{(IV.4)}$$

where $A^{\mu K} = A_a^{\mu K} K_a$, with K_a being the generators of K, and $R(K_a)_{ij}$ being the matrix representation of K defined by the P_i's, i.e.

$$[K_a , P_i] = P_j R(K_a)_{ji} \tag{IV.5}$$

Therefore, the zero curvature condition is only determined by the representation of K defined by the subspace P. Such a representation is in general reducible, and we shall denote the branching as

$$R = \sum_l R_l^K \tag{IV.6}$$

with R_l^K being irreducible representations of K. Suppose now that a given representation R^λ of $\mathcal{G}$ presents a branching rule

$$R^\lambda = \sum_l R_l^K + \text{something} \tag{IV.7}$$

Then we can introduce an operator

$$\tilde{B}_\mu^\lambda \equiv \tilde{B}_{\mu,i} P_i^\lambda \tag{IV.8}$$

where $\tilde{B}_{\mu,i}$ are the same coefficients as in the expansion of the original $\tilde{B}_\mu$ in terms of the basis P_i of the representation R of $\mathcal{G}$, i.e. $\tilde{B}_\mu = \tilde{B}_{\mu,i} P_i$. In addition, P_i^λ are the generators of $\mathcal{G}_{R^\lambda}$ associated to the representation R^λ, corresponding to the subspace $\sum_l R_l^K$, and transforming exactly as the P_i's under the subalgebra K, i.e.

$$[K_a , P_i^\lambda] = P_j^\lambda R(K_a)_{ji} \tag{IV.9}$$

Therefore, one gets that

$$\begin{aligned} D^\mu \tilde{B}_\mu^\lambda &= \partial^\mu \tilde{B}_\mu^\lambda + \left[A^\mu , \tilde{B}_\mu^\lambda \right] \\ &= \left(\partial^\mu \tilde{B}_{\mu,i} + A_a^{\mu K} \tilde{B}_{\mu,j} R(K_a)_{ij} \right) P_i^\lambda \\ &+ \left[A^{\mu S} , \tilde{B}_\mu^\lambda \right] \end{aligned} \tag{IV.10}$$

Notice that the first term after the last equality in (IV.10) corresponds to the zero curvature for the theory under consideration, namely (IV.4). Consequently, if one imposes the constraint

$$\left[A^{\mu S} , \tilde{B}_\mu^\lambda \right] = A_r^{\mu S} \tilde{B}_\mu^i [S_r , P_i^\lambda] = 0 \tag{IV.11}$$

with S_r being the generators of the subspace S, one gets a submodel with the conserved currents given by (see (III.9))

$$J_\mu^\lambda \equiv W^{-1} \tilde{B}_\mu^\lambda W \tag{IV.12}$$

Suppose now that there exists an infinite number of representations, like R^λ, satisfying (IV.7) such that the conditions (IV.11) impose the same set of constraints on the model. Then the submodel defined by the equations (IV.4) and the constraints (IV.11) possesses an infinite number of local conserved currents. Several examples fullfilling those requirements were constructed in [1,7,6,10].

An important case where the above construction works is when the representation of $\mathcal{G}$ defined by the P_i's, possesses at least one charge zero singlet of the subalgebra K, i.e. there exists in the abelian subalgebra P a generator P_Λ, such that [6]

$$[K_a , P_\Lambda] = 0 \tag{IV.13}$$

Then one can easily construct representations satisfying (IV.7) by considering tensor products of the representation R, defined by the P_i's, with itself. The subsapces given by the tensor products of R with the singlet

P_Λ, transform like $\sum_l R_l^K$. For instance, in the case of $R \otimes R$ one has that $P_\Lambda \otimes R$ (equivalently $R \otimes P_\Lambda$) satisfies

$$[1 \otimes K_a + K_a \otimes 1 \,,\, P_\Lambda \otimes P_i] = P_\Lambda \otimes P_j \;\; R(K_a)_{ji} \qquad \text{(IV.14)}$$

For the case of $(\otimes R)^n$ any representation of the form $(\otimes P_\Lambda)^l \otimes R\,(\otimes P_\Lambda)^{n-l-1}$ is equivalent to R (viewed as representations of the subalgebra K) . Therefore, one introduces the potentials

$$A_\mu^{(n)} \equiv A_\mu^\alpha \sum_{l=0}^{n-1} (\otimes 1)^l \otimes T_\alpha \, (\otimes 1)^{n-l-1}$$

$$\tilde{B}_\mu^{(n)} \equiv \tilde{B}_{\mu,i} \sum_{l=0}^{n-1} c_{n,l} \;\, (\otimes P_\Lambda)^l \otimes P_i \, (\otimes P_\Lambda)^{n-l-1} \qquad \text{(IV.15)}$$

where we have denoted $A_\mu = A_\mu^\alpha T_\alpha$, with T_α being the generators of $\mathcal{G}$, and where $c_{n,l}$ are constants. We introduce such constants because one can rescale the basis of each irreducible component of the representations of K independently, without affecting the equations (IV.4). Only the constraints, defining the submodel, are affected by the constants $c_{n,l}$.

Notice that the curvature $F_{\mu\nu}^{(n)}$ associated to the connection $A_\mu^{(n)}$ vanishes as a consquence of the fact that A_μ is flat. Therefore, the first condition in (III.4) leads in this case, to the same equations as following equations as (IV.4), i.e.

$$\partial^\mu \tilde{B}_{\mu,i} + A_a^{\mu K} \tilde{B}_{\mu,j} R\,(K_a)_{ij} = 0 \qquad \text{(IV.16)}$$

and the constraints

$$A_\mu^{S,r} \tilde{B}_i^\mu \left[\left(\sum_{m=0}^{n-1} (\otimes 1)^m \otimes S_r \, (\otimes 1)^{n-m-1} \right) , \left(\sum_{l=0}^{n-1} c_{n,l} \;\, (\otimes P_\Lambda)^l \otimes P_j \, (\otimes P_\Lambda)^{n-l-1} \right) \right] = 0 \qquad \text{(IV.17)}$$

Therefore, since (IV.16) are the same equations as (IV.4) we have a submodel, and the subclass of solutions is determined by the constraints (IV.17).

The conserved currents obtained from the zero curvature are (see (III.9))

$$J_\mu^{\lambda(n)} \equiv \left(\otimes W^{-1}\right)^n \, \tilde{B}_\mu^{(n)} \, (\otimes W)^n \qquad \text{(IV.18)}$$

Therefore, if we can choose the constants $c_{n,l}$ in such way that (IV.17) imposes the same set of constraints for any n, then one gets that the corresponding submodel possesses an infinity number of conserved currents.

V APPLICATIONS AND OUTLOOK

The first applications of our zero curvature generalization have been to produce new geometrical formulations of well known theories in higher dimensions. The most obvious way to have flatness locally is by requiring each component of $B_{\mu\nu}$ to be covariantly constant and to commute at the same point. This simplest possibility, which has not been reviewd here, produces topological theories like Chern-Simons [11] . Since these theories are rather well understood in the quantum case, a possible application is to compare to the classical solution. Self dual Yang-Mills or BPS are also easily incorporated and were also given in [1] as 4d examples.

The case which has been studied in more detail is when the local equation are based on a non-semisimple Lie algebra, which has produced many new results, specially obtaining reductions with infinite number of conserved currents. First it was applied to CP^1 in [1], and to the chiral ($su(2)$) model in [7], later generalized to Grassmannian models in [12] and then to homogeneous spaces in general in [6], including symmetric spaces (compact and noncompact). In this last general formulation, new models and many classes of the previous cases where treated with new results and insights. In particular, the reason behind the infinity of conserved currents, discussed only heuristically in [1], has been fully understood in [6] with the coset construction G/K in terms of the branching of representations of G in those of K with some singlets states playing a special and important role. This has been explained in detail here (see section IV) directly in terms of the decomposition of the algebra of G in a subalgebra K and its complement.

One of the most relevant aspects of our approach is the reduction to submodels possessing infinite number of conserved charges. Although that may resemble the BPS condition, it apparently does not involve saturation of any bound. Such methods have also been applied to Skyrme and Skyrme-type models [10,13]. The existence of solutions for the constraints is highly non trivial, specially for the hedgehog for Skyrme [13]. Indeed it is a non trivial fact that there are solutions to the constraints, like the special one found in [10] with the rational map, and the meromorphic ones for adjusted babyskyrme with an old trick due to Smirnov and Sobolev in [14].

Although there is a possibility to implement dressing, as mentioned in [1], the systematic method to construct solutions, which come naturally in $2d$ from the affine algebraic structures, has still to be developed.

ACKNOWLEDGMENTS

JSG would like to thank the La Plata colleagues for the perfect organization and all the participants for the friendly and rewarding atmosphere. This work is partially supported by grants from from CNPq (Brazil), NSF (PHY-9507829), EC (TMRERBFMRXCT960012) and DGICYT (PB 96-0960).

REFERENCES

1. Orlando Alvarez, Luiz A. Ferreira and J. Sánchez Guillén *Nucl. Phys.* **B529** (1998) 689.
2. D.I. Olive, N. Turok, *Nucl. Phys.* **B220** (1983) 491, Nucl. Phys. **B257** (1985) 277.
3. L.A.Ferreira, J.L.Miramontes, J. Sánchez Guillén *J. Math. Phys.* **38** (1997) 882; J.L.Miramontes, hep-th/9809052.
4. R.Donagi, E.Witten, *Nucl. Phys.* **B460** (1996) 97; J. Edelstein and J.Mas, these Proceedings, hep-th/9902161.
5. Ya. Aref'eva *Theor. Mat. Phys.* **43** (1980) 353, N.E.Bralic *Phys. Rev.* **D22** (1980) 3090.
6. L.A. Ferreira and E. Leite, *Integrable theories in any dimension and homogeneous spaces*, hep-th/9810067, to appear in *Nuclear Physics B.*
7. D. Gianzo, J.O. Madsen and J. Sánchez Guillén, *Nucl. Phys.* **B537** (1999) 586.
8. V.E. Zakharov and A.B. Shabat, *Zh. Exp. Teor. Fiz.* **61** (1971) 118-134; english transl. *Soviet Phys. JETP* **34** (1972) 62-69; P. Lax, *Comm. Pure Appl. Math.* **21** (1968) 467-490.
9. P.G.O. Freund and R. Nepomechie, *Nucl. Phys.* **B199** (1982) 482.
10. H.Aratyn, L.A. Ferreira and A. Zimerman, *Toroidal solitons in* 3 + 1 *dimensional integrable theories*, hep-th/9902141.
11. J.M.F. Labastida, these proceedings.
12. K. Fuji, Y. Homma and T. Suzuki,*Phys. Lett.* **438B** (1998) 290
13. J.O. Madsen and J. Sánchez Guillén, in preparation.
14. K. Fuji, Y. Homma and T. Suzuki, hep-th/9809149.

WIMPZILLAS!

Edward W. Kolb[†‡1],
Daniel J. H. Chung[§2],
Antonio Riotto[¶3]

[†] *Theoretical Astrophysics*
Fermi National Accelerator Laboratory
Batavia, Illinois 60510.

[‡] *Department of Astronomy and Astrophysics*
Enrico Fermi Institute
The University of Chicago
Chicago, Illinois 60637.

[§] *Department of Physics*
The University of Michigan
Ann Arbor, Michigan 48109.

[¶] *Theory Division*
CERN
CH-1211 Geneva 23, Switzerland.

Abstract. There are many reasons to believe the present mass density of the universe is dominated by a weakly interacting massive particle (WIMP), a fossil relic of the early universe. Theoretical ideas and experimental efforts have focused mostly on production and detection of *thermal* relics, with mass typically in the range a few GeV to a hundred GeV. Here, I will review scenarios for production of *nonthermal* dark matter. Since the masses of the nonthermal WIMPS are in the range 10^{12} to 10^{16} GeV, much larger than the mass of thermal wimpy WIMPS, they may be referred to as WIMPZILLAS. In searches for dark matter it may be well to remember that "size does matter."

I INTRODUCTION

There is conclusive evidence that the dominant component of the matter density in the universe is dark. The most striking indication of the existence of dark matter is the dynamical motions of astronomical objects. Observations of flat rotation curves for spiral galaxies [1] indicates that the dark component of galactic halos is about ten times the luminous component. Dynamical evidence for DM in galaxy clusters from the velocity dispersion of individual galaxies, as well as from the large x-ray temperatures of clusters, is also compelling [2]. Bulk flows, as well as the peculiar motion of our own local group, also implies a universe dominated by dark matter [3].

The mass of galaxy clusters inferred by their gravitational lensing of background images is consistent with the large dark-to-visible mass ratios determined by dynamical methods [4].

There is also compelling evidence that the bulk of the dark component must be nonbaryonic. The present baryonic density is restricted by big-bang nucleosynthesis to be less than that inferred by the methods discussed above [5]. The theory of structure formation from the gravitational instability of small initial seed inhomogeneities requires a significant nonbaryonic component to the mass density [6].

---

1) E-mail: rocky@rigoletto.fnal.gov
2) E-mail: djchung@feynman.physics.lsa.umich.edu
3) E-mail: riotto@nxth04.cern.ch

CP484, *Trends in Theoretical Physics II*, edited by H. Falomir, R. E. Gamboa Saraví, and F. A. Schaposnik

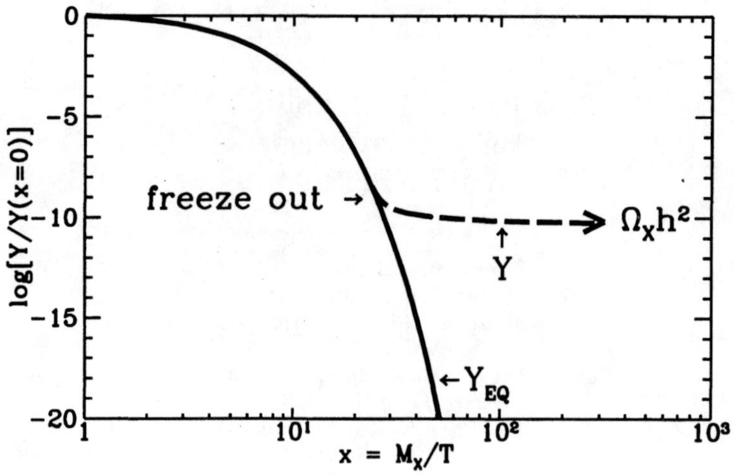

FIGURE 1. A thermal relic starts in LTE at $T \gg M_X$. When the rates keeping the relic in chemical equilibrium become smaller than the expansion rate, the density of the relic relative to the entropy density freezes out.

In terms of the critical density, $\rho_C = 3H_0^2 M_{\rm Pl}^2/8\pi = 1.88 \times 10^{-29}$g cm^{-3} with Hubble constant $H_0 \equiv 100h$ km sec^{-1}Mpc^{-1} and Planck mass $M_{\rm Pl}$, the dark-matter density inferred from dynamics is $\Omega_{\rm DM} \equiv \rho_{\rm DM}/\rho_C \gtrsim 0.3$. In addition, the most natural inflation models predict a flat universe, *i.e.*, $\Omega_0 = 1$, while standard big-bang nucleosynthesis implies that ordinary baryonic matter can contribute at most 10% to Ω_0. This means that about 90% of the matter in our universe may be dark.

II THERMAL RELICS—WIMPY WIMPS

It is usually assumed that the dark matter consists of a species of a new, yet undiscovered, massive particle, traditionally denoted by X. It is also often assumed that the dark matter is a thermal relic, *i.e.*, it was in chemical equilibrium in the early universe.

A thermal relic is assumed to be in local thermodynamic equilibrium (LTE) at early times. The *equilibrium* abundance of a particle, say relative to the entropy density, depends upon the ratio of the mass of the particle to the temperature. Define the variable $Y \equiv n_X/s$, where n_X is the number density of WIMP X with mass M_X, and $s \sim T^3$ is the entropy density. The equilibrium value of Y, Y_{EQ}, is proportional to $\exp(-x)$ for $x \gg 1$, while $Y_{EQ} \sim$ constant for $x \ll 1$, where $x = M_X/T$.

A particle will track its equilibrium abundance as long as reactions which keep the particle in chemical equilibrium can proceed rapidly enough. Here, rapidly enough means on a timescale more rapid than the expansion rate of the universe, H. When the reaction rate becomes smaller than the expansion rate, then the particle can no longer track its equilibrium value, and thereafter Y is constant. When this occurs the particle is said to be "frozen out." A schematic illustration of this is given in Fig. 1.

The more strongly interacting the particle, the longer it stays in LTE, and the smaller its eventual freeze-out abundance. Conversely, the more weakly interacting the particle, the larger its present abundance. The freeze-out value of Y is related to the mass of the particle and its annihilation cross section (here characterized by σ_0) by [7]

$$Y \propto \frac{1}{M_X m_{Pl} \sigma_0} \; . \tag{1}$$

Since the contribution to Ω is proportional to $M_X n_X$, which in turn is proportional to $M_X Y$, the present contribution to Ω from a thermal relic roughly is *independent* of its mass,[1] and depends only upon the annihilation cross section. The cross section that results in $\Omega_X h^2 \sim 1$ is of order 10^{-37}cm^2, of the order the weak scale.

[1]) To first approximation the relic dependence depends upon the mass only indirectly through the dependence of the annihilation cross section on the mass.

This is one of the attractions of thermal relics. The scale of the annihilation cross section is related to a known mass scale.

The simple assumption that dark matter is a thermal relic is surprisingly restrictive. The largest the annihilation cross section can be is roughly M_X^{-2}. This implies that large-mass WIMPS would have such a small annihilation cross section that their present abundance would be too large. Thus one expects a maximum mass for a thermal WIMP, which turns out to be a few hundred TeV [8].

The standard lore is that the hunt for dark matter should concentrate on particles with mass of the order of the weak scale and with interaction with ordinary matter on the scale of the weak force. This has been the driving force behind the vast effort in dark matter direct detection described in this meeting by Cabrera [9], Liubarsky [10], Bernabei [11], Ramachers [12], and Baudis [13].

In view of the unitarity argument, in order to consider *thermal* WIMPZILLAS, one must invoke, for example, late-time entropy production to dilute the abundance of these supermassive particles [14], rendering the scenario unattractive.

III NONTHERMAL RELICS—WIMPZILLAS

There are two necessary conditions for the WIMPZILLA scenario. First, the WIMPZILLA must be stable, or at least have a lifetime much greater than the age of the universe. This may result from, for instance, supersymmetric theories where the breaking of supersymmetry is communicated to ordinary sparticles via the usual gauge forces [15]. In particular, the secluded and the messenger sectors often have accidental symmetries analogous to baryon number. This means that the lightest particle in those sectors might be stable and very massive if supersymmetry is broken at a large scale [16]. Other natural candidates arise in theories with discrete gauge symmetries [17] and in string theory and M theory [18,19].

It is useful here to note that WIMPZILLA decay might be able to account for ultra-high energy cosmic rays above the Greisen–Zatzepin–Kuzmin cutoff [20,21]. A wimpy little thermal relic would be too light to do the job, a WIMPZILLA is needed.

The second condition for a WIMPZILLA is that it must not have been in equilibrium when it froze out (*i.e.*, it is not a thermal relic), otherwise $\Omega_X h^2$ would be much larger than one. A sufficient condition for nonequilibrium is that the annihilation rate (per particle) must be smaller than the expansion rate: $n_X\sigma|v| < H$, where $\sigma|v|$ is the annihilation rate times the Møller flux factor, and H is the expansion rate. Conversely, if the dark matter was created at some temperature T_* *and* $\Omega_X h^2 < 1$, then it is easy to show that it could not have attained equilibrium. To see this, assume X's were created in a radiation-dominated universe at temperature T_*. Then $\Omega_X h^2$ is given by

$$\Omega_X h^2 = \Omega_\gamma h^2 (T_*/T_0) m_X n_X(T_*)/\rho_\gamma(T_*) \ , \tag{2}$$

where T_0 is the present temperature. Using the fact that $\rho_\gamma(T_*) = H(T_*)M_{Pl}T_*^2$, $n_X(T_*)/H(T_*) = (\Omega_X/\Omega_\gamma)T_0 M_{Pl} T_*/M_X$. One may safely take the limit $\sigma|v| < M_X^{-2}$, so $n_X(T_*)\sigma|v|/H(T_*)$ must be less than $(\Omega_X/\Omega_\gamma)T_0 M_{Pl} T_*/M_X^3$. Thus, the requirement for nonequilibrium is

$$\left(\frac{200\ \mathrm{TeV}}{M_X}\right)^2 \left(\frac{T_*}{M_X}\right) < 1 \ . \tag{3}$$

This implies that if a nonrelativistic particle with $M_X \gtrsim 200$ TeV was created at $T_* < M_X$ with a density low enough to result in $\Omega_X \lesssim 1$, then its abundance must have been so small that it never attained equilibrium. Therefore, if there is some way to create WIMPZILLAS in the correct abundance to give $\Omega_X \sim 1$, nonequilibrium is automatic.

Any WIMPZILLA production scenario must meet these two criteria. Before turning to several WIMPZILLA production scenarios, it is useful to estimate the fraction of the total energy density of the universe in WIMPZILLAS at the time of their production that will eventually result in $\Omega \sim 1$ today.

The most likely time for WIMPZILLA production is just after inflation. The first step in estimating the fraction of the energy density in WIMPZILLAS is to estimate the total energy density when the universe is "reheated" after inflation.

Consider the calculation of the reheat temperature, denoted as T_{RH}. The reheat temperature is calculated by assuming an instantaneous conversion of the energy density in the inflaton field into radiation when the decay width of the inflaton energy, Γ_ϕ, is equal to H, the expansion rate of the universe.

The reheat temperature is calculated quite easily [7]. After inflation the inflaton field executes coherent oscillations about the minimum of the potential. Averaged over several oscillations, the coherent oscillation energy density redshifts as matter: $\rho_\phi \propto a^{-3}$, where a is the Robertson–Walker scale factor. If ρ_I and a_I denotes the total inflaton energy density and the scale factor at the initiation of coherent oscillations, then the Hubble expansion rate as a function of a is

$$H(a) = \sqrt{\frac{8\pi}{3}\frac{\rho_I}{M_{Pl}^2}\left(\frac{a_I}{a}\right)^3} . \tag{4}$$

Equating $H(a)$ and Γ_ϕ leads to an expression for a_I/a. Now if all available coherent energy density is instantaneously converted into radiation at this value of a_I/a, one can define the reheat temperature by setting the coherent energy density, $\rho_\phi = \rho_I(a_I/a)^3$, equal to the radiation energy density, $\rho_R = (\pi^2/30)g_* T_{RH}^4$, where g_* is the effective number of relativistic degrees of freedom at temperature T_{RH}. The result is

$$T_{RH} = \left(\frac{90}{8\pi^3 g_*}\right)^{1/4} \sqrt{\Gamma_\phi M_{Pl}} = 0.2\left(\frac{200}{g_*}\right)^{1/4} \sqrt{\Gamma_\phi M_{Pl}} . \tag{5}$$

The limit from gravitino overproduction is $T_{RH} \lesssim 10^9$ to 10^{10} GeV.

Now consider the WIMPZILLA density at reheating. Suppose the WIMPZILLA never attained LTE and was nonrelativistic at the time of production. The usual quantity $\Omega_X h^2$ associated with the dark matter density today can be related to the dark matter density when it was produced. First write

$$\frac{\rho_X(t_0)}{\rho_R(t_0)} = \frac{\rho_X(t_{RH})}{\rho_R(t_{RH})}\left(\frac{T_{RH}}{T_0}\right), \tag{6}$$

where ρ_R denotes the energy density in radiation, ρ_X denotes the energy density in the dark matter, T_{RH} is the reheat temperature, T_0 is the temperature today, t_0 denotes the time today, and t_{RH} denotes the approximate time of reheating.[2] To obtain $\rho_X(t_{RH})/\rho_R(t_{RH})$, one must determine when X particles are produced with respect to the completion of reheating and the effective equation of state between X production and the completion of reheating.

At the end of inflation the universe may have a brief period of matter domination resulting either from the coherent oscillations phase of the inflaton condensate or from the preheating phase [22]. If the X particles are produced at time $t = t_e$ when the de Sitter phase ends and the coherent oscillation period just begins, then both the X particle energy density and the inflaton energy density will redshift at approximately the same rate until reheating is completed and radiation domination begins. Hence, the ratio of energy densities preserved in this way until the time of radiation domination is

$$\frac{\rho_X(t_{RH})}{\rho_R(t_{RH})} \approx \frac{8\pi}{3}\frac{\rho_X(t_e)}{M_{Pl}^2 H^2(t_e)}, \tag{7}$$

where $M_{Pl} \approx 10^{19}$ GeV is the Planck mass and most of the energy density in the universe just before time t_{RH} is presumed to turn into radiation. Thus, using Eq. 6, one may obtain an expression for the quantity $\Omega_X \equiv \rho_X(t_0)/\rho_C(t_0)$, where $\rho_C(t_0) = 3H_0^2 M_{Pl}^2/8\pi$ and $H_0 = 100\,h$ km sec^{-1} Mpc^{-1}:

$$\Omega_X h^2 \approx \Omega_R h^2 \left(\frac{T_{RH}}{T_0}\right)\frac{8\pi}{3}\left(\frac{M_X}{M_{Pl}}\right)\frac{n_X(t_e)}{M_{Pl}H^2(t_e)}. \tag{8}$$

Here $\Omega_R h^2 \approx 4.31 \times 10^{-5}$ is the fraction of critical energy density in radiation today and n_X is the density of X particles at the time when they were produced.

Note that because the reheating temperature must be much greater than the temperature today ($T_{RH}/T_0 \gtrsim 4.2 \times 10^{14}$), in order to satisfy the cosmological bound $\Omega_X h^2 \lesssim 1$, the fraction of total WIMPZILLA energy density at the time when they were produced must be extremely small. One sees from Eq. 8 that $\Omega_X h^2 \sim 10^{17}(T_{RH}/10^9\text{GeV})(\rho_X(t_e)/\rho(t_e))$. It is indeed a very small fraction of the total energy density extracted in WIMPZILLAS.

---

[2] More specifically, this is approximately the time at which the universe becomes radiation dominated after inflation.

This means that if the WIMPZILLA is extremely massive, the challenge lies in creating very few of them. Gravitational production discussed in Section IV A naturally gives the needed suppression. Note that if reheating occurs abruptly at the end of inflation, then the matter domination phase may be negligibly short and the radiation domination phase may follow immediately after the end of inflation. However, this does not change Eq. 8.

IV WIMPZILLA PRODUCTION

A Gravitational Production

First consider the possibility that WIMPZILLAS are produced in the transition between an inflationary and a matter-dominated (or radiation-dominated) universe due to the "nonadiabatic" expansion of the background spacetime acting on the vacuum quantum fluctuations [23].

The distinguishing feature of this mechanism is the capability of generating particles with mass of the order of the inflaton mass (usually much larger than the reheating temperature) even when the particles only interact extremely weakly (or not at all) with other particles and do not couple to the inflaton. They may still be produced in sufficient abundance to achieve critical density today due to the classical gravitational effect on the vacuum state at the end of inflation. More specifically, if $0.04 \lesssim M_X/H_I \lesssim 2$, where $H_I \sim m_\phi \sim 10^{13}$GeV is the Hubble constant at the end of inflation (m_ϕ is the mass of the inflaton), WIMPZILLAS produced gravitationally can have a density today of the order of the critical density. This result is quite robust with respect to the "fine" details of the transition between the inflationary phase and the matter-dominated phase, and independent of the coupling of the WIMPZILLA to any other particle.

Conceptually, gravitational WIMPZILLA production is similar to the inflationary generation of gravitational perturbations that seed the formation of large scale structures. In the usual scenarios, however, the quantum generation of energy density fluctuations from inflation is associated with the inflaton field that dominated the mass density of the universe, and not a generic, sub-dominant scalar field. Another difference is that the usual density fluctuations become larger than the Hubble radius, while most of the WIMPZILLA perturbations remain smaller than the Hubble radius.

There are various inequivalent ways of calculating the particle production due to interaction of a classical gravitational field with the vacuum (see for example [24], [25], and [26]). Here, I use the method of finding the Bogoliubov coefficient for the transformation between positive frequency modes defined at two different times. For $M_X/H_I \lesssim 1$ the results are quite insensitive to the differentiability or the fine details of the time dependence of the scale factor. For $0.04 \lesssim M_X/H_I \lesssim 2$, all the dark matter needed for closure of the universe can be made gravitationally, quite independently of the details of the transition between the inflationary phase and the matter dominated phase.

Start with the canonical quantization of the X field in an action of the form (with metric $ds^2 = dt^2 - a^2(t)d\mathbf{x}^2 = a^2(\eta)\left[d\eta^2 - d\mathbf{x}^2\right]$ where η is conformal time)

$$S = \int dt \int d^3x \, \frac{a^3}{2}\left(\dot{X}^2 - \frac{(\nabla X)^2}{a^2} - M_X^2 X^2 - \xi R X^2\right) \tag{9}$$

where R is the Ricci scalar. After transforming to conformal time coordinate, use the mode expansion

$$X(\mathbf{x}) = \int \frac{d^3k}{(2\pi)^{3/2}a(\eta)}\left[a_k h_k(\eta)e^{i\mathbf{k}\cdot\mathbf{x}} + a_k^\dagger h_k^*(\eta)e^{-i\mathbf{k}\cdot\mathbf{x}}\right], \tag{10}$$

where because the creation and annihilation operators obey the commutator $[a_{k_1}, a_{k_2}^\dagger] = \delta^{(3)}(\mathbf{k}_1 - \mathbf{k}_2)$, the h_ks obey a normalization condition $h_k h_k^{\prime *} - h_k^\prime h_k^* = i$ to satisfy the canonical field commutators (henceforth, all primes on functions of η refer to derivatives with respect to η). The resulting mode equation is

$$h_k''(\eta) + w_k^2(\eta)h_k(\eta) = 0, \tag{11}$$

where

$$w_k^2 = k^2 + M_X^2 a^2 + (6\xi - 1)a''/a \,. \tag{12}$$

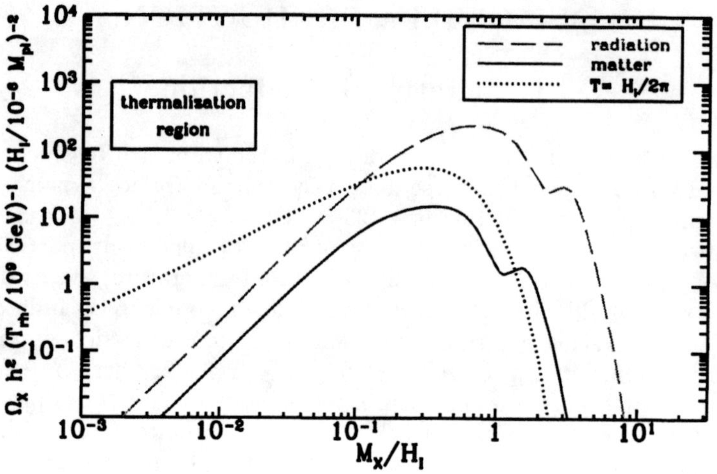

FIGURE 2. The contribution of gravitationally produced WIMPZILLAS to $\Omega_X h^2$ as a function of M_X/H_I. The shaded area is where thermalization *may* occur if the annihilation cross section is its maximum value. Also shown is the contribution assuming that the WIMPZILLA is present at the end of inflation with a temperature $T = H_I/2\pi$.

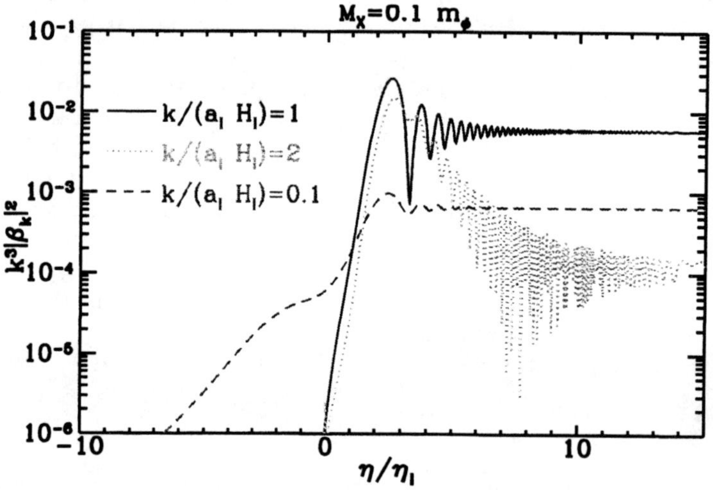

FIGURE 3. The evolution of the Bogoliubov coefficient with conformal time for several wavenumbers. $\eta = \eta_I$ corresponds to the end of the inflationary era.

The parameter ξ is $1/6$ for conformal coupling and 0 for minimal coupling. From now on, $\xi = 1/6$ for simplicity but without much loss of generality. By a change in variable $\eta \to k/a$, one can rewrite the differential equation such that it depends only on $H(\eta)$, $H'(\eta)/k$, $k/a(\eta)$, and M_X. Hence, the parameters H_I and a_I correspond to the Hubble parameter and the scale factor evaluated at an arbitrary conformal time η_I, which can be taken to be the approximate time at which Xs are produced (*i.e.*, η_I is the conformal time at the end of inflation).

One may then rewrite Eq. 11 as

$$h''_{\tilde{k}}(\tilde{\eta}) + \left(\tilde{k}^2 + \frac{M_X^2}{H_I^2}\tilde{a}^2\right) h_{\tilde{k}}(\tilde{\eta}) = 0 \; , \tag{13}$$

where $\tilde{\eta} = \eta a_I H_I$, $\tilde{a} = a/a_I$, and $\tilde{k} = k/(a_I H_I)$. For simplicity of notation, drop all the tildes. This differential equation can be solved once the boundary conditions are supplied.

The number density of the WIMPZILLAS is found by a Bogoliubov transformation from the vacuum mode solution with the boundary condition at $\eta = \eta_0$ (the initial time at which the vacuum of the universe is determined) into the one with the boundary condition at $\eta = \eta_1$ (any later time at which the particles are no longer being created). η_0 will be taken to be $-\infty$ while η_1 will be taken to be at $+\infty$. Defining the Bogoliubov transformation as $h_k^{\eta_1}(\eta) = \alpha_k h_k^{\eta_0}(\eta) + \beta_k h_k^{*\eta_0}(\eta)$ (the superscripts denote where the boundary condition is set), the energy density of produced particles is

$$\rho_X(\eta_1) = M_X n_X(\eta_1) = M_X H_I^3 \left(\frac{1}{\tilde{a}(\eta_1)}\right)^3 \int_0^\infty \frac{d\tilde{k}}{2\pi^2}\tilde{k}^2 |\beta_{\tilde{k}}|^2, \tag{14}$$

where one should note that the number operator is defined at η_1 while the quantum state (approximated to be the vacuum state) defined at η_0 does not change in time in the Heisenberg representation.

As one can see from Eq. 13, the input parameter is M_X/H_I. One must also specify the behavior of $a(\eta)$ near the end of inflation. In Fig. 2 (from [23]), I show the resulting values of $\Omega_X h^2$ as a function of M_X/H_I assuming the evolution of the scale factor smoothly interpolates between exponential expansion during inflation and either a matter-dominated universe or radiation-dominated universe. The peak at $M_X/H_I \sim 1$ is similar to the case presented in Ref. [27]. As expected, for large M_X/H_I, the number density falls off faster than any inverse power of M_X/H_I.

Now most of the action occurs around the transition from inflation to the matter-dominated or radiation-dominated universe. This is shown in Fig. 3. Also from Fig. 3 one can see that most of the particles are created with wavenumber of order H_I.

To conclude, there is a significant mass range ($0.1H_I$ to H_I, where $H_I \sim 10^{13}$GeV) for which WIMPZILLAS will have critical density today regardless of the fine details of the transition out of inflation. Because this production mechanism is inherent in the dynamics between the classical gravitational field and a quantum field, it needs no fine tuning of field couplings or any coupling to the inflaton field. However, only if the particles are stable (or sufficiently long lived) will these particles give contribution of the order of critical density.

B Production during Reheating

Another attractive origin for WIMPZILLAS is during the defrosting phase after inflation. It is important to recall that it is not necessary to convert a significant fraction of the available energy into massive particles; in fact, it must be an infinitesimal amount. I now will discuss how particles of mass much greater than T_{RH} may be created in the correct amount after inflation in reheating [28].

In one extreme is the assumption that the vacuum energy of inflation is immediately converted to radiation resulting in a reheat temperature T_{RH}. In this case Ω_X can be calculated by integrating the Boltzmann equation with initial condition $N_X = 0$ at $T = T_{RH}$. One expects the X density to be suppressed by $\exp(-2M_X/T_{RH})$; indeed, one finds $\Omega_X \sim 1$ for $M_X/T_{RH} \sim 25 + 0.5\ln(m_X^2\langle\sigma|v|\rangle)$, in agreement with previous estimates [20] that for $T_{RH} \sim 10^9$GeV, the WIMPZILLA mass would be about 2.5×10^{10}GeV.

A second (and more plausible) scenario is that reheating is not instantaneous, but is the result of the slow decay of the inflaton field. The simplest way to envision this process is if the comoving energy density in the zero mode of the inflaton decays into normal particles, which then scatter and thermalize to form a thermal background. It is usually assumed that the decay width of this process is the same as the decay width of a free inflaton field.

There are two reasons to suspect that the inflaton decay width might be small. The requisite flatness of the inflaton potential suggests a weak coupling of the inflaton field to other fields since the potential is renormalized by the inflaton coupling to other fields [29]. However, this restriction may be evaded in supersymmetric theories where the nonrenormalization theorem ensures a cancelation between fields and their superpartners. A second reason to suspect weak coupling is that in local supersymmetric theories gravitinos are produced during reheating. Unless reheating is delayed, gravitinos will be overproduced, leading to a large undesired entropy production when they decay after big-bang nucleosynthesis [30].

It is simple to calculate the WIMPZILLA abundance in the slow reheating scenario. It will be important to keep in mind that what is commonly called the reheat temperature, T_{RH}, is not the maximum temperature obtained after inflation. The maximum temperature is, in fact, much larger than T_{RH}. The reheat temperature is best regarded as the temperature below which the universe expands as a radiation-dominated universe, with the scale factor decreasing as $g_*^{-1/3}T^{-1}$. In this regard it has a limited meaning [7,31]. One implication of this is that it is incorrect to assume that the maximum abundance of a massive particle species produced after inflation is suppressed by a factor of $\exp(-M/T_{RH})$.

To estimate WIMPZILLA production in reheating, consider a model universe with three components: inflaton field energy, ρ_ϕ, radiation energy density, ρ_R, and WIMPZILLA energy density, ρ_X. Assume that the decay rate of the inflaton field energy density is Γ_ϕ. Also assume the WIMPZILLA lifetime is longer than any timescale in the problem (in fact it must be longer than the present age of the universe). Finally, assume that the light degrees of freedom are in local thermodynamic equilibrium.

With the above assumptions, the Boltzmann equations describing the redshift and interchange in the energy density among the different components is

$$\begin{aligned} &\dot\rho_\phi + 3H\rho_\phi + \Gamma_\phi\rho_\phi = 0 \\ &\dot\rho_R + 4H\rho_R - \Gamma_\phi\rho_\phi - \frac{\langle\sigma|v|\rangle}{m_X}\left[\rho_X^2 - \left(\rho_X^{EQ}\right)^2\right] = 0 \\ &\dot\rho_X + 3H\rho_X + \frac{\langle\sigma|v|\rangle}{m_X}\left[\rho_X^2 - \left(\rho_X^{EQ}\right)^2\right] = 0 \ , \end{aligned} \tag{15}$$

where dot denotes time derivative. As already mentioned, $\langle\sigma|v|\rangle$ is the thermal average of the X annihilation cross section times the Møller flux factor. The equilibrium energy density for the X particles, ρ_X^{EQ}, is determined by the radiation temperature, $T = (30\rho_R/\pi^2 g_*)^{1/4}$.

It is useful to introduce two dimensionless constants, α_ϕ and α_X, defined in terms of Γ_ϕ and $\langle\sigma|v|\rangle$ as

$$\Gamma_\phi = \alpha_\phi M_\phi \qquad \langle\sigma|v|\rangle = \alpha_X M_X^{-2} \ . \tag{16}$$

For a reheat temperature much smaller than M_ϕ, Γ_ϕ must be small. From Eq. (5), the reheat temperature in terms of α_X and M_X is $T_{RH} \simeq \alpha_\phi^{1/2}\sqrt{M_\phi M_{Pl}}$. For $M_\phi = 10^{13}$GeV, α_ϕ must be smaller than of order 10^{-13}. On the other hand, α_X may be as large as of order unity, or it may be small also.

It is also convenient to work with dimensionless quantities that can absorb the effect of expansion of the universe. This may be accomplished with the definitions

$$\Phi \equiv \rho_\phi M_\phi^{-1} a^3 \ ; \quad R \equiv \rho_R a^4 \ ; \quad X \equiv \rho_X M_X^{-1} a^3 \ . \tag{17}$$

It is also convenient to use the scale factor, rather than time, for the independent variable, so one may define a variable $x = aM_\phi$. With this choice the system of equations can be written as (prime denotes d/dx)

$$\begin{aligned} \Phi' &= -c_1 \frac{x}{\sqrt{\Phi x + R}} \Phi \\ R' &= c_1 \frac{x^2}{\sqrt{\Phi x + R}} \Phi + c_2 \frac{x^{-1}}{\sqrt{\Phi x + R}} \left(X^2 - X_{EQ}^2\right) \\ X' &= -c_3 \frac{x^{-2}}{\sqrt{\Phi x + R}} \left(X^2 - X_{EQ}^2\right) \ . \end{aligned} \tag{18}$$

The constants c_1, c_2, and c_3 are given by

$$c_1 = \sqrt{\frac{3}{8\pi}\frac{M_{Pl}}{M_\phi}}\alpha_\phi \qquad c_2 = c_1 \frac{M_\phi}{M_X}\frac{\alpha_X}{\alpha_\phi} \qquad c_3 = c_2 \frac{M_\phi}{M_X} \ . \tag{19}$$

X_{EQ} is the equilibrium value of X, given in terms of the temperature T as (assuming a single degree of freedom for the X species)

$$X_{EQ} = \frac{M_X^3}{M_\phi^3}\left(\frac{1}{2\pi}\right)^{3/2} x^3 \left(\frac{T}{M_X}\right)^{3/2} \exp(-M_X/T) . \tag{20}$$

The temperature depends upon R and g_*, the effective number of degrees of freedom in the radiation:

$$\frac{T}{M_X} = \left(\frac{30}{g_*\pi^2}\right)^{1/4} \frac{M_\phi}{M_X}\frac{R^{1/4}}{x} . \tag{21}$$

It is straightforward to solve the system of equations in Eq. (18) with initial conditions at $x = x_I$ of $R(x_I) = X(x_I) = 0$ and $\Phi(x_I) = \Phi_I$. It is convenient to express $\rho_\phi(x = x_I)$ in terms of the expansion rate at x_I, which leads to

$$\Phi_I = \frac{3}{8\pi}\frac{M_{Pl}^2}{M_\phi^2}\frac{H_I^2}{M_\phi^2} x_I^3 . \tag{22}$$

The numerical value of x_I is irrelevant.

Before numerically solving the system of equations, it is useful to consider the early-time solution for R. Here, early times means $H \gg \Gamma_\phi$, *i.e.*, before a significant fraction of the comoving coherent energy density is converted to radiation. At early times $\Phi \simeq \Phi_I$, and $R \simeq X \simeq 0$, so the equation for R' becomes $R' = c_1 x^{3/2}\Phi_I^{1/2}$. Thus, the early time solution for R is simple to obtain:

$$R \simeq \frac{2}{5}c_1\left(x^{5/2} - x_I^{5/2}\right)\Phi_I^{1/2} \qquad (H \gg \Gamma_\phi) . \tag{23}$$

Now express T in terms of R to yield the early-time solution for T:

$$\frac{T}{M_\phi} \simeq \left(\frac{12}{\pi^2 g_*}\right)^{1/4} c_1^{1/4}\left(\frac{\Phi_I}{x_I^3}\right)^{1/8} \times \left[\left(\frac{x}{x_I}\right)^{-3/2} - \left(\frac{x}{x_I}\right)^{-4}\right]^{1/4} \qquad (H \gg \Gamma_\phi) . \tag{24}$$

Thus, T has a maximum value of

$$\frac{T_{MAX}}{M_\phi} = 0.77\left(\frac{12}{\pi^2 g_*}\right)^{1/4} c_1^{1/4}\left(\frac{\Phi_I}{x_I^3}\right)^{1/8} = 0.77\alpha_\phi^{1/4}\left(\frac{9}{2\pi^3 g_*}\right)^{1/4}\left(\frac{M_{Pl}^2 H_I}{M_\phi^3}\right)^{1/4} , \tag{25}$$

which is obtained at $x/x_I = (8/3)^{2/5} = 1.48$. It is also possible to express α_ϕ in terms of T_{RH} and obtain

$$\frac{T_{MAX}}{T_{RH}} = 0.77\left(\frac{9}{5\pi^3 g_*}\right)^{1/8}\left(\frac{H_I M_{Pl}}{T_{RH}^2}\right)^{1/4} . \tag{26}$$

For an illustration, in the simplest model of chaotic inflation $H_I \sim M_\phi$ with $M_\phi \simeq 10^{13}$GeV, which leads to $T_{MAX}/T_{RH} \sim 10^3(200/g_*)^{1/8}$ for $T_{RH} = 10^9$GeV.

We can see from Eq. (23) that for $x/x_I > 1$, in the early-time regime T scales as $a^{-3/8}$, which implies that entropy is created in the early-time regime [31]. So if one is producing a massive particle during reheating it is necessary to take into account the fact that the maximum temperature is greater than T_{RH}, and that during the early-time evolution, $T \propto a^{-3/8}$.

An example of a numerical evaluation of the complete system in Eq. (18) is shown in Fig. 4 (from [28]). The model parameters chosen were $M_\phi = 10^{13}$GeV, $\alpha_\phi = 2 \times 10^{-13}$, $M_X = 1.15 \times 10^{12}$GeV, $\alpha_X = 10^{-2}$, and

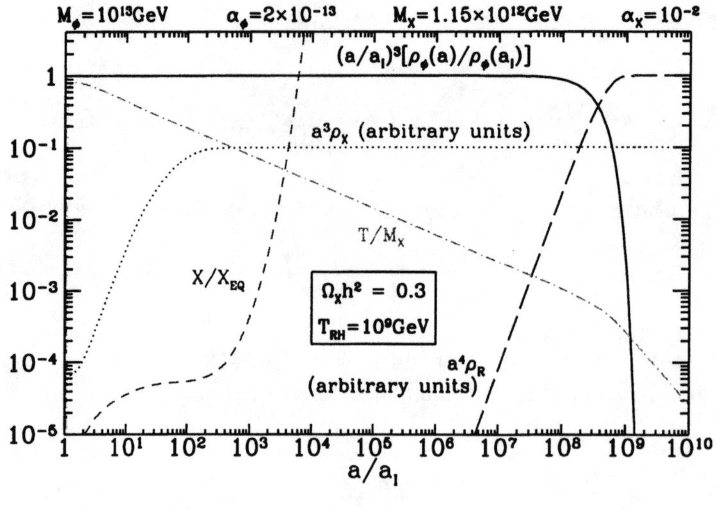

FIGURE 4. The evolution of energy densities and T/M_X as a function of the scale factor. Also shown is X/X_{EQ}.

$g_* = 200$. The expansion rate at the beginning of the coherent oscillation period was chosen to be $H_I = M_\phi$. These parameters result in $T_{RH} = 10^9$GeV and $\Omega_X h^2 = 0.3$.

Figure 4 serves to illustrate several aspects of the problem. Just as expected, the comoving energy density of ϕ (*i.e.*, $a^3\rho_\phi$) remains roughly constant until $\Gamma_\phi \simeq H$, which for the chosen model parameters occurs around $a/a_I \simeq 5 \times 10^8$. But of course, that does not mean that the temperature is zero. Notice that the temperature peaks well before "reheating." The maximum temperature, $T_{MAX} = 10^{12}$GeV, is reached at a/a_I slightly larger than unity (in fact at $a/a_I = 1.48$ as expected), while the reheat temperature, $T_{RH} = 10^9$GeV, occurs much later, around $a/a_I \sim 10^8$. Note that $T_{MAX} \simeq 10^3 T_{RH}$ in agreement with Eq. (26).

From the figure it is clear that $X \ll X_{EQ}$ at the epoch of freeze out of the comoving X number density, which occurs around $a/a_I \simeq 10^2$. The rapid rise of the ratio after freeze out is simply a reflection of the fact that X is constant while X_{EQ} decreases exponentially.

A close examination of the behavior of T shows that after the sharp initial rise of the temperature, the temperature decreases as $a^{-3/8}$ [as follows from Eq. (24)] until $H \simeq \Gamma_\phi$, and thereafter $T \propto a^{-1}$ as expected for the radiation-dominated era.

For the choices of M_ϕ, α_ϕ, g_*, and α_X used for the model illustrated in Fig. 4, $\Omega_X h^2 = 0.3$ for $M_X = 1.15 \times 10^{12}$GeV, in excellent agreement with the mass predicted by using an analytic estimate for the result [28]

$$\Omega_X h^2 = M_X^2 \langle\sigma|v|\rangle \left(\frac{g_*}{200}\right)^{-3/2} \left(\frac{2000 T_{RH}}{M_X}\right)^7 . \tag{27}$$

Here again, the results have also important implications for the conjecture that ultra-high cosmic rays, above the Greisen-Zatsepin-Kuzmin cut-off of the cosmic ray spectrum, may be produced in decays of superheavy long-living particles [19–21,32]. In order to produce cosmic rays of energies larger than about 10^{13} GeV, the mass of the X-particles must be very large, $M_X \gtrsim 10^{13}$ GeV and their lifetime τ_X cannot be much smaller than the age of the Universe, $\tau_X \gtrsim 10^{10}$ yr. With the smallest value of the lifetime, the observed flux of ultra-high energy cosmic rays will be reproduced with a rather low density of X-particles, $\Omega_X \sim 10^{-12}$. It has been suggested that X-particles can be produced in the right amount by usual collisions and decay processes taking place during the reheating stage after inflation if the reheat temperature never exceeded M_X [32]. Again, assuming naively that that the maximum number density of a massive particle species X produced after inflation is suppressed by a factor of $(M_X/T_{RH})^{3/2}\exp(-M_X/T_{RH})$ with respect to the photon number density, one concludes that the reheat temperature T_{RH} should be in the range 10^{11} to 10^{15}GeV [20]. This is a rather high value and leads to the gravitino problem in generic supersymmetric models. This is one reason alternative production mechanisms of these superheavy X-particles have been proposed [23,33,34]. However, our analysis show that the situation is much more promising. Making use of Eq. (27), the right amount of X-particles to explain the observed ultra-high energy cosmic rays is produced for

$$\left(\frac{T_{RH}}{10^{10}\,\mathrm{GeV}}\right) \simeq \left(\frac{g_*}{200}\right)^{3/14} \left(\frac{M_X}{10^{15}\,\mathrm{GeV}}\right), \tag{28}$$

where it has been assumed that $\langle\sigma|v|\rangle \sim M_X^{-2}$. Therefore, particles as massive as 10^{15} GeV may be generated during the reheating stage in abundances large enough to explain the ultra-high energy cosmic rays even if the reheat temperature satisfies the gravitino bound.

C Production During Preheating

Another way to produce WIMPZILLAS after inflation is in a preliminary stage of reheating called "preheating" [22], where nonlinear quantum effects may lead to an extremely effective dissipational dynamics and explosive particle production.

Particles can be created in a broad parametric resonance with a fraction of the energy stored in the form of coherent inflaton oscillations at the end of inflation released after only a dozen oscillation periods. A crucial observation for our discussion is that particles with mass up to 10^{15} GeV may be created during preheating [33,35,36], and that their distribution is nonthermal. If these particles are stable, they may be good candidates for WIMPZILLAS [37].

The main ingredient of the preheating scenario introduced in the early 1990s is the nonperturbative resonant transfer of energy to particles induced by the coherently oscillating inflaton fields. It was realized that this nonperturbative mechanism can be much more efficient than the usual perturbative mechanism for certain parameter ranges of the theory [22]. The basic picture can be seen as follows. Suppose there is a scalar field X with a coupling $g^2\phi^2X^2$ where ϕ is a homogeneous classical inflaton field. The mode equation for X field then can be written in terms of a redefined variable $\chi_k \equiv X_k a^{3/2}$ as

$$\ddot{\chi}_k(t) + [A + 2q\cos(2t)]\chi_k(t) = 0 \tag{29}$$

where A depends on the energy of the particle and q depends on the inflaton field oscillation amplitude. When A and q are constants, this equation is usually referred to as the Mathieu equation which exhibits resonant mode instability for certain values of A and q. In an expanding universe, A and q will vary in time, but if they vary slowly compared to the frequency of oscillations, the effects of resonance will remain. If the mode occupation number for the X particles is large, the number density per mode of the X particles will be proportional to $|\chi_k|^2$. If A and q have the appropriate values for resonance, χ_k will grow exponentially in time, and hence the number density will attain an exponential enhancement above the usual perturbative decay. This period of enhanced rate of energy transfer has been called preheating primarily because the particles that are produced during this period have yet to achieve thermal equilibrium.

This resonant amplification leads to an efficient transfer of energy from the inflaton to other particles which may have stronger coupling to other particles than the inflaton, thereby speeding up the reheating process and leading to a higher reheating temperature than in the usual scenario. Another interesting feature is that particles of mass larger than the inflaton mass can be produced through this coherent resonant effect. This has been exploited to construct a baryogenesis scenario [35] in which the baryon number violating bosons with masses larger than the inflaton mass are created through the resonance mechanism. A natural variation on this idea is to produce WIMPZILLAS by the same resonance mechanism.

Interestingly enough, what was found [37] is that in the context of a slow-roll inflation with the potential $V(\phi) = m_\phi^2\phi^2/2$ with the inflaton coupling of $g^2\phi^2X^2/2$, the resonance phenomenon is mostly irrelevant to WIMPZILLA production because too many particles would be produced if the resonance is effective. For the tiny amount of energy conversion needed for WIMPZILLA production, the coupling g^2 must be small enough (for a fixed M_X) such that the motion of the inflaton field at the transition out of the inflationary phase generates just enough nonadiabaticity in the mode frequency to produce WIMPZILLAS . The rest of the oscillations, damped by the expansion of the universe, will not contribute significantly to WIMPZILLA production as in the resonant case. In other words, the quasi-periodicity necessary for a true resonance phenomenon is not present in the case when only an extremely tiny fraction of the energy density is converted into WIMPZILLAS. Of course, if the energy scales are lowered such that a fair fraction of the energy density can be converted to WIMPZILLAS without overclosing the universe, this argument may not apply.

The main finding of a detailed treatment [37] is that WIMPZILLAS with a mass as large as $10^3 H_I$, where H_I is the value of the Hubble expansion rate at the end of inflation, can be produced in sufficient abundance to be cosmologically significant today.

If the WIMPZILLA is coupled to the inflaton ϕ by a term $g^2\phi^2X^2/2$, then the mode equation in Eq. 12 is now changed to

$$\omega_k^2 + k^2 + \left(M_X^2 + g^2\phi^2\right) a^2 \ , \tag{30}$$

again taking $\xi = 1/6$.

The procedure to calculate the WIMPZILLA density is the same as in Section IV A. Now, in addition to the parameter M_X/H_I, there is another parameter gM_{Pl}/H_I. Now in large-field models $H_I \sim 10^{13}$GeV, so M_{Pl}/H_I might be as large as 10^6. The choice of $g = 10^{-3}$ would yield $gM_{Pl}/H_I = 10^3$.

Fig. 5 (from [37]) shows the dependence of the WIMPZILLA density upon M_X/H_I for the particular choice $gM_{Pl}/H_I = 10^6$. This would correspond to $g \sim 1$ in large-field inflation models where $M_{Pl}/H_I = 10^6$, about the largest possible value. Note that $\Omega_X \sim 1$ obtains for $M_X/H_I \approx 10^3$. The dashed and dotted curves are two analytic approximations discussed in [37], while the solid curve is the numerical result. The approximations are in very good agreement with the numerical results.

Fig. 6 (also from [37]) shows the dependence of the WIMPZILLA density upon gM_{Pl}/H_I. For this graph M_X/H_I was chosen to be unity. This figure illustrates the fact that the dependence of $\Omega_X h^2$ on gM_{Pl}/H_I is not monotonic. For details, see [37].

D Production in Bubble Collisions

WIMPZILLAS may also be produced [34] in theories where inflation is completed by a first-order phase transition [38], in which the universe exits from a false-vacuum state by bubble nucleation [39]. When bubbles of true vacuum form, the energy of the false vacuum is entirely transformed into potential energy in the bubble walls. As the bubbles expand, more and more of their energy becomes kinetic as the walls become highly relativistic.

In bubble collisions the walls oscillate through each other [40] and their kinetic energy is dispersed into low-energy scalar waves [40,41]. We are interested in the potential energy of the walls, $M_P = 4\pi\eta R^2$, where η is the energy per unit area of a bubble wall of radius R. The bubble walls can be visualized as a coherent state of inflaton particles, so the typical energy E of the products of their decays is simply the inverse thickness of the wall, $E \sim \Delta^{-1}$. If the bubble walls are highly relativistic when they collide, there is the possibility of quantum production of nonthermal particles with mass well above the mass of the inflaton field, up to energy $\Delta^{-1} = \gamma M_\phi$, with γ the relativistic Lorentz factor.

Suppose for illustration that the WIMPZILLA is a fermion coupled to the inflaton field by a Yukawa coupling $g\phi\overline{X}X$. One can treat ϕ (the bubbles or walls) as a classical, external field and the WIMPZILLA as a quantum field in the presence of this source. The number of WIMPZILLAS created in the collisions from the wall potential energy is $N_X \sim f_X M_P/M_X$, where f_X parametrizes the fraction of the primary decay products in WIMPZILLAS. The fraction f_X will depend in general on the masses and the couplings of a particular theory in question. For the Yukawa coupling g, it is $f_X \simeq g^2 \ln(\gamma M_\phi/2M_X)$ [41,42]. WIMPZILLAS may be produced in bubble collisions out of equilibrium and never attain chemical equilibrium. Even with T_{RH} as low as 100 GeV, the present WIMPZILLA abundance would be $\Omega_X \sim 1$ if $g \sim 10^{-5}\alpha^{1/2}$. Here $\alpha^{-1} \ll 1$ is the fraction of the bubble energy at nucleation in the form of potential energy at the time of collision. This simple analysis indicates that the correct magnitude for the abundance of WIMPZILLAS may be naturally obtained in the process of reheating in theories where inflation is terminated by bubble nucleation.

V CONCLUSIONS

In this talk I have pointed out several ways to generate nonthermal dark matter. All of the methods can result in dark matter much more massive than the feeble little weak-scale mass thermal relics. The nonthermal dark matter may be as massive as the GUT scale, truly in the WIMPZILLA range.

The mass scale of the WIMPZILLAS is determined by the mass scale of inflation, more exactly, the expansion rate of the universe at the end of inflation. For large-field inflation models, that mass scale is of order 10^{13}GeV. For small-field inflation models, it may be less, perhaps much less.

The mass scale of inflation may one day be measured! In addition to scalar density perturbations, tensor perturbations are produced in inflation. The tensor perturbations are directly proportional to the expansion rate during inflation, so determination of a tensor contribution to cosmic background radiation temperature

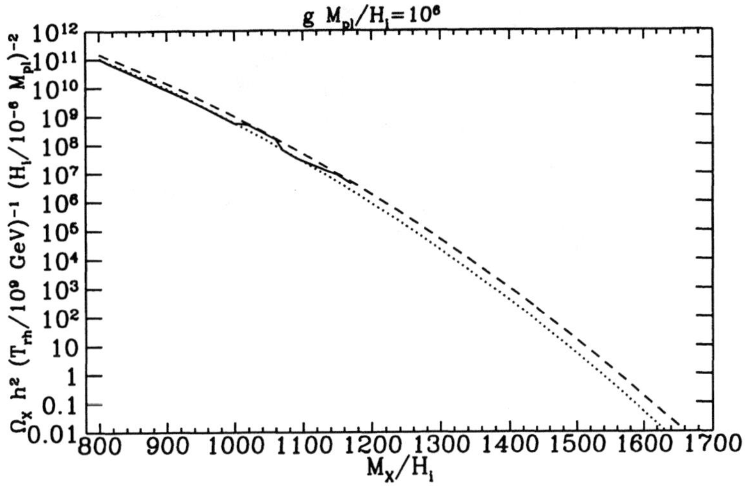

FIGURE 5. A graph of $\Omega_X h^2$ versus M_X/H_I for $gM_{Pl}/H_I = 10^6$. The solid curve is a numerical result, while the dashed and dotted curves are analytic approximations.

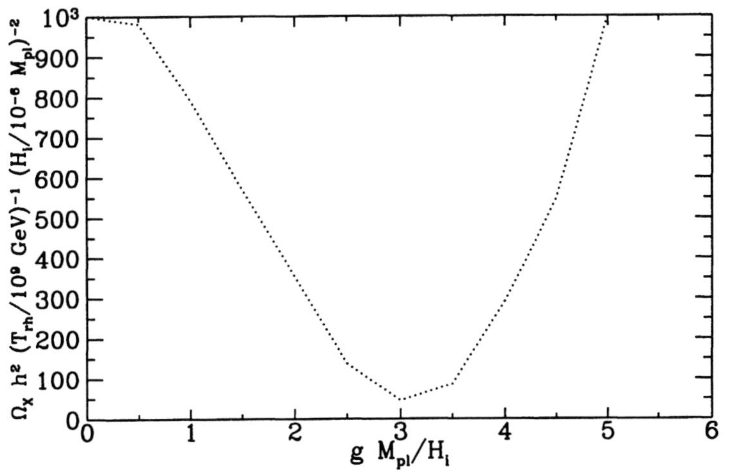

FIGURE 6. An illustration of the nonmonotonic behavior of the particle density produced with the variation of the coupling constant. The value of M_X/H_I is set to unity.

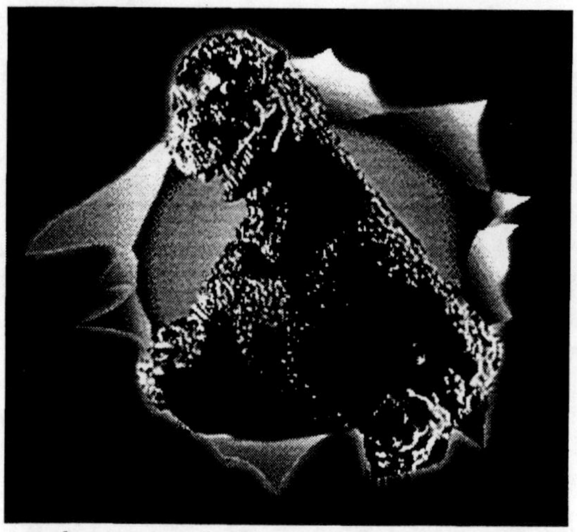

FIGURE 7. Dark matter may be much more massive than usually assumed, much more massive than wimpy WIMPS, perhaps in the WIMPZILLA class.

fluctuations would give the value of the expansion rate of the universe during inflation and set the scale for the mass of the WIMPZILLA.

Undoubtedly, other methods for WIMPZILLA production will be developed. But perhaps even with the present scenarios one should start to investigate methods for WIMPZILLA detection. While wimpy WIMPS must be color singlets and electrically neutral, WIMPZILLAS may be endowed with color and electric charge. This should open new avenues for detection and exclusion of WIMPZILLAS.

The lesson of the talk is illustrated in Fig. 7. WIMPZILLAS may surprise and be the dark matter, and we may learn that size does matter!

ACKNOWLEDGEMENTS

This work was supported by the Department of Energy and NASA (grant number NAG5-7092).

REFERENCES

1. B. Fuchs, "DARK98 Proceedings of the Second International Conference on Dark Matter in Astro and Particle Physics," eds. H V Klapdor-Kleingrothaus and L. Baudis, (Institute of Physics Publishing, Bristol and Philadelphia, 1999).
2. G. Evrard, "DARK98 Proceedings."
3. See, *e.g.*, A. Dekel, in Proceedings of the 3rd ESO-VLT Workshop on "Galaxy Scaling Relations: Origins, Evolution and Applications," ed. L. da Costa (Springer).
4. See, *e.g.*, J. A. Tyson, G. P. Kochanski, and I. P.. Dell'Antonio, astro-ph/9801193.
5. S. Sarkar, "DARK98 Proceedings."
6. G. Böerner, "DARK98 Proceedings."
7. E. W. Kolb and M. S. Turner, *The Early Universe*, (Addison-Wesley, Menlo Park, Ca., 1990).
8. K. Griest and M. Kamionkowski, Phys. Rev. Lett. **64**, 615 (1990).
9. B. Cabrera, "DARK98 Proceedings."
10. I. Liubarsky, "DARK98 Proceedings."
11. R. Bernabei, "DARK98 Proceedings."
12. Y. Ramachers, "DARK98 Proceedings."
13. L. Baudis, "DARK98 Proceedings."
14. J. Ellis, J.L. Lopez and D.V. Nanopoulos, Phys. Lett. **B247**, 257 (1990).
15. See, for instance, G. F. Giudice and R. Rattazzi, hep-ph/9801271.
16. S. Raby, Phys. Rev. D **56**, (1997).
17. K. Hamaguchi, Y. Nomura and T. Yanagida, hep-ph/9805346.
18. K. Benakli, J. Ellis and D.V. Nanopoulos, hep-ph/9803333.

19. D. V. Nanopoulos, "DARK98 Proceedings."
20. V. A. Kuzmin and V. A. Rubakov, Phys. Atom. Nucl. **61**, 1028 (1998).
21. M. Birkel and S. Sarkar, hep-ph/9804285.
22. L. A. Kofman, A. D. Linde and A. A. Starobinsky, Phys. Rev. Lett. **73**, 3195 (1994); S. Yu. Khlebnikov and I. I. Tkachev, Phys. Rev. Lett. **77**, 219 (1996); Phys. Lett. **B390**, 80 (1997); Phys. Rev. Lett. **79**, 1607 (1997); Phys. Rev. D **56**, 653 (1997); G. W. Anderson, A. Linde and A. Riotto, Phys. Rev. Lett. **77**, 3716 (1996); see L. Kofman, *The origin of matter in the Universe: reheating after inflation*, astro-ph/9605155, UH-IFA-96-28 preprint, 16pp., to appear in *Relativistic Astrophysics: A Conference in Honor of Igor Novikov's 60th Birthday*, eds. B. Jones and D. Markovic for a more recent review and a collection of references; see also L. Kofman, A. D. Linde and A. A. Starobinsky, Phys. Rev. D **56**, 3258 (1997); J. Traschen and R. Brandenberger, Phys. Rev. D **42**, 2491 (1990); Y. Shtanov, J. Traschen, and R. Brandenberger, Phys. Rev. D **51**, 5438 (1995).
23. D. J. H. Chung, Edward W. Kolb, and A. Riotto, hep-ph/9802238.
24. S. Fulling, Gen. Rel. and Grav. **10**, 807 (1979).
25. N. D. Birrell and P. C. W. Davies, *Quantum Fields in Curved Space* (Cambridge University Press, Cambridge, 1982).
26. D. M. Chitre and J. B. Hartle, Phys. Rev. D **16**, 251 (1977); D. J. Raine and C. P. Winlove, Phys. Rev. D **12**, 946 (1975); G. Schaefer and H. Dehnen, Astron. Astrophys. **54**, 823 (1977).
27. N. D. Birrell and P. C. W. Davies, J. Phys. A:Math. Gen. **13**, 2109 (1980).
28. D. J. H. Chung, E. W. Kolb, and A. Riotto, hep-ph/9809453.
29. D. H. Lyth and A. Riotto, hep-ph/9807278.
30. J. Ellis, J. Kim and D. V. Nanopoulos, Phys. Lett. **B145**, 181 (1984); L. M. Krauss, Nucl. Phys. **B227**, 556 (1983); M. Yu. Khlopov and A. D. Linde, Phys. Lett. **138B**, 265 (1984); J. Ellis, D. V. Nanopoulos, and S. Sarkar, Nucl. Phys. **B461**, 597 (1996).
31. R. J. Scherrer and M. S. Turner, Phys. Rev. **D31**, 681 (1985).
32. V. Berezinsky, M. Kachelriess and A. Vilenkin, Phys. Rev. Lett. **79**, 4302 (1997).
33. V. Kuzmin and I. I. Tkachev, hep-ph/9802304.
34. D. J. H. Chung, E. W. Kolb and A. Riotto, hep-ph/9805473.
35. E. W. Kolb, A. D. Linde and A. Riotto, Phys. Rev. Lett. **77**, 4290 (1996).
36. E. W. Kolb, A. Riotto and I. I. Tkachev, *Phys. Lett.* **B423**, 348 (1998).
37. D. J. H. Chung, hep-ph/9809489.
38. D. La and P. J. Steinhardt, Phys. Rev. Lett. **62**, 376 (1989).
39. A. H. Guth, Phys. Rev. D **23**, 347 (1981).
40. S. W. Hawking, I. G. Moss and J. M. Stewart, Phys. Rev. D **26**, 2681 (1982).
41. R. Watkins and L. Widrow, Nucl. Phys. **B374**, 446 (1992).
42. A. Masiero and A. Riotto, Phys. Lett. **B289**, 73 (1992).

Spectral functions in mathematics and physics

Klaus Kirsten[1]
Universität Leipzig
Fakultät für Physik und Geowissenschaften
Institut für Theoretische Physik
Augustusplatz 10/11
04109 Leipzig
Germany

Abstract.
Spectral functions relevant in the context of quantum field theory under the influence of spherically symmetric external conditions are analysed. Examples comprise heat-kernels, determinants and spectral sums needed for the analysis of Casimir energies. First, we summarize that a convenient way of handling them is to use the associated zeta function. A way to determine all its needed properties is derived. Using the connection with the mentioned spectral functions, we provide: i.) a method for the calculation of heat-kernel coefficients of Laplace-like operators on Riemannian manifolds with smooth boundaries and ii.) an analysis of vacuum energies in the presence of spherically symmetric boundaries and external background potentials.

I INTRODUCTION

There is a notorious appearance of spectral functions in many branches of mathematics and physics. These functions are associated with suitable sequences of numbers $\{\lambda_k\}_{k\in\mathbb{N}}$, which for most applications are eigenvalues of certain interesting, in most applications Laplace-like, operators. A rich source of problems where spectral functions are encountered is quantum field theory under the influence of "external conditions". External means that the condition is assumed to be known (as a function of space and time) and only appears in the equation of motion of other fields, which are to be quantized under these conditions. Given that the focus of interest is on the influence of the external conditions, these other fields are often assumed to be non-selfinteracting. It is in this setting that the present work is to be viewed.

Under the described circumstances the action of the considered theory will be quadratic in the quantized field. Using a path integral formulation to describe the underlying quantized theory, one encounters Gaussian functional integrals which lead to functional determinants formally defined as $\prod_{k=1}^{\infty}\lambda_k$. Making sense out of this kind of expressions is one of the basic themes of quantum field theory. The present contribution deals with scalar fields and will develop and apply techniques to analyse determinants arising when the external conditions are described by boundary conditions (Casimir effect) or when external scalar background fields are present. We restrict to scalar fields because the application of the provided ideas to spinor fields or electromagnetic fields is then immediate (see f.e. [1,2]).

In some more detail the situation considered will be described by the action (we assume an Euclidean formulation)

$$S[\Phi] = -\frac{1}{2}\int_{\mathcal{M}} dx\Phi(x)(\Box_E - V(x))\Phi(x), \tag{1.1}$$

[1] e-mail:Klaus.Kirsten@itp.uni-leipzig.de
Present address: Department of Physics and Astronomy, The University of Manchester, Oxford Road, Manchester UK M139PL

CP484, *Trends in Theoretical Physics II*, edited by H. Falomir, R. E. Gamboa Saraví, and F. A. Schaposnik

for a scalar field Φ in the background potential $V(x)$. Here, $\mathcal{M}$ is a D-dimensional Riemannian manifold and dx its volume element.

The corresponding field equation is

$$(\Box_E - V(x))\Phi(x) = 0. \tag{1.2}$$

If boundaries are present, the equations of motion are supplemented by boundary conditions to be specified later.

Physical properties of the systems are conveniently described by means of the path-integral functionals (an infinite normalization constant is neglected)

$$\mathcal{Z}[V] = \int D\Phi e^{-S[\Phi]}, \tag{1.3}$$

where the functionals are taken over all fields satisfying, if applicable, the boundary conditions imposed.

Under the circumstances described, the effective action can (at least formally) easily be computed to be (assume that there are no zero modes; if these are present they have to be omitted because otherwise the determinant is trivially zero),

$$\Gamma[V] = -\ln \mathcal{Z}[V] = \frac{1}{2}\ln\det\left[(-\Box_E + V(x))/\mu^2\right].$$

Here, μ is an arbitrary parameter with dimension of a mass, to make the argument of the logarithm dimensionless. The operators involved are thus Laplace-like operators, extensively dealt with afterwards. Let us write them in the unified form also used later, namely

$$P = -g^{\rho\nu}\nabla_\rho\nabla_\nu - E, \tag{1.4}$$

where $g^{\rho\nu}$ is the Riemannian metric of the manifold $\mathcal{M}$, ∇ is a connection and E an endomorphism defined on $\mathcal{M}$. We are thus confronted with the task of calculating expressions of the type

$$\Gamma[V] = \frac{1}{2}\ln[\det(P/\mu^2)]. \tag{1.5}$$

Clearly, expression (1.5) is not defined because the eigenvalues λ_n of P,

$$P\phi_n = \lambda_n\phi_n, \tag{1.6}$$

grow without bound for $n \to \infty$. Of course, there are various possible regularization procedures; let us mention only Pauli-Villars, dimensional regularization and zeta function regularization. Here, we will use zeta function regularization [3,4] because it is mathematically appealing as well as (probably) the most convenient one in the context of our work. The basic idea is to generalize the identity, valid for a $(N \times N)$-matrix P,

$$\ln\det P = \sum_{n=1}^{N}\ln\lambda_n = -\frac{d}{ds}\sum_{n=1}^{N}\lambda_n^{-s}|_{s=0} = -\frac{d}{ds}\zeta_P(s)|_{s=0},$$

with the zeta function

$$\zeta_P(s) = \sum_{n=1}^{N}\lambda_n^{-s},$$

to the differential operator P appearing in (1.6) by

$$\ln\det P = -\zeta_P'(0), \tag{1.7}$$

with

$$\zeta_P(s) = \sum_{n=1}^{\infty}\lambda_n^{-s}. \tag{1.8}$$

That this definition is in fact sensible is a result of deep mathematical theorems on the analytical structure of $\zeta_P(s)$ (explained in the following).

First of all, due to a classical theorem of Weyl [5], which says that for a second order elliptic differential operator the eigenvalues behave asymptotically for $n \to \infty$ as

$$\lambda_n^{D/2} \sim \frac{2^{D-1}\pi^{D/2} D\Gamma(D/2)}{\mathrm{vol}(\mathcal{M})} n,$$

the representation (1.8) of $\zeta_P(s)$ is valid for $\Re s > D/2$. In order to use definition (1.7), the question arises of how to analytically continue $\zeta_P(s)$ to the left and to determine its analytic structure. This is very elegantly done by using an integral representation of the Γ-function to write (see f.e. [6]), still for $\Re s > D/2$,

$$\zeta_P(s) = \frac{1}{\Gamma(s)} \int_0^\infty t^{s-1} K(t), \tag{1.9}$$

with the heat-kernel

$$K(t) = \sum_{n=1}^{\infty} e^{-\lambda_n t}.$$

For $t \to \infty$ the integral is well behaved due to the exponential damping coming from $K(t)$ (for simplicity we assume a positive definite operator P). Possible residues only arise from the $t \to 0$ behaviour of the integrand, and we need information on $K(t)$ for $t \to 0$. This is given by the heat-kernel expansion [7–9]

$$K(t) \sim \sum_{l=0,1/2,1,\dots}^{\infty} a_l(P) t^{l-D/2}, \tag{1.10}$$

with the heat-kernel coefficients $a_l(P)$ depending, of course, explicitly on the operator P together with the boundary conditions chosen. If the manifold has no boundary, the coefficients with half-integer index vanish. Splitting the integral for example into $\int_0^1 dt + \int_1^\infty dt$ the following connection is obtained [10],

$$\mathrm{Res}\ (\zeta_P(s)\Gamma(s))|_{s=D/2-l} = a_l(P), \tag{1.11}$$

or, showing the information contained more clearly, for $z = D/2, (D-1)/2, ..., 1/2, -(2n+1)/2, n \in \mathbb{N}_0$,

$$\mathrm{Res}\ \zeta_P(z) = \frac{a_{D/2-z}(P)}{\Gamma(z)}, \tag{1.12}$$

and for $q \in \mathbb{N}_0$,

$$\zeta_P(-q) = (-1)^q q! a_{D/2+q}(P). \tag{1.13}$$

This clearly shows that, for manifolds without boundaries, the poles are located at $z = D/2, D/2-1, ..., 1$ for D even, and $z = D/2, D/2-1, ..., 1/2, -(2n+1)/2, n \in \mathbb{N}_0$, for D odd. In addition, for D odd and $q \in \mathbb{N}_0$, one gets $\zeta_P(-q) = 0$. For manifolds with boundary, additional possible poles appear and here one finds poles at $D/2, (D-1)/2, ..., 1/2, -(2n+1)/2, n \in \mathbb{N}_0$. In all cases, $\zeta_P(s)$ is an analytical function in a neighbourhood of $s = 0$ such that eq. (1.7) is well defined. This definition was first used by the mathematicians Ray and Singer [11], when trying to give a definition of the Reidemeister-Franz torsion [12].

Let us stress that the expansion (1.10) of the heat-kernel and the connection (1.11) with the zeta function strongly depend on the assumptions made, and strictly they hold only if we are dealing with a *second order elliptic* differential operator on a *smooth compact Riemannian* manifold with a *smooth* boundary and *local elliptic* boundary conditions [9]. For example, in the context of admissible pseudo differential operators and global spectral boundary conditions, different powers of t can be involved in (1.10) and, in addition, $(\ln t)$-terms might appear. This was noted starting with [13] and later, for example, in [14–16], where the asymptotic expansions for these cases can be found. That indeed generically all possible terms are present (what means generically can be made precise) has been shown recently [17]. As a consequence, the simple poles of the zeta function may well be located at other points depending on the power of t appearing in the asymptotic small-t

expansion of $K(t)$, and, if there are $\ln t$-terms, double poles may be present as well. Also in this more general context, analogous relations to equation (1.11) may be stated [18], but here we will make no use of these and, therefore, we won't bother to state more details.

A case of particular interest is the manifold $\mathcal{M} = S^1 \times M_s$ with the operator

$$P = -\frac{\partial^2}{\partial \tau^2} + P_s, \tag{1.14}$$

and P_s Laplace-like. Imposing periodic boundary conditions in the τ-variable this is finite temperature quantum field theory for a scalar field and the perimeter β of the circle plays the role of the inverse temperature. Under the assumption that the potential V and the boundary conditions are static, P_s is a purely spatial operator and depends only on coordinates on M_s. In this context, it is known that the energy of the system is by definition

$$E = -\frac{\partial}{\partial \beta} \ln \mathcal{Z} = -\frac{1}{2}\frac{\partial}{\partial \beta}\zeta'_{P/\mu^2}(0),$$

and it seems natural to use [19]

$$E_{vac} = \lim_{\beta\to\infty} E = \frac{1}{2} FP\ \zeta_{P_s}(-1/2) - \frac{1}{2\sqrt{4\pi}} a_{D/2}(P_s) \ln \tilde{\mu}^2, \tag{1.15}$$

with the scale $\tilde{\mu} = (\mu e/2)$ as the definition of the vacuum energy. Here, FP means finite part for cases where $\zeta_{P_s}(-1/2)$ has a pole. Due to the presence of the arbitrary scale μ, it is seen that the vacuum energy is ambiguous, which generally causes problems to achieve a physically sensible answer. However, this is the way the Casimir energy is usually defined (see for example [20–24,19,25–30]), the idea for the derivation going back already to Gibbons [31]. The ambiguity may be discussed within renormalization group equations found by demanding [32,33]

$$\mu \frac{d}{d\mu} \Gamma[V] = 0.$$

Eq. (1.15) then clearly shows that the total energy of the system must contain all terms present in $a_{D/2}(P_s)$ in order to define the needed counterterms and running coupling constants. These terms describe the energy of the external fields or are the energy needed to get a model for the boundary conditions. If $a_{D/2}(P_s) \neq 0$, the Casimir energy is determined only up to terms proportional to $a_{D/2}(P_s)$ and this finite ambiguity can (in principle) only be eliminated by experiments [28]. A possible way to fix the ambiguity for massive fields will be discussed in Section IV. If, on the contrary, $a_{D/2}(P_s) = 0$, eq. (1.15) gives a unique answer for the energy.

Eq. (1.15) is also the definition one would naively use. To see this, write the Hamilton operator formally as

$$H = \sum_k E_k \left(N_k + \frac{1}{2} \right),$$

with N_k the number operator, to obtain for the vacuum energy

$$E_{vac} =< 0|H|0 >= \frac{1}{2} \sum_k E_k. \tag{1.16}$$

The regularization

$$\begin{aligned} E_{vac} &= \frac{\mu^{2s}}{2} \sum_k (E_k^2)^{1/2-s}|_{s=0} = \frac{\mu^{2s}}{2} \zeta_{P_s}(s-1/2)|_{s=0} \\ &= \frac{1}{2} FP\ \zeta_{P_s}(-1/2) + \frac{1}{2}\left(\frac{1}{s} + \ln \mu^2\right) \text{Res}\ \zeta_{P_s}(-1/2) \\ &= \frac{1}{2} FP\ \zeta_{P_s}(-1/2) - \left(\frac{1}{s} + \ln \mu^2\right) \frac{1}{2\sqrt{4\pi}} a_{D/2}(P_s) \end{aligned} \tag{1.17}$$

is clearly equivalent to eq. (1.15), the only difference being that renormalization now involves infinities.

Let us stress that, although we have focused on non-selfinteracting scalar fields, the calculation of determinants like in eq. (1.5) is of great relevance in several areas of modern theoretical physics. In many situations the potential is provided by classical solutions to nonlinear field equations. Quantizing a theory about these solutions leads to the same kind of determinants as discussed above [34]. The classical solutions involved may be monopoles [35,36], sphalerons [37] or electroweak Skyrmions [38–46]. The determinant in these external fields enters semiclassical transition rates as well as the nucleation of bubbles or droplets [47,48]. In general, the classical fields are inhomogeneous configurations and, as a rule, the effective potential approximation to the effective action, where quantum fluctuations are integrated out about a constant classical field, is not expected to be adequate. The derivative expansion [49] improves on this by accounting for spatially varying background fields. Being a perturbative approximation it has however its own limitations. Having in mind that even the classical solutions are often known only numerically, it is clear that it is desirable to have a numerical procedure to determine the quantum corrections. Some contributions in this direction are [50–53].

To summarize, as we have briefly described, the most relevant spectral functions in quantum field theory under external conditions are determinants, eq. (1.5), respectively the spectral sum, eq. (1.16), and the heat-kernel, eq. (1.10), describing the ambiguity in Casimir and ground state energy calculations. It is the aim of the present work, to provide and apply techniques for the analysis of all these spectral functions. As we have seen in the general introduction above, it is the zeta function associated with the spectrum which can be used as the organizing quantity for all the calculations to be done.

The starting point of our investigation was the aim to develop a machinery adequate to deal with spherically symmetric external conditions, the motivation being that many of the classical solutions have this property. In all spherically symmetric problems, the total angular momentum is a conserved quantity and, as a result, eigenvalues will be labelled by this angular momentum and an additional main quantum number. Compared to one-dimensional problems, where many calculations can be done exactly (see f.e. [54–57]), one encounters here the common technical complication of angular momentum sums. Any spherically symmetric example can help to understand the difficulties arising therefrom.

Intuitively, to impose boundary conditions on a spherical shell seems simpler than dealing with an arbitrary spherically symmetric external field, because eigenfunctions are known explicitly. However, let us stress that analytical expressions for the eigenvalues are not available which complicates the analysis considerably. Based on the knowledge of the eigenfunctions, an important achievement of our consideration is that various properties of spectral functions associated with a spectrum which is not known explicitly can be determined. We explain this procedure in detail for the spectrum of the Laplace operator on the three dimensional ball [58].

Having a very good (if not complete) knowledge of the zeta function at hand, and using the connections explained above, the obtained results are applied to the calculation of the associated heat-kernel coefficients and Casimir energies. The applications are grouped according to their complexity: i.) the heat-kernel coefficients, which are "local" quantities and whose determination is purely analytical, ii.) Casimir energies, which are non-local and where additional numerical work is needed.

For the mentioned reason we start, in Section 3, with the consideration of heat-kernel coefficients. An extremely important aspect of this calculation is that it is *not* just a special case calculation, but very rich information about heat-kernel coefficients for Laplace-like operators on arbitrary compact Riemannian manifolds evolves. Supplemented by other (already known) techniques involving conformal variations and the application of index theorems [9], it allowed for the determination of new coefficients for all types of boundary conditions [59–63] (including Dirichlet, Robin, mixed, oblique and spectral boundary conditions).

In Section IV we consider Casimir energies in the presence of spherically symmetric boundaries. Electromagnetic field fluctuations in a spherical shell were considered already a long time ago in the context of a simple model for an electron [64–67], where the zero-point energy of the electromagnetic field was supposed to stabilize the classical electron. However, as was first shown by Boyer [65], a repulsive force is the result and the simple model fails.

Our focus here will be on the influence of a mass m of the quantum field, and we will consider a scalar field with Dirichlet boundary conditions. Based on the only previously existing complete calculation for planar boundaries at a distance R [68], the general believe is that for $mR \gg 1$ the Casimir energy is exponentially small and thus of very short range. As we will see and explain, this is due to the planar boundaries and does not hold if the boundaries are curved. It might even happen that the Casimir energy changes sign as a function of the mass. This surprising result justifies to stand the additional complication due to the non-vanishing mass. Although we restrict for simplicity to scalar fields with Dirichlet boundary conditions, higher spin fields with various boundary conditions as f.e. MIT bag-boundary conditions can be treated as well along these lines (see f.e. [69–71]).

Afterwards, we analyze the influence of external background fields [72,73]. Although, in general, not even the eigenfunctions are explicitly given, this knowledge is replaced by the information available from scattering theory. So we will express the ground state energy by the Jost function, which is known (at least) numerically by solving the Lippmann-Schwinger equation. Applying the formalism developed in Section II for the treatment of the angular momentum sums, quite explicit results for the ground state energies are obtained. An example shows the typical features of the dependence of the ground state energy on the external potential.

Section II provides the basis for the rest of this work. However, the sections on the determination of heat-kernel coefficients and the calculation of Casimir and ground state energies are completely independent and can be read separately.

II DEVELOPING THE FORMALISM: SCALAR FIELD ON THE THREE DIMENSIONAL BALL

To start we focus our interest on the zeta function of the operator $(-\Delta+m^2)$ on the three dimensional ball $B^3=\{x\in \mathbb{R}^3; |x|\leq a\}$ endowed with Dirichlet boundary conditions. The zeta function is formally defined as

$$\zeta(s)=\sum_k \lambda_k^{-s}, \tag{2.1}$$

with the eigenvalues λ_k being determined through

$$(-\Delta+m^2)\phi_k(x)=\lambda_k\phi_k(x) \tag{2.2}$$

(k is in general a multiindex here), together with Dirichlet boundary conditions. It is convenient to introduce a spherical coordinate basis, with $r=|x|$ and the angles $\Omega=(\theta,\varphi)$. In these coordinates, a complete set of solutions of eq. (2.2) together with the boundary conditions may be given in the form

$$\phi_{l,m,n}(r,\Omega)=r^{-1/2}J_{l+1/2}(w_{l,n}r)Y_{lm}(\Omega), \tag{2.3}$$

with the Bessel function $J_{l+1/2}$ and $Y_{lm}(\Omega)$ the spherical surface harmonics [74]. The $w_{l,n}$ (>0) are determined through the boundary condition by

$$J_{l+1/2}(w_{l,n}a)=0. \tag{2.4}$$

In this notation, using $\lambda_{l,n}=w_{l,n}^2+m^2$, the zeta function can be given in the form

$$\zeta(s)=\sum_{n=0}^\infty\sum_{l=0}^\infty(2l+1)(w_{l,n}^2+m^2)^{-s}, \tag{2.5}$$

where $w_{l,n}$ is defined as the n-th root of the l-th equation (2.4). Here, the sum over n is extended over all possible roots $w_{l,n}$ on the positive real axis, and $(2l+1)$ is the number of independent harmonic polynomials, which defines the degeneracy of each value of l and n in three dimensions.

Eq. (2.5) may be written under the form of a contour integral on the complex plane,

$$\zeta(s)=\sum_{l=0}^\infty(2l+1)\int_\gamma \frac{dk}{2\pi i}\,(k^2+m^2)^{-s}\frac{\partial}{\partial k}\ln J_{l+1/2}(ka), \tag{2.6}$$

where the contour γ runs counterclockwise and must enclose all the solutions of (2.4) on the positive real axis, see Fig. 1 (for a similar treatment of the zeta function as a contour integral see [75,76,56]). Clearly, $(\partial/\partial k)\ln J_{l+1/2}(ka)=aJ'_{l+1/2}(ka)/J_{l+1/2}(ka)$, having simple poles at the solutions of eq. (2.4) and, by the residue theorem, definition (2.5) is reproduced. This representation of the zeta function in terms of a contour integral around some circuit γ on the complex plane, eq. (2.6), is the first step of our procedure.

As it stands, the representation (2.6) is valid for $\Re s>3/2$. However, we are especially interested in the properties of $\zeta(s)$ in the range $\Re s<3/2$ and thus, we need to perform the analytical continuation to the left. Before considering in detail the l-summation, we will first proceed with the k-integral alone.

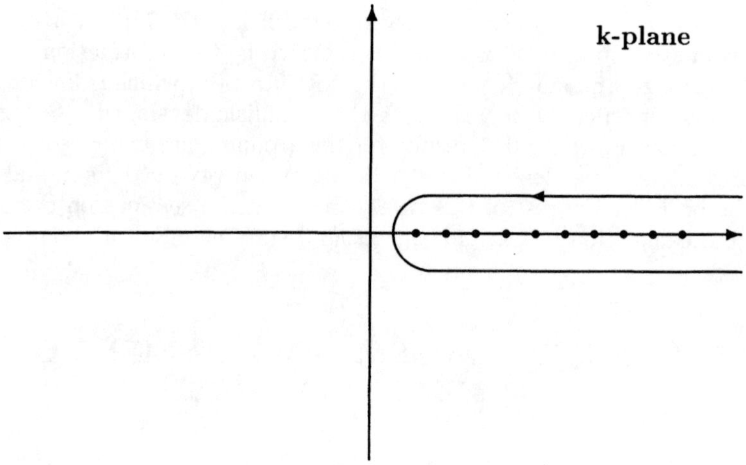

FIGURE 1. Contour γ

The first specific idea is to shift the integration contour and place it along the imaginary axis. For $k \to 0$, to leading order, one has the behaviour $J_\nu(k) \sim k^\nu/(2^\nu\Gamma(\nu+1))$ and, in order to avoid contributions coming from the origin $k=0$, we will consider (with $\nu = l + 1/2$) the expression

$$\zeta^\nu(s) = \int_\gamma \frac{dk}{2\pi i}\,(k^2+m^2)^{-s}\frac{\partial}{\partial k}\ln\left(k^{-\nu}J_\nu(ka)\right), \tag{2.7}$$

where the additional factor $k^{-\nu}$ in the logarithm does not change the result, for no additional pole is enclosed. Using the relations $J_\nu(ik) = e^{i\pi\nu}J_\nu(-ix)$ and $I_\nu(x) = e^{-ix\pi/2}J_\nu(ix)$ [77], one then easily obtains

$$\zeta^\nu(s) = \frac{\sin(\pi s)}{\pi}\int_m^\infty dk\,[k^2-m^2]^{-s}\frac{\partial}{\partial k}\ln\left(k^{-\nu}I_\nu(ka)\right) \tag{2.8}$$

valid in the strip $1/2 < \Re s < 1$. Here, the upper restriction, $\Re s < 1$, is imposed by the behaviour $(k-m)^{-s}$ of the integrand at the lower integration bound $k \to m$. For $k \to \infty$ one uses instead $I_\nu(k) \sim e^k/\sqrt{2\pi k}$ to find the behaviour k^{-2s} and, thus, the restriction $1/2 < \Re s$. It gets clear that, with $m = 0$, this representation is defined for no value of s and a slightly more difficult procedure has to be used [78,79]. We prefer to include the mass m and consider the limit $m \to 0$ (whenever needed) at a stage of the calculation where the limit will be well defined.

Given that the interesting properties of the zeta function (namely nearly all heat-kernel coefficients of $(-\Delta)$, the determinant and the Casimir energy) are encoded at the left of the strip $1/2 < \Re s < 1$, how can we find its analytical continuation to the left? As explained, the restriction $1/2 < \Re s$ is a result of the behaviour of the integrand for $k \to \infty$. If we subtract this asymptotic behaviour from the integrand in (2.8), the strip of convergence will certainly move to the left, and the hope will be that the asymptotic terms alone can be treated analytically, whereas $\zeta^\nu(s)$ with the asymptotic terms subtracted can be treated numerically. In detail, one needs to make use of the uniform asymptotic expansion of the Bessel function $I_\nu(k)$ for $\nu \to \infty$ as $z = k/\nu$ fixed [80] which, as we will see, guarantees that integration as well as summation lead to analytical expressions. One has

$$I_\nu(\nu z) \sim \frac{1}{\sqrt{2\pi\nu}}\frac{e^{\nu\eta}}{(1+z^2)^{\frac{1}{4}}}\left[1 + \sum_{k=1}^{\infty}\frac{u_k(t)}{\nu^k}\right], \tag{2.9}$$

with $t = 1/\sqrt{1+z^2}$ and $\eta = \sqrt{1+z^2} + \ln[z/(1+\sqrt{1+z^2})]$. The first few coefficients are listed in [80]; higher coefficients are immediate to obtain by using the recursion [80]

$$u_{k+1}(t) = \frac{1}{2}t^2(1-t^2)u_k'(t) + \frac{1}{8}\int_0^t d\tau\ (1-5\tau^2)u_k(\tau), \tag{2.10}$$

starting with $u_0(t) = 1$. As is clear, all the $u_k(t)$ are polynomials in t. Furthermore, the coefficients $D_n(t)$ defined by

$$\ln\left[1 + \sum_{k=1}^{\infty} \frac{u_k(t)}{\nu^k}\right] \sim \sum_{n=1}^{\infty} \frac{D_n(t)}{\nu^n}, \tag{2.11}$$

are easily found with the help of a simple computer program. For example one has

$$\begin{aligned} D_1(t) &= \frac{1}{8}t - \frac{5}{24}t^3, \\ D_2(t) &= \frac{1}{16}t^2 - \frac{3}{8}t^4 + \frac{5}{16}t^6. \end{aligned} \tag{2.12}$$

By adding and subtracting N leading terms of the asymptotic expansion, eq. (2.11), for $\nu \to \infty$, eq. (2.8) may be split into the following pieces,

$$\zeta^\nu(s) = Z^\nu(s) + \sum_{i=-1}^{N} A_i^\nu(s), \tag{2.13}$$

with the definitions

$$\begin{aligned} Z^\nu(s) = \frac{\sin(\pi s)}{\pi} \int_{ma/\nu}^{\infty} dz\ \left[\left(\frac{z\nu}{a}\right)^2 - m^2\right]^{-s} \frac{\partial}{\partial z} \{\ln\left[z^{-\nu} I_\nu(z\nu)\right] \\ - \ln\left[\frac{z^{-\nu}}{\sqrt{2\pi\nu}} \frac{e^{\nu\eta}}{(1+z^2)^{\frac{1}{4}}}\right] - \sum_{n=1}^{N} \frac{D_n(t)}{\nu^n}\right\}, \end{aligned} \tag{2.14}$$

and

$$A_{-1}^\nu = \frac{\sin(\pi s)}{\pi} \int_{ma/\nu}^{\infty} dz\ \left[\left(\frac{z\nu}{a}\right)^2 - m^2\right]^{-s} \frac{\partial}{\partial z} \ln\left(z^{-\nu} e^{\nu\eta}\right), \tag{2.15}$$

$$A_0^\nu = \frac{\sin(\pi s)}{\pi} \int_{ma/\nu}^{\infty} dz\ \left[\left(\frac{z\nu}{a}\right)^2 - m^2\right]^{-s} \frac{\partial}{\partial z} \ln(1+z^2)^{-\frac{1}{4}}, \tag{2.16}$$

$$A_i^\nu = \frac{\sin(\pi s)}{\pi} \int_{ma/\nu}^{\infty} dz\ \left[\left(\frac{z\nu}{a}\right)^2 - m^2\right]^{-s} \frac{\partial}{\partial z} \left(\frac{D_i(t)}{\nu^i}\right). \tag{2.17}$$

The essential idea is conveyed here by the fact that the representation (2.13) has the following anticipated properties. First, by considering the asymptotics of the integrand in eq. (2.14) for $z \to ma/\nu$ and $z \to \infty$, and by considering the behaviour of $Z^\nu(s)$ for $\nu \to \infty$, which is $\nu^{-2s-N-1}$, it can be seen that the function

$$Z(s) = \sum_{l=0}^{\infty} (2l+1) Z^\nu(s) \tag{2.18}$$

is analytic on the half plane $(1-N)/2 < \Re s$. (The integral alone has poles at $s = k \in \mathbb{N}$ coming from the $z \to ma/\nu$ region which is seen by writing $[(z\nu/a)^2 - m^2]^{-s} = (-1)^j(\Gamma(1-s)/\Gamma(j+1-s))(d^j/d(m^2)^j)[(z\nu/a)^2 - m^2]^{-s+j}$. But these are cancelled by the zeroes of the prefactor.) For this reason, it gives no contribution to the residue of $\zeta(s)$ in that range. Furthermore, for $s = -k$, $k \in \mathbb{N}_0$, $k < (-1+N)/2$, we have $Z(s) = 0$ and, thus, it also yields no contribution to the values of the zeta function at these points. This result means that the heat kernel coefficients, see eqs. (1.12) and (1.13), are just determined by the terms $A_i(s)$ with

$$A_i(s) = \sum_{l=0}^{\infty} (2l+1) A_i^\nu(s). \tag{2.19}$$

As they stand, the $A_i^\nu(s)$ in eqs. (2.15), (2.16) and (2.17) are well defined on the strip $1/2 < \Re s < 1$ (at least). We will now show that the analytic continuation in the parameter s to the whole of the complex plane, in terms of known functions, can be performed. Keeping in mind that $D_i(t)$ is a polynomial in t (see the examples in eq. (2.12)), all the $A_i^\nu(s)$ are in fact hypergeometric functions, which is seen by means of the basic relation [77]

$${}_2F_1(a,b;c;z) = \frac{\Gamma(c)}{\Gamma(b)\Gamma(c-b)} \int_0^1 dt\ t^{b-1}(1-t)^{c-b-1}(1-tz)^{-a}. \tag{2.20}$$

Let us consider first in detail $A_{-1}^\nu(s)$, $A_0^\nu(s)$, and the corresponding $A_{-1}(s)$, $A_0(s)$. One finds immediately that

$$A_{-1}^\nu(s) = \frac{m^{-2s}}{2\sqrt{\pi}} am \frac{\Gamma\left(s-\frac{1}{2}\right)}{\Gamma(s)}\ {}_2F_1\left(-\frac{1}{2}, s-\frac{1}{2}; \frac{1}{2}; -\left(\frac{\nu}{ma}\right)^2\right)$$
$$-\frac{\nu}{2} m^{-2s}, \tag{2.21}$$

$$A_0^\nu(s) = -\frac{1}{4} m^{-2s}\ {}_2F_1\left(1, s; 1; -\left(\frac{\nu}{ma}\right)^2\right)$$
$$= -\frac{1}{4} m^{-2s} \left[1 + \left(\frac{\nu}{ma}\right)^2\right]^{-s}, \tag{2.22}$$

where in the last equality we have used that ${}_2F_1(b,s;b;x) = (1-x)^{-s}$. These representations show the meromorphic structure of $A_{-1}^\nu(s), A_0^\nu(s)$ for all values of s.

The next step is to consider the summation over l. For $A_{-1}^\nu(s)$ this is best done using a Mellin-Barnes type integral representation of the hypergeometric functions, namely

$${}_2F_1(a,b;c;z) = \frac{\Gamma(c)}{\Gamma(a)\Gamma(b)} \frac{1}{2\pi i} \int_C dt\ \frac{\Gamma(a+t)\Gamma(b+t)\Gamma(-t)}{\Gamma(c+t)} (-z)^t, \tag{2.23}$$

where the contour is such that the poles of $\Gamma(a+t)\Gamma(b+t)/\Gamma(c+t)$ lie to the left of it and the poles of $\Gamma(-t)$ to the right [77]. The contour involved in eq. (2.21) is shown in the Fig. 2.

It is clear that one wishes to interchange the summation over l and the integration in (2.23) in order to arrive at a Hurwitz zeta function, which is defined as

$$\zeta_H(s;v) = \sum_{l=0}^{\infty} (l+v)^{-s}, \qquad \Re s > 1. \tag{2.24}$$

However, as is well known, one has to be very careful with this kind of manipulations, what has been realized and explained with great detail in [81–83]. Before interchanging the summation and integration one has to ensure that the resulting sum will be absolutely convergent along the contour C. Applying this criterium to $A_{-1}(s)$,

$$A_{-1}(s) = \sum_{l=0}^{\infty} (2l+1) \left[\frac{m^{-2s}}{2\sqrt{\pi}} am \frac{\Gamma\left(s-\frac{1}{2}\right)}{\Gamma(s)}\ {}_2F_1\left(-\frac{1}{2}, s-\frac{1}{2}; \frac{1}{2}; -\left(\frac{l+\frac{1}{2}}{ma}\right)^2\right)\right.$$
$$\left. -\frac{l+\frac{1}{2}}{2} m^{-2s}\right],$$

it turns out that we may interchange the $\sum_l$ and the integral in eq. (2.23) only if for the real part $\Re C$ of the contour the condition $\Re C < -1$ is satisfied. However, the integrand $\Gamma(-1/2+t)\Gamma(s-1/2+t)/\Gamma(1/2+t)$ has

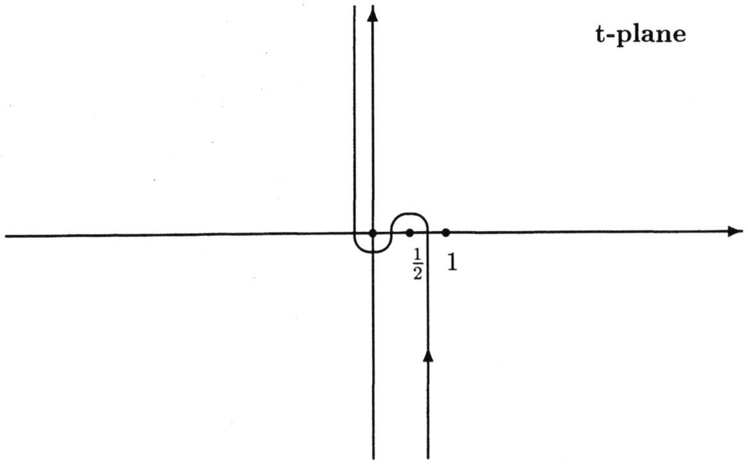

FIGURE 2. Contour C for eq. (2.21)

a pole at $t = 1/2$ (assume for the moment $\Re s \gg 1$ so that all arguments go through). Thus the contour C coming from $-i\infty$ must cross the real axis to the right of $t = 1/2$, and then once more between 0 and 1/2 (in order that the pole $t = 0$ of $\Gamma(-t)$ lies to the right of it), before going to $+i\infty$ (see Fig. 2). That is, before interchanging the sum and the integral we have to shift the contour C over the pole at $t = 1/2$ to the left, which cancels the (potentially divergent) second piece in $A_{-1}(s)$. This term, $-(l+1/2)m^{-2s}/2$, is the result of the factor $\ln k^{-\nu}$ introduced in eq. (2.7) to avoid contributions coming from the origin, and its crucial importance is made explicit in the above step.

Closing then the contour to the left, we end up with the following expression in terms of Hurwitz zeta functions (the contour at infinity does not contribute, which is seen by considering the asymptotics of the integrand)

$$A_{-1}(s) = \frac{a^{2s}}{2\sqrt{\pi}\Gamma(s)} \sum_{j=0}^{\infty} \frac{(-1)^j}{j!} (ma)^{2j} \frac{\Gamma\left(j+s-\frac{1}{2}\right)}{s+j} \zeta_H(2j+2s-2;1/2). \tag{2.25}$$

For $A_0(s)$, one only needs to use the binomial expansion in order to find

$$A_0(s) = -\frac{a^{2s}}{2\Gamma(s)} \sum_{j=0}^{\infty} \frac{(-1)^j}{j!} (ma)^{2j} \Gamma(s+j) \zeta_H(2j+2s-1;1/2). \tag{2.26}$$

The series are convergent for $|ma| < 1/2$. Finally, we need to obtain analytic expressions for the $A_i(s)$, $i \in \mathbb{N}$. As is easy to see, they are similar to the ones for $A_{-1}(s)$ and $A_0(s)$ above. We need to recall only that $D_i(t)$, eq. (2.11), is a polynomial in t,

$$D_i(t) = \sum_{b=0}^{i} x_{i,b} t^{i+2b}, \tag{2.27}$$

which coefficients $x_{i,a}$ are easily found by using eqs. (2.10) and (2.11) directly. Thus the calculation of $A_i^\nu(s)$ is essentially solved through the identity

$$\int_{ma/\nu}^{\infty} dz \left[\left(\frac{z\nu}{a}\right)^2 - m^2\right]^{-s} \frac{\partial}{\partial z} t^n = -m^{-2s} \frac{n}{2(ma)^n} \frac{\Gamma\left(s+\frac{n}{2}\right)\Gamma(1-s)}{\Gamma\left(1+\frac{n}{2}\right)} \times \nu^n \left[1+\left(\frac{\nu}{ma}\right)^2\right]^{-s-\frac{n}{2}}. \tag{2.28}$$

The remaining sum may be done as mentioned for $A_0(s)$, and we end up with

$$A_i(s) = -\frac{2a^{2s}}{\Gamma(s)} \sum_{j=0}^{\infty} \frac{(-1)^j}{j!} (ma)^{2j} \zeta_H(-1+i+2j+2s;1/2) \times \sum_{b=0}^{i} x_{i,b} \frac{\Gamma\left(s+b+j+\frac{i}{2}\right)}{\Gamma\left(b+\frac{i}{2}\right)}, \tag{2.29}$$

convergent once more for $|ma| < 1/2$. Restricting attention to the massless field, the asymptotic contributions take the surprisingly simple form

$$\begin{aligned} A_{-1}(s) &= \frac{a^{2s}}{2\sqrt{\pi}} \frac{\Gamma(s-1/2)}{\Gamma(s+1)} \zeta_H(2s-2;1/2), \\ A_0(s) &= -\frac{a^{2s}}{2} \zeta_H(2s-1;1/2), \\ A_i(s) &= -\frac{2a^{2s}}{\Gamma(s)} \zeta_H(i-1+2s;1/2) \sum_{b=0}^{i} x_{i,b} \frac{\Gamma\left(s+b+j+\frac{i}{2}\right)}{\Gamma\left(b+\frac{i}{2}\right)}. \end{aligned} \tag{2.30}$$

In summary, the analytical structure of the zeta function for the problem considered is made completely explicit. Eq. (2.30) allows for the direct determination of residues, function values and derivatives at whatever values of s needed. Clearly, all ingredients can be easily dealt with by Mathematica and calculations can be easily automized. The remaining piece to obtain the full zeta function, eq. (2.18) and (2.14), is suitable for numerical evaluation for values of s in the half plane mentioned. Furthermore, an analytical treatment of $Z'(0)$ is possible, which leads to the determinant $\zeta'(0)$ of the Laplacian on the ball in terms of elementary Hurwitz zeta functions [84,59]. For various different approaches in a similar context see also [85,86,76,87,88], with possible applications in quantum cosmology [89,90].

III CALCULATION OF HEAT-KERNEL COEFFICIENTS VIA SPECIAL CASES

Let us now come to the first application of our analysis in Section II, namely the calculation of heat-kernel coefficients for Laplace-like operators on smooth manifolds with smooth boundaries. Our main emphasis is on the determination of the boundary contribution to the heat kernel coefficients. This is motivated because it is here that the special case of the ball gives rich information. But it is also justified because the calculation of the volume part is nowadays nearly automatic [91–93] and these terms do not depend on the boundary conditions [94] and are thus known already for all problems to be dealt with.

Concerning the relevance of heat-kernel coefficients, in the Introduction we have already mentioned that a knowledge of these coefficients is equivalent to a knowledge of the one-loop renormalization equations in various theories [32], which provides one reason for the consideration of heat-kernel coefficients in physics. In addition, if an exact evaluation of relevant quantities is not possible, asymptotic expansions (with respect to inverse masses, slowly varying background fields, high temperature,...) are often very useful and naturally given in terms of heat-kernel coefficients. An especially important link between physics and mathematics is provided by index theorems [9], again, with the well-known connection to the heat equation. But, in mathematics, the interest extends to basically all of Geometric Analysis, including analytic torsion [12,11], characteristic classes [9], sharp inequalities of borderline Sobolev and Moser-Trudinger type [95], etc.

We will start the Section by giving what can be named the general form of the heat-kernel coefficients for Dirichlet and Robin boundary conditions. (We will explain the approach for these boundary conditions; comments on various other boundary conditions are given at the end of the Section.) As we will see these are built from certain geometrical invariants with unknown numerical coefficients. Relations between the unknown coefficients can be derived by conformal transformation techniques most systematically used by Branson and Gilkey [96]. However, in order to determine the numerical coefficients, additional information is needed. The product formula gives a certain subset of the numerical coefficients but in general by far not enough to complete the calculation by using the conformal techniques [96]. Especially the group of terms containing the extrinsic curvature is not even touched by the product formula and our calculation on the ball will turn out to be very valuable. Combined with the conformal techniques, the application of index theorems, and additional examples, the conglomerate of all methods allows for the determination of (at least) the leading heat-kernel coefficients for all classical boundary conditions.

Let us clearly state that the coefficients we are going to "determine", namely the coefficients up to $a_{3/2}$ for Dirichlet and Robin boundary conditions, are known already for a long time, see. f.e. [96]. It is just to explain the ideas in detail that we have chosen the first coefficients and the simplest boundary conditions. But in fact, the method has been successfully applied to $a_{5/2}$ containing more than 100 terms [97] (see also [98]), showing the effectiveness of the procedure employed.

A General form of the coefficients for Dirichlet and Robin boundary conditions

In order to be as self-contained as possible, in the following we are going to summarize some basic properties of heat-kernel coefficients on manifolds with boundary. We follow [96].

Let $\mathcal{M}$ be a compact $D = (d+1)$-dimensional Riemannian manifold with boundary $\partial\mathcal{M}$. Let V be a smooth vector bundle over $\mathcal{M}$ equipped with a connection ∇^V and finally let E be an endomorphism of V. Our interest is then in Laplace-type operators of the form (1.4),

$$P = -g^{ij}\nabla_i^V \nabla_j^V - E. \tag{3.1}$$

Let us mention that every second order elliptic differential operator on $\mathcal{M}$ with leading symbol given by the metric can be put in this form. We will see this explicitly below, starting with eq. (3.16).

If there is no boundary, it is well known that P defines a symmetric operator. If, however, there is a boundary present, Green's theorem says

$$(v, Pw)_{L^2} - (Pv, w)_{L^2} = \int_{\mathcal{M}} dx (v^\dagger Pw - (Pv)^\dagger w) = \int_{\partial\mathcal{M}} dy (v^\dagger_{;m} w - v^\dagger w_{;m}), \tag{3.2}$$

with dx and dy the volume elements on $\mathcal{M}$ and $\partial\mathcal{M}$, and $v_{;m}$ the normal covariant derivative of v with respect to the *exterior* normal N to the boundary $\partial\mathcal{M}$. In order that P be symmetric one has to impose boundary conditions. Obvious possibilities are the classical Dirichlet or Robin boundary conditions,

$$\mathcal{B}^- \phi \equiv \phi|_{\partial\mathcal{M}} \quad \text{and} \quad \mathcal{B}^+_S \phi \equiv (\phi_{;m} - S\phi)\,|_{\partial\mathcal{M}}, \tag{3.3}$$

with S a hermitian endomorphism of V defined on $\partial\mathcal{M}$. Clearly, in these cases the argument of the integral over the boundary in (3.2) vanishes and, with these boundary conditions, a symmetric operator P is defined.

Another possibility is to impose a mixture of Dirichlet and Robin boundary conditions. In order to do this, a suitable splitting of $V = V_- \oplus V_+$ is needed and in V_- one imposes Dirichlet and in V_+ Robin boundary conditions. Also here, the integrand itself vanishes.

However, a further possibility is that the integrand in (3.2) does not vanish but equals a boundary divergence. This condition involves tangential derivatives and, in the mathematical literature, it is sometimes referred to as oblique. Although these last boundary conditions have been the subject of classical analysis (see, f.e. [99–101]), very little is known about the associated heat equation asymptotics. We will comment on the application of the techniques to all these boundary conditions at the end of the section, but concentrate now on conditions (3.3).

To have a uniform notation, we set $S = 0$ for Dirichlet boundary conditions and write $B^{\mp}_S$. If F is a smooth function on $\mathcal{M}$, there is an asymptotic series as $t \to 0$ of the form

$$\mathrm{Tr}_{L^2}\left(Fe^{-tP}\right) \approx \sum_{n\geq 0} t^{n-\frac{D}{2}} a_n(F, P, \mathcal{B}^{\mp}_S), \tag{3.4}$$

where the $a_n(F, P, \mathcal{B}^{\mp}_S)$ are locally computable [9].

The use of the smearing function F is important for at least three reasons. First, near the boundary the heat-kernel behaves like a distribution and, by studying $\mathrm{Tr}_{L^2}(Fe^{-tP})$, this local behaviour is recovered. As an example consider $\partial\mathcal{M} = \emptyset$. For $F = 1$ volume divergences are integrated away, which is not possible if a smearing function is present. A second reason is that for the functorial formalism to be described [96], the smearing function is at the heart of this method. Finally, it is exactly this *smeared* coefficient appearing in the integration of conformal anomalies that is relevant for several physical applications [102–111].

To state the general form of the coefficients in (3.4), let us introduce some notation. Here and in the following $F[\mathcal{M}] = \text{Tr} \int_{\mathcal{M}} dx\, F(x)$ and $F[\partial\mathcal{M}] = \text{Tr} \int_{\partial\mathcal{M}} dyF(y)$, with Tr the fiber trace. In addition, ";" denotes differentiation with respect to the Levi-Civita connection of $\mathcal{M}$ and ":" covariant differentiation tangentially with respect to the Levi-Civita connection of the boundary. Furthermore, Ω is the curvature of the connection ∇^V, $[\nabla_i^V, \nabla_j^V] = \Omega_{ij}$, and R_{ijkl}, R_{ij}, R, are as usual Riemann tensor, Ricci tensor and Riemann scalar. Finally let $N^\nu(F) = F_{;m\ldots}$ be the νth normal covariant derivative. Then there exist local formulae $a_n(x,P)$ and $a_{n,\nu}(y,P,\mathcal{B}_S^\mp)$ so that [112]

$$a_n(F,P,\mathcal{B}_S^\mp) = \{Fa_n(x,P)\}[\mathcal{M}] + \{\sum_{\nu=0}^{2n-1} N^\nu(F)a_{n,\nu}(y,P,\mathcal{B}_S^\mp)\}[\partial\mathcal{M}]. \tag{3.5}$$

Important homogeneity properties follow from the Seeley calculus [94], so one has for $0 < c \in \mathbb{R}$ [112]

$$a_n(x,c^{-2}P) = c^{-2n}a_n(x,P), \quad a_{n,\nu}(y,c^{-2}P,c^{-1}\mathcal{B}_S^\mp) = c^{-(2n-\nu)}a_{n,\nu}(y,P,\mathcal{B}_S^\mp). \tag{3.6}$$

Physicists would say that P carries dimension *length*$^{-2}$, thus in order that e^{-tP} makes sense, t has to have dimension of *length*^{2}. Then, eq. (3.4) is dimensionless and $a_n(F,P,\mathcal{B}_S^\mp)$ must have dimension of *length*$^{-D+2n}$. As a result, $a_n(x,P_\mathcal{B})$ has dimension of length to the power $2n$ and $a_{n,\nu}(y,P_\mathcal{B},\mathcal{B}_S^\mp)$ to the power $2n-\nu$ which is equivalent to the above.

The interior invariants $a_n(x,P)$ are built universally and polynomially from the metric tensor, its inverse, and the covariant derivatives of R, Ω, and E. By Weyl's work on the invariants of the orthogonal group [113], these polynomials can be formed using only tensor products and contraction of tensor arguments (indices). If A is a monomial term of $a_n(x,P)$ of degree (k_R, k_Ω, k_E) in (R,Ω,E), and if k_∇ explicit covariant derivatives appear in A, then by the homogeneity property of $a_n(x,P)$,

$$2(k_R + k_\Omega + k_E) + k_\nabla = 2n.$$

When considering the boundary invariants $a_{n,\nu}(y,P,\mathcal{B}_S^\mp)$ we must also introduce the second fundamental form $K_{ab} = (\nabla_{e_a}e_b, N)$, $K = K_a^a$, where $\{e_1,\ldots,e_d\}$ is an orthonormal frame of $T(\partial\mathcal{M})$, the tangent bundle of the boundary and, when considering Robin boundary conditions, we must also consider the tensor S. Given that these are defined only at the boundary, we only differentiate $\{K,S\}$ tangentially. We use Weyl's [113] theorem again to construct invariants. The structure group now is $O(D-1)$, and the normal N plays a distinguished role. If A is a monomial term of $a_{n,\nu}(y,P,\mathcal{B}_S^\mp)$ of degree $(k_R,k_\Omega,k_E,k_K,k_S)$ in (R,Ω,E,K,S), and if k_∇ explicit covariant derivatives appear in A, then once more by homogeneity

$$2(k_R + k_\Omega + k_E) + k_K + k_S + k_\nabla = 2n - \nu.$$

By constructing a basis for the space of invariants of a given homogeneity, we write down the following general form of the heat-kernel coefficients [96],

$$a_0(F,P,\mathcal{B}_S^\mp) = (4\pi)^{-D/2}F[\mathcal{M}], \tag{3.7}$$

$$a_{1/2}(F,P,\mathcal{B}_S^\mp) = \delta(4\pi)^{-d/2}F[\partial\mathcal{M}], \tag{3.8}$$

$$a_1(F,P,\mathcal{B}_S^\mp) = (4\pi)^{-D/2}6^{-1}\left\{(6FE + FR)[\mathcal{M}] + (b_0FK + b_1F_{;m} + b_2FS)[\partial\mathcal{M}]\right\}, \tag{3.9}$$

$$\begin{aligned} a_{3/2}(F,P,\mathcal{B}_S^\mp) = \frac{\delta}{96(4\pi)^{d/2}}\{F\big(c_0E + c_1R + c_2R_{mm} + c_3K^2 + c_4K_{ab}K^{ab} + c_7SK + c_8S^2\big) \\ + F_{;m}(c_5K + c_9S) + c_6F_{;mm}\}[\partial\mathcal{M}]. \end{aligned} \tag{3.10}$$

All numerical constants involved have the very important property of being independent of the dimension D [112]. A direct proof has been given in [96] and this goes as follows. First one shows a product formula for heat-kernel coefficients. Let $\mathcal{M} = \mathcal{M}_1 \times \mathcal{M}_2$ and $P = P_1 \otimes 1 + 1 \otimes P_2$ and $\partial\mathcal{M}_2 = \emptyset$ and let S only depend on coordinates in $\mathcal{M}_1$. Then

$$\begin{aligned} a_n(x,P) &= \sum_{p+q=n} a_p(x_1,P_1)a_q(x_2,P_2), \\ a_{n,\nu}(y,P,\mathcal{B}_S^\mp) &= \sum_{p+q=n} a_{p,\nu}(y_1,P_1,\mathcal{B}_S^\mp)a_q(x_2,P_2). \end{aligned} \tag{3.11}$$

This is a purely formal computation because by separation of variables the heat-kernel of the operator P gets the product of the heat-kernels of P_1 and P_2 and the result follows by just comparing powers of t.

To avoid the appearance of factors of $\sqrt{4\pi}$ normalize for the moment $a_0(x,P) = 1$. Application of formula (3.11) to $(\mathcal{M}_2, P_2) = (S^1, -\partial^2/\partial\theta^2)$ leads to $a_n(x_1, P_1) = a_n((x_1,\theta), P)$ and $a_{n,\nu}(y_1, P_1, \mathcal{B}_S^{\mp}) = a_{n,\nu}((y_1,\theta), P, \mathcal{B}_S^{\mp})$, because $a_0(\theta, P_2) = 1$ and $a_q(\theta, P_2) = 0$ for $q > 0$. However, invariants formed by contractions of indices are restricted from $\mathcal{M}_1 \times S^1$ to $\mathcal{M}_1$ by restricting the range of summation, but have the same appearance. This shows that the numerical constants are independent of the dimension.

We are thus left with the task of the determination of the universal numerical constants by whatever method. An obvious (rich) source of information are special case calculations. Let us discuss this in detail for the example of the ball in order to motivate the calculation of the coefficients for this setting. The way to deal with the three dimensional ball and Dirichlet boundary conditions has been shown in Section II. In Section III B we will show how these results can be generalized to arbitrary dimensions $D = d+1$. In addition, the way to deal with Robin boundary conditions with a constant endomorphism S is explained. For these manifolds we will have $K_a^b = \delta_a^b$. As a result we get $K = d$, $K_{ab}K^{ab} = d$, $K^2 = d^2$, and so on. The polynomials, traces and contractions of K_{ab} give a polynomial in the dimension d. In addition, for the example of the ball, we have $R_{abce} = 0$, and we have chosen $P = -\Delta_{\mathcal{M}}$, thus $E = 0$. Finally, we included no smearing function and have $F = 1$. In this setting, restricting (3.8) to the ball, we have

$$a_{1/2}(1, -\Delta_{\mathcal{M}}, \mathcal{B}_S^{\mp}) = \delta (4\pi)^{-d/2} |S^d|,$$

with the volume $|S^d| = 2\pi^{(d+1)/2}/\Gamma((d+1)/2)$ of the d-sphere. Calculating the heat-kernel coefficients explicitly one will find the "unknown" numerical constants δ and then knows $a_{1/2}(F, P, \mathcal{B}_S^{\mp})$ for an arbitrary manifold by eq. (3.8). Passing on to a_1, on the ball we have

$$a_1(1, -\Delta_{\mathcal{M}}, \mathcal{B}_S^{\mp}) = (4\pi)^{-D/2} 6^{-1} |S^d| \left(b_0 d + b_2 S\right).$$

Just by comparing powers of d and S one can determine b_0 and b_2 from the explicit $a_1(1, -\Delta_{\mathcal{M}}\ \mathcal{B}_S^{\mp})$ on the ball. Let us stress that, if we include a smearing function $F(r)$ into the formalism, then $F_{;m} = (d/dr)F(r)$ is the derivative with respect to the exterior derivative, and we would have determined also b_1. By application of (3.9) we see that one can get $a_1(F, P, \mathcal{B}_S^{\mp})$ by just having the result on the ball.

Continuing with the same argumentation, in eq. (3.10) one can determine c_3, c_4, c_7 and c_8 for $F = 1$ and furthermore c_5, c_9 and c_6 including an $F(r)$. Let us stress that c_3 and c_4 both can be determined only because we will perform our calculation in arbitrary dimension (this observation gets more important for the higher coefficients). Thus only 3 of 10 unknowns are left and it gets clear that the special case calculation chosen contains rich information. However, one also realizes, and this was, of course, clear from the beginning that the example can not determine the full coefficients, because E and the Riemann tensor vanishes. Both aspects can be improved a bit by including a mass (as we have done) and by dealing with a so-called bounded generalized cone [59]. But the information obtained thereby is also very easily obtained by an application of the product formula (3.11) and we will not give details of this generalization here. However, product manifolds have vanishing normal components of the Riemann tensor (their appearance starts with $a_{3/2}$) and the corresponding universal constants have to be determined by different means.

An extremely effective method to continue at this point is to study the functorial properties of the heat-kernel coefficients [114,96]. We summarize here the main points in order to allow for a self-consistent reading. Consider the one-parameter family of differential operators

$$P(\epsilon) = e^{-2\epsilon F} P \tag{3.12}$$

and boundary operators

$$\mathcal{B}_S^{\mp}(\epsilon) = e^{-\epsilon F} \mathcal{B}_S^{\mp}. \tag{3.13}$$

Here, ϵ is a real-valued parameter and F, as before, is a function. Eq. (3.13) guarantees that the boundary condition remains invariant along the one-parameter family of operators (see below, eq. (3.19)). Then one might ask what is the dependence of the heat-kernel coefficients on the parameter ϵ. The relevant transformation behaviour is described by the following

Lemma:

$$(a)\frac{d}{d\epsilon}\mid_{\epsilon=0} a_n(1,P(\epsilon),\mathcal{B}_S^{\mp}(\epsilon)) = (D-2n)a_n(F,P,\mathcal{B}_S^{\mp}), \tag{3.14}$$

$$(b)\text{If } D=2n+2, \text{ then } \frac{d}{d\epsilon}\mid_{\epsilon=0} a_n\left(e^{-2\epsilon f}F,P(\epsilon),\mathcal{B}_S^{\mp}(\epsilon)\right) = 0. \tag{3.15}$$

Proceeding formally, (a) is proven by considering

$$\frac{d}{d\epsilon}\mid_{\epsilon=0} \mathrm{Tr}_{L^2}\left(e^{-tP(\epsilon)}\right) = -t\mathrm{Tr}_{L^2}\left(\left[\frac{d}{d\epsilon}\mid_{\epsilon=0} P(\epsilon)\right]e^{-tP}\right)$$

$$= 2t\mathrm{Tr}_{L^2}\left(FPe^{-tP}\right) = -2t\frac{\partial}{\partial t}\mathrm{Tr}_{L^2}\left(Fe^{-tP}\right),$$

and comparing powers of t in the asymptotic expansion. For the necessary justification of the analytic steps see [112]. Part (b) is much the same by starting with $(d/d\epsilon)|_{\epsilon=0}\mathrm{Tr}_{L^2}(e^{-2\epsilon f}Fe^{-tP(\epsilon)})$.

This Lemma will be applied in Section III C; let us here only explain the way it determines universal constants. First, we obviously need the heat-kernel coefficients for the operator $P(\epsilon)$, so let us see how the family (3.12) of operators can be generated. Let P be an arbitrary second order differential operator with leading symbol given by the metric tensor. In local coordinates we have

$$P = -\left(g^{ij}\frac{\partial^2}{\partial x^i \partial x^j} + P^k\frac{\partial}{\partial x^k} + Q\right). \tag{3.16}$$

Obviously, $P(\epsilon)$ is obtained by defining

$$g^{ij}(\epsilon) = e^{-2\epsilon F}g^{ij}, \quad P^k(\epsilon) = e^{-2\epsilon F}P^k, \quad Q(\epsilon) = e^{-2\epsilon F}Q.$$

In order to obtain the heat-kernel coefficients of $P(\epsilon)$ in the form (3.7)—(3.10), write P invariantly in the form (3.1), $P = -(g^{ij}\nabla_i^V\nabla_j^V + E)$. Let ω_l be the connection 1 form and consequently $\Omega_{ij} = [\nabla_i^V,\nabla_j^V] = \omega_{j,i} - \omega_{i,j} + \omega_i\omega_j - \omega_j\omega_i$ with "," the partial derivative. Comparing the two different representations of P one finds

$$\omega_l = \frac{1}{2}g_{il}\left(P^i + g^{jk}\Gamma^i_{jk}\right), \tag{3.17}$$

$$E = Q - g^{ij}\left(\omega_{i,j} + \omega_i\omega_j - \omega_k\Gamma^k_{ij}\right), \tag{3.18}$$

with the Christoffel symbols Γ^i_{jk}. As a result, this defines

$$\omega_l(\epsilon) = \omega_l + \frac{1}{2}\epsilon(2-D)F_{;l} \tag{3.19}$$

$$\Omega_{ij}(\epsilon) = \Omega_{ij} \tag{3.20}$$

$$E(\epsilon) = e^{-2\epsilon F}(E + \frac{1}{2}(D-2)\epsilon\Delta_{\mathcal{M}}F + \frac{1}{4}(D-2)^2\epsilon^2F_{;k}F^{k}_{;}). \tag{3.21}$$

This shows that the leading $a_n(f,P(\epsilon),\mathcal{B}_S^{\mp}(\epsilon))$ are given by eqs. (3.7)—(3.10) once the above definitions are used and once all geometrical tensors and covariant derivatives are calculated with respect to the metric $g_{ij}(\epsilon)$ (for many useful relations see [96]). Let us here only mention that Dirichlet boundary conditions are obviously conformally invariant, and that with

$$S(\epsilon) = e^{-\epsilon F}\left(S - \epsilon\frac{D-2}{2}F_{;m}\right)$$

the same holds for Robin conditions. This is seen immediately from eq. (3.19) and the boundary condition (3.3). As a result, $(d/d\epsilon)|_{\epsilon=0}a_k(1,P(\epsilon),\mathcal{B}_S^{\mp}(\epsilon))$ will have the same appearance as $a_k(F,P,\mathcal{B}_S^{\mp})$, and Lemma (3.14) will give relations among the universal constants, as well as (3.15) does. To make it very clear, look at $a_{3/2}(F,P,\mathcal{B}_S^{\mp})$. Then $(d/d\epsilon)E(\epsilon)$ contains a term $F_{;mm}$ being part of ΔF, see (3.21). Eq. (3.14) then states that the numerical constant c_0 is connected with c_6 (more invariants are involved however). Let us stress already here, that due care must be taken that only *independent* terms are compared in eqs. (3.14) and (3.15) and that partial integrations (or more involved manipuliations) may be necessary to see that apparently independent terms are actually dependent.

On its own, the functorial method is unable to determine the coefficients fully. But, given a subset of numerical coefficients, found f.e. by special case calculations, the method provides the required information with relative ease. For that reason, we start by applying the product Lemma (3.11) and, by calculating the heat-kernel coefficients for the Laplacian on the ball, determine (a subset of the) universal constants and complete the calculation by use of the functorial properties.

In the last step it will turn out that the generalization of the calculations in Section III B to the smeared zeta function is essential. This is, as seen in eq. (3.14), because the functorial techniques (apart from other things) yield relations between the smeared and non-smeared case; the information one can get on the "smeared side" is crucial to find the full "non-smeared" side. The way this can actually be done is explained in [61].

B Scalar field on the D-dimensional ball

As briefly explained, for the applications of our calculations to the heat-equation asymptotics, it will be very important that results in arbitrary dimensions are available. For that reason, let us briefly explain how the three dimensional calculation of Section II is generalized to arbitrary dimension $D = d+1$ [59]. We put the radius of the ball $a = 1$, the dependence of the results on the radius are easily recovered by dimensional arguments.

Instead of (2.3), the nonzero eigenmodes of Δ that are finite at the origin have eigenvalues $-\alpha^2$ and are of the form

$$\frac{J_\nu(\alpha r)}{r^{(d-1)/2}} Y_{l+D/2}(\Omega), \tag{3.22}$$

where the spherical harmonics satisfy

$$\Delta_{\mathcal{N}} Y_{l+D/2}(\Omega) = -\lambda_l^2 Y_{l+D/2}(\Omega) \tag{3.23}$$

with $\Delta_{\mathcal{N}}$ the Laplacian on the sphere, and

$$\nu^2 = \lambda_l^2 + (d-1)^2/4. \tag{3.24}$$

The eigenvalues are [74]

$$\lambda_l^2 = l(l+d-1) = \left(l + \frac{d-1}{2}\right)^2 - \left(\frac{d-1}{2}\right)^2, \quad l \in \mathbb{N}_0,$$

such that

$$\nu^2 = \left(l + \frac{d-1}{2}\right)^2,$$

and they have degeneracy

$$d(l) = (2l+d-1)\frac{(l+d-2)!}{l!(d-1)!}.$$

Let us next impose the boundary conditions and this time let us consider Dirichlet as well as generalized Neumann (or Robin) boundary conditions. In the notation of, for example, [115], these read explicitly

$$J_\nu(\alpha) = 0 \tag{3.25}$$

for Dirichlet and

$$uJ_\nu(\alpha) + \alpha J_\nu'(\alpha) = 0 \tag{3.26}$$

for Robin, with $u \in \mathbb{R}$. In the following we will write $u = 1 - \frac{D}{2} - \beta$, such that $\beta = 0$ corresponds to Neumann boundary conditions and β is to be identified with the endomorphism S.

As we have seen in the three dimensional calculation, the angular momentum sum leads to a zeta function of the type,

$$\zeta_{\mathcal{N}}(s) = \sum d(\nu)\nu^{-2s}, \tag{3.27}$$

and we anticipate this to be the central object to state the asymptotic contributions. Its definition is clearly motivated by the calculation in 3 dimensions where $2\zeta_H(2s-1;1/2)$ plays the role of $\zeta_{\mathcal{N}}(s)$.

Our first aim will be to express the whole zeta function on the ball

$$\zeta(s) = \sum \alpha^{-2s},$$

as far as possible in terms of this quantity. That is, we seek to replace analysis on the ball by that on the sphere.

Following the analysis of the previous section for Dirichlet boundary conditions, the starting point is the representation of the zeta function in terms of a contour integral

$$\zeta(s) = \sum d(\nu) \int\limits_\gamma \frac{dk}{2\pi i} \, (k^2 + m^2)^{-s} \frac{\partial}{\partial k} \ln J_\nu(k), \tag{3.28}$$

where the anticlockwise contour γ must enclose all the solutions of (3.25) on the positive real axis.

As we have already seen, it is very useful to split the zeta function into two parts,

$$\zeta(s) = Z(s) + \sum_{i=-1}^{N} A_i(s). \tag{3.29}$$

Performing identical steps as before, the different pieces are determined to be

$$Z(s) = \frac{\sin(\pi s)}{\pi} \sum d(\nu) \int\limits_0^\infty dz \, (z\nu)^{-2s} \frac{\partial}{\partial z} \Bigg(\ln \left(z^{-\nu} I_\nu(z\nu)\right) - \ln \left[\frac{z^{-\nu}}{\sqrt{2\pi\nu}} \frac{e^{\nu\eta}}{(1+z^2)^{\frac{1}{4}}} \right] - \sum_{n=1}^{N} \frac{D_n(t)}{\nu^n} \Bigg), \tag{3.30}$$

and

$$A_{-1}(s) = \frac{1}{4\sqrt{\pi}} \frac{\Gamma\left(s - \frac{1}{2}\right)}{\Gamma(s+1)} \zeta_{\mathcal{N}}\left(s - 1/2\right), \tag{3.31}$$

$$A_0(s) = -\frac{1}{4} \zeta_{\mathcal{N}}(s), \tag{3.32}$$

$$A_i(s) = -\frac{1}{\Gamma(s)} \zeta_{\mathcal{N}}\left(s + i/2\right) \sum_{b=0}^{i} x_{i,b} \frac{\Gamma\left(s + b + i/2\right)}{\Gamma\left(b + i/2\right)}. \tag{3.33}$$

These results can be read off from eqs. (2.14) and (2.18) once the definition (3.24) for ν^2 is used and $2\zeta_H(2s-1;1/2)$ is replaced by $\zeta_{\mathcal{N}}(s)$, eq. (3.27). The function $Z(s)$ is analytic on the strip $(d-1-N)/2 < \Re s$, which may be seen by considering the asymptotics of the integrand in eq. (3.30).

As is clearly apparent in eq. (3.31)–(3.33), sphere contributions are separated from radial ones. Again, the analytical structure is made very explicit and the result is very well organized by the introduction of the zeta function $\zeta_{\mathcal{N}}(s)$.

In order to treat Robin boundary conditions, only a few changes are necessary. In addition to expansion (2.9) we need [116,80]

$$I'_\nu(\nu z) \sim \frac{1}{\sqrt{2\pi\nu}} \frac{e^{\nu\eta}(1+z^2)^{1/4}}{z} \left[1 + \sum_{k=1}^{\infty} \frac{v_k(t)}{\nu^k} \right], \tag{3.34}$$

with the $v_k(t)$ determined by

$$v_k(t) = u_k(t) + t(t^2-1)\left[\frac{1}{2}u_{k-1}(t) + tu'_{k-1}(t)\right]. \tag{3.35}$$

The relevant polynomials analogous to the $D_n(t)$, eq. (2.11), are defined by

$$\ln\left[1+\sum_{k=1}^{\infty}\frac{v_k(t)}{\nu^k} + \frac{1-D/2-\beta}{\nu}t\left(1+\sum_{k=1}^{\infty}\frac{u_k(t)}{\nu^k}\right)\right] \sim \sum_{n=1}^{\infty}\frac{M_n(t)}{\nu^n} \tag{3.36}$$

and have the same structure,

$$M_n(t) = \sum_{b=0}^{n} z_{n,b}\, t^{n+2b}. \tag{3.37}$$

One may again introduce a split as in eq. (3.29) with $A_{-1}^R(s) = A_{-1}(s)$ and $A_0^R(s) = -A_0(s)$, where the upper index R indicates that these are the results for Robin boundary conditions. The $A_i^R(s)$ are given by eq. (3.33) once the $x_{i,a}$ is replaced by $z_{i,a}$. In addition one finds

$$\begin{aligned} Z^R(s) = \frac{\sin(\pi s)}{\pi}\sum d(\nu)\int_0^{\infty} dz\ (z\nu)^{-2s}\frac{\partial}{\partial z}\Bigg(&\ln\left((1-D/2-\beta)I_\nu(z\nu) + z\nu I'_\nu(z\nu)\right) \\ &-\ln\left[\sqrt{\frac{\nu}{2\pi}}e^{\nu\eta}(1+z^2)^{\frac{1}{4}}\right] - \sum_{n=1}^{N}\frac{M_n(t)}{\nu^n}\Bigg). \end{aligned} \tag{3.38}$$

The remaining task is the analysis of the meromorphic structure of the zeta function $\zeta_{\mathcal{N}}(s)$. It reads explicitly

$$\zeta_{\mathcal{N}}(s) = \sum_{l=0}^{\infty}(2l+d-1)\frac{(l+d-2)!}{l!(d-1)!}\left(l+\frac{d-1}{2}\right)^{-2s}. \tag{3.39}$$

Writing

$$d(l) = \binom{l+d-1}{d-1} + \binom{l+d-2}{d-1},$$

it is seen immediately that this is a sum of Barnes zeta functions [117–119] defined as

$$\zeta_{\mathcal{B}}(s,a)\sum_{\vec{m}=0}^{\infty}\frac{1}{(a+m_1+...+m_d)^s} = \sum_{l=0}^{\infty}\binom{l+d-1}{d-1}(l+a)^{-s}. \tag{3.40}$$

In detail, one finds

$$\zeta_{\mathcal{N}}(s) = \zeta_{\mathcal{B}}\left(2s, \frac{d+1}{2}\right) + \zeta_{\mathcal{B}}\left(2s, \frac{d-1}{2}\right). \tag{3.41}$$

Its residues are determined in the following way. Using the integral representation

$$\zeta_{\mathcal{B}}(s,c) = \frac{i\Gamma(1-s)}{2\pi}\int_L dz\ \frac{e^{z(d/2-c)}(-z)^{s-1}}{2^d\sinh^d(z/2)}, \tag{3.42}$$

where L is the Hankel contour, one immediately finds for the base function

$$\begin{aligned} \zeta_{\mathcal{N}}(s) &= \frac{i\Gamma(1-2s)}{2\pi}2^{2s+1-d}\int_L dz\ (-z)^{2s-1}\frac{\cosh z}{\sinh^d z} \\ &= \frac{i\Gamma(2-2s)}{2\pi(d-1)}2^{2s+1-d}\int_L dz\ (-z)^{2s-2}\frac{1}{\sinh^{d-1} z}. \end{aligned} \tag{3.43}$$

For the residues this yields $(m = 1, 2, ..., d)$

$$\text{Res } \zeta_{\mathcal{N}}(m/2) = \frac{2^{m-d} D_{d-m}^{(d-1)}}{(d-1)(m-2)!(d-m)!}, \tag{3.44}$$

with the $D_\nu^{(d-1)}$ defined through (*cf* [120])

$$\left(\frac{z}{\sinh z}\right)^{d-1} = \sum_{\nu=0}^{\infty} D_\nu^{(d-1)} \frac{z^\nu}{\nu!}. \tag{3.45}$$

Obviously $D_\nu^{(d-1)} = 0$ for ν odd, so there are actually poles only for $m = 1, 2, ..., d$ with $d - m$ even. The advantage of this approach is that known recursion formulas allow for an efficient evaluation of the $D_\nu^{(n)}$ as polynomials in d, [121].

Using eq. (3.44) in eqs. (3.31)—(3.33) we find, by the connection (1.12) for the heat-kernel coefficients $a_{k/2}$ with Dirichlet boundary conditions,

$$\begin{aligned}\frac{(4\pi)^{D/2}}{|S^d|} a_{k/2} = & \frac{(d-k-1)}{(d-1)(d-k+1)k!} \left(\frac{d+1-k}{2}\right)_{k/2} D_k^{(d-1)} \\ & - \frac{(d-k)}{4(d-1)(k-1)!} \left(\frac{d+2-k}{2}\right)_{(k-1)/2} D_{k-1}^{(d-1)} \\ & - \frac{2\sqrt{\pi}}{(d-1)} \sum_{i=1}^{k-1} \frac{d+i-k}{(k-1-i)!} \left(\frac{d+2-k+i}{2}\right)_{(k-i-1)/2} \times \\ & \sum_{b=0}^{i} \frac{x_{i,b}}{\Gamma(b+i/2)} \left(\frac{d+1-k+i}{2}\right)_b D_{k-1-i}^{d-1},\end{aligned} \tag{3.46}$$

where $(y)_n = \Gamma(y+n)/\Gamma(y)$ is the Pochhammer symbol. Eq.(3.46) exhibits the heat-kernel coefficients as explicit functions of the dimension d (partly encoded in $D_\nu^{(n)}$). Clearly, evaluation of eq. (3.46) is a simple routine machine matter, because all ingredients can be found by simple algebraic computer programs.

For later use, the polynomials up to $a_{3/2}$ are listed in the following,

$$\frac{(4\pi)^{d/2}}{|S^d|} a_{1/2} = -\frac{1}{4} \tag{3.47}$$

$$\frac{(4\pi)^{D/2}}{|S^d|} a_1 = \frac{d}{3} \tag{3.48}$$

$$\frac{(4\pi)^{d/2}}{|S^d|} a_{3/2} = \frac{(10-7d)d}{384}. \tag{3.49}$$

For Robin boundary conditions one has to make the modifications outlined above eq. (3.34). The results, up to $a_{3/2}$ are listed below,

$$\frac{(4\pi)^{d/2}}{|S^d|} a_{1/2} = \frac{1}{4} \tag{3.50}$$

$$\frac{(4\pi)^{D/2}}{|S^d|} a_1 = \frac{d}{3} + 2\beta \tag{3.51}$$

$$\frac{(4\pi)^{d/2}}{|S^d|} a_{3/2} = \frac{192\beta^2 + 96\beta d + d(2+13d)}{384} \tag{3.52}$$

Further coefficients could be calculated with ease, f.e. for the first 20 coefficients the program needs about 2 minutes.

C Determination of the general heat-kernel coefficients

We now compare, one by one, the general form of the coefficients with our special case evaluation. The coefficient a_0 is, by normalization,

$$a_0(F, P, \mathcal{B}_S^{\mp}) = (4\pi)^{-D/2} F[\mathcal{M}].$$

The next one is

$$a_{1/2}(F, P, \mathcal{B}_S^{\mp}) = \delta (4\pi)^{-d/2} F[\partial\mathcal{M}].$$

For the ball this means

$$a_{1/2}(F, -\Delta_{\mathcal{M}}, \mathcal{B}_S^{\mp}) = \delta (4\pi)^{-d/2} F(1) |S^d|.$$

Using the relations (3.47) and (3.50) we can immediately determine δ,

$$\delta = \left(-\frac{1^-}{4}, \frac{1^+}{4} \right).$$

The coefficient $a_{1/2}$ is thus given for a general manifold from the result on the ball (which was clear of course). Passing on to a_1, the general form is

$$a_1(F, P, \mathcal{B}_S^{\mp}) = (4\pi)^{-D/2} 6^{-1} \{(6FE + FR)[\mathcal{M}] + (b_0 FK + b_1 F_{;m} + b_2 FS)[\partial\mathcal{M}]\}$$

In our special case on the ball, $K_a^b = \delta_a^b$ and thus

$$a_1(F, -\Delta_{\mathcal{M}}, \mathcal{B}_S^{\mp}) = (4\pi)^{-D/2} 6^{-1} \mathrm{vol}(S^d) \{b_0 F(1) d + b_1 F_{,r}(1) + b_2 F(1) S\}.$$

Comparing with the results (3.48), (3.51), one finds

$$b_0 = 2, \qquad b_2 = 12.$$

Taking into account also the smeared calculation [61], in addition one can find

$$b_1 = (3^-, -3^+).$$

Thus, our special case also gives the entire a_1 coefficient without any further information being needed. It is very important that the calculation can be performed for an arbitrary ball dimension D, and also for a smearing function $F(r)$. This allows one to just compare polynomials in d with the associated extrinsic curvature terms in the general expression and simply to read off the universal constants in this expression.

We continue with the next higher coefficient, with the general form,

$$a_{3/2}(F, P, \mathcal{B}_S^{\mp}) = \frac{\delta}{96(4\pi)^{d/2}} \Big(F(c_0 E + c_1 R + c_2 R_{mm} + c_3 K^2 + c_4 K_{ab} K^{ab} c_7 SK + c_8 S^2) + F_{;m}(c_5 K + c_9 S) + c_6 F_{;mm} \Big).$$

The (smeared) ball calculation immediately gives 7 of the 10 unknowns,

$$c_3 = (7^-, 13^+), \quad c_4 = (-10/, 2^+), \quad c_5 = (30^-, -6^+),$$
$$c_6 = 24, \quad c_7 = 96, \quad c_8 = 192, \quad c_9 = -96.$$

Next, apply the lemma on product manifolds (3.11) [96]. For $a_{3/2}$ this means

$$a_{3/2}(y, P, \mathcal{B}_S^{\mp}) = a_{3/2}(y_1, P_1, \mathcal{B}_S^{\mp}) a_0(x_2, P_2) + a_{1/2}(y_1, P_1, \mathcal{B}_S^{\mp}) a_1(x_2, P_2).$$

We will choose $P_1 = -\Delta_1$ and $P_2 = -\Delta_2 + E(x_2)$, with obvious notation, to obtain

$$\delta 96^{-1}(c_0 E + c_1 R(\mathcal{M}_2)) = \delta 6^{-1}(6E + R(\mathcal{M}_2)),$$

where we used in addition $R(\mathcal{M}_1 \times \mathcal{M}_2) = R(\mathcal{M}_1) + R(\mathcal{M}_2)$. This gives

$$c_0 = 96, \quad c_1 = 16.$$

It is seen that the determination of $a_{3/2}$ is relatively simple, once the ball result is at hand. The lemma on product manifolds is also very easily applied and already only one of the universal constants c_i, namely c_2, is missing.

The remaining information is obtained using the relations between the heat- kernel coefficients under conformal rescaling, (3.14). Setting to zero the coefficients of all terms in (3.14) gives several relations between the universal constants c_i. We will need only one of them. Thus, setting to zero the coefficient of $F_{;mm}$ gives

$$\frac{1}{2}(D-2)c_0 - 2(D-1)c_1 - (D-1)c_2 - (D-3)c_6 = 0$$

and so $c_2 = -8$ for Dirichlet and Robin boundary conditions. This completes the calculation of $a_{3/2}$.

It gets clear, already, that the combination of the different methods is extremely effective in obtaining heat-kernel coefficients and, as mentioned, $a_{5/2}$ has been obtained proceeding in this way. For other boundary conditions, the general form of the heat-kernel coefficients can contain several other geometrical quantities, as projectors for mixed boundary conditions (for a discussion of their physical relevance see [122]) or tangential vector fields for oblique boundary conditions. Although the analysis gets partly more involved, the ideas described seem to apply generally and, as mentioned, many new results have already been obtained [123–125,62,63].

IV CASIMIR AND GROUND STATE ENERGIES

In this section we apply the techniques developed in Section II to the calculation of vacuum energies when spherically symmetric boundaries or external potentials are present. The boundary calculation is an immediate continuation of Section II. When external potentials are present, the implicit eigenvalue equation (2.4) is replaced by an asymptotic equivalent, see eq. (4.32), which allows the energy to be expressed by quantum mechanical scattering dates.

A Massive scalar field with Dirichlet boundary conditions

So let us start to analyse the influence the mass of a field has on the Casimir energy. We have seen that the Casimir energy is generally plagued by ambiguities. For a massive field, however, there is the possibility to separate (at least in principle) the classical part and the quantum part of the energy. The idea is that, if the mass of the quantum field tends to infinity, quantum fluctuations should die out and as a result, in this limit, the quantum contributions to the Casimir energy should vanish. As we will see, this provides a unique definition [126]. Before we proceed, let us first discuss this in more detail.

To actually impose the condition, the $m \to \infty$ behaviour of E_{vac} is needed. This is easily derived using the heat-kernel expansion. For the scalar field, the heat-kernel of $(-\Delta + m^2)$ clearly is

$$K(t) = e^{-m^2 t} K_{-\Delta}(t),$$

with $K_{-\Delta}(t)$ the heat-kernel of minus the Laplacian on the ball. The asymptotic expansion for $m \to \infty$ is then found by employing eq. (1.9),

$$\begin{aligned} \zeta(\alpha) &= \frac{1}{\Gamma(\alpha)} \int_0^\infty dt\, t^{\alpha-1} e^{-m^2 t} K_{-\Delta}(t) \\ &\sim \frac{1}{\Gamma(\alpha)} \sum_{l=0,1/2,1,\dots}^{\infty} a_l(-\Delta) \frac{\Gamma(\alpha+l-3/2)}{m^{2(\alpha+l-3/2)}}. \end{aligned} \tag{4.1}$$

About $\alpha = -1/2$ it reads,

$$\zeta(-1/2+s) = -\frac{m^4}{4\sqrt{\pi}}a_0\left(\frac{1}{s}-\frac{1}{2}+\ln\left[\frac{4\mu^2}{m^2}\right]\right) - \frac{2m^3}{3}a_{1/2}$$
$$+\frac{m^2}{2\sqrt{\pi}}a_1\left(\frac{1}{s}-1+\ln\left[\frac{4\mu^2}{m^2}\right]\right)+ma_{3/2} \qquad (4.2)$$
$$-\frac{1}{2\sqrt{\pi}}a_2\left(\frac{1}{s}-2+\ln\left[\frac{4\mu^2}{m^2}\right]\right)+\mathcal{O}(1/m)+\mathcal{O}(s). \qquad (4.3)$$

Thus, in order to impose the normalization condition

$$\lim_{m\to\infty} E_{vac}^{ren} = 0, \qquad (4.4)$$

this defines the terms to be subtracted. As is seen, the zeta functional regularization used leaves the contributions of the coefficients with half integer index finite in the limit $s \to 0$. This is a specific feature of this regularization. However, in other regularizations, as for example the proper time cutoff [28] or the exponential cutoff [127], these contributions are divergent when the cutoff is removed. For this reason, and in order to impose the normalization condition (4.4), we include them among the "divergent" terms suffering renormalization. By this, we are led to the following definition of the renormalized Casimir energy,

$$E_{vac}^{ren} = E_{vac} - E_{vac}^{div} \qquad (4.5)$$

with

$$E_{vac}^{div} = -\frac{m^4}{8\sqrt{\pi}}a_0\left(\frac{1}{s}-\frac{1}{2}+\ln\left[\frac{4\mu^2}{m^2}\right]\right) - \frac{m^3}{3}a_{1/2}$$
$$+\frac{m^2}{4\sqrt{\pi}}a_1\left(\frac{1}{s}-1+\ln\left[\frac{4\mu^2}{m^2}\right]\right)+\frac{1}{2}ma_{3/2}$$
$$-\frac{1}{4\sqrt{\pi}}a_2\left(\frac{1}{s}-2+\ln\left[\frac{4\mu^2}{m^2}\right]\right). \qquad (4.6)$$

To interpret the subtraction in (4.5) as a renormalization of "bare" parameters, we need to consider the following physical system composed of two parts:

1. A classical system consisting of a spherical surface of radius a. Its energy reads:

$$E_{class} = pV + \sigma S + Fa + k + \frac{h}{a}, \qquad (4.7)$$

   where $V = \frac{4}{3}\pi a^3$ and $S = 4\pi a^2$ are the volume and surface, respectively. This energy is determined by the parameters p = pressure, σ = surface tension, and F, k, and h, which do not have special names.

2. A quantized field $\hat{\varphi}(x)$ whose classical counterpart obeys the Klein-Gordon equation

$$(\Box + m^2)\varphi(x) = 0, \qquad (4.8)$$

   as well as suitable boundary conditions on the surface ensuring selfadjointness of the corresponding elliptic operator on perturbations. We choose Dirichlet boundary conditions as the easiest to handle.

For this system one can consider three models, which will behave in a different way [126]. These models consist of the classical part given by the surface and

(i) the quantized field in the interior of the surface,

(ii) the quantized field in the exterior of the surface,

(iii) the quantized field in both regions together,

respectively. We will give here the analysis for model (i); models (ii) and (iii) can be treated along the same lines [126] and only a few comments will be added.

The heat-kernel coefficients needed to impose the normalization (4.4) are well known (see for instance [128], or, alternatively, the calculation provided in Section III B), and read

$$a_0 = \frac{1}{6\sqrt{\pi}}a^3, \quad a_{1/2} = -\frac{1}{4}a^2, \quad a_1 = \frac{1}{3\sqrt{\pi}}a,$$
$$a_{3/2} = -\frac{1}{24}, \quad a_2 = \frac{2}{315\sqrt{\pi}a}.$$

Thus, we have five divergent contributions. This remains true for model (ii); instead, for model (iii), only two divergent terms remain due to a cancellation of poles which is easily understood by realizing that the extrinsic curvature seen from the interior and exterior has opposite sign.

As the physical system we consider, as described, the classical part (4.7) and the ground state energy of the quantum field together, and write for the complete energy

$$E = E_{class} + E_{vac}. \tag{4.9}$$

In this context, the renormalization can be achieved by shifting the parameters in E_{class} by an amount which cancels the divergent contributions and removes completely the contribution of the corresponding heat kernel coefficients. In detail we have

$$\begin{aligned} p &\to p - \frac{m^4}{64\pi^2}\left(\frac{1}{s} - \frac{1}{2} + \ln\left[\frac{4\mu^2}{m^2}\right]\right), \quad \sigma \to \sigma + \frac{m^3}{48\pi}, \\ F &\to F + \frac{m^2}{12\pi}\left(\frac{1}{s} - 1 + \ln\left[\frac{4\mu^2}{m^2}\right]\right), \quad k \to k - \frac{m}{96}, \\ h &\to h + \frac{1}{630\pi}\left(\frac{1}{s} - 2 + \ln\left[\frac{4\mu^2}{m^2}\right]\right). \end{aligned} \tag{4.10}$$

After the subtraction of these contributions from E_{vac} we denote it by E_{vac}^{ren}, see eq. (4.5), and the complete energy becomes

$$E = E_{class} + E_{vac}^{ren}, \tag{4.11}$$

with the bare parameters in E_{class} replaced by the renormalized ones using eq. 4.10). In our renormalization scheme, we have defined a unique renormalized groundstate energy E_{vac}^{ren}.

Now, having discussed in detail the subtraction procedure and the structure of divergences, let us come to a full evaluation of E_{vac}. The equations of Section II serve as a starting point. Choosing $N = 3$ in eq. (2.13), $Z(s)$ is finite at $s = -1/2$ and can be used for the numerical evaluation of E_{vac} for (in principle) arbitrary mass m. For $A_i(s)$, $i = -1, 0, 1, 2, 3$, a representation valid for $|ma| < 1/2$ has been given, see eqs. (2.25), (2.26) and (2.29), but here we need a representation of $A_i(s)$ beyond that range. This is a purely technical complication explained in Appendix A. The final result for $A_{-1}(s)$ and $A_0(s)$ about $s = -1/2$ is given in eqs. (A.9) and (A.10). The remaining $A_i(s)$ have the form, see eq. (2.28),

$$A_i(s) = -\frac{2m^{-2s}}{\Gamma(s)}\sum_{c=0}^{i}\frac{x_{i,c}}{(ma)^{i+2c}}\frac{\Gamma(s+c+i/2)}{\Gamma(c+i/2)}f(s;1+2c;c+i/2), \tag{4.12}$$

with

$$f(s;c;b) = \sum_{\nu=1/2,3/2,\ldots}^{\infty} \nu^c\left(1+\left(\frac{\nu}{ma}\right)^2\right)^{-s-b}. \tag{4.13}$$

The remaining task is to calculate $f(s;c;b)$ for the relevant values of c and b about $s = -1/2$. This is a systematic calculation sketched also in Appendix A. It is slightly simplified by realizing the recurrence

$$f(s;c;b) = (ma)^2\left[f(s;c-2;b-1) - f(s;c-2;b)\right]. \tag{4.14}$$

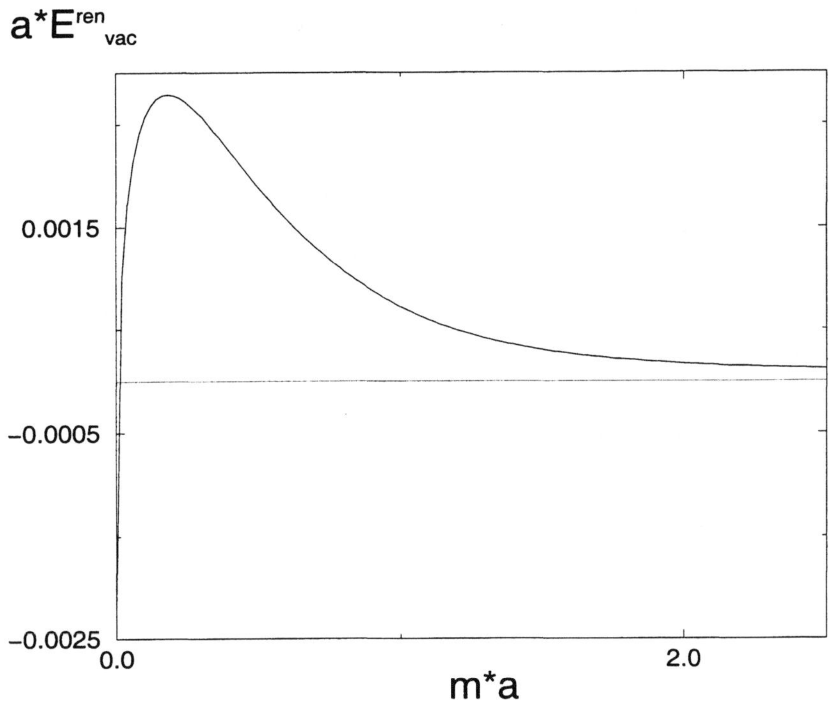

FIGURE 3. Plot of the renormalized vacuum energy E_{vac}^{ren} measured in units of the inverse of the radius.

In summary, all analytic expressions needed for the numerical evaluation have been provided.

The result for the Casimir energy for the interior region of the ball is shown in Fig. 3 for $a = 1$ as a function of m.

For very small values of the argument, ma, the function takes negative values, whereas for larger values it is positive and tends to zero for $(ma) \to \infty$. Thus, as a function of the mass, attractive as well as repulsive forces are possible and the influence of the mass is by no means negligeable. This is different from the example of parallel plates [21,68], where for large mass the influence of the mass is exponentially damped away and is of a very short range. The origin of this different behaviour is intimately connected with the extrinsic curvature being non-vanishing or vanishing. This is already seen from the heat-kernel expansion (4.1), where the leading contribution to E_{vac}^{ren} is proportional to $a_{5/2}(-\Delta)/m$. For the ball, this coefficient is non-vanishing; however, for parallel plates this one as well as all the following coefficients are vanishing, what explains the very different behaviour of the renormalized Casimir energy as a function of the mass for this two configurations.

B Scalar field in the presence of an external background field

In the above described situation, "empty space" is supplemented by boundary conditions. However, clearly (perfect) boundaries are an idealization and no boundary made of matter can be perfectly smooth. Seen as modeling some distribution of matter which interacts with the quantum field, it might be more justified to introduce an external potential and to quantize a theory within this potential. This will change the quantum modes and the associated spectrum and, as a result, the energy of the vacuum. For certain types of external confinement this situation has been called Casimir effect with soft or semihard boundaries [129,130]. As mentioned, apart from this type of setting, in many situations the potential is provided by classical solutions to nonlinear field equations.

Again, we will start the consideration by describing our concrete model and its renormalization, introducing also various notations used in the following. We will consider the Lagrangian

$$L = \frac{1}{2}\Phi(\Box - M^2 - \lambda\Phi^2)\Phi + \frac{1}{2}\varphi(\Box - m^2 - \lambda'\Phi^2)\varphi. \tag{4.15}$$

Here, the field Φ is a classical background field. By means of

$$V(x) = \lambda'\Phi^2 \tag{4.16}$$

it defines the potential in (4.15) for the field $\varphi(x)$, which should be quantized in the background of $V(x)$. As explained below, the embedding into this external system is necessary in order to guarantee the renormalizability of the ground state energy. Actually, this is clear already from the beginning, because the external system needs to comprise of the counterterms of a $(\lambda\varphi^4)$-theory.

The complete energy

$$E[\Phi] = E_{class}[\Phi] + E_{vac}[\Phi] \tag{4.17}$$

of the system consists, as before, of the classical part and the contributions resulting from the ground state energy of the quantum field φ in the background of the field Φ. The classical part reads

$$E_{class}[\Phi] = \frac{1}{2}V_g + \frac{1}{2}M^2V_1 + \lambda V_2, \tag{4.18}$$

with the definitions $V_g = \int d^3x(\nabla\Phi)^2$, $V_1 = \int d^3x\Phi^2$ and $V_2 = \int d^3x\Phi^4$. Here M^2 and λ are the bare mass, respectively coupling constant, which need renormalization. Continuing as in Section IV, for the ground state energy one defines [28]

$$E_{vac}[\Phi] = \frac{1}{2}\sum_{(n)}(\lambda_{(n)}^2 + m^2)^{1/2-s}\mu^{2s}, \tag{4.19}$$

where μ is the arbitrary mass parameter and s is the regularization parameter, which has to be put to zero after renormalization. Furthermore, $\lambda_{(n)}$ are the eigenvalues of the corresponding Laplace equation

$$(-\Delta + V(x))\phi_{(n)}(x) = \lambda_{(n)}^2\phi_{(n)}(x). \tag{4.20}$$

For the moment, we assume the space to be a large ball of radius R as an intermediate step to have discrete eigenvalues and, thus, a discrete multiindex (n) and, in addition, to avoid pure volume divergences. For simplicity we impose Dirichlet boundary conditions and explain below under which conditions on the potential $V(x)$ the final result does not receive boundary contributions.

Expressing the ground state energy (4.19) in terms of the zeta function

$$\zeta_V(s) = \sum_{(n)}(\lambda_{(n)}^2 + m^2)^{-s} \tag{4.21}$$

of the wave operator with potential $V(x)$ as defined in (4.20), one has

$$E_{vac}[\Phi] = \frac{1}{2}\zeta_V(s - 1/2)\mu^{2s}. \tag{4.22}$$

The residue of $\zeta_V(-1/2)$ is determined by the a_2 heat-kernel coefficient associated with the spectrum $\lambda_{(n)}^2 + m^2$, and the classical energy (4.18) provides another example for our previous viewpoint, that it should consist of the terms contained in a_2. In order to fix the finite renormalizations, we impose again the normalization condition

$$\lim_{m\to\infty} E_{vac}^{ren}[\Phi] = 0, \tag{4.23}$$

which is implemented along the lines leading to eq. (4.6). For convenience we state it again,

$$\begin{aligned} E_{vac}^{div}[\Phi] = &-\frac{m^4}{8\sqrt{\pi}}\left(\frac{1}{s} + \ln\frac{4\mu^2}{m^2} - \frac{1}{2}\right)a_0 - \frac{m^3}{3}a_{1/2} \\ &+\frac{m^2}{4\sqrt{\pi}}\left(\frac{1}{s} + \ln\frac{4\mu^2}{m^2} - 1\right)a_1 + \frac{1}{2}a_{3/2} \\ &-\frac{1}{4\sqrt{\pi}}\left(\frac{1}{s} + \ln\frac{4\mu^2}{m^2} - 2\right)a_2, \end{aligned} \tag{4.24}$$

with $a_i, i = 0, 1/2, ..., 2$ the heat-kernel coefficients of the Laplace equation (4.20). This is a good moment to consider the limiting behaviour for $R \to \infty$ of the boundary terms. Before actually performing the limit $R \to \infty$ we must clearly state what the ground state energy in the given context is meant to be. It is a quantity that describes the influence of the background potential $V(x)$ on the vacuum energy and, as such, is to be compared with the field free case $V(x) = 0$. Thus, from definition (4.19), we will subtract the Casimir energy of a free field inside a large ball of radius R in order to normalize $E_{vac}[\Phi = 0] = 0$. In eq. (4.24), this subtraction cancels the contributions of a_0 and $a_{1/2}$, in a_1 it cancels the potential independent boundary terms. In $a_{3/2}$ the term proportional to $V(R)R^2$ survives, see eq. (3.10), and, in the limit $R \to \infty$ vanishes only if $V(r) \sim r^{-2-\epsilon}$, $\epsilon > 0$, for $r \to \infty$. As is easily seen by dimensional arguments, the boundary contributions of the higher coefficients depending on $V(r)$ will then vanish too.

In summary, under the assumption that $V(r) \sim r^{-2-\epsilon}$ for $r \to \infty$, in the limit $R \to \infty$ no boundary contributions will appear and the normalization condition (4.23) is achieved by a renormalization of the mass M of the background field,

$$M^2 \to M^2 + \frac{\lambda' m^2}{16\pi^2}\left(-\frac{1}{s} + 1 + \ln\frac{m^2}{4\mu^2}\right), \tag{4.25}$$

and the coupling constant λ by

$$\lambda \to \lambda + \frac{\lambda'^2}{64\pi^2}\left(-\frac{1}{s} + 2 + \ln\frac{m^2}{4\mu^2}\right). \tag{4.26}$$

This is an immediate result of the volume contributions to the heat-kernel coefficients, $a_1 = -(4\pi)^{-3/2}\int d^3x V(x)$ and $a_2 = (1/2)(4\pi)^{-3/2}\int d^3x V^2(x)$. The kinetic term V_g in $E[\Phi]$ suffers no renormalization.

By defining

$$E_{vac}^{ren} = E_{vac}[\Phi] - E_{vac}^{div} \tag{4.27}$$

one obtains the finite groundstate energy, which is normalized in a way that the functional dependence on Φ^2 present in the classical energy is now absent in the quantum corrections E_{vac}^{ren}.

Let us note that this is just the well known general renormalization scheme written down here explicitly in the notations needed in our case.

For the calculation of $E_{vac}[\Phi]$ let us take further advantage of the background field $\Phi(r)$ being spherically symmetric. The multiindex $(n) \to n, l, m$ consists of the main quantum number n, the angular momentum number l and the magnetic quantum number m. In polar coordinates the ansatz for a solution of the wave equation (4.20) reads

$$\phi_{(n)}(x) = \frac{1}{r}\phi_{n,l}(r)Y_{lm}(\theta, \varphi) \tag{4.28}$$

where the radial wave equation takes the form

$$\left[\frac{d^2}{dr^2} - \frac{l(l+1)}{r^2} - V(r) + \lambda_{n,l}^2\right]\phi_{n,l}(r) = 0. \tag{4.29}$$

Now we use the standard scattering theory within $r \in [0, \infty)$ and take the momentum p instead of the discrete $\lambda_{n,l}$. Let $\phi_{p,l}(r)$ be the so called regular solution which is defined so as to have the same behaviour at $r \to 0$ as the solution without potential

$$\phi_{p,l}(r) \underset{r\to 0}{\sim} j_l(pr) \tag{4.30}$$

with j_l the spherical Bessel function [131]. This regular solution defines the Jost function f_l through its asymptotics as $r \to \infty$,

$$\phi_{l,p}(r) \underset{r\to\infty}{\sim} \frac{i}{2}\left[f_l(p)\hat{h}_l^-(pr) - f_l^*(p)\hat{h}_l^+(pr)\right], \tag{4.31}$$

where $\hat{h}_l^-(pr)$ and $\hat{h}_l^+(pr)$ are the Riccati-Hankel functions [131]. The analytic properties of the Jost function $f_l(p)$ strongly depend on the properties of the potential $V(r)$. If in addition to $V(r) \sim r^{-2-\epsilon}$ for $r \to \infty$, we impose $V(r) \sim r^{-2+\epsilon}$ for $\epsilon \to 0$ and continuity of $V(r)$ in $0 < r < \infty$ (except maybe at a finite number of finite discontinuities), one may show that the Jost function is an analytic function of p for $\Im p > 0$. It vanishes in the finite set of points $p = i\kappa_{n,l}$ of the positive imaginary half axis, corresponding to the bound states with energy $-\kappa_{n,l}^2$.

Now we use the Jost function to transform the frequency sum in eq. (4.21) in a contour integral. Let us assume that the support of the potential is contained in the cavity of radius R. Then the above eq. (4.31) gets exact at $r = R$ and may be interpreted as an implicit equation for the eigenvalues $p = \lambda_{n,l}$. Choosing Dirichlet boundary conditions at $r = R$, $\phi_{p,l}(R) = 0$, it reads explicitly,

$$f_l(p)\hat{h}_l^-(pR) - f_l^*(p)\hat{h}_l^+(pR) = 0. \tag{4.32}$$

As already mentioned, ultimately we are interested in the limit $R \to \infty$ and in that limit the results will not receive boundary contributions, once we assume that $V(r) \sim r^{-2-\epsilon}$ for $r \to \infty$.

Let us now consider the ground state energy associated with the eigenvalues determined by (4.32). As described in detail in Section II, we represent the frequency sum in (4.19) by a contour integral. Using eq. (4.32), one immediately finds

$$\begin{aligned} E_{vac}[\Phi] &= \mu^{2s}\sum_{l=0}^{\infty}(l+1/2)\int_\gamma \frac{dp}{2\pi i}\,(p^2+m^2)^{1/2-s}\frac{\partial}{\partial p}\ln\left[\frac{f_l(p)\hat{h}_l^-(pR) - f_l^*(p)\hat{h}_l^+(pR)}{h_l^-(pR) - \hat{h}_l^+(pR)}\right] \\ &+\mu^{2s}\sum_{l=0}^{\infty}(l+1/2)\sum_n (m^2-\kappa_{n,l}^2)^{1/2-s}, \end{aligned} \tag{4.33}$$

with $-\kappa_{n,l}^2$ the energy eigenvalues of the bound states with given orbital momentum l. The denominator in the logarithm provides the subtraction of the Minkowski space contribution, where $f_l(p) = 1$. The contour γ is chosen counterclockwise enclosing all real solutions of eq. (4.32) on the positive real axis. In the limit of the infinite space the negative eigenvalues become the usual boundstates and the $\lambda_{n,l} > 0$ turn into the scattering states.

For the calculation of (4.33), as the next step, one deforms the contour γ to the imaginary axis. A contour coming from $i\infty + \epsilon$, crossing the imaginary axis at some positive value smaller than the smallest κ_n and going to $i\infty - \epsilon$ results first. Shifting the contour over the bound state values κ_n, which are the zeroes of the Jost function on the imaginary axis, the bound state contributions in eq. (4.33) are cancelled and in the limit $R \to \infty$ one finds

$$E_{vac}[\Phi] = -\frac{\cos\pi s}{\pi}\mu^{2s}\sum_{l=0}^{\infty}(l+1/2)\int_m^\infty dk\,[k^2-m^2]^{\frac{1}{2}-s}\,\frac{\partial}{\partial k}\ln f_l(ik). \tag{4.34}$$

For details of the contour deformation see Fig. 4.

In the first step of the deformation one makes use of the following properties [131],

$$\begin{aligned} f_l(-p) &= f_l^*(p), \\ \hat{h}_l^{\pm}(-z) &= (-1)^l\hat{h}_l^{\mp}(z). \end{aligned}$$

This is the representation of the ground state energy (and by means of (4.22) as well of the zeta function) in terms of the Jost function, which is the starting point of our following analysis. It has the nice property, that the dependence on the bound states is not present explicitly, being however contained in the Jost function by its properties on the positive imaginary axis. It is connected with the more conventional representations by means of the analytic properties of the Jost function. Expressing them by the dispersion relation

$$f_l(ik) = \prod_n\left(1-\frac{\kappa_{n,l}^2}{k^2}\right)\exp\left(-\frac{2}{\pi}\int_0^\infty\frac{dq\,q}{q^2+k^2}\delta_l(q)\right),$$

where $\delta_l(q)$ is the scattering phase, we obtain from (4.34)

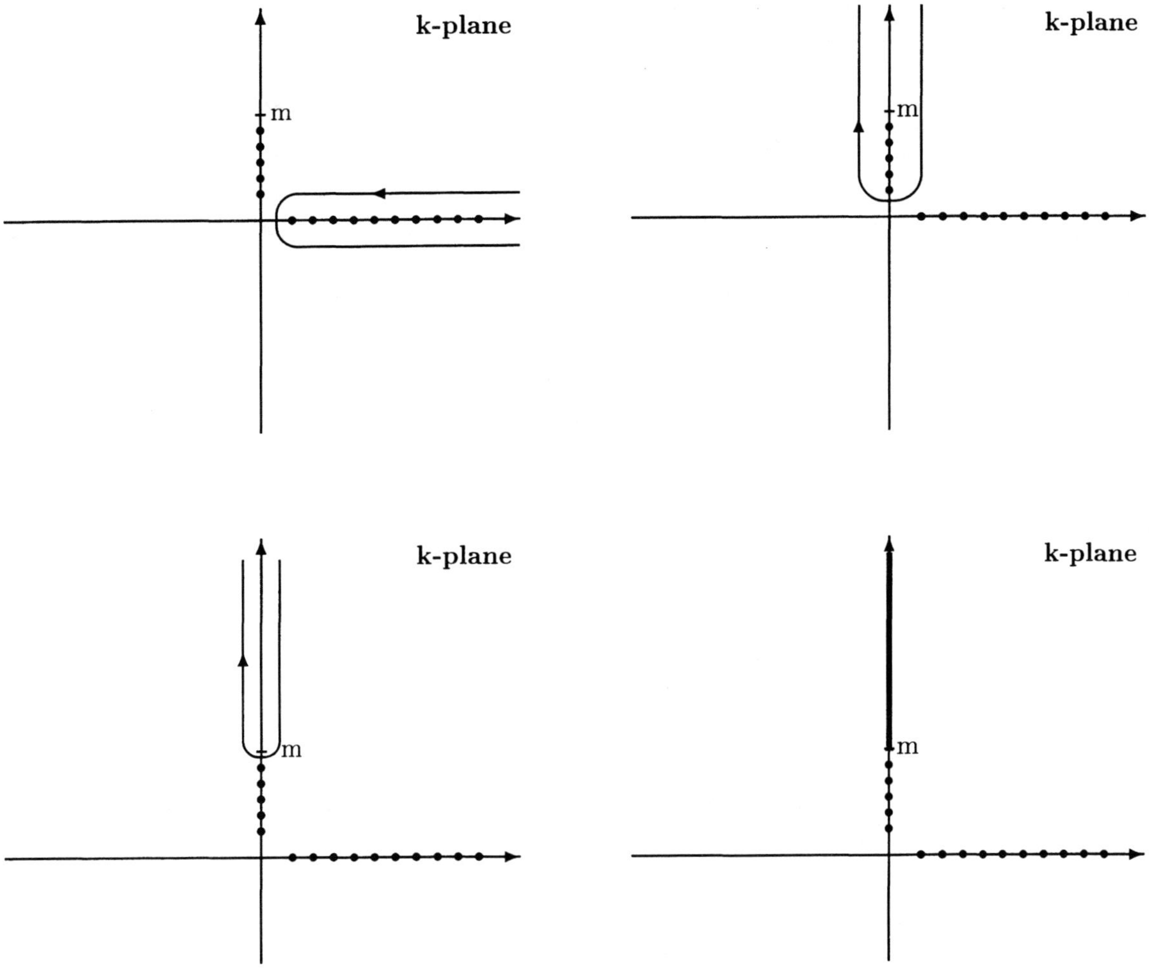

FIGURE 4. Deformation of the contour γ

$$E_{vac}[\Phi] = \mu^{2s} \sum_{l=0}^{\infty} \left(l + \frac{1}{2}\right) \left\{ -\sum_{n} \left(m^{1-2s} - \sqrt{m^2 - \kappa_{n,l}^2}^{\,1-2s} \right) - \frac{1-2s}{\pi} \int_0^{\infty} dq \, \frac{q}{\sqrt{q^2 + m^2}^{\,1-2s}} \, \delta_l(q) \right\}, \tag{4.35}$$

which gives the expression of the ground state energy through the scattering phase. From here one can pass to the representation through the mode density by integrating by parts.

Note that this representation can be obtained also directly from (4.33) in the limit $R \to \infty$ by deforming the contour γ.

Let us add a discussion on the sign of the ground state energy. In (4.35) the first contribution results from the bound states and is completely negative. The second contribution, which contains the scattering phase $\delta_l(q)$, is positive (negative) for an attractive (repulsive) potential, i.e., for $V(r) < 0$ ($V(r) > 0$) for all r [131]. So the regularized (still not renormalized) ground state energy $E_{vac}[\Phi]$ (4.35) is positive for a potential which is repulsive for all r (there are no bound states in this case) and it is negative for a potential which is attractive for all r. Now, if we perform the renormalization in accordance with (4.25) and (4.26), we obtain

$$E_{vac}^{ren} = E_{vac}[\Phi] + \frac{m^2}{8\pi}\left(\frac{1}{s} + \log\frac{4\mu^2}{m^2} - 1\right)\int_0^\infty dr\, r^2\, V(r)$$
$$+\frac{1}{16\pi}\left(\frac{1}{s} + \log\frac{4\mu^2}{m^2} - 2\right)\int_0^\infty dr\, r^2\, V(r)^2\,. \qquad (4.36)$$

This expression is finite for $s \to 0$, i.e., when removing the regularization. But due to the subtracted terms there is no longer any definite result on the sign. Note that this is in contrast to the case of a one-dimensional potential, where it had been possible to express the subtracted terms through the scattering phase [56].

Let us continue with a detailed analysis of the ground state energy, eq. (4.34). As we have easily seen using heat-kernel techniques, the non-renormalized vacuum energy $E_{vac}[\Phi]$ contains divergencies in $s = 0$, see eq. (4.24), which are removed by the renormalization prescription given in eqs. (4.25) – (4.27), resp. (4.36). The poles present are by no means obvious in the representation (4.34) of $E_{vac}[\Phi]$. However, in order to actually perform the renormalization, eq. (4.27), it is necessary to represent the groundstate energy eq. (4.34) in a form which makes the explicit subtraction of the divergencies visible. This will be our first task.

As is known from general zeta function theory as well as one sees from simply counting the large momentum behaviour, the representation eq. (4.34) of $E_{vac}[\Phi]$ will be convergent for $\mathrm{Re}\, s > 2$. However, for the calculation of the ground state energy we need the value of eq. (4.34) at $s = 0$; thus, an analytical continuation to the left has to be constructed. The basic idea is the same as before: adding and subtracting the leading uniform asymptotics of the integrand in eq. (4.34) [72]. Let

$$E_{vac}[\Phi] = E_f + E_{as}, \qquad (4.37)$$

where

$$E_f = -\frac{\cos(\pi s)}{\pi}\mu^{2s}\sum_{l=0}^{\infty}(l+1/2)\int_m^\infty dk\; [k^2 - m^2]^{\frac{1}{2}-s}\frac{\partial}{\partial k}[\ln f_l(ik) - \ln f_l^{asym}(ik)] \qquad (4.38)$$

and

$$E_{as} = -\frac{\cos(\pi s)}{\pi}\mu^{2s}\sum_{l=0}^{\infty}(l+1/2)\int_m^\infty dk\; [k^2 - m^2]^{\frac{1}{2}-s}\frac{\partial}{\partial k}\ln f_l^{asym}(ik). \qquad (4.39)$$

As described in detail, the idea is that as many asymptotic terms are subtracted as necessary to take $s = 0$ in the integrand of E_f. This term will then (in general) be evaluated numerically.

In E_{as} the analytic continuation to $s = 0$ can be done explicitly, showing that the pole contributions cancel when subtracting $E_{vac}^{div}[\Phi]$, eq. (4.24).

The first task, thus, is to obtain the asymptotics of the Jost functions. Contrary to the case with boundary conditions, the asymptotics cannot be taken just from tables, but need to be calculated itself. Fortunately, this may be done by using the integral equation (Lippmann-Schwinger equation) known from scattering theory [131]. For the Jost function one has ($\nu \equiv l + 1/2$)

$$f_l(ik) = 1 + \int_0^\infty dr\; r\, V(r)\phi_{l,ik}(r)K_\nu(kr), \qquad (4.40)$$

with the regular solution given by the integral equation

$$\phi_{l,ik}(r) = I_\nu(kr) + \int_0^r dr'\; r'\; [I_\nu(kr)K_\nu(kr') - I_\nu(kr')K_\nu(kr)]V(r')\phi_{l,ik}(r'). \qquad (4.41)$$

General zeta function theory tells us that the divergence at $s = 0$ contains at most terms of order V^2. Thus, one might expand $\ln f_l(ik)$ in powers of V and take into account only the asymptotics of terms up to $\mathcal{O}(V^2)$. The expansion in powers of V is easily obtained. Using eqs. (4.40) and (4.41) one finds

$$\ln f_l(ik) = \int_0^\infty dr\ rV(r)K_\nu(kr)I_\nu(kr)$$
$$- \int_0^\infty dr\ rV(r)K_\nu^2(kr)\int_0^r dr'\ r'V(r')I_\nu^2(kr')$$
$$+\mathcal{O}(V^3). \tag{4.42}$$

Now, the uniform asymptotics for $l \to \infty$ of $\ln f_l(ik)$ is essentially reduced to the well known uniform asymptotics of the modified Bessel functions K_ν and I_ν, (9.7.7) and (9.7.8) in [80]. With the notation $t = 1/\sqrt{1+(kr/\nu)^2}$ and $\eta(k) = \sqrt{1+(kr/\nu)^2} + \ln[(kr/\nu)/(1+\sqrt{1+(kr/\nu)^2})]$, one finds for $\nu \to \infty$, $k \to \infty$ with k/ν fixed,

$$I_\nu(kr)K_\nu(kr) \sim \frac{1}{2\nu t} + \frac{t^3}{16\nu^3}\left(1 - 6t^2 + 5t^4\right) + \mathcal{O}(1/\nu^4)$$
$$I_\nu(kr')K_\nu(kr) \sim \frac{1}{2\nu}\frac{e^{-\nu(\eta(k)-\eta(kr'/r))}}{(1+(kr/\nu)^2)^{1/4}(1+(kr'/\nu)^2)^{1/4}}\left[1+\mathcal{O}(1/\nu)\right].$$

Using these terms in the rhs of eq. (4.42) we define

$$\ln f_l^{asym}(ik) = \frac{1}{2\nu}\int_0^\infty dr\ \frac{r\,V(r)}{\left[1+\left(\frac{kr}{\nu}\right)^2\right]^{1/2}}$$
$$+\frac{1}{16\nu^3}\int_0^\infty dr\ \frac{r\,V(r)}{\left[1+\left(\frac{kr}{\nu}\right)^2\right]^{3/2}}\left[1 - \frac{6}{\left[1+\left(\frac{kr}{\nu}\right)^2\right]} + \frac{5}{\left[1+\left(\frac{kr}{\nu}\right)^2\right]^2}\right]$$
$$-\frac{1}{8\nu^3}\int_0^\infty dr\ \frac{r^3\,V^2(r)}{\left[1+\left(\frac{kr}{\nu}\right)^2\right]^{3/2}}. \tag{4.43}$$

Thereby the r'-integration in the term quadratic in V has been performed by the saddlepoint method using the monotony of $\eta(k)$. Now, by means of (4.43) the limit $s \to 0$ can be performed in eq. (4.38) and we obtain

$$E_f = -\frac{1}{\pi}\sum_{l=0}^\infty (l+1/2)\int_m^\infty dk\ \sqrt{k^2-m^2}\frac{\partial}{\partial k}\left(\ln f_l(ik) - \ln f_l^{asym}(ik)\right), \tag{4.44}$$

a form which is suited for a numerical evaluation.

For E_{as} at $s = 0$ one might explicitly find the analytical continuation. First of all, the k-integrals may be done using

$$\int_m^\infty dk\ [k^2-m^2]^{\frac{1}{2}-s}\frac{\partial}{\partial k}\left[1+\left(\frac{kr}{\nu}\right)^2\right]^{-\frac{n}{2}} = -\frac{\Gamma(s+\frac{n-1}{2})\Gamma(\frac{3}{2}-s)}{\Gamma(n/2)}\frac{\left(\frac{\nu}{mr}\right)^n m^{1-2s}}{\left(1+\left(\frac{\nu}{mr}\right)^2\right)^{s+\frac{n-1}{2}}}.$$

This naturally leads to the functions encountered already in the case of boundary conditions,

$$f(s;c;b) = \sum_{\nu=1/2,3/2,\dots}^\infty \nu^c\left(1+\left(\frac{\nu}{mr}\right)^2\right)^{-s-b}. \tag{4.45}$$

In terms of these functions, for E_{as} we obtain

$$E_{as}[\Phi] = -\frac{\Gamma(s)}{2\sqrt{\pi}\Gamma(s-1/2)}\left(\frac{\mu}{m}\right)^{2s}\int_0^\infty dr\ V(r) f(s-1/2;1;1/2)$$
$$+\frac{\Gamma(s+1)}{4\sqrt{\pi}m^2\Gamma(s-1/2)}\left(\frac{\mu}{m}\right)^{2s}\int_0^\infty dr\ \left[V^2(r)-\frac{V(r)}{2r^2}\right] f(s-1/2,1,3/2)$$
$$+\frac{\Gamma(s+2)}{2\sqrt{\pi}m^4\Gamma(s-1/2)}\left(\frac{\mu}{m}\right)^{2s}\int_0^\infty dr\ \frac{V(r)}{r^4} f(s-1/2;3;5/2)$$
$$-\frac{\Gamma(s+3)}{6\sqrt{\pi}m^6\Gamma(s-1/2)}\left(\frac{\mu}{m}\right)^{2s}\int_0^\infty dr\ \frac{V(r)}{r^6} f(s-1/2;5;7/2).$$

The relevant expansions about $s=0$ of the $f(s;c;b)$ are found in Appendix A. All divergences are made explicit, and

$$E_{as}^{ren}[\Phi] = E_{as}[\Phi] - E_{vac}^{div}[\Phi]$$

takes the compact form [73]

$$\begin{aligned}
E_{as}^{ren}[\Phi] = &-\frac{1}{8\pi}\int_0^\infty dr\ r^2V^2(r)\ln(mr) \\
&-\frac{1}{2\pi}\int_0^\infty dr\ V(r)\int_0^\infty d\nu \frac{\nu}{1+e^{2\pi\nu}}\ln|\nu^2-(mr)^2| \\
&-\frac{1}{8\pi}\int_0^\infty dr\ \left[r^2V^2(r)-\frac{1}{2}V(r)\right]\int_0^\infty d\nu\ \left(\frac{d}{d\nu}\frac{1}{1+e^{2\pi\nu}}\right)\ln|\nu^2-x^2| \\
&-\frac{1}{8\pi}\int_0^\infty dr\ V(r)\int_0^\infty d\nu\ \left[\frac{d}{d\nu}\left(\frac{1}{\nu}\frac{d}{d\nu}\frac{\nu^2}{1+e^{2\pi\nu}}\right)\right]\ln|\nu^2-x^2| \\
&+\frac{1}{48\pi}\int_0^\infty dr\ V(r)\int_0^\infty d\nu\ \left[\frac{d}{d\nu}\left(\frac{1}{\nu}\frac{d}{d\nu}\frac{1}{\nu}\frac{d}{d\nu}\frac{\nu^4}{1+e^{2\pi\nu}}\right)\right]\ln|\nu^2-x^2|,
\end{aligned} \tag{4.46}$$

valid for any potential with the above mentioned properties and very suitable for numerical evaluation.

The remaining task for the analysis of the renormalized ground state energy

$$E_{vac}^{ren}[\Phi] = E_f[\Phi] + E_{as}^{ren}[\Phi]$$

in the presence of a spherically symmetric potential is the numerical analysis of E_f. To achieve that, a (as a rule numerical) knowledge of the Jost function $f_l(ik)$ is necessary. This will be obtained from a numerical knowledge of the regular solution at a single point [73].

Under the assumption that the potential has compact support, let's say $V(r)=0$ for $r\geq R$, the regular solution may be written as

$$\phi_{l,p}(r) = u_{l,p}(r)\Theta(R-r)+\frac{i}{2}\left[f_l(p)\hat{h}_l^-(pr)-f_l^*(p)\hat{h}_l^+(pr)\right]\Theta(r-R). \tag{4.47}$$

Assuming continuity of the regular solution and its derivative, the matching conditions are

$$u_{l,p}(R) = \frac{i}{2}\left[f_l(p)\hat{h}_l^-(pR)-f_l^*(p)\hat{h}_l^+(pR)\right],$$
$$u'_{l,p}(R) = \frac{i}{2}p\left[f_l(p)\hat{h}_l^{-\prime}(pR)-f_l^*(p)\hat{h}_l^{+\prime}(pR)\right].$$

These conditions provide already the Jost function in terms of the regular solutions,

$$f_l(p) = -\frac{1}{p}\left(pu_{l,p}(R)\hat{h}_l^{+\prime}(pR)-u'_{l,p}(R)\hat{h}_l^+(pR)\right), \tag{4.48}$$

where we used that the Wronskian determinant of $\hat{h}_l^\pm$ is $2i$.

Compared to integral representations of the Jost function [131],

$$f_l(ik) = 1 + \int_0^\infty dr\; r\, V(r)\phi_{l,ik}(r) K_\nu(kr), \tag{4.49}$$

eq. (4.48) has the advantage that the solution is only needed at the point $r = R$. This simplifies the numerical procedure considerably.

In order to determine a unique solution of the differential equation (4.29), we need to pose an initial value problem. Given that $\phi_{l,p}(r)$ is the regular solution, we have for $r \to 0$

$$\phi_{l,p}(r) = u_{l,p}(r) \sim \hat{j}_l(pr) \sim \frac{\sqrt{\pi}}{\Gamma(l+3/2)} \left(\frac{z}{2}\right)^{l+1}.$$

The natural ansatz thus is

$$u_{l,p}(r) = \frac{\sqrt{\pi}}{\Gamma(l+3/2)} \left(\frac{pr}{2}\right)^{l+1} g_{l,p}(r),$$

with the inital value $g_{l,p}(0) = 1$. The differential equation for $g_{l,p}(r)$ reads

$$\left\{\frac{d^2}{dr^2} + 2\frac{l+1}{r}\frac{d}{dr} - V(r) + p^2\right\} g_{l,p}(r) = 0,$$

or, going to the needed imaginary p-axis and writing it as a first order differential equation with $(\partial/\partial r) g_{l,ip}(r) = v_{l,ip}(r)$,

$$\frac{d}{dr}\begin{pmatrix} g_{l,ip}(r) \\ v_{l,ip}(r) \end{pmatrix} = \begin{pmatrix} 0 & 1 \\ V(r) + p^2 & -\frac{2}{r}(l+1) \end{pmatrix} \begin{pmatrix} g_{l,ip}(r) \\ v_{l,ip}(r) \end{pmatrix}. \tag{4.50}$$

To fix the solution uniquely we only need to fix $v_{l,ip}(0)$. A power series ansatz for $g_{l,ip}(r)$ about $r = 0$ shows that, for $V(r) = \mathcal{O}(r^{-1+\epsilon})$, the condition reads $v_{l,ip}(0) = 0$. With this unique solution of (4.50), the Jost function takes the form

$$f_l(ip) = \frac{2}{\Gamma(l+3/2)} \left(\frac{pR}{2}\right)^{l+3/2} \left\{ g_{l,ip}(R) K_{l+3/2}(pR) + \frac{1}{p} g'_{l,ip}(R) K_{l+1/2}(pR) \right\}. \tag{4.51}$$

Finally, doing a partial integration and the substitution $q = \sqrt{k^2 - m^2}$ in eq. (4.39), the starting point for the numerical evaluation used for E_f is

$$E_f = \frac{1}{\pi} \sum_{l=0}^\infty (l+1/2) \int_0^\infty dq\; \left[\ln f_l(i\sqrt{q^2+m^2}) - \ln f_l^{asym}(i\sqrt{q^2+m^2})\right]. \tag{4.52}$$

Now we are prepared to use eqs. (4.46) and (4.52) for the calculation of E_{vac}^{ren}. We choose a potential with compact support $\Phi(r \geq R) = 0$. For the numerical analysis to come, we have used the dimensionless quantities

$$\epsilon = E_{vac}^{ren} R, \quad \mu = mR, \quad \rho = \frac{r}{R},$$
$$V(r) = \lambda' \Phi^2(r) = \frac{\lambda'}{R^2} \varphi^2(\rho).$$

The example we consider here is [73]

$$\varphi(\rho) = \frac{16 a \rho^2 (1-\rho)^2}{a + (1-2\rho)^2},$$

which is a kind of a spherical wall; the parameter a allows to vary the shape of the potential (see Fig. 5).

The most determining property of the potential $V(r)$ is the number and depth of boundstates. The parameter dependence on λ' of the $l = 0$ boundstates, determined as zeroes of the Jost function, are given in Fig. 5. As

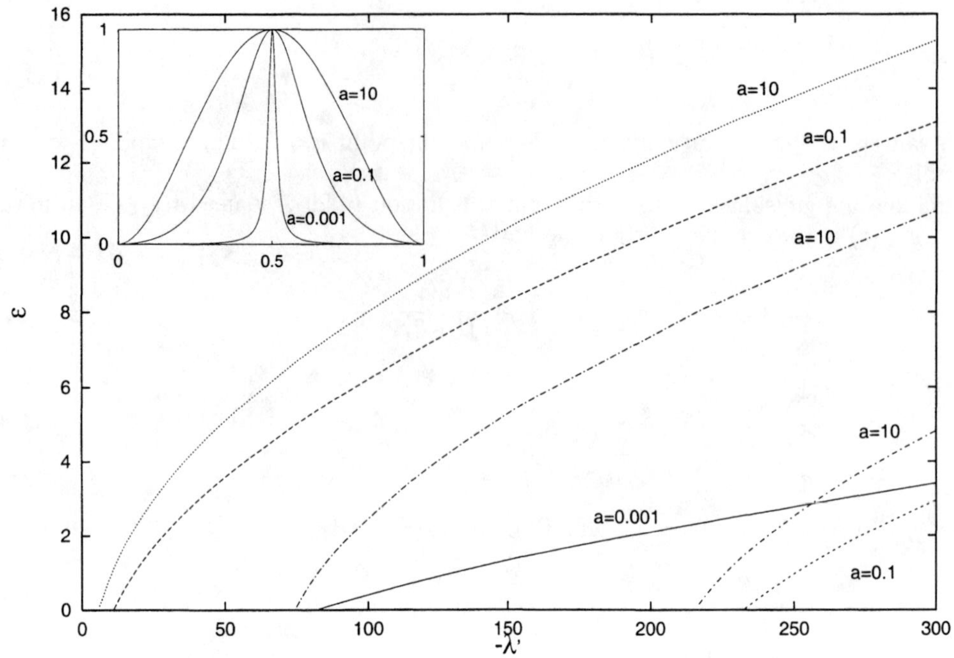

FIGURE 5. Energy of bound states for several shapes of the potential and negative λ'. (The inset shows $\varphi(\rho)$.)

is clear, number as well as depth of the boundstates increases with increasing $(-\lambda')$, and, as is seen, with increasing a.

To show clearly the way this influences the vacuum energy, we consider the dependence of the energy for fixed λ' as a function of the parameter a, Fig. 6, as well as for fixed $a = 1$ as a function of λ', Fig. 7.

Let us start with a description of Fig. 6. For a large enough, the contribution of the bound state(s) to the vacuum energy is large enough to overcompensate the scattering state contributions and the energy is negative. At some "critical value" of a, the energy of the only remaining bound state is so small, that the positive contributions of the scattering states get the upper hand and the vacuum energy is positive. Finally, for $a \to 0$ we have normalized $E_{vac}^{ren}[\Phi = 0] = 0$, such that at some point, E_{vac}^{ren} starts to decrease again with decreasing a.

The same features are clearly recovered in Fig. 7. For large enough $(-\lambda')$ a negative vacuum energy is obtained. With decreasing $(-\lambda')$ the energy increases and can be positive or negative depending on the parameter μ; for fixed value of R this means the mass m. Again, at some point the scattering states get the upper hand, the energy gets positive and tends to zero as λ' tends to zero. For positive λ' the energy starts to increase again, being virtually identical for $\lambda = \pm 2$, and it seems that E_{vac}^{ren} is a monotonically increasing function of λ' for $\lambda' \geq 0$.

Exactly the same features are observed if instead of the chosen spherical wall a lump-like potential concentrated around $r = 0$ is chosen, and it is expected that these features hold for any kind of potential [73]. In addition, the numerical analysis seems to indicate that, without bound states, the vacuum energy is always positive, indicating that scattering states contribute positively to the vacuum energy (a property we actually used already in the description of the figures), a result for which no analytical proof exists in three dimensions.

V CONCLUSIONS

In this contribution we have provided methods for the analysis of zeta functions associated with second order differential operators, when spherically symmetric external conditions are present. Basic analytical skills applied are contour and Mellin-Barnes integral representations and, at a second stage, the Barnes zeta function. Due to the close connection between zeta functions and various other spectral functions as the heat-kernel and determinants, applications to the calculation of heat-kernel coefficients and ground state energies have been given.

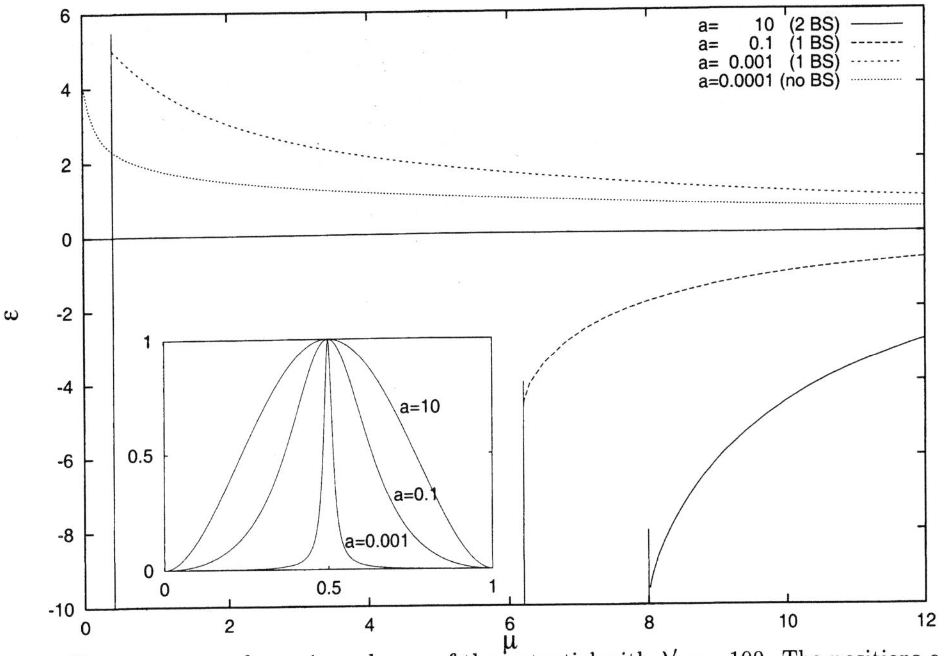

FIGURE 6. Vacuum energy for various shapes of the potential with $\lambda' = -100$. The positions of bound states at the μ axis are shown as vertical lines and the inset shows $\varphi(\rho)$.

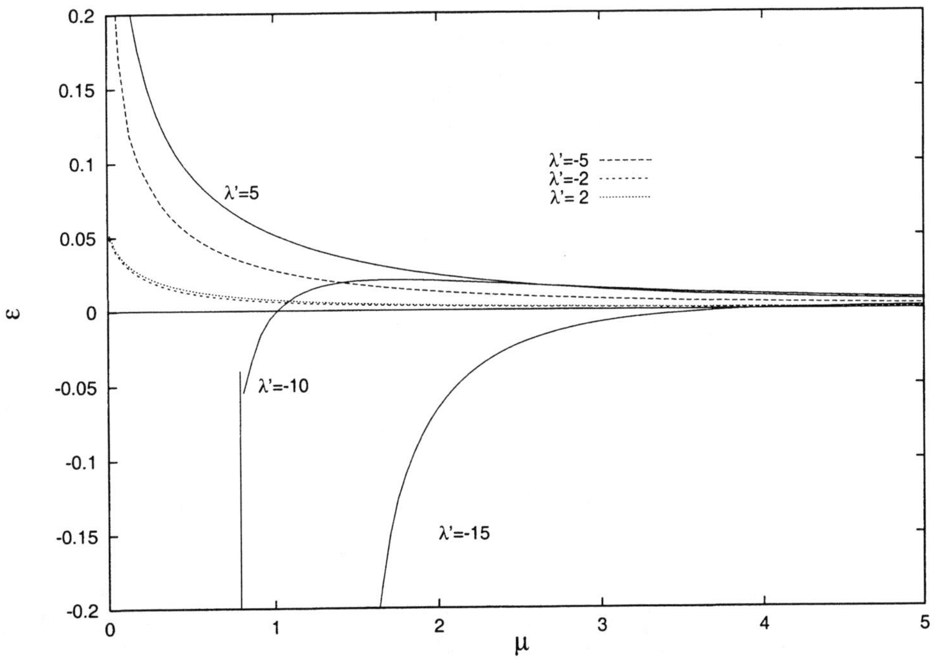

FIGURE 7. Vacuum energy for different magnitudes λ' with equal shape parameter $a = 1$.

The determination of heat-kernel coefficients is achieved by a combination of different methods, see Section III. As a first step write down the general form of the coefficients in terms of geometrical invariants. The possible invariants strongly depend on the boundary conditions considered as has been briefly described [9,123]. Next one can consider the special case calculations of the ball given in Section III B with this general form and fit unknown numerical constants, as has been explained in detail in Section III C. Together with information obtained from the product formula, this gives a very good starting point to apply the functorial techniques developed in [96]. Experience shows that, seemingly, any boundary condition can be successfully attacked in that way.

Afterwards, we concentrated on the analysis of vacuum energies. For the case of a massive scalar field inside a spherical shell we calculated the Casimir energy for the case when the field satisfies Dirichlet boundary conditions. Contrary to what one expects from the example of parallel plates, the energy depends strongly on the mass and, as a function of it, attraction as well as repulsion is possible. Mathematically, the calculation of the vacuum energy amounts to an evaluation of the associated zeta function at the specific point $s=-1/2$. With the help of the analytical approach provided in Section II this reduces to a numerical analysis of convergent integrals.

If an external potential is present we have expressed E_{vac}^{ren} through the associated quantum mechanical scattering dates, namely the Jost function. This is explicitly calculated by solving numerically a first order linear differential equation and by performing numerical integrations. The vacuum energy behaves in a way one can understand from physical reasons and the definition introduced by the normalization condition (4.23) seems to be very reasonable.

Let us emphasize that the external potential can be a solution of some classical nonlinear equation. But to apply the method one needs just any spherically symmetric potential of otherwise whatever shape to come immediately to a determination of E_{vac}^{ren}. This is hoped to find further applications in the context of the fields mentioned briefly in the Introduction.

ACKNOWLEDGMENTS

I am indebted to the institute of physics of the university of La Plata for the invitation to the conference and for the very kind hospitality during my stay (partially supported by ANPCyT of Argentina, under grant PICT 00039). Several discussions with the members of the theory group, especially Mariel Santangelo and Horacio Falomir, have been influential on the presentation as well as the content of the contribution. Finally, let me thank Michael Bordag, Stuart Dowker and Emilio Elizalde for the pleasant and fruitful scientific collaboration on the subjects presented. The work is supported by the DFG under contract number Bo 1112/4-2.

A REPRESENTATIONS FOR THE ASYMPTOTIC CONTRIBUTIONS

In this Appendix we derive explicit representations of the asymptotic contributions in the Casimir and ground state energies for massive fields. Let us start with $A_{-1}(s)$ for the massive scalar field, eq. (2.15), which, a little bit more explicit reads

$$A_{-1}(s)=2\frac{\sin(\pi s)}{\pi}\sum_{l=0}^{\infty}\nu^2\int_{ma/\nu}^{\infty}\left[\left(\frac{x\nu}{a}\right)^2-m^2\right]^{-s}\frac{\sqrt{1+x^2}-1}{x}, \tag{A.1}$$

of which we need the analytical continuation to $s=-1/2$. With the substitution $t=(x\nu/a)^2-m^2$, this expression results into the following one

$$\begin{aligned}A_{-1}(s)&=\frac{\sin(\pi s)}{\pi}\sum_{l=0}^{\infty}\nu\int_0^{\infty}dt\,\frac{t^{-s}}{t+m^2}\left\{\sqrt{\nu^2+a^2(t+m^2)}-\nu\right\}\\&=-\frac{1}{2\sqrt{\pi}}\frac{\sin(\pi s)}{\pi}\sum_{l=0}^{\infty}\nu\int_0^{\infty}dt\,t^{-s}\int_0^{\infty}d\alpha\,e^{-\alpha(t+m^2)}\end{aligned} \tag{A.2}$$

$$\times \int_0^\infty d\beta\, \beta^{-3/2} \left\{ e^{-\beta(\nu^2+a^2[t+m^2])} - e^{-\beta\nu^2} \right\},$$

where the Mellin integral representation for the single factors has been used. As we see, the β-integral is well defined. Introducing a regularization parameter δ, $A_{-1}(s)$ can then be written as

$$A_{-1}(s) = \lim_{\delta\to 0} \left[A_{-1}^1(s,\delta) + A_{-1}^2(s,\delta) \right], \tag{A.3}$$

with

$$A_{-1}^1(s,\delta) = -\frac{1}{2\sqrt{\pi}} \frac{\sin(\pi s)}{\pi} \sum_{l=0}^\infty \nu \int_0^\infty d\alpha\, e^{-\alpha m^2} \int_0^\infty d\beta\, \beta^{-3/2+\delta} e^{-\beta(\nu^2+a^2m^2)} \int_0^\infty dt\, t^{-s} e^{-t(\alpha+\beta a^2)}$$

and

$$A_{-1}^2(s,\delta) = \frac{1}{2\sqrt{\pi}} \Gamma(1-s) \frac{\sin(\pi s)}{\pi} \sum_{l=0}^\infty \nu \int_0^\infty d\alpha\, e^{-\alpha m^2} \alpha^{s-1} \int_0^\infty d\beta\, \beta^{-3/2+\delta} e^{-\beta\nu^2}.$$

Let us proceed with the remaining pieces. In $A_{-1}^1(s,\delta)$ two of the integrals can be done, yielding

$$A_{-1}^1(s,\delta) = -\frac{a^{1-2\delta}}{2\sqrt{\pi}\Gamma(s)} \Gamma(s+\delta-1/2) \times \sum_{l=0}^\infty \nu \int_0^\infty dy\, y^{\delta-3/2} \left[m^2 + y\left(\frac{\nu}{a}\right)^2 \right]^{1/2-s-\delta}. \tag{A.4}$$

For $A_{-1}^2(s,\delta)$, one gets

$$\begin{aligned} A_{-1}^2(s,\delta) &= \frac{m^{-2s}}{2\sqrt{\pi}} \Gamma(\delta-1/2) \sum_{l=0}^\infty \nu^{2-2\delta} \\ &= \frac{a^{1-2\delta}}{2\sqrt{\pi}\Gamma(s)} \Gamma(s+\delta-1/2) \sum_{l=0}^\infty \nu \int_0^\infty dx\, x^{s-1} \left[m^2 x + \left(\frac{\nu}{a}\right)^2 \right]^{1/2-s-\delta}. \end{aligned} \tag{A.5}$$

Adding up (A.4) and (A.5) yields

$$\begin{aligned} A_{-1}(s) &= \frac{a}{2\sqrt{\pi}\Gamma(s)} \Gamma(s-1/2) \sum_{l=0}^\infty \nu \int_0^1 dx\, x^{s-1} \left[m^2 x + \left(\frac{\nu}{a}\right)^2 \right]^{1/2-s} \\ &= \frac{a^{2s}}{2\sqrt{\pi}} \frac{\Gamma(s-1/2)}{\Gamma(s+1)} \sum_{l=0}^\infty \nu \left\{ \frac{1}{[\nu^2+(ma)^2]^{s-1/2}} \right. \\ &\quad \left. + \left(s-\frac{1}{2}\right)(ma)^2 \int_0^1 dx\, \frac{x^s}{(\nu^2+(ma)^2 x)^{s+1/2}} \right\}, \end{aligned} \tag{A.6}$$

where in the last step a partial integration has been performed such that the x-integral is well behaved at $s=-1/2$.

This is a form suited for the treatment of the angular momentum sum, which is performed using

$$\sum_{\nu=1/2}^\infty f(\nu) = \int_0^\infty d\nu\, f(\nu) - i \int_0^\infty d\nu\, \frac{f(i\nu+\epsilon) - f(-i\nu+\epsilon)}{1+e^{2\pi\nu}}, \tag{A.7}$$

where $\epsilon \to 0$ is understood and appropriate analytic properties of the function $f(\nu)$ are assumed. This is often used in finite temperature field theory and is equivalent there to switching from imaginary time to real time [132]. When expanding the function $f(\nu)$ in a Taylor series, one arrives at the well known Euler-Maclaurin summation formula (a thorough treatment of the Euler-Maclaurin summation formula can be found in Ref. [133]).

For $A_{-1}(s)$ the relevant application of eq. (A.7) is

$$\sum_{\nu=1/2,3/2,\ldots}^{\infty} \nu^{2n+1}\left(1+\left(\frac{\nu}{x}\right)^2\right)^{-s} = \frac{1}{2}\frac{n!\Gamma(s-n-1)}{\Gamma(s)}x^{2n+2}$$
$$+(-1)^n 2\int_0^x d\nu\, \frac{\nu^{2n+1}}{1+e^{2\pi\nu}}\left(1-\left(\frac{\nu}{x}\right)^2\right)^{-s}$$
$$+(-1)^n 2\cos(\pi s)\int_x^\infty d\nu\, \frac{\nu^{2n+1}}{1+e^{2\pi\nu}}\left(\left(\frac{\nu}{x}\right)^2-1\right)^{-s}. \qquad \text{(A.8)}$$

Together with the integral

$$\int_0^\infty \frac{\nu^n}{1+e^{2\pi\nu}} = \frac{n!}{(2\pi)^{n+1}}\eta(n+1),$$

where $\eta(s)$ is the eta-function,

$$\eta(s) = \sum_{k=1}^\infty \frac{(-1)^{k+1}}{k^s},$$

the following final form of $A_{-1}(s)$ is obtained,

$$A_{-1}(-1/2+s) = \left(\frac{1}{s}+\ln a^2\right)\left(\frac{7}{1920\pi a}+\frac{m^2 a}{48\pi}-\frac{m^4a^3}{24\pi}\right)$$
$$+\ln 4\left(\frac{7}{1920\pi a}+\frac{m^2a}{48\pi}-\frac{m^4a^3}{24\pi}\right) \qquad \text{(A.9)}$$
$$+\frac{7}{1920\pi a}-\frac{m^2a}{48\pi}+\frac{m^4a^3}{48\pi}(1+4\ln(ma))$$
$$-\frac{1}{\pi a}\int_0^\infty d\nu\,\frac{\nu}{1+e^{2\pi\nu}}(\nu^2-m^2a^2)\ln|\nu^2-m^2a^2|$$
$$-\frac{2m^2a}{\pi}\int_0^\infty d\nu\,\frac{\nu}{1+e^{2\pi\nu}}\left(\ln|\nu^2-m^2a^2|+\frac{\nu}{ma}\ln\left|\frac{ma+\nu}{ma-\nu}\right|\right).$$

In the same way, for

$$A_0(s) = -\frac{m^{-2s}}{2}\sum_{l=0}^\infty \nu\left[1+\left(\frac{\nu}{ma}\right)^2\right]^{-2s},$$

using (A.8), it follows

$$A_0(s) = \frac{1}{6}a^2m^3 - m\int_0^{ma} d\nu\,\frac{\nu}{1+e^{2\pi\nu}}\sqrt{1-\left(\frac{\nu}{ma}\right)^2}. \qquad \text{(A.10)}$$

For the remaining asymptotic contributions, the central spectral function is, see eq. (4.13),

$$f(s;c;b)=\sum_{\nu=1/2,3/2,\ldots}^{\infty}\nu^c\left(1+\left(\frac{\nu}{ma}\right)^2\right)^{-s-b}.$$

To deal with all values of c and b needed, in addition to eq. (A.8), we need

$$\sum_{\nu=1/2,3/2,\ldots}^{\infty}\nu^{2n}\left(1+\left(\frac{\nu}{x}\right)^2\right)^{-s}=\frac{1}{2}\frac{\Gamma(n+1/2)\Gamma(s-n-1/2)}{\Gamma(s)}x^{2n+1}$$
$$-(-1)^n 2\sin(\pi s)\int_x^\infty d\nu\,\frac{\nu^{2n}}{1+e^{2\pi\nu}}\left(\left(\frac{\nu}{x}\right)^2-1\right)^{-s}. \tag{A.11}$$

Using partial integrations one can obtain representations valid for values of s needed for the asymptotic contributions. To simplify notation, in the following we shall use x for (ma) and $f(c;b)=f(-1/2;c;b)$. A full list of ingredients needed is the following:

$$f(1;1/2)=-\frac{1}{2}x^2+\frac{1}{24},$$
$$\frac{d}{ds}\Big|_{s=-1/2}f(s;1;1/2)=-\frac{1}{2}x^2-2\int_0^\infty d\nu\frac{\nu}{1+e^{2\pi\nu}}\ln\left|1-\left(\frac{\nu}{x}\right)^2\right|,$$
$$f(1;3/2)=\frac{x^2}{2(s+1/2)}+x^2\ln x+x^2\int_0^\infty d\nu\left(\frac{d}{d\nu}\frac{1}{1+e^{2\pi\nu}}\right)\ln\left|\nu^2-x^2\right|,$$
$$f(1;1)=2x^2\int_0^x d\nu\left(\frac{d}{d\nu}\frac{1}{1+e^{2\pi\nu}}\right)\left[1-\left(\frac{\nu}{x}\right)^2\right]^{1/2},$$
$$f(1;2)=-2x^2\int_0^x d\nu\left(\frac{d}{d\nu}\frac{1}{1+e^{2\pi\nu}}\right)\left|1-\left(\frac{\nu}{x}\right)^2\right|^{-1/2},$$
$$f(3;2)=\frac{5}{2}x^4+2x^4\int_0^x d\nu\left[\frac{d}{d\nu}\left(\frac{1}{\nu}\frac{d}{d\nu}\frac{\nu^2}{1+e^{2\pi\nu}}\right)\right]\left[1-\left(\frac{\nu}{x}\right)^2\right]^{1/2},$$
$$f(3;5/2)=\frac{x^4}{2(s+1/2)}+(\ln x-1/2)x^4+\frac{x^4}{2}\int_0^\infty d\nu\left[\frac{d}{d\nu}\left(\frac{1}{\nu}\frac{d}{d\nu}\frac{\nu^2}{1+e^{2\pi\nu}}\right)\right]\ln\left|\nu^2-x^2\right|,$$
$$f(3;3)=-\frac{2}{3}x^4\int_0^x d\nu\left[\frac{d}{d\nu}\left(\frac{1}{\nu}\frac{d}{d\nu}\frac{\nu^2}{1+e^{2\pi\nu}}\right)\right]\left[1-\left(\frac{\nu}{x}\right)^2\right]^{-1/2},$$
$$f(5;7/2)=\frac{x^6}{2(s+1/2)}+(\ln x-3/4)x^6$$
$$+\frac{x^6}{8}\int_0^\infty d\nu\left[\frac{d}{d\nu}\left(\frac{1}{\nu}\frac{d}{d\nu}\frac{1}{\nu}\frac{d}{d\nu}\frac{\nu^4}{1+e^{2\pi\nu}}\right)\right]\ln\left|\nu^2-x^2\right|,$$
$$f(7;9/2)=\frac{x^8}{2(s+1/2)}+(\ln x-11/12)x^8$$
$$+\frac{x^8}{48}\int_0^\infty d\nu\left[\frac{d}{d\nu}\left(\frac{1}{\nu}\frac{d}{d\nu}\frac{1}{\nu}\frac{d}{d\nu}\frac{1}{\nu}\frac{d}{d\nu}\frac{\nu^6}{1+e^{2\pi\nu}}\right)\right]\ln\left|\nu^2-x^2\right|. \tag{A.12}$$

REFERENCES

1. M. Bordag and K. Kirsten. *The ground state energy of a spinor field in the background of a finite radius flux tube,* hep–th/9812060.
2. M. Bordag, K. Kirsten and D. Vassilevich. Phys. Rev. D, to appear, *On the ground state energy for a penetrable sphere and for a dielectric ball,* hep–th/9811015.
3. J.S. Dowker and R. Critchley. Phys. Rev. **D 13**, 3224 (1976).
4. S.W. Hawking. Commun. Math. Phys. **55**, 133 (1977).

5. H. Weyl. Math. Ann. **71**, 441 (1912).
6. A. Voros. Commun. Math. Phys. **110**, 439 (1987).
7. S. Minakshisundaram and A. Pleijel. Can. J. Math. **1**, 242 (1949).
8. S.J. Minakshisundaram. Indian Math. Soc. **17**, 158 (1953).
9. P.B. Gilkey. *Invariance Theory, The Heat Equation and the Atiyah-Singer Index Theorem, 2nd. Edn.* CTC Press, Boca Raton, (1995).
10. R. Seeley. *Complex Powers of an Elliptic Operator.* Singular Integrals (Proc. Symp. Pure Math., Chicago), (1966). Amer. Math. Soc., Providence, R.I.
11. D.B. Ray and I.M. Singer. Advances in Math. **7**, 145 (1971).
12. W. Franz. J. Reine Angew. Math. **173**, 245 (1935).
13. J. Duistermaat and V. Guilleimin. Invent. Math. **29**, 39 (1975).
14. J. Bruning and R. Seeley. Adv. in Maths. **58**, 133 (1985).
15. G. Grubb. *Functional Calculus of Pseudo-Differential Boundary Problems.* In *Progress in Mathematics.* J. Coates and S. Helgason, editors. Birkhauser, Boston, (1986).
16. N. Kurokawa. Proc. Japan Acad., Ser. A **64**, 21 (1988).
17. P.B. Gilkey and G. Grubb. Commun. Part. Diff. Equat. **23**, 777 (1998).
18. J. Jorgenson and S. Lang. *Basic Analysis of regularized Series and Products.* Springer, Berlin, (1993). Lecture Notes in Mathematics 1564.
19. G. Cognola, L. Vanzo and S. Zerbini. J. Math. Phys. **33**, 222 (1992).
20. J.S. Dowker and G. Kennedy. J. Phys. **A 11**, 895 (1978).
21. J. Ambjørn and S. Wolfram. Ann. Phys. **147**, 1 (1983).
22. J.S. Dowker. Class. Quantum Grav. **1**, 359 (1984).
23. B.P. Dolan and C. Nash. Commun. Math. Phys. **148**, 139 (1992).
24. F. Caruso, N.P. Neto, B.F. Svaiter and N.F. Svaiter. Phys. Rev. D **43**, 1300 (1991).
25. A.A. Bytsenko, G. Cognola, L. Vanzo and S. Zerbini. Phys. Rep. **266**, 1 (1996).
26. E. Elizalde, S. D. Odintsov, A. Romeo, A.A. Bytsenko and S. Zerbini. *Zeta Regularization Techniques with Applications.* World Scientific, Singapore, (1994).
27. E. Elizalde. *Ten Physical applications of Spectral Zeta Functions.* Springer-Verlag, (1995). Berlin.
28. S.K. Blau, M. Visser and A. Wipf. Nucl. Phys. **B310**, 163 (1988).
29. L.C. De Albuquerque. Phys. Rev. D **55**, 7754 (1997).
30. C.G. Beneventano, M. De Francia and E.M. Santangelo Dirac fields in the background of a magnetic flux string and spectral boundary conditions, hep–th/9809081.
31. G.W. Gibbons. Phys. Lett. **A 60**, 385 (1977).
32. C. Wiesendanger and A. Wipf. Ann. Phys. **233**, 125 (1994).
33. G. Cognola, K. Kirsten and S. Zerbini. Phys. Rev. **D 48**, 790 (1993).
34. R. Rajaraman. *Solitons and Instantons.* North-Holland Publishing Company, (1982). Amsterdam.
35. G. t'Hooft. Nucl. Phys. B **79**, 276 (1974).
36. A.M. Polyakov. JETP Lett. **20**, 194 (1974).
37. F.R. Klinkhammer and N.S. Manton. Phys. Rev. D **30**, 2212 (1984).
38. J.M. Gipson and H.Ch. Tze. Nucl. Phys. B **183**, 524 (1981).
39. J.M. Gipson. Nucl. Phys. B **231**, 365 (1984).
40. J. Ambjorn and V.A. Rubakov. Nucl. Phys. B **256**, 434 (1985).
41. G. Eilam, D. Klabucar and A. Stern. Phys. Rev. Lett. **56**, 1331 (1986).
42. R. Friedberg and T.D. Lee. Phys. Rev. D **15**, 1694 (1977).
43. R. Friedberg and T.D. Lee. Phys. Rev. D **16**, 1096 (1977).
44. T.H.R. Skyrme. Proc. R. Soc. A **260**, 127 (1961).
45. T.H.R. Skyrme. Nucl. Phys. B **31**, 556 (1972).
46. G.S. Adkins, C.R. Nappi and E. Witten. Nucl. Phys. B **228**, 552 (1993).
47. S. Coleman. Phys. Rev. D **15**, 2929 (1977).
48. C.G. Callan and S. Coleman. Phys. Rev. D **16**, 1762 (1977).
49. L.-H. Chan. Phys. Rev. Lett. **54**, 1222 (1985).
50. J. Baacke. Z. Phys. C **47**, 263 (1990).
51. J. Baacke and V.G. Kiselev. Phys. Rev. D **48**, 5648 (1993).
52. C.L.Y. Lee. Phys. Rev. D **49**, 4101 (1994).
53. D.E. Brahm and C.L.Y. Lee. Phys. Rev. D **49**, 4094 (1994).
54. A.W. Wipf. Helvetica Physica Acta **58**, 531 (1985).
55. A.I. Bochkarev. Phys. Rev. D **46**, 5550 (1992).
56. M. Bordag. J. Phys. **A28**, 755 (1995).
57. H. Nastase, M. Stephanov, P. van Nieuwenhuizen and A. Rebhan. *Topological boundary conditions, the BPS bound,*

and elimination of ambiguities in the quantum mass of solitons, hep-th/9802074.
58. M. Bordag, E. Elizalde and K. Kirsten. J. Math. Phys. **37**, 895 (1996).
59. M. Bordag, K. Kirsten and S. Dowker. Commun. Math. Phys. **182**, 371 (1996).
60. J.S. Dowker and K. Kirsten. Communications in Analysis and Geometry, to appear.
61. J.S. Dowker and K. Kirsten. *Smeared heat-kernel coefficients on the ball and generalized cone*, hep-th/9803094.
62. J.S. Dowker and K. Kirsten. *The $a_{3/2}$ heat kernel coefficients for oblique boundary conditions*, hep-th/9806168.
63. S. Dowker, P. Gilkey and K. Kirsten. Geometric Aspects of PDE, Roskilde Conference, Sept. 1998.
64. H.B.G. Casimir. Physica **19**, 846 (1953).
65. T.H. Boyer. Phys. Rev. **174**, 1764 (1968).
66. K.A. Milton, L.L. DeRaad Jr. and J. Schwinger. Ann. Phys. **115**, 388 (1978).
67. R. Balian and B. Duplantier. Ann. Phys. **112**, 165 (1978).
68. G. Plunien, B. Müller and W. Greiner. Phys. Rep. **134**, 87 (1986).
69. M. de Francia, H. Falomir and M. Loewe. Phys. Rev. D **55**, 2477 (1997).
70. M. de Francia. Phys. Rev. D **50**, 2908 (1994).
71. E. Elizalde, M. Bordag and K. Kirsten. J. Phys. A: Math. Gen. **31**, 1743 (1998).
72. M. Bordag and K. Kirsten. Phys. Rev. D **53**, 5753 (1996).
73. M. Bordag, M. Hellmund and K. Kirsten. *Dependence of the vacuum energy on spherically symmetric background fields*, in preparation.
74. Staff of the Bateman Manuscript Project. *Higher Trascendental Functions.* In *A. Erderlyi.* McGraw-Hill Book Company, New York, (1955). Based on the notes of Harry Bateman.
75. A.Yu. Kamenshchik and I.V. Mishakov. Int. J. Mod. Phys. **A7**, 3713 (1992).
76. A.O. Barvinsky, A.Yu. Kamenshchik and I.P. Karmazin. Ann. Phys. **219**, 201 (1991).
77. I.S. Gradshteyn and I.M. Ryzhik. *Table of Integrals, Series and Products.* Academic Press, New York, (1965).
78. E. Elizalde, S. Leseduarte and A. Romeo. J. Phys. **A26**, 2409 (1993).
79. S. Leseduarte and A. Romeo. J. Phys. **A27**, 2483 (1994).
80. M. Abramowitz and I.A. Stegun. *Handbook of Mathematical Functions.* Dover, New York, (1970).
81. H.A. Weldon. Nucl. Phys. **B270**, 79 (1986).
82. E. Elizalde and A. Romeo. Phys. Rev. **D 40**, 436 (1989).
83. E. Elizalde, K. Kirsten and S. Zerbini. J. Phys. **A28**, 617 (1995).
84. M. Bordag, E. Elizalde, B. Geyer and K. Kirsten. Commun. Math. Phys. **179**, 215 (1996).
85. J.S. Dowker. Class. Quantum Grav. **13**, 1 (1996).
86. J.S. Dowker. Phys. Lett. B **366**, 89 (1996).
87. H. Falomir, E. Gamboa Saraví, M.A. Muschietti, E.M. Santangelo and J. Solomin. J. Math. Phys. **37**, 5805 (1996).
88. H. Falomir, R.E. Gamboa Saravi and E.M. Santangelo. J. Math. Phys. **39**, 532 (1998).
89. G. Esposito. *Quantum Gravity, Quantum Cosmology and Lorentzian Geometries.* Lecture Notes in Physics, Monographs, Vol. m12, (1994). Berlin: Springer-Verlag.
90. G. Esposito, A.Yu. Kamenshchik and G. Pollifrone. *Euclidean Quantum Gravity on Manifolds with Boundary.* Dordrecht: Kluwer, (1997). Fundamental Theories of Physics 85.
91. I.G. Avramidi. Nucl. Phys. **B355**, 712 (1991).
92. S.A. Fulling and G. Kennedy. Amer. Math. Soc. **310**, 583 (1988).
93. A. Van den Ven. Class. Quantum Grav. **15**, 2311 (1998).
94. R. Seeley. Amer. J. Math. **91**, 889 (1969).
95. T. Branson. Trans. Amer. Math. Soc. **347**, 3671 (1995).
96. T.P. Branson and P.B. Gilkey. Commun. Part. Diff. Equat. **15**, 245 (1990).
97. K. Kirsten. Class. Quantum Grav. **15**, L5 (1998).
98. T.P. Branson, P.B. Gilkey and D.V. Vassilevich. Bollettin U.M.I. **11-B**, 39 (1997).
99. F. Treves. *Introduction to Pseudodifferential and Fourier Integral Operators, Vol. 1.* Plenum, (1980). New York.
100. Yu.V. Egorov and M.A. Shubin. *Partial Differential Equations.* Springer, (1991). Berlin.
101. S.G. Krantz. *Partial Differential Equations and Complex Analysis.* Boca Raton, FL: CRC, (1992).
102. S.K. Blau, M. Visser and A. Wipf. Int. J. Mod. Phys. A **4**, 1467 (1989).
103. J.S. Dowker. Phys. Rev. D **33**, 3150 (1986).
104. J.S. Dowker and J.P. Schofield. Phys. Rev. **D 38**, 3327 (1988).
105. J.S. Dowker and J.P. Schofield. Nucl. Phys. **B327**, 267 (1989).
106. J.S. Dowker and J.P. Schofield. J. Math. Phys. **31**, 808 (1990).
107. S.K. Blau, M. Visser and A. Wipf. Phys. Lett. B **209**, 209 (1988).
108. I.L. Buchbinder, V.P. Gusynin and P.I. Fomin. J. Nucl. Phys. **44**, 828 (1986).
109. V.P. Gusynin and V.V. Romankov. J. Nucl. Phys. **46**, 1832 (1987).
110. A. Wipf and S. Duerr. Nucl. Phys. B **443**, 201 (1995).
111. R.E. Gamboa Saravi, M.A. Muschietti, F.A. Schaposnik and J.E. Solomin. Ann. Phys. **157**, 360 (1984).

112. P. Gilkey and L. Smith. Comm. on Pure and Appl. Math. **XXXVI**, 83 (1983).
113. H. Weyl. Rendiconti del Circolo Mat. di Palermo **39**, 1 (1915).
114. T. Branson, P. Gilkey and B. Ørsted. Proc. Amer. Math. Soc. **109**, 437 (1990).
115. M. Levitin. Differential Geometry and its Application **8**, 35 (1998).
116. F.W. Olver. Phil. Trans. Roy. Soc. A **247**, 328 (1954).
117. E.W. Barnes. Trans. Camb. Philos. Soc **19**, 374, 426 (1903).
118. E.W. Barnes. Trans. Camb. Philos. Soc. **19**, 426 (1903).
119. J.S. Dowker. J. Math. Phys. **35**, 4989 (1994).
120. A. Chodos and E. Myers. Ann. Phys. **156**, 412 (1984).
121. N.E. Nörlund. Acta Math. **43**, 21 (1922).
122. H.C. Luckock. J. Math. Phys. **32**, 1755 (1991).
123. I.G. Avramidi and G. Esposito. Class. Quantum Grav. **15**, 281 (1998).
124. I.G. Avramidi and G. Esposito. Commun. Math. Phys. **200**, 495 (1999).
125. J.S. Dowker and K. Kirsten. Class. Quantum Grav. **14**, L169 (1997).
126. M. Bordag, E. Elizalde, K. Kirsten and S. Leseduarte. Phys. Rev. D **56**, 4896 (1997).
127. C.G. Beneventano and E.M. Santangelo. Int. J. Mod. Phys. **11**, 2871 (1996).
128. G. Kennedy. J. Phys. **A 11**, 173 (1978).
129. A.A. Actor and I. Bender. Phys. Rev. D **52**, 3581 (1995).
130. A.A. Actor and I. Bender. Casimir Effect with a Semihard Boundary September 21, 1995.
131. J.R. Taylor. *Scattering Theory.* Wiley, (1972). New York.
132. J.I. Kapusta. *Finite-Temperature Field Theory.* Cambridge University Press, Cambridge, (1989).
133. R. Wong. *Asymptotic Approximations of Integrals.* Academic Press, (1989). New York.

Three-dimensional quantum geometry and black holes

Máximo Bañados

Departamento de Física Teórica, Universidad de Zaragoza, Ciudad Universitaria 50009, Zaragoza, Spain.

Abstract. We review some aspects of three-dimensional quantum gravity with emphasis in the 'CFT → Geometry' map that follows from the Brown-Henneaux conformal algebra. The general solution to the classical equations of motion with anti-de Sitter boundary conditions is displayed. This solution is parametrized by two functions which become Virasoro operators after quantisation. A map from the space of states to the space of classical solutions is exhibited. Some recent proposals to understand the Bekenstein-Hawking entropy are reviewed in this context. The origin of the boundary degrees of freedom arising in 2+1 gravity is analysed in detail using a Hamiltonian Chern-Simons formalism.

I INTRODUCTION

General relativity is a highly complicated non-linear field theory both classically and quantum mechanically. Even though a large number of classical solutions exists, a general classification of the space of solutions has never been achieved. The non-renormabizibility of quantum gravity is not related to this issue, but if the general structure of the space of solutions of the Einstein equations was known, then the quantum version of phase space perhaps would be more manageable.

It is this aspect of three-dimensional gravity that makes it attractive because the general solution to the equations of motion can be written down. In this paper, we shall exploit this fact trying to formulate a quantum theory of black holes by quantizing the space of solutions directly.

Another important aspect of three-dimensional gravity is its formulation as a Chern-Simons theory [1]. Quantum Chern-Simons theory is well understood for compact groups [2–4]. However, we shall be interested in Euclidean gravity with a negative cosmological constant whose associated group is $SL(2,C)$, which is not compact. The quantization is then not straightforward. We shall follow an alternative route by first solving the equations of motion with prescribed boundary conditions and then quantise. We shall see that the boundary conditions will play an important role in making the quantum theory well-defined.

A Brief description of the results contained in this article

Let us start by briefly mentioning, without proofs, the main results which will be of interest for us here. The relevant proofs will be given below. We should remark at this point that most of the results presented here are known in the literature in various contexts (the relevant quotations will be given in the main text). The aim of this article is to put things together in a self-contained framework, and to explore some aspects of quantum black holes in three dimensions.

Let M be a three dimensional manifold with a boundary denoted by ∂M. We assume that ∂M has the topology of a 2-torus. Let $\{w, \bar{w}, \rho\}$ coordinates on M such that the boundary is located at $e^\rho =: lr \to \infty$, and $w = \varphi + it$, $\bar{w} = \varphi - it$ are complex coordinates on the torus. The three-dimensional metric [1],

$$ds^2 = 4Gl(Ldw^2 + \bar{L}d\bar{w}^2) + \left(l^2 e^{2\rho} + 16G^2 L\bar{L}e^{-2\rho}\right) dw d\bar{w} + l^2 d\rho^2, \quad (1)$$

[1]) G represents Newton's constant and l is the anti-de Sitter radius related to a negative cosmological constant by $\Lambda = -1/l^2$. We set $\hbar = 1$ throughout the paper.

CP484, *Trends in Theoretical Physics II*, edited by H. Falomir, R. E. Gamboa Saraví, and F. A. Schaposnik

where $L = L(w)$ and $\bar{L} = \bar{L}(\bar{w})$ are arbitrary functions of their arguments satisfies the following properties:

(i) Exact solution

The metric (1) is an exact solution to the three-dimensional vacuum Einstein equations with a negative cosmological constant $\Lambda = -1/l^2$. The leading and first subleading terms of (1) (in powers of $r = e^\rho$) are, of course, the ones dictated by the general analysis of [5] for asymptotically anti-de Sitter spacetimes. What is perhaps not so well-known is that adding the term $e^{-2\rho}L\bar{L}$, the metric becomes an *exact* solution. Most importantly, (1) is the most general solution, up to trivial diffeomorphisms, which is asymptotically anti-de Sitter. Note that since (1) contains two arbitrary functions, it gives rise to an infinite number of solutions.

We shall work in the Euclidean sector of the theory. This means that w is a complex coordinate related to the spacetime coordinates as $w = \varphi + it$. The metric (1) is then complex. As we shall see, this will not bring in any problems in the quantization. For real values of w, the metric (1) is a solution to the Minkowskian equations of motion.

(ii) Physical degrees of freedom

Two solutions of the form (1) with different values for L and $\bar{L}$ represent physically different configurations which cannot be connected via a gauge transformation. In the quantum theory, where L and $\bar{L}$ will become operators, different expectations values for them will be associated to different solutions.

This is a non-trivial statement. Since (1) is a solution to the three-dimensional Einstein equations it has constant curvature and then, locally, is isometric to anti-de Sitter space (see Eq. (11) below). The point here is that the coordinate transformations which change the values of L and $\bar{L}$ are not generated by constraints and therefore they are not gauge symmetries. This point will be analysed in detail in the Chern-Simons formulation in section II C, and in the metric formulation in Sec. IV B.

(iii) Residual conformal symmetry

The metric (1) has a residual conformal symmetry. There exists a change of coordinates $\{w, \bar{w}, \rho\} \to \{w', \bar{w}', \rho'\}$ such that the new metric looks exactly like (1) with new functions L' and $\bar{L}'$. See Sec. IV B for the proof of this statement. This change of coordinates is parametrized by two functions $\varepsilon(w)$ and $\bar{\varepsilon}(\bar{w})$. The new function L' is related to the old one via $L' = L + \delta L$ with,

$$\delta L = i(\varepsilon\partial L + 2\partial\varepsilon L - \frac{c}{12}\partial^3\varepsilon) \tag{2}$$

where c is given by

$$c = \frac{3l}{2G}. \tag{3}$$

The same transformation holds for $\bar{L}$. Thus, under this symmetry, L and $\bar{L}$ are quasi-primary fields of conformal dimension 2. This symmetry, properly defined acting on the gravitational variables, can be shown to be also a global symmetry of the action [5]. The canonical generators are the functions L and $\bar{L}$ themselves and the associated algebra is the Virasoro algebra [5],

$$[L_n, L_m] = (n-m)L_{n+m} + \frac{c}{12}n^3\delta_{n+m}, \tag{4}$$

where the central charge is defined in (3) and,

$$L(w) = \sum_{n\in Z} L_n e^{inw}. \tag{5}$$

We are using here a non-standard form of the central term. The usual $n(n^2-1)$ form can be obtained simply by shifting the L_0 mode as $L_0 \to L_0 - c/24$. This convention is appropriated to the black hole background which has an exact $SO(2) \times SO(2)$ invariance. See [6] for a discussion on this point in the supergravity context.

(iv) Asymptotic conformal symmetry

The residual symmetry (2) is not an exact symmetry of any background metric. Rather it is a symmetry of the space of solutions described by (1); it maps one solution into another one. However, this symmetry can be regarded as an *asymptotic* symmetry of anti-de Sitter space because the $r \to \infty$ form of (1) is asymptotic Euclidean adS_3 space (note the redefinition of coordinates: $w = \varphi + it$, $r = le^{\rho}$),

$$ds^2_{(r\to\infty)} \to r^2(dt^2 + d\varphi^2). \tag{6}$$

(We have kept here only the leading terms in powers of r, but note that there is also a term $2i(L - \bar{L})dtd\varphi$ of order one which is allowed by the boundary conditions [5].) Since the asymptotic behaviour (6) does not see L and $\bar{L}$ it is invariant under the transformation (2). This symmetry was discovered in [5]. See [7,8] for recent discussions.

(v) Basic dynamical variables and induced Poisson brackets

Up to some global issues, the Virasoro algebra (4) can be regarded as the basic Poisson bracket algebra of the gauge-fixed residual variables. In other words, the functions $L(w)$ and $\bar{L}(\bar{w})$ appearing in (1) are the part of the metric field $g_{\mu\nu}(x^\mu)$ which survives after the gauge is fixed (i.e., after gauge conditions are imposed and the constraints solved), with anti-de Sitter boundary conditions. The equal-time Poisson bracket of general relativity, $\{\pi^{ij}, g_{kl}\} = \delta^{ij}_{kl}$, induces the Virasoro algebra (4) on the residual dynamical functions L and $\bar{L}$. A technical note is convenient here. From the dynamical point of view, L and $\bar{L}$ both depend on w and $\bar{w}$. However, the gauge-fixed equations of motion read $\partial_{\bar{w}}L = 0$ and $\partial_w\bar{L} = 0$ leading to $L = L(w)$ and $\bar{L} = \bar{L}(\bar{w})$.

The identification of the Virasoro operators as basic variables is not natural from the point of view of conformal field theory. In the standard situation, the Virasoro algebra is associated to a symmetry rather than to the basic commutator. (A good analogy is to consider the angular momentum components L_i as basic variables, satisfying $[L_i, L_j] = i\epsilon_{ijk}L_k$, without knowing the existence of q^i, p_j.) One of the main problems of three-dimensional quantum gravity is to identify what is the conformal field theory behind the Brown-Henneaux conformal symmetry. Classically, Liouville theory [9] seems to be a good candidate, however, its quantisation does not give the right counting for the black hole entropy degeneracy [10]. Treating the Virasoro algebra as basic Poisson algebra is also well-motivated classically but it does not give the right counting either (see Sec. IV D). We shall discuss this issues in Sec. IV D, as well as two possible modifications of the boundary dynamics which do provide the right counting of states.

(vi) Black holes and adS space

The space of solutions described by (1) contains black holes. If L and $\bar{L}$ are constants (no $w, \bar{w}$ dependence) with only L_0, $\bar{L}_0$ different from zero and parametrized as

$$Ml = L_0 + \bar{L}_0, \qquad J = L_0 - \bar{L}_0, \tag{7}$$

then the metric (1) is globally isometric to the Euclidean three-dimensional black hole [11,12] of mass M and angular momentum J. Eq. (7) means that the Virasoro operators vanish on the vacuum black hole. The corresponding algebra is (4). The Euclidean black hole metric in Schwarzschild coordinates reads

$$ds^2 = l^2N^2dt^2 + N^{-2}dr^2 + r^2(d\varphi + iN^\varphi dt)^2, \tag{8}$$

with

$$N^2(r) = -8MG + \frac{r^2}{l^2} + \frac{16G^2J^2}{r^2}, \tag{9}$$

$$N^\varphi(r) = \frac{4GJ}{r^2}. \tag{10}$$

See Sec. IV C for the explicit transition from (1) to (8).

For $M > |J|$, this metric has two horizons which are the solutions to the equation $N^2(r_\pm) = 0$. It is often convenient to define the Euclidean angular momentum J_E as $J_E = iJ$ and then the i in (8) does not occur. Note also that in the Euclidean sector, the black hole manifold does not see the interior $r < r_+$. See Sec. IV C for more details on the relation between (8) and (1). Note that since the coordinates w and $\bar{w}$ are defined on a torus, the only globally well-defined solutions are the ones with constant L and $\bar{L}$.

For $J = 0$ and $8MG = -1$ the metric (8) reduces to Euclidean anti-de Sitter space in three dimensions,

$$ds^2_{adS} = l^2\left(1 + \frac{r^2}{l^2}\right)dt^2 + \left(1 + \frac{r^2}{l^2}\right)^{-1}dr^2 + r^2d\varphi^2. \tag{11}$$

(vii) A quantum metric, the 'CFT→ Geometry' map and black hole entropy

Once the functions L and $\bar{L}$ are promoted to be operators acting on Fock space, the metric (1) becomes a well-defined operator, denoted as $d\hat{s}^2$, on that space. We then find a map from Fock's space (representations of the Virasoro algebra) into the independent classical solutions of Einstein's equations. Let $|\Psi>$ a state in Fock's space, we have

$$|\Psi> \rightarrow \; ds^2_\Psi =< \Psi|d\hat{s}^2|\Psi> \tag{12}$$

with $L_\Psi =< \Psi|L|\Psi>$ and $\bar{L}_\psi =< \Psi|\bar{L}|\Psi>$. By construction ds^2_Ψ is a solution to the classical Einstein equations because (1) is a solution for arbitrary functions L and $\bar{L}$. It also follows that the full set of states $|\Psi>$ generates the full space of classical solutions. Note that this map is valid independently of the structure of the conformal field theory generating the Virasoro algebra. We can then ask the question of how many states are there in Fock space such that they induce through (12) a black hole of a mass M and angular momentum J. The answer to this question of course depends on the structure of the Hilbert space. We shall study this point in detail in Sec. IV D.

(viii) Relation to 2d induced gravity

Finally, note that for $\bar{L} = 0$ and fixed r, the metric (1) is equal to Polyakov's [13] 2d lightlike metric which yields an $SL(2,\Re)$ algebra. Since 3d gravity is known to induce 2d gravity at the boundary (r fixed), the understanding of the quantum properties of (1) may yield new information about 2d gravity.

B Organisation of the article

The goal of this article is to discuss and provide the proofs for the above properties of the metric (1). We have written (i)-(viii) in a metric formulation of gravity because our final target is quantum gravity. However, the explicit proofs will be given in terms of the Chern-Simons formulation [1] of three-dimensional gravity because they are simpler and provide a rich mathematical structure.

In Sec. II we give a short introduction to Chern-Simons gravity and its phase space. A detailed discussion about boundary degrees of freedom is included in that section. In Sec. III the explicit solution to the equations of motion, with two different classes of boundary conditions, is written down (in terms of the Chern-Simons fields) and their induced Poisson brackets are displayed. Finally, in Sec. IV, we go back to the metric formulation and apply the results to quantum three-dimensional gravity.

II CHERN-SIMONS GRAVITY AND GLOBAL DEGREES OF FREEDOM

In this section we shall first briefly describe the Chern-Simons formulation of three-dimensional gravity. Then we analyse the issue of global degrees of freedom associated to the presence of boundaries. We shall also show in this section (see Sec. II F) how the boundary conditions solve part of the unitary problems of three-dimensional gravity.

A Chern-Simons gravity, its equations of motion and their solutions

In our approach to the quantum black hole problem, the Chern-Simons formulation of 2+1 gravity will be of great help. This formulation was discovered in [1] and its quantum properties (for closed manifolds) were explored in [14]. An extensive treatment can be found in [15]. In a few words, the Chern-Simons formulation is a field redefinition that simplifies the equations and introduces a rich mathematical structure.

The basic variables of general relativity in the tetrad formalism are the triad e^a and the spin connection[2] ω^a.

---

[2] In three dimensions one defines $\omega^a = (-1/2)\epsilon^a{}_{bc}\omega^{bc}$. It follows that the 2-form curvature $R^{ab} = d\omega^{ab} + \omega^a{}_c{}_\wedge\omega^{cb}$ can be written in the form $R^{ab} = -\epsilon^{ab}{}_c R^c$ with $R^a = d\omega^a + (1/2)\epsilon^a{}_{bc}\omega^b{}_\wedge\omega^c$. In the same way, the torsion $T^a = de^a + \omega^a{}_b{}_\wedge e^b$ reads $T^a = de^a + \epsilon^a{}_{bc}\omega^b{}_\wedge e^c$. These definitions depend on the signature. The formulae displayed here are appropriated to Euclidean signature.

The equations of motion of three dimensional gravity with a negative cosmological constant in these variables are simply

$$R^a = \frac{1}{2l^2}\epsilon^a{}_{bc}\, e^b{}_\wedge e^c, \qquad T^a = 0. \tag{13}$$

We define now two new fields according to,

$$A^a = \omega^a + \frac{i}{l}e^a, \qquad \bar{A}^a = \omega^a - \frac{i}{l}e^a. \tag{14}$$

The 1-form A^a is an $SL(2,C)$ Yang-Mills gauge field. Let F^a and $\bar{F}^a$ the curvatures associated to A^a and $\bar{A}^a$. The discovery of Achúcarro and Townsend [1] is that the equations,

$$F^a = 0, \qquad \bar{F}^a = 0, \tag{15}$$

are exactly equivalent to the three-dimensional Einstein equations (13). Furthermore, the Einstein-Hilbert action is equal to the combination,

$$I[A, \bar{A}] = I[A] - I[\bar{A}], \tag{16}$$

where $I[A]$ is the Chern-Simons action,

$$I[A] = \frac{k}{4\pi}\int \mathrm{Tr}(AdA + \frac{2}{3}A^3). \tag{17}$$

To determine the Chern-Simons coupling constant, or level, k as a function of the gravitational constants G and l we need to fix the representation of A and $\bar{A}$. We use the anti-Hermitian $SU(2)$ generators,

$$J_1 = \frac{i}{2}\begin{pmatrix} 0 & 1 \\ 1 & 0 \end{pmatrix}, \quad J_2 = \frac{1}{2}\begin{pmatrix} 0 & -1 \\ 1 & 0 \end{pmatrix}, \quad J_3 = \frac{i}{2}\begin{pmatrix} 1 & 0 \\ 0 & -1 \end{pmatrix}, \tag{18}$$

which satisfy $[J_a, J_b] = \epsilon_{ab}{}^c J_c$ and $\mathrm{Tr}(J_a J_b) = -(1/2)\delta_{ab}$ and define

$$A = A^a J_a, \quad \bar{A} = \bar{A}^a J_a. \tag{19}$$

Note that we use the same J's in both cases. This means that $\bar{A}$ is not the complex conjugate of A. With these conventions, comparing the Chern-Simons action with the Einstein-Hilbert action one finds

$$k = -\frac{l}{4G}. \tag{20}$$

The sign of k depends on the identity $\sqrt{g} = \pm e$ where e is the determinant of the triad. This sign determines the relative orientation of the coordinate and orthonormal basis. We have chosen here the plus sign which means that we work with $e > 0$.

Note that the gauge field A is complex and thus the relevant group is $SL(2,C)$ which is non-compact. This means that we cannot apply in a straightforward way the quantization of Chern-Simons theory described in [2]. Our prescription to define the quantum theory will be to first find the general solution the classical equations of motion, under prescribed boundary conditions, and then quantize that space. As we shall see, the boundary conditions will play a key role in making the quantum theory unitary (see Sec. II F).

The convenience of the Chern-Simons formulation is evident. Instead of working with a second order action in terms of the metric, we work with two flat Yang-Mills fields. The equations (15) show clearly that 2+1 gravity does not have any local degrees of freedom. This means that all dynamics is contained in the holonomies [14] and boundary degrees of freedom [16,17].

The general solution to the equations (15) can be written in the form,

$$A = g^{-1}dg + g^{-1}Hg, \tag{21}$$

where H is also flat ($dH + H{}_\wedge H = 0$) but cannot be written as $u^{-1}du$ with u single valued. Similar arguments hold for $\bar{A}$. The group element $g(x)$ is a single valued map from the manifold to the group. The space of solutions (21) is invariant under

$$A \to A' = U^{-1}AU + U^{-1}dU \tag{22}$$

where U is another map from the manifold to the group. In principle (see below for a detailed discussion), we can use this symmetry to set $g = 1$ in (21) and thus all solutions are classified only by the independent values of holonomy H. The quantization of this sector of phase space was first discussed in [14]. Its dimensionality is finite and cannot account for the large black hole degeneracy. For this reason we shall not consider them here anymore. However, it is important to stress that the black hole gauge field does have non-trivial holonomies. Indeed, it can be shown that the gauge field corresponding to a black hole satisfies [18],

$$\mathrm{P}\exp\oint A =: \exp(w), \qquad \mathrm{P}\exp\oint \bar{A} =: \exp(\bar{w}) \tag{23}$$

where

$$\mathrm{Tr}(w^2 + \bar{w}^2) = 32\pi^2 MG, \qquad \mathrm{Tr}(w^2 - \bar{w}^2) = \frac{32\pi^2 JG}{l}, \tag{24}$$

and M and J are the black hole mass and angular momentum, respectively. Only for $8GM = -1$ and $J = 0$ these holonomies are trivial. The corresponding solution is anti-de Sitter space (11).

B The strategy

Since Chern-Simons theory does not have any local excitations, all relevant degrees of freedom are global. We shall consider here the situation on which the topology is fixed and thus all relevant states come from the presence of boundaries. The existence of boundary degrees of freedom has been analysed in great detail by Carlip [19] using a covariant formalism and path integrals (see also [17] for an approach similar to ours). Here, we shall describe an equivalent procedure based on the Hamiltonian formalism in the form discussed by Regge and Teitelboim [20]. This approach can be summarised in the following steps. Given an action $I[\phi]$ with a gauge symmetry $\delta_G\phi$ we need to:

- Impose boundary conditions on the fields such that $\delta I[\phi]/\delta\phi$ exists. These boundary conditions are not unique and their election represent an important physical input into the theory. [In practice, one first decides the boundary conditions and then add to the action the necessary boundary terms to make it differentiable.]
- Find the sub-group of gauge transformations that leave the boundary conditions and the action invariant.
- Find the canonical generators which generate the symmetries of the action. If a generator is a constraint, we shall call the associated symmetry a *gauge* symmetry. Configurations which differ by gauge symmetries are identified and represent the same physical state. Conversely, if a generator is different from zero (even on-shell), we call the associated symmetry a *global* symmetry. Note that according to this definition, global symmetries do not need to be rigid. Global symmetries map the space of physical states into itself.

We shall see that a proper distinction between global and gauge symmetries is crucial to understand the boundary degrees of freedom in Chern-Simons theories.

C Global symmetries and boundary degrees of freedom

The appearance of boundary degrees of freedom can be summarized as follows. Chern-Simons theory has $3N$ fields A^a_μ ($\mu = 0, 1, 2;\ a = 1, ..., N$). However, the gauge symmetries of the action tells us that they do not represent independent physical degrees of freedom. In fact, locally, using the symmetry (22) one can kill all of them (the temporal component A^a_0 is a Lagrange multiplier, while the spatial components A^a_i are $2N$ fields subject to N constraints $F^a_{ij} = 0$ plus N gauge conditions).

The question we want to address here is whether the symmetry (22) is really a gauge symmetry, in the sense that two fields related by it are to be considered the same, or not. After properly defining what a gauge transformation is, we shall see that at the boundary the transformation (22) is not a gauge symmetry, although it is still a symmetry of the action. Then, two solutions of the form (21) with g and g' such that, at the boundary, $g \neq g'$ represent two different physical configurations. This boundary effect can give rise to an infinite number of degrees of freedom (independent solutions to the equations of motion). At this point we can make contact with Carlip's would-be-gauge degrees of freedom approach: the field g, at the boundary, is dynamical and its dynamics is governed by a WZW action [19].

In the presence of boundaries the definition of a gauge symmetry becomes delicate because not all the transformations encoded in (22) are generated by constraints. Indeed, if U does not approach the identity map at the boundary, then the associated canonical generator is a non-zero quantity and hence that transformation is not a gauge symmetry. See Sec. II E for a proof of this statement in Chern-Simons theory.

Following Dirac's quantization procedure (see [21] for an extensive treatment), we define a *gauge* transformation as a symmetry generated by a (first class) constraint. On the contrary, a symmetry of the action generated by a non-zero quantity is called *global*, even if it is not rigid. By definition, the space of physical states, or phase space, is the set of fields which satisfy the equations of motion, modulo gauge transformations. Let us ignore the holonomies for a moment. The general solution (21) then reduces to $A = g^{-1}dg$. If no boundaries are present, this space of solutions is trivial containing only one element $A = 0$ because the transformations (22) are generated by constraints (hence, they represent gauge symmetries) and one can use (22) to set $g = 1$ and thus $A = 0$.

On the contrary, if there is a boundary, part of the symmetry (22) is not generated by a constraint (to be proved below). Therefore, while it is still true that we can transform any flat A to 0 using (22), it is not true that the state A and the state 0 represent the same physical configuration. Both states, $A \neq 0$ and $A = 0$ (at the boundary), are solutions to the equations of motion and they are related by a symmetry of the action. However, they are physically distinguishable. Indeed, there is a gauge invariant conserved charge which takes different values in each state. Our main problem is then to determine the set of fields $\hat{A}$ which solve the equations of motion and cannot be set to zero by the action of a constraint. As we shall see, in Chern-Simons theory there is an infinite number of them.

In a quantum mechanical notation, the above discussion can be summarized as follows. Denote by G_0 the set of transformations which are true gauge symmetries generated by constraints, and by Q those which are not. Physical states satisfy $G_0|\Psi>= 0$. On the other hand, Q generates a symmetry of the space of physical states, that is $Q|\Psi>= |\Psi'>$. We shall prove explicitly (at least in Chern-Simons theory; for a general discussion see [22]) that G_0 and Q satisfy an algebra of the form,

$$[G_0, G_0] = G_0, \tag{25}$$
$$[G_0, Q] = G_0, \tag{26}$$
$$[Q, Q] = Q + c \tag{27}$$

where c represents (schematically) a possible central term. Eq. (25) is the definition of first class constraints. Eq. (26) means that if $|\Psi>$ is physical ($G_0|\Psi>= 0$) then $Q|\Psi>$ is also physical (Q generates a global symmetry of the Hilbert space). Finally, Eq. (27) is the algebra of the globally symmetry. The appearance of central terms in (27) cannot be discarded by a general principle [22]. Note however that since Q does not generate a gauge symmetry and it is different from zero, the central term does not represent any trouble after quantization. An interesting and important example on which the central term is present was discovered in [5].

The above discussion is a quick summary of the results presented in [20,23,22,5], and many other papers that have followed this work. The nice property of Chern-Simons theory is that these ideas can be tested with minimum calculations. Another system which is simple to analyse is Yang-Mills theory on which the above analysis leads to the definition of global colour charges [24]. However, in that case, the resulting global algebra is finite dimensional and does not have any central terms.

D Boundary conditions in Chern-Simons gravity

1 *Making the action differentiable*

The black hole manifold is asymptotically anti-de Sitter and then it has a boundary. In the Euclidean sector, the boundary has the topology of a torus with compact coordinates φ and t. It is convenient to define the complex coordinates on the torus

$$w = \varphi + it, \qquad \bar{w} = \varphi - it, \tag{28}$$

and then $A_\varphi d\varphi + A_t dt = A_w dw + A_{\bar{w}} d\bar{w}$.

Boundary conditions are necessary in order to ensure that the action principle has well defined variations. As discussed above, all the dynamics of 2+1 gravity is contained in the boundary conditions. For this reason, it is a key problem to choose them judiciously. In particular, if they are too strong there will be no dynamics left in the theory. For the black hole problem (which is asymptotically anti-de Sitter) there is a natural choice of boundary conditions first discussed in [9] in the Minkowskian signature and extended to Euclidean signature in [25]. In the coordinates (28) they read simply

$$A^a_{\bar{w}} = 0, \quad \bar{A}^a_w = 0 \quad \text{(at the boundary)}. \tag{29}$$

A quick way to convince ourselves that the black hole satisfies this condition is to consider the constant curvature metric

$$ds^2 = e^{2\rho}(dx^2 + dy^2) + l^2 d\rho^2. \tag{30}$$

A natural election for the triads is $\{e^1 = e^\rho dx, e^2 = e^\rho dy, e^3 = l d\rho\}$. The torsion equation $de^a + \epsilon^a{}_{bc}\omega^b \wedge e^c = 0$ yields for the components of ω^a $\{\omega^1 = -(1/l)e^\rho dy,\ \omega^2 = (1/l)e^\rho dx, \omega^3 = 0\}$. Defining $w = x + iy$ it is clear that $A^a = \omega^a + (i/l)e^a$ and $\bar{A}^a = \omega^a - (i/l)e^a$ satisfy (29). It can be shown that the black hole metric (8) which is also of constant curvature satisfies (29) as well [25]. See [26] for the explicit transition from (30) to (8).

This example also illustrates the choice of orientation. Suppose we choose a new set of triads given by $\tilde{e}^1 = -e^\rho dx$, $\tilde{e}^2 = -e^\rho dy$ and $\tilde{e}^3 = -l d\rho$. These new fields satisfy the torsion equation because is homogeneos in e, and Einstein equations because they are quadratic in e. However, the determinant of $\tilde{e}^a_\mu$ is negative. As we remarked before, the value of k given in (20) depends on the orientation of the orthonormal basis, and the identity $\sqrt{g} = \pm e$. We have chosen $e > 0$ and then the election $\tilde{e}^a$ is not allowed.

Let us check that (29) are enough to make the action differentiable. The variation of the Chern-Simons action gives a term proportional to the equations of motion plus a boundary term,

$$\begin{aligned}\delta I_{CS} &= \int_M (\text{eom})_a \delta A^a + \frac{k}{4\pi}\int_{\partial M} g_{ab} A^a \wedge \delta A^b \\ &= \int_M (\text{eom})_a \delta A^a + \frac{k}{4\pi}\int_{\partial M} g_{ab}(A^a_w \delta A^b_{\bar{w}} - A^a_{\bar{w}}\delta A^b_w) \\ &= \int_M (\text{eom})_a \delta A^a + 0.\end{aligned} \tag{31}$$

The boundary term vanishes due to (29). Thus, the variation of the action under the boundary condition (29) is well defined. Later we will restrict further the values of the gauge field at then boundary, but for the purposes of this discussion the above boundary conditions are very useful.

In summary, we work with the Chern-Simons action (16) supplemented with the boundary conditions (29) and no added boundary terms. [Note that when passing to the Hamiltonian formalism there will be a boundary term [25].] As a further check that this action is appropriated to the black hole problem, one can prove [27] that its value on the Euclidean black hole solution is finite and gives the right canonical free energy (Gibbons-Hawking approximation).

2 *The chiral boundary group*

The second step in the Regge-Teitelboim procedure is to determined how the gauge symmetries are affected by the boundary conditions, i.e., to determine the residual group of transformations that preserves (29). This is actually very simple. We look for the set of parameters λ^a satisfying

$$\delta A^a_{\bar{w}} = D_{\bar{w}}\lambda^a = 0 \qquad \text{(at the boundary)}. \tag{32}$$

Since by (29) $A_{\bar{w}} = 0$ this condition simply imply that $\partial_{\bar{w}}\lambda^a = 0$. The subset of gauge transformations leaving (29) invariant are then those whose parameters at the boundary are chiral, only depend on w.

Let us now check that this group leaves the action invariant. The variation of the Chern-Simons action under $\delta A^a = D\lambda^a$ gives a boundary term,

$$\begin{aligned}\delta I_{CS} &= \frac{k}{4\pi}\int_{\partial M} A^a \wedge D\lambda_a \\ &= \frac{k}{4\pi}\int_{\partial M} (A^a_w D_{\bar{w}}\lambda_a - A^a_{\bar{w}} D_w \lambda_a) \\ &= 0\end{aligned} \tag{33}$$

which vanishes thanks to (29) and (32). There is an important point to be stressed here. It is often said in the literature that the Chern-Simons action is invariant under $\delta A^a = D\lambda^a$ only if $\lambda = 0$ at the boundary. This is, as we have just shown, not true. The right statement is that λ cannot be completely arbitrary at the boundary but it can be different from zero. Under the boundary condition (29), the action is invariant under transformations with non-zero values of λ^a at the boundary provided that parameter is chiral ($\lambda^a = \lambda^a(w)$). This gives rise to an infinite dimensional symmetry.

E Affine (Kac-Moody) algebras

Let us briefly describe the main steps leading to (25-27) in Chern-Simons theory. For more details, the reader is referred to [28,29].

In the 2+1 decomposition of the gauge field $A^a = A^a_0 dt + A^a_i dx^i$, the Chern-Simons action reads,

$$I[A_i, A_0] = \frac{k}{8\pi}\int dt \int_\Sigma \epsilon^{ij}\delta_{ab}(A^a_i \dot{A}^b_i - A^a_0 F^b_{ij}) + B, \tag{34}$$

where B is a boundary term. Here we have used that $\mathrm{Tr}(J_a J_b) = -(1/2)\delta_{ab}$. The coordinates x^i are local coordinates on the spatial surface denoted by Σ. This action has $2N$ dynamical fields A^a_i ($a = 1, ..., N$; $i = 1, 2$) and N Lagrange multipliers A^a_0. The dynamical fields satisfy the basic equal-time Poisson bracket algebra,

$$\{A^a_i(x), A^b_j(y)\} = \frac{4\pi}{k}\epsilon_{ij}\delta^{ab}\delta^2(x, y). \tag{35}$$

The Poisson bracket of two functions $F(A_i)$ and $H(A_i)$ is computed as

$$\{F, H\} = \frac{4\pi}{k}\int_\Sigma d^2z \frac{\delta F}{\delta A^a_i(z)}\epsilon_{ij}\delta^{ab}\frac{\delta H}{\delta A^b_j(z)}. \tag{36}$$

The functionals F and H need to be differentiable with respect to A_i.

The equation of motion with respect to A_0 leads to the constraint equation,

$$G^a_0 = \frac{k}{8\pi}\epsilon^{ij}F^a_{ij} \approx 0, \tag{37}$$

which, we expect, will be the canonical generator of the gauge transformations $\delta A^a_i = D_i\lambda^a$. This is indeed true but only for those transformation whose parameters vanish at the boundary. Indeed, define $G_0(\lambda) = \int_\Sigma \lambda_a G^a_0$ and compute $\delta_\lambda A_i(x) = [A^a_i(x), G_0(\lambda)]$. It is direct to see that the functional derivative of $G_0(\lambda)$ with respect to A_i is well-defined only if λ^a vanishes at the boundary. In that case, one does find $[A^a_i(x), G_0(\lambda)] = D_i\lambda^a$ and thus $G_0(\lambda)$ generates the correct gauge transformation. (We stress here that $G_0(\lambda)$ should not be identified with the constraint: $G_0(\lambda)$ is the constraint smeared with a parameter that vanishes at the boundary.)

However, as we discussed in Sec. II D 2, the Chern-Simons action with the boundary condition (29) is also invariant under transformations whose parameters at the boundary are chiral $\lambda^a = \lambda^a(w)$ but different from zero. What is then the generator of those transformations? Consider

$$Q(\lambda) = \int_\Sigma \lambda_a G^a_0 - \frac{k}{4\pi}\int_{\partial\Sigma}\lambda_a A^a. \tag{38}$$

It is easy to check that the boundary term arising when varying the bulk part of (38) is cancelled by the boundary term, without imposing any conditions over λ. The combination (38) then has well defined variations even if λ does not vanish at the boundary. Furthermore, one can check that $[A_i^a(x), Q(\lambda)] = D_i\lambda^a$ and therefore Q indeed generates those transformations whose parameters do not vanish at the boundary.

The key point here is that Q is no longer a combination of the constraints (for $\lambda^a|_{\partial\Sigma} \neq 0$) and thus it is different from zero, even on-shell. According to the previous discussion, Q generates a global symmetry of the action. Two configurations which differ by a transformation generated by it represent physically different states. As a direct application of this result, we find that two flat connections A and A' whose values at the boundary differ, $(A - A')|_{\partial\Sigma} \neq 0$, cannot be connected by the action of a constraint. Thus, as we have anticipated, the values of A at the boundary represent the physically relevant degrees of freedom. The next step is to prove that there exists solutions to the equations of motion, satisfying the boundary conditions (29), with different values for A at the boundary. This is done in the next section.

By direct application of the Poisson bracket (36) one can find the algebra of two transformations with parameters η and λ not vanishing at the boundary,

$$[Q(\eta), Q(\lambda)] = Q([\eta,\lambda]) + \frac{k}{4\pi}\int_{\partial\Sigma} \eta_a d\lambda^a \tag{39}$$

where $[\eta,\lambda]^a = \epsilon^a{}_{bc}\eta^b\lambda^c$. This equation should be compared with (27). Also, note that if λ vanishes at the boundary then $Q(\lambda) = G_0(\lambda)$. One can then easily see that (39) reproduces (25) and (26) as well.

The algebra (39) provides the simplest way to determine the Poisson bracket structure on the space of functions which cannot be set to zero by the action of a constraint. We shall do this explicitly in next section.

F Unitarity. An $SU(2)$ field

To end this section, we mention an important consequence of the boundary conditions (29). Namely, they provide a simple solution to one of the problems with unitarity in Chern-Simons gravity. As we have mentioned above, the gauge field $A = \omega^a + (i/l)e^a$ is complex and therefore the relevant group is $SL(2,C)$ which is non-compact. It has been argued in [30], and explicitly used for example in [31] and [25], that under some conditions in the path integral one can set $e^a = 0$ and work with the $SU(2)$ gauge field $A^a = \omega^a$. For closed manifolds, this has been shown to give a good prescription [30], but it is not the case for manifolds with a boundary.

The boundary conditions (29) lead to a simple solution to part of this problem. Indeed, expressing (29) explicitly in terms the triad and spin connection one finds,

$$\omega_{\bar{w}}^a + \frac{i}{l}e_{\bar{w}} = 0, \qquad \omega_w^a - \frac{i}{l}e_w = 0. \tag{40}$$

This means that the real and imaginary parts of A_z and $\bar{A}_{\bar{z}}$ are not independent. Indeed, using these equations, the non-zero components of A_μ and $\bar{A}_\mu$ at the boundary, namely A_w and $\bar{A}_{\bar{w}}$, can be written in terms of the spin connection as,

$$A_w^a = 2\omega_w^a, \qquad \bar{A}_{\bar{w}}^a = 2\omega_{\bar{w}}^a. \tag{41}$$

This shows that the non-zero components of the gauge field at the boundary, which in fact carry all the dynamics, are real $SU(2)$ currents.

Since all the dynamics of Chern-Simons theory will be defined at the boundary, this simple observation means that we can indeed work with two $SU(2)$ currents and forget about the non-compact nature of $SL(2,C)$. Of course in the bulk we still have an $SL(2,C)$ field and this makes the statement "all the dynamics is contained at the boundary" delicate. We shall not discussed this point anymore in this paper. Our prescription will be to treat the gauge degrees of freedom classically and to quantise the reduced phase space, after the gauge has been fixed. In the language of [3], the Chiral WZW action arises classically (by varying with respect to A_0 instead of integrating over it) and we work with its basic Poisson bracket which is the $SU(2)$ affine algebra. Actually, we shall not make explicit use of [3], but rederive the same algebras by studying global symmetries of the Chern-Simons action. Some comments on the relation between both methods will be given in Sec. III C.

For later convenience, we mention here that using (40) the formulae (41) can also be rewritten in terms of the triad as

$$A_w^a = \frac{2i}{l} e_w^a, \qquad \bar{A}_{\bar{w}}^a = -\frac{2i}{l} e_{\bar{w}}^a. \tag{42}$$

These formulae are more useful when constructing the metric out of the connection via $g_{\mu\nu} = e_\mu^a e_\nu^b \delta_{ab}$.

Note finally that the equations (40) make a link between the i appearing in $A = w + (i/l)e$ and the i appearing in the complex structure given to the torus through $w = \varphi + \tau t$ (we are using here $\tau = i$).

III THE AFFINE AND ANTI-DE SITTER SOLUTIONS

As discussed in detail in the last section, the presence of a boundary means that not all the values of the field A are related by proper gauge transformations. Two solutions of the equations of motion A and A' whose values at the boundary differ by a chiral non-zero transformation are physically distinguishable solutions.

This means that the space of solutions is not trivial. In this section we shall explicitly solve the equations of motion with the boundary conditions (29) and isolate the variables which are physically relevant. We shall also find the induced Poisson bracket acting on the space of dynamical (gauge-fixed residual functions) degrees of freedom.

A The affine solution

We work on the solid torus with coordinates $\{w, \bar{w}, \rho\}$ and $A = A_w dw + A_{\bar{w}} d\bar{w} + A_\rho d\rho$. Our goal in this section is to find the general solution to the equations of motion which satisfies the boundary condition (29).

The first step is to fix the gauge and eliminate the redundant degrees of freedom. We impose the gauge condition,

$$A_\rho = iJ_3, \tag{43}$$

which implies that ρ is a proper radial coordinate (see below). Our conventions for the matrices J_a are given in (18). Note that the i present in (43) means that $e_\rho \neq 0$. This is necessary because otherwise the triad would be degenerate. Next we impose the boundary condition $A_{\bar{w}} = 0$ [see (29)]. The equations of motion $F = 0$ in the coordinates $\{w, \bar{w}, \rho\}$ read explicitly,

$$\begin{aligned}
\partial_\rho A_w - \partial_w A_\rho + [A_\rho, A_w] &= 0, \\
\partial_\rho A_{\bar{w}} - \partial_{\bar{w}} A_\rho + [A_\rho, A_{\bar{w}}] &= 0, \\
\partial_w A_{\bar{w}} - \partial_{\bar{w}} A_w + [A_w, A_{\bar{w}}] &= 0.
\end{aligned} \tag{44}$$

The general solution to these equations in the gauge (43) and satisfying $A_{\bar{w}} = 0$ at ∂M is,

$$\begin{aligned}
A_w &= b^{-1} \hat{A}(w) b, \\
A_{\bar{w}} &= 0, \\
A_\rho &= b^{-1} \partial_\rho b
\end{aligned} \tag{45}$$

where $\hat{A}$ is a chiral function, $\hat{A} = \hat{A}(w)$, but otherwise arbitrary and

$$b = e^{i\rho J_3} = \begin{pmatrix} e^{-\rho/2} & 0 \\ 0 & e^{\rho/2} \end{pmatrix}. \tag{46}$$

Since $\hat{A}(w)$ is arbitrary, the space of solutions (45) is infinite dimensional. The question is whether different solutions with different values for $\hat{A}$ are related by gauge transformations or not. Consider a configuration of the form (45) and act on it with the transformation,

$$\delta A_\mu = D_\mu \eta, \qquad \eta = b^{-1} \hat{\eta}(w) b. \tag{47}$$

It is direct to see that the effect of this transformation on the solution is to produce another solution of the form (45) with $\hat{A}' = \hat{A} + \hat{D}_w \hat{\eta}$ (here $\hat{D}_w \hat{\eta}$ denotes covariant derivative in $\hat{A}$, $\hat{D}_w \hat{\eta} = \partial_w \hat{\eta} + [\hat{A}, \hat{\eta}]$). Thus, by acting on (45) with (47) we move around on the space of solutions. Actually, the transformations (47) are the most general set of transformations that leave the boundary condition and gauge fixing conditions invariant.

Since the parameter $\hat{\eta}$ appearing in (47) does not vanish at the boundary, the canonical generator of (47) is a non-zero quantity of the form (38). We then conclude that different values of the function $\hat{A}$ are connected by global transformations generated by (38) and not by the action of constraints. The function $\hat{A}$ then represents dynamical degrees of freedom. Even more, since (45) is the most general solution to the equations of motion with the boundary condition (29), the function $\hat{A}$ generates the full space of non-trivial solutions with those boundary conditions.

As it was pointed out in [25], this analysis is also valid if the boundary is located at a finite value of the radial coordinate. This observation is important, for example, if one expects to find a conformal field theory at the black hole horizon. The main difference between the metric and Chern-Simons approaches to global symmetries is that in the former one works with diffeomorphisms while in the latter with gauge transformations. Contrary to gauge transformations, diffeomorphisms are not local and move the position of the boundary. For example, the residual diffeomorphisms associated to anti-de Sitter space [5] have a non-zero radial component.

Our next step is to determine the Poisson bracket structure acting on the space of solutions, that is, the induced Poisson bracket acting on the functions $\hat{A}$. This is very easy thanks to a general theorem proved in [22]. First, we note that after the gauge is fixed (the constraint is solved and (43) is imposed) the value of Q given in (38) reduces to the boundary term,

$$\hat{Q}(\hat{\eta}) = -\frac{k}{4\pi}\int \hat{\eta}_a \hat{A}^a, \tag{48}$$

which is, as we have emphasized, different from zero. The theorem [22] states that after the gauge is fixed and one works with the induced Poisson bracket (or Dirac bracket), the charge $\hat{Q}$ satisfies the same algebra (39) as it did the full charge Q,

$$[\hat{Q}(\hat{\eta}), \hat{Q}(\hat{\lambda})]^* = \hat{Q}([\hat{\eta}, \hat{\lambda}]) + \frac{k}{4\pi}\int_{\partial\Sigma} \hat{\eta}_a d\hat{\lambda}^a \tag{49}$$

This algebra can be put in a more explicit form by defining,

$$\hat{A}^a(w) = \frac{2}{k}\sum_{n\in Z} T^a_n e^{inw}. \tag{50}$$

One finds

$$[T^a_n, T^b_m]^* = -\epsilon^{ab}{}_c T^c_{n+m} + \frac{ink}{2}\delta^{ab}\delta_{n+m,0}. \tag{51}$$

The algebra (51) is called Kac-Moody or affine algebra and represents an infinite dimensional symmetry of the space of solutions. The quantum version is obtained simply by replacing the Dirac bracket by $-i$ times the commutator,

$$[T^a_n, T^b_m] = i\epsilon^{ab}{}_c T^c_{n+m} + \frac{nk}{2}\delta^{ab}\delta_{n+m,0}. \tag{52}$$

This equation represents the algebra of the gauge-fixed basic variables (analogous to $[q,p] = i$) of Chern-Simons theory with the boundary condition (29). There are various ways to see this explicitly, which also provide alternative derivations of (52). Conceptually, the most direct derivation of this result is by starting with the three-dimensional Poisson bracket (35). Then we fix the gauge as in (43) and solve the constraint $F_{\rho\varphi} = 0$. The solutions to the constraint equation in this gauge are parametrized by the function $\hat{A}$. One can then compute the Dirac bracket of $\hat{A}$ with itself and find the affine algebra (52). Other methods yielding the same result are the WZW approach followed in [3], and the symplectic method [32]. The idea of looking at first class quantities and their algebra in Chern-Simons theory was first discussed in [28]. The presence of central terms in the algebra of global charges in Chern-Simons theory was first discussed in [33].

Note that in our applications to general relativity, the value of k is negative (see Eq. (20)). This can be cured simply by replacing[3] $T^a_n \to T^a_{-n}$, or in other words, by replacing the maximum weight condition $T^a_{-n}|0>= 0$

---

3) We thank M. Asorey and F. Falceto for useful conversation on this point.

by $T_n^a|0>= 0$ $(n > 0)$. In any case, we shall not consider (52) as the starting point for quantization but rather a restriction of it which yields the Virasoro algebra with central charge $c = -6k$. For this reason, in this particular calculation, it seems convenient to work with a negative k. Incidentally, note that the Virasoro algebra is also invariant under $L_n \to -L_{-n}$, $c \to -c$.

Because of its affine symmetry we shall call the solution (45) the affine solution to the equations of motion.

B The anti-de Sitter solution

In principle we could consider the algebra (52) as our definition for the gauge-fixed basic quantum commutator. As we have pointed out in Sec. II F, the gauge field $\hat{A}$ is a real $SU(2)$ current and therefore the modes T_n^a satisfy the Hermitian condition $(T_n^a)^\dagger = T_{-n}^a$. We shall explore the quantum properties of this general metric elsewhere. Our goal here is to describe the space of solutions to general relativity with anti-de Sitter boundary conditions. The metric which follows from the general affine solution does not satisfy this requirement and thus we need to impose further restrictions on it [9].

In the conventions displayed in (18), the solution (45) for A_w can be written as,

$$A_w = \frac{i}{2} b^{-1} \begin{pmatrix} \hat{A}^3 & \hat{A}^+ \\ \hat{A}^- & -\hat{A}^3 \end{pmatrix} b \tag{53}$$

with $\hat{A}^\pm = \hat{A}^1 \pm i\hat{A}^2$ and b is defined in (46). We remind that the other components of the solution are $A_{\bar{w}} = 0$ and $A_\rho = iJ_3$, and similar expressions hold for the anti-holomorphic field.

The extra boundary conditions follow from looking at the form of the gauge field associated to Euclidean anti-de Sitter space which satisfies

$$\hat{A}^+ = -2, \qquad \hat{A}^3 = 0. \tag{54}$$

The anti-holomorphic field satisfies $\hat{A}^- = -2, \hat{A}^3 = 0$. These conditions were first discussed in [34], and their relation to anti-de Sitter spaces was realised in [9]. In the WZW approach, they can be incorporated in the action by adding Lagrange multipliers. This leads to a gauged WZW action whose conformal generators follow from the GKO coset construction.

We shall impose (54) as part of the boundary data. Since (45) solves the equations of motion for arbitrary values of $\hat{A}^+, \hat{A}^-$ and $\hat{A}^3$, the gauge field will still be a solution after imposing (54). This means that among the three components of the gauge field $\hat{A}^a$ only one component, $\hat{A}^-$, remains as an arbitrary function. It is convenient to rename this function as,

$$L(w) = -\frac{k}{2}\hat{A}^-(w) \tag{55}$$

and thus the solution we are interested in has the form,

$$A_w(L) = -ib^{-1} \begin{pmatrix} 0 & 1 \\ (1/k)L(w) & 0 \end{pmatrix} b, \tag{56}$$

where $L(w)$ is an arbitrary function of w, and b is given in (46). As we shall see, this solution is appropriated to anti-de Sitter spacetimes.

The boundary group leaving (54) invariant is no longer the Kac-Moody algebra (52) because that algebra does not preserve (54). Let us find the group of transformations leaving (54) invariant. First, we look for those gauge transformations $\delta A = D\lambda$ which preserve the form (56) changing only the values of L. These transformations are generated by parameters of the form [34],

$$\lambda = b^{-1} \begin{pmatrix} (i/2)\partial\varepsilon & \varepsilon \\ (1/k)\varepsilon L + (1/2)\partial^2\varepsilon & -(i/2)\partial\varepsilon \end{pmatrix} b \tag{57}$$

where $\varepsilon = \varepsilon(w)$ is an arbitrary function of w, and b is given in (46). Acting with (57) on A_w we find,

$$A_w(L) + D_w\lambda = A_w(L + \delta L), \tag{58}$$

with

$$\delta L = i(\varepsilon\partial L + 2\partial\varepsilon L + \frac{k}{2}\partial^3\varepsilon). \tag{59}$$

This shows that L is a quasi-primary field of dimension two under the residual group of transformations leaving (56) invariant.

The next step is to determine the algebra associated to these transformations. This can be done by imposing the reduction conditions (54) in the algebra (52) and computing the induced algebra. Geometrically speaking, given a Poisson bracket structure of the form $[x^a, x^b] = J^{ab}(x^a)$ (J^{ab} invertible) one defines the symplectic form σ_{ab} as the inverse of J^{ab}. The antisymmetry and Jacobi identity satisfied by J^{ab} imply that σ_{ab} is a closed 2-form. Now, let $\chi_\alpha(x^a) = 0$ a set of constraints on phase space such that $C_{\alpha\beta} := [\chi_\alpha, \chi_\beta]$ is invertible. The surface defined by $\chi_\alpha(x^a) = 0$ will be called Σ. Let σ^* the pull-back of σ into Σ. It follows that the induced Poisson bracket structure on Σ is simply the inverse of σ^* (the invertibility of σ^* is guaranteed by the invertibility of $C_{\alpha\beta}$). See, for example, [21] (chapter 2) for more details on this construction and in particular its relation with the Dirac bracket.

In terms of the modes T^a_n defined in (50), conditions (54) read

$$T^+_n = -k\delta^0_n, \quad T^3_n = 0 \tag{60}$$

we then need to consider the matrix (evaluated on the surface (60)),

$$C := \begin{pmatrix} [T^+_n, T^+_m] & [T^+_n, T^3_m] \\ [T^3_n, T^+_m] & [T^3_n, T^3_m] \end{pmatrix} = \begin{pmatrix} 0 & k\delta_{n+m} \\ -k\delta_{n+m} & (kn/2)\delta_{n+m} \end{pmatrix} \tag{61}$$

which is indeed invertible. We remind here the form of the algebra (52) in the basis $\{T^\pm = T^1 \pm iT^2, T^3_n\}$,

$$[T^+_n, T^-_m] = 2T^3_{n+m} + nk\delta_{n+m}, \tag{62}$$

$$[T^3_n, T^\pm_m] = \pm T^\pm_{n+m}, \tag{63}$$

$$[T^3_n, T^3_m] = \frac{kn}{2}\delta_{n+m}. \tag{64}$$

By an straightforward application of the method explained above, the induced Poisson structure $[\ ,\ ]^*$ (or Dirac bracket) acting on the surface (54) can be written in terms of the original bracket $[\ ,\ ]$ as (sum over $n \in Z$ is assumed),

$$[a,b]^* = [a,b] + \frac{n}{2k}[a,T^+_n][T^+_{-n},b] + \frac{1}{k}[a,T^+_n][T^3_{-n},b] - \frac{1}{k}[a,T^3_n][T^+_{-n},b]. \tag{65}$$

This bracket, by definition, satisfies $[a,T^3_n]^* = 0 = [a,T^+_n]^*$ for any function a, on the surface (60). Now we compute the algebra of the remaining component T^-_n. As before, it is convenient to define $L_n = -T^-_n$. The L_n's are then the Fourier modes of the function L defined in (55),

$$L(w) = \sum_{n\in Z} L_n e^{inw} \tag{66}$$

and satisfy the Virasoro algebra,

$$[L_n, L_m]^* = (n-m)L_{n+m} - \frac{k}{2}n^3\delta_{n+m} \tag{67}$$

with a central charge

$$c = -6k. \tag{68}$$

Note that since k is negative (see (20)), the central charge in (67) is positive and thus highest weight unitary representations exist.

The form of the central term in (67) is not the standard one. One could shift L_0 in order to find the usual $n(n^2-1)$ term. However, for the black hole whose exact isometries are $SO(2) \times SO(2)$ (due to the identifications) it is more natural to leave the central term as in (67). This is also natural from the point of view of supergravity since the vacuum black hole Killing spinors are periodic [6].

The space of solutions (56) is invariant under conformal transformations generated by $L(w)$. Note that in the Chern-Simons formulation of three-dimensional gravity the Chern-Simons coupling k was related to Newton's constant G as $k = -l/4G$ (see (20)). This means that the central charge in the Virasoro algebra $c = -6k$ coincides with the Brown-Henneaux [5] central charge $c = 3l/2G$. This is not a coincidence [9]. As we shall see, the above conformal algebra represents exactly the Brown-Henneaux conformal symmetry of three-dimensional adS gravity. The relation between the reduction conditions (54) and the conformal symmetry found in [5] was stablished in [9]. A previous calculation of the central charge using the Chern-Simons formulation of three-dimensional gravity and a twisted Sugawara construction was presented in [29].

What we have done for the holomorphic sector can be repeated for the anti-holomorphic sector. The reduction conditions in this case read $\bar{T}^-_n = -k\delta^0_n$ and $\bar{T}^3_n = 0$. Since the affine $SU(2)_k$ algebra is invariant under the change $T^+_n \leftrightarrow T^-_n$, $T^3 \to -T^3_n$, one finds the same induced algebra. The anti-de Sitter solution for the anti-holomorphic part reads

$$\bar{A}_{\bar{w}}(\bar{L}) = -ib \begin{pmatrix} 0 & (1/k)\bar{L}(\bar{w}) \\ 1 & 0 \end{pmatrix} b^{-1}, \tag{69}$$

plus $\bar{A}_w = 0$ and $\bar{A}_\rho = -iJ_3$. The residual gauge transformations are

$$\bar{\lambda} = b \begin{pmatrix} -(i/2)\partial\bar{\varepsilon} & (1/k)\bar{\varepsilon}\bar{L} + (1/2)\partial^2\bar{\epsilon} \\ \bar{\varepsilon} & (i/2)\partial\bar{\varepsilon} \end{pmatrix} b^{-1}, \tag{70}$$

where $\bar{\varepsilon} = \bar{\varepsilon}(\bar{w})$ is an arbitrary function of $\bar{w}$. Again $\bar{L}$ is a Virasoro operator and the central charge is $c = -6k$. Note that the effective coupling in the anti-holomorphic Chern-Simons theory is $-k$ because the Chern-Simons action $I[\bar{A}]$ has a minus sign in front. However, the non-zero current is now $\bar{A}_{\bar{w}}$ instead of $\bar{A}_w$. The change $k \to -k$ is compensated by $w \to \bar{w}$ and we find the same Virasoro algebra with the same central charge in both cases.

We shall call the gauge fields (56) and (69) the anti-de Sitter solution of the equations of motion because they are appropriated to anti-de Sitter spacetimes.

C The WZW and Liouville actions

We have displayed in this section two solutions to the Chern-Simons equations of motion with an affine (Sec. III A) and Virasoro (Sec. III B) symmetries. We have also argued that, at least classically, the corresponding algebras can be interpreted as basic Poisson brackets acting on the corresponding phase spaces. A natural and powerful way to justify this point is by studying the induced theories at the boundary for the corresponding boundary conditions. It is known [3] that under the boundary condition (29), the Chern-Simons action reduces to a WZW action at the boundary whose basic Poisson bracket, first calculated in [35], is the affine algebra (52). Reducing the affine solution via (54) then gives (67). At this point, a natural question to ask is what is the boundary action (analogue to the WZW action) which would give rise directly to the Virasoro algebra (67) as its basic Poisson bracket. We do not know the answer to this question. However, an alternative route can be taken. It was shown in [9] that the two chiral WZW actions arising in Chern-Simons gravity can be combined into a single non-Chiral action via $g = g_1^{-1}g_2$. Furthermore, the reduction conditions (54) applied to the non-Chiral theory lead to a Liouville action [36] which has the expected conformal symmetry with a central charge equal to $c = -6k$. The solutions of three-dimensional gravity can be classified in terms of the solutions of Liouville theory. One finds that the different monodromy conditions (elliptic, parabolic and hyperbolic) led to the three classes of solutions: conical singularities, extreme, and black holes [37]. See [8] for a direct relation between the Liouville energy momentum tensor and the anti-de Sitter boundary conditions, without using the Chern-Simons formalism. The Liouville field has also appeared in [38] in the study of solutions to the 2+1 classical equations with a positive cosmological constant.

The Liouville action is certainly a good candidate to describe the dynamics of 2+1 gravity with anti-de Sitter boundary conditions. However, it should be kept in mind that its derivation from the WZW model is not unique and a better control on some global issues is necessary. First, merging the two Chiral WZW actions into a single one through $g = g_1^{-1}g_2$ is not unique because g is invariant under $g_1 \to Ag_1$, $g_2 \to Ag_2$ with A arbitrary. Second, the Liouville action arises in terms of a Gauss decomposition of the group element which is not global. For these reasons we have not committed ourselves to any particular form for the boundary action but instead we have treated the basic Poisson bracket algebras as the starting point for quantization.

IV A QUANTUM SPACETIME

We have found in the last section a general solution for the classical equations of motion with prescribed boundary conditions. We have also found the induced Poisson bracket structure acting on the gauge-fixed dynamical functions. Our aim in this section is to quantise those spaces and apply the results to three-dimensional gravity.

However, an important warning is necessary here. The space of solutions that we have displayed (affine and anti-de Sitter solutions) are explicitly not coordinate invariant. This means that we could have chosen other coordinates to describe that space and it is not guaranteed that the corresponding quantum versions would be equivalent. For example, we know that a full quantization of the Chern-Simons action with compact groups induces a shift in the coupling constant [2–4] which is not seen in our gauge-fixed approach (although it is suggested by the Sugawara form of the conformal generators; see [29] for a construction of the Virasoro generators using a twisted Sugawara operator). We shall not attack this problem here in the belief that at least in the large k limit our results can be trusted.

It is worth stressing here that the fact that we have found non-Abelian Poisson structures (Virasoro and Kac-Moody algebras) is a consequence of the self-interacting character of gravity. The non-Abelian pieces in those algebras are measured by the coupling k which in the semiclassical limit, when gravity becomes linearized, goes to infinity.

A The metric

The general solution to Einstein equations in three dimensions with anti-de Sitter boundary conditions can be found using the results of last section together with the correspondence between metrics $g_{\mu\nu}$ and connections $A^a_\mu, \bar{A}^b_\nu$ discovered in [1]. Given the connections $A^a_\mu, \bar{A}^a_\mu$, one constructs the Lorentz vector $e^a_\mu = (l/2i)(A^a_\mu - \bar{A}^a_\mu)$ and then the metric $g_{\mu\nu} = e^a_\mu e^b_\nu \delta_{ab}$. It follows from the analysis of [1] that if A^a_μ and $\bar{A}^a_\mu$ satisfy the Chern-Simons equations of motion, then $g_{\mu\nu}$ satisfies the three-dimensional gravitational equations.

For our calculations it will be useful to define the matrix,

$$e_\mu = \frac{l}{2i}(A_\mu - \bar{A}_\mu) \tag{71}$$

where $A = A^a J_a$ and $\bar{A} = \bar{A}^a J_a$. The spacetime metric is then given by

$$g_{\mu\nu} = -2\mathrm{Tr}(e_\mu e_\nu). \tag{72}$$

The conventions for the J_a matrices are displayed in (18).

The affine solution (45) (and its anti-holomorphic part) is certainly a very general and interesting solution for the Chern-Simons equations of motion, however, the induced metric does not satisfy the anti-de Sitter boundary conditions prescribed in [5]. This is the reason that we have also considered the reduction of (45) via (54) which leaves an asymptotically anti-de Sitter spacetime, with a conformal symmetry.

Consider the anti-de Sitter solutions (56) and (69) for the Chern-Simons equations of motion and let us compute the associated metric. Using (71) we compute the components of the triad,

$$e_w = -\frac{l}{2}\begin{pmatrix} 0 & e^\rho \\ e^{-\rho}L/k & 0 \end{pmatrix}, \tag{73}$$

$$e_{\bar{w}} = \frac{l}{2}\begin{pmatrix} 0 & e^{-\rho}\bar{L}/k \\ e^\rho & 0 \end{pmatrix}, \tag{74}$$

$$e_\rho = lJ_3, \tag{75}$$

and then using (72) we find the metric,

$$ds^2 = 4Gl(Ldw^2 + \bar{L}d\bar{w}^2) + \left(l^2e^{2\rho} + 16G^2L\bar{L}e^{-2\rho}\right) dwd\bar{w} + l^2d\rho^2. \tag{76}$$

Here we have used the value of k given in (20). This is the metric that was displayed in Sec. I A. We can now go through the properties listed in that section and check their validity.

First, by construction (76) solves the Einstein equations because the corresponding gauge fields solve the Chern-Simons equations. This, of course, can be proved explicitly by checking that (76) has constant curvature. Since $L(w)$ and $\bar{L}(\bar{w})$ are arbitrary functions, the metric (76) provides an infinite number of solutions to Einstein equations with a negative cosmological constant in three dimensions. These solutions represent different physical states because two metrics with different values of $L, \bar{L}$ are related by a global diffeomorphism (global diffeomorphism here is what was called "improper" in [23]). At this point it is necessary to prove that the notion of global diffeomorphism has an analogue within general relativity. Since (76) has constant curvature, there exists a change of coordinates mapping (76) into the anti-de Sitter metric (11). The question is whether that change of coordinates is generated by one of the constraints of general relativity or not.

In complete analogy with the gauge case, we define global diffeomorphisms as coordinate transformations which are not generated by the constraints of general relativity [20]. Rather, global diffeomorphisms are generated by non-zero quantities which enter as boundary terms in their canonical generators. This has been analysed in detail in [5] from where we conclude that two metrics of the form (76), which only differ of the values of L and $\bar{L}$, are connected by a global diffeomorphism. The functions L and $\bar{L}$ then represent physical degrees of freedom from the gravitational point of view.

B Diffeomorphisms in Chern-Simons gravity

It is interesting and instructive to prove explicitly that there exists a change of coordinates $\{w, \bar{w}, \rho\} \to \{w', \bar{w}', \rho'\}$ which preserve the form of (76) changing only the values of L and $\bar{L}$. This is the goal of this paragraph. To find these transformations we can either do it by brute force acting with Lie derivatives on (76), or by using the results of the last sections making a dictionary between gauge transformations and diffeomorphisms. We shall follow this last procedure.

It is well known that, due to the flatness of the gauge field, in Chern-Simons theory the diffeomorphism invariance is not an independent symmetry. Indeed, a diffeomorphism along a vector field ξ^μ can be written as a gauge transformation with a parameter $\lambda^a = A^a_\mu \xi^\mu$ [14]. The converse is, in general, not true. However, in the case of Chern-Simons gravity where the relevant group is $SL(2,C)$ and an invertible triad exists, one can prove that all gauge transformations act on the metric via diffeomorphisms.

More explicitly, let $g_{\mu\nu}$ the metric associated to a particular configuration $A^a_\mu, \bar{A}^b_\nu$ through (71) and (72). Now, consider an arbitrary gauge transformation with parameters $\lambda^a, \bar{\lambda}^a$ acting on $A^a, \bar{A}^a$. It follows that the transformed metric (associated to the transformed fields) is related to the original one by a diffeomorphism generated by a vector field $\xi^\mu_{(\lambda,\bar{\lambda})}$.

To prove this statement, and find the explicit formula for $\xi^\mu_{(\lambda,\bar{\lambda})}$, consider the action of the gauge group on the triad. From (71) we find that under a gauge transformation $\delta A_\mu = D_\mu \lambda$, $\delta \bar{A}_\mu = \bar{D}_\mu \bar{\lambda}$ the triad changes according to,

$$\delta e_\mu = \frac{l}{2i} D^{(w)}_\mu (\lambda - \bar{\lambda}) - \frac{1}{2}[e_\mu, \lambda + \bar{\lambda}]. \tag{77}$$

Here we have used that $A_\mu + \bar{A}_\mu = 2\omega_\mu$ where $\omega_\mu = \omega^a_\mu J_a$ is the spin connection, and $D^{(w)}_\mu$ denotes its associated covariant derivative. The second term in (77) is a Lorentz rotation of the triad which does not change the metric. We then concentrate on the first term. Let us define the $SO(3)$ vector ρ^a and its associated vector field $\xi^\mu_{(\rho)}$ by,

$$\rho^a = \frac{l}{2i}(\lambda^a - \bar{\lambda}^a), \tag{78}$$

$$\xi^\mu = e^\mu_a \rho^a. \tag{79}$$

We assume that e^a_μ is invertible then (79) does make sense. We now study how does the transformation,

$$\delta e^a_\mu = D^{(w)}_\mu \rho^a = D^{(w)}_\mu (e^a_\nu \xi^\nu), \tag{80}$$

change the metric. Define the Christoffel symbols in the standard way by $D^{(w)}_\mu e^a_\nu = \Gamma^\sigma_{\nu\mu} e^a_\sigma$ [4]. The transformation (80) becomes $\delta e^a_\mu = e^a_\nu \xi^\nu{}_{;\mu}$ where the semicolon denotes standard covariant derivative. The action of this transformation on the metric is,

$$\begin{aligned} \delta g_{\mu\nu} &= \delta e^a_\mu e_{a\nu} + e^a_\mu \delta e_{a\nu}, \\ &= (e^a_\sigma \xi^\sigma{}_{;\mu}) e_{a\nu} + e^a_\mu (e_{a\sigma} \xi^\sigma{}_{;\nu}), \\ &= \xi_{\mu;\nu} + \xi_{\nu;\mu}, \end{aligned} \tag{81}$$

where in the last line we have used the definition of the metric tensor and the identity $g_{\mu\nu;\sigma} = 0$. Thus, a transformation in the triad of the form (80) is indeed seen in the metric as a diffeomorphism. Since the gauge transformations acting on $A, \bar{A}$ produce (up to a Lorentz rotation) a transformation of the form (80) with ρ^a given in (78), we conclude that the gauge group acts on the metric via a diffeomorphism with a parameter defined in (79).

Now we apply this result to the particular case of the residual gauge transformations (57) and (70). The residual vector field $\xi^\mu_{(\varepsilon,\bar{\varepsilon})}$ associated to those transformations is computed directly from (57) and (70) plus (78) and (79). The formulae for the triad are given in (73)-(75).

It should be clear from the above analysis that $\xi^\mu_{\varepsilon,\bar{\varepsilon}}$ generates a residual symmetry of the metric (76). This can be summarised in the following table. $A_\mu(L)$ and $\bar{A}^a_\nu(\bar{L})$ represent the residual connections (56) and (69), and $g_{\mu\nu}(L, \bar{L})$ the associated metric (76). Under the residual gauge transformations (57) and (70) the gauge field and metric transform according to:

$$\begin{array}{|ccccc|} A_\mu(L) & \multirow{2}{*}{$\rightarrow$} & A_\mu(L) + \delta A_\mu & = & A_\mu(L + \delta L) \\ \bar{A}_\nu(\bar{L}) & & \bar{A}_\nu(\bar{L}) + \delta \bar{A}_\nu & = & \bar{A}_\nu(\bar{L} + \delta\bar{L}) \\ \Downarrow & & \Downarrow & & \Downarrow \\ g_{\mu\nu}(L,\bar{L}) \rightarrow & & g_{\mu\nu}(L,\bar{L}) + \mathcal{L}_{\xi_{(\varepsilon,\bar{\varepsilon})}} g_{\mu\nu} & = & g_{\mu\nu}(L + \delta L, \bar{L} + \delta \bar{L}) \end{array}$$

where δL is given in (59), and the same expression holds for $\delta \bar{L}$. The equalities in the first two lines simply express the fact that the residual gauge transformations leave the gauge field invariant changing only the values of L and $\bar{L}$. The third line contains non-trivial information. First, as we discussed above, the metric associated to the transformed gauge fields is related to the original metric via a diffeomorphism parametrised with the residual vector field $\xi^\mu_{(\varepsilon,\bar{\varepsilon})}$. Then, since the transformed metric can also be written in terms of $A(L + \delta L), \bar{A}(\bar{L} + \delta \bar{L})$, we find the last equality and conclude that the vector field $\xi^\mu_{(\varepsilon,\bar{\varepsilon})}$ generates a residual diffeomorphism of the metric (76). This analysis exhibit the power of the Chern-Simons formalism. To discover that the metric (76) has a residual conformal invariance –not only asymptotically– using Lie derivatives would have been extremely complicated.

Let us work out explicitly the case on which $\bar{L} = 0$. The metric (76) reduces to the simple form

$$ds^2 = 4GlL(w)dw^2 + l^2 e^{2\rho} dw d\bar{w} + l^2 d\rho^2. \tag{82}$$

We act on this metric with the holomorphic residual transformation generated by (57). The associated residual vector $\xi^\mu = \delta x^\mu$ is computed from (57) and the triad (73)-(75) with $\bar{L} = 0$. In the coordinates $\{w, \bar{w}, \rho\}$ it reads,

$$\delta\rho = -\frac{i}{2}\partial\varepsilon, \qquad \delta w = i\varepsilon, \qquad \delta\bar{w} = -\frac{i}{2} e^{-2\rho}\partial^2\varepsilon. \tag{83}$$

Transforming the metric (82) with this vector one finds the same metric with L replaced by $L' = L + \delta L$, and

$$\delta L = i(\varepsilon\partial L + 2\partial\varepsilon L - \frac{l}{8G}\partial^3\varepsilon). \tag{84}$$

---

[4] The meaning of this definition can be uncovered by writing it in the form $\Gamma^\sigma_{\mu\nu} = e^\sigma_a \omega^a{}_{b\nu} e^b_\mu + e^\sigma_a e^a_{\mu,\nu}$. This is the transformation law of a connection, $\omega^a{}_{b\nu} \rightarrow \Gamma^\sigma{}_{\mu\nu}$, under the change of basis from the coordinate basis ∂_μ to the orthonormal frame $\vec{v}_a$ described by the matrix e^a_μ: $\partial_\mu = e^a_\mu \vec{v}_a$.

Since $l/8G = c/12 = -k/2$ with c and k given respectively in (3) and (20), we find consistency with (59), as expected. As a further check, we can now transform (82) with the anti-holomorphic residual transformation generated by (70). Since $\bar{L}$ is a quasi-primary field, we expect that this transformation will not preserve $\bar{L} = 0$ and thus the metric (82) will be transformed into (76) with $\bar{L} = \delta\bar{L}$. Indeed, from (70) and (73)-(75) we find the associated transformation,

$$\delta\rho = -\frac{i}{2}\bar{\partial}\bar{\varepsilon}, \qquad \delta w = -\frac{i}{2}e^{-2\rho}\bar{\partial}^2\bar{\varepsilon}, \qquad \delta\bar{w} = i\varepsilon + \frac{2iGL}{l}e^{-4\rho}\bar{\partial}^2\bar{\varepsilon}. \tag{85}$$

We act on (82) with this vector and find a metric of the form (76) with $\bar{L} = (-il/8G)\bar{\partial}^3\bar{\varepsilon}$. This is exactly the right transformation, in accordance with (84) applied to the anti-holomorphic field $\bar{L}(\bar{w})$.

As a last example of the conformal residual symmetries of (76) we mention here the case of the finite (holomorphic) exponential map. To simplify the notation we set here $4G = 1$ and $l = 1$. Let us make the finite change of coordinates on (82) $\{w, \bar{w}, \rho\} \to \{z, \bar{w}', \rho'\}$ defined by

$$z = e^{-iw}, \qquad \bar{w}' = \bar{w} + (i/2)e^{-2\rho}, \qquad e^{2\rho'} = ie^{2\rho+iw} \tag{86}$$

This transformation maps the cylinder w into the plane z, but leaves $\bar{w}$ in the cylinder. Note that here we are using explicitly the independence of w and $\bar{w}$. We act on the metric (82) with this change of coordinates and find again the same metric with $L(w)$ replaced by $T(z) = -z^{-2}(L(w) + 1/2)$. The shift $1/2 = 6/12$, of course, corresponds to the Schwarztian derivative of the map (86). (In the conventions $4G = 1$ and $l = 1$ the central charge (3) is equal to 6.)

C Black holes

As we have mentioned above, the metric (76) reduces to a three-dimensional black hole [11,12] when $L = L_0$ and $\bar{L} = \bar{L}_0$ are constants. This can be proved as follows. First, we define the constants M,J and $r_\pm$ by

$$L_0 + \bar{L}_0 = Ml = \frac{r_+^2 + r_-^2}{8Gl}, \tag{87}$$

$$L_0 - \bar{L}_0 = J = \frac{2r_+r_-}{8Gl}. \tag{88}$$

Next we define the real coordinates φ, t and r by,

$$w = \varphi + it, \tag{89}$$

and

$$r^2 = r_+^2\cosh^2(\rho - \rho_0) - r_-^2\sinh^2(\rho - \rho_0). \tag{90}$$

The constant ρ_0 is given by $e^{2\rho_0} = (r_+^2 - r_-^2)/(4l^2)$. This radial definition has the property $l^2d\rho^2 = N^{-2}dr^2$ where N^2 is the lapse function appearing in the black hole metric (8). The constants $r_\pm$ are the solutions of the equation $N^2(r_\pm) = 0$. After a long but direct calculation one can prove that the metric (76), in the coordinates $\{t, r, \varphi\}$, is exactly equal to the metric (8) with mass M and angular momentum J.

Since we are working in the Euclidean sector, the coordinate t appearing in (89) and (8) is periodic, $0 \le t < \beta$, with $\beta = 2\pi l^2 r_+/(r_+^2 - r_-^2)$. In order to fix the period of the time coordinate to be independent of the black hole parameters (and thus fix the complex structure of the torus), one can define $z = \varphi + \tau x^0$ with $0 \le x^0 < 2\pi$ and $\tau = i\beta/2\pi$. Since φ and x^0 are periodic, the complex coordinate z is defined on a torus,

$$z \sim z + 2\pi n + 2\pi\tau m, \qquad n, m \in Z. \tag{91}$$

with τ its modular parameter. Introducing τ is particularly convenient when studying modular invariance on the black hole manifold [25,39]. We shall not deal with this issue here, so we use (89).

D The quantum space of metrics. State counting

We have described in the last paragraph a set of metrics parametrized by two functions whose induced Poisson bracket yield the Virasoro algebra with a non-zero central charge. We shall now promote the algebra (67) to be a quantum algebra and study the properties of the associated quantum metric.

The unitary representations of the Virasoro algebra for a given positive central charge c are parametrized by a single real positive number h. In the semiclassical limit with $-k$ large, the Virasoro central charge $c=-6k$ is then large. Under these conditions, there exists one unitary representation for each conformal dimension h. We start with the vacuum state $|h>$ satisfying $L_0|h>=h|h>$ and $L_n|h>=0$ $(n>0)$. The excited states are constructed with the negative modes L_{-n} acting on $|h>$. The full representation, for a given h, is spanned by the vectors $|n_1,...,n_r;h>:=L_{-n_1}\cdots L_{-n_r}|h>$ with $r=1,2,....$ The same construction has to be repeated for the other Virasoro algebra $\bar{L}_n$.

In standard conformal field theory, the values of h are not arbitrary. They are equal to the conformal dimensions of the primary fields ϕ_h of the theory. The state $|h>$ is created by ϕ_h via $|h>=\phi_h(0)|0>$ where $|0>$ is the true conformal vacuum. In our situation, we do not have a field theory at the boundary (of course it could be Liouville theory, see Sec. III C for a discussion on this point) but only the Virasoro algebra. The usual state-operator map will then be missing until we decide which is right conformal field theory.

Once we promote the modes L_n and $\bar{L}_m$ to be operators acting on Fock space, the metric (76) becomes an operator that we shall denote by $d\hat{s}^2$. Note that since the metric (76) is an algebraic function in L and $\bar{L}$ which does not involve products of non-commuting operators, $d\hat{s}^2$ is well defined in the operator sense.

We can now define a natural map from Fock space to the space of classical solutions. For each state $|\Psi>$ in Fock space, we associate a classical solution to the equations of motion given by,

$$ds^2_\psi=<\psi|d\hat{s}^2|\psi> \tag{92}$$

where ds^2_ψ is the metric (76) with $L=<\psi|\hat{L}|\psi>$ and $\bar{L}=<\psi|\hat{\bar{L}}|\psi>$. Since (76) is a solution for arbitrary values of L and $\bar{L}$, the metric (92) is a solution for any state $|\psi>$. More interesting, the full set of solutions (76) can be generated by the above map.

According to (92), every state $|\psi>$ induces a unique classical solution. The converse is not true. For a given classical solution there may be many associated states. In particular, there are many states associated to a given black hole of mass M and angular momentum J. As explained before, the metric (76) gives rise to a black hole when L and $\bar{L}$ are constants and related to M and J by (87) and (88). Let $|M,J;\lambda>$ the set of states in the Hilbert space such that they satisfy

$$\begin{aligned}<M,J;\lambda|(L_n+\bar{L}_n)|M,J;\lambda> &= lM\,\delta^0_n,\\ <M,J;\lambda|(L_n-\bar{L}_n)|M,J;\lambda> &= J\,\delta^0_n,\end{aligned} \tag{93}$$

for all $\lambda=1,2,3,...,\rho(M,J)$. These states generate through (92) a black hole of mass M and angular momentum J. We can then formulate the problem of black hole degeneracy as whether the logarithm of the number of these states, $\ln\rho(M,J)$, is equal to the Bekenstein-Hawking entropy of the corresponding black hole of mass M and angular momentum J or not. The answer to this question depends on the structure of the Hilbert space.

Let us first work under the assumption that the Virasoro algebra is the basic quantum commutator of the theory. In this case, the counting is very simple. The states $|n_1,...,n_r;h>$, properly normalized, precisely have the property,

$$<n_1,...,n_r;h|L_n|n_1,...,n_r;h>=L_0\delta^0_n \quad \text{with} \quad L_0=h+\sum_{i=1}^{r}n_i. \tag{94}$$

The number of these states, $\rho(L_0,\bar{L}_0)$, is then equal to number of ways that one can write an integer as a sum of integers. For large values of L_0 and $\bar{L}_0$ this number is approximated by the well-known Ramanujan formula,

$$\rho_{c'}(L_0,\bar{L}_0)=e^{2\pi\sqrt{c'L_0/6}+2\pi\sqrt{c'\bar{L}_0/6}}, \tag{95}$$

with $c'=1$. Unfortunately, this naive counting does not give the right result. Inserting $lM=L_0+\bar{L}_0$ and $J=L_0-\bar{L}_0$ in (95) gives an entropy equal to $S=c^{-1/2}A/4G$, where c is the central charge (3) and $A=2\pi r_+$

is the perimeter of the horizon. (The relation between the different parameters is given in (87) and (88).) The prefactor $c^{-1/2}$ shows that our naive procedure is not yet correct because we would expect the degeneracy of states to be equal to the Bekenstein-Hawking value.

If we do not regard (67) as the basic algebra but only as representing the symmetry algebra of some underlying conformal field theory, then an elegant and striking way to relate (67) with the correct Bekenstein-Hawking entropy is available [7]. Suppose that the algebra (67) represents the Virasoro algebra associated to some conformal field theory with central charge c. Suppose also that this CFT is unitary, in the sense that $L_0, \bar{L}_0 > -c/24$ (note that we are using the Virasoro generators which vanish for the vacuum black hole), and that the partition function,

$$Z[\tau] = \mathrm{Tr}\, e^{2\pi i\tau L_0 - 2\pi i\bar{\tau}\bar{L}_0}, \tag{96}$$

is modular invariant. This means

$$Z[\tau'] = Z[\tau], \qquad \tau' = \frac{a\tau + b}{c\tau + d}, \tag{97}$$

for any $a, b, c, d \in Z$ and $ad - bc = 1$. Then, it follows [40] that that number of states with L_0 and $\bar{L}_0$ fixed is again given by (95) but this time with $c' = c$. The associated entropy is then exactly equal to the Bekenstein-Hawking value $S = A/4G$ with $A = 2\pi r_+$ equal to the perimeter of the horizon [7].

As stressed in [10], this result is too beatiful to be wrong. Even more, recently [41,42], it has been shown that under some boundary conditions at the horizon, similar results can be applied to higher dimensions. These are exiting results which bring closer the long standing dream of a statistical mechanical description for the Bekenstein-Hawking entropy. However, there remains to find the conformal field theory responsible for the degrees of freedom and, most importantly, to determine whether general relativity is enough to describe that CFT, or other degrees of freedom like string theory are necessary.

An alternative route to get the right counting was suggested in [43]. For integer values of the central charge c, there is a natural way to add degrees of freedom to the theory in such a way that the counting yields the right result. The idea is that the Virasoro algebra (67) can be regarded as a sub-algebra of another Virasoro algebra, with central charge 1 and generators Q_n, via the formula,

$$L_n = \frac{1}{c} Q_{cn}. \tag{98}$$

See [44] and references therein for a detailed description of this embedding. (The formula (98) has also appeared in [45].) The number of states associated to the representations of the operators Q_n is again given by (95) with $c' = 1$ and L_0 replaced by Q_0. Since by (98) $Q_0 = cL_0$, this yields the right result when using (87) and (88). The main problem with this approach is that we do not know how to relate the gravitational degrees of freedom to the generators Q_n. Perhaps one should look for other boundary conditions, generalising (54), which may give other conformal structures, generalising (67). This issue is presently under investigation.

Whether the Virasoro operators are fundamental variables or not, this will not change our quantum geometry picture. The microscopical origin of the black hole degeneracy is associated to different states (living in the correct CFT) which generate the same classical metric through (92).

V FINAL REMARKS

Maldacena [46] has conjectured a duality between large N super-conformal field theory in four dimensions and Type IIB string theory compactified on $\mathrm{adS}_5 \times S_5$. This relation has become known as adS/CFT correspondence due to the relation between the symmetry groups in each theory. The result of Brown and Henneaux [5] relating adS_3 and a conformal algebra in 1+1 dimensions can also be regarded as an adS/CFT correspondence. Note however that contrary to the higher dimensional case, this relation involves only asymptotic adS space whose isometry group is infinite dimensional. In [37,47,48] the relation between these two aspects of the adS/CFT correspondence has been explored.

Finally, we would like to mention here a surprising motivation to study three-dimensional gravity. It has been shown in [49] and [50] (see also the recent review [51]) that there exists duality transformations relating five-dimensional black holes with three-dimensional ones. This means that everything we can learn about three-dimensional quantum gravity can be useful to higher dimensional situations.

ACKNOWLEDGMENTS

The author would like to thank H. Falomir, R.E. Gamboa Saraví and F.A. Schaposnik for the invitation to the Buenos Aires' Meeting "Trends in Theoretical Physics II". Useful discussions with M.Asorey and F. Falceto are acknowledged. Financial support from CICYT (Spain) grant AEN-97-1680, and the Spanish postdoctoral program of Ministerio de Educación y Ciencia is also acknowledged.

REFERENCES

1. A. Achúcarro and P.K. Townsend, Phys. Lett. **B180**, 89 (1986).
2. E. Witten, Commun. Math. Phys. **121**, 351 (1989).
3. G. Moore and N. Seiberg, Phys. Lett. **B220**, 422 (1989); S. Elitzur, G. Moore, A. Schwimmer and N. Seiberg, Nucl. Phys. **B326**, 108 (1989)
4. J.M.F. Labastida and A.V. Ramallo, Phys.Lett. **B227**, 92 (1989).
5. J.D. Brown and M. Henneaux, Commun. Math. Phys. **104**, 207 (1986).
6. O. Coussaert, M. Henneaux, Phys. Rev. Lett. **72**, 183 (1994).
7. A. Strominger, High Energy Phys. **02** 009 (1998).
8. P. Navarro and J. Navarro-Salas, Phys. Lett. **B439**, 262 (1998).
9. O. Coussaert, M. Henneaux, P. van Driel, Class. Quant. Grav. **12**, 2961 (1995).
10. S. Carlip, Class. Quant. Grav. **15** 3609 (1998).
11. M. Bañados, C. Teitelboim and J.Zanelli, Phys. Rev. Lett. **69**, 1849 (1992).
12. M. Bañados, M. Henneaux, C. Teitelboim and J.Zanelli, Phys. Rev. **D48**, 1506 (1993).
13. A.M. Polyakov, Mod. Phys. Lett. **A2**, 893 (1987)
14. E. Witten, Nucl. Phys. **B 311**, 4 (1988).
15. S. Carlip, *Quantum Gravity in 2+1 Dimensions*, Cambridge University Press (1998).
16. S. Carlip, Phys. Rev. **D51**, 632 (1995).
17. A.P. Balachandran, L. Chandar and A. Momen, Nucl. Phys. **B461**, 581 (1996)
18. D. Cangemi, M. Leblanc and R.B. Mann, Phys. Rev. **D48**, 3606 (1993).
19. S. Carlip, Nucl.Phys.Proc.Suppl. **57**, 8 (1997); S. Carlip, gr-qc/9603049; gr-qc/9509024
20. T. Regge and C. Teitelboim, Ann. Phys. (N.Y.) **88**, 286 (1974).
21. M. Henneaux and C. Teitelboim, *Quantization of Gauge Systems* (Princeton University Press, Princeton, 1992).
22. J.D. Brown and M. Henneaux, Journ. Math. Phys. **27**, 489 (1986).
23. R. Benguria, P. Cordero and C.Teitelboim, Nucl. Phys. **B122**, 61 (1977).
24. L.F. Abbott and S. Deser, Phys. Lett. **116B**, 259 (1982)
25. M. Bañados, T. Brotz and M. Ortiz, "Boundary dynamics and the statistical mechanics of the 2+1 dimensional black hole", hep-th/9802076. To appear in Nucl.Phys.B.
26. S. Carlip, C. Teitelboim, Phys. Rev. **D51** 622 (1995).
27. M. Bañados and F. Mendez, Phys. Rev. **D58** 104014 (1998)
28. A. P. Balachandran, G. Bimonti, K.S. Gupta, A. Stern, Int. Jour. Mod. Phys. **A7**, 4655 (1992).
29. M. Bañados, Phys. Rev. **D52**, 5816 (1995).
30. E. Witten, Commun. Math. Phys. **137**, 29 (1991).
31. S. Carlip, Phys. Rev. **D55**, 878 (1997)
32. P. Oh and M.I. Park, "Symplectic reduction and symmetry algebra in boundary Chern-Simons theory", hep-th/9805178. M.I. Park, hep-th/9811033 (to appear in Nucl. Phys. B)
33. J. Mickelsson, Lett.Math.Phys. **7**, 45 (1983)
34. A.M. Polyakov, Int. J. Mod. Phys. **A5** (1990) 833.
35. E. Witten, Commun. Math. Phys. **92**, 455 (1984).
36. A. Alekseev and S. Shatashvili, Nucl. Phys. **B323**, 719 (1989). See also, P. Forgács, A. Wipf, J. Balog, L. Fehér and L. O'Raifeartaigh, Phys. Lett. **227 B** (1989) 214.
37. E. Martinec, "Conformal field theory, geometry, and entropy", hep-th/9809021
38. S. Deser and R. Jackiw, Ann. Phys. (NY), **153**, 405 (1984)
39. J. Maldacena, A. Strominger, "AdS(3) black holes and a stringy exclusion principle", hep-th/9804085
40. J. A. Cardy, Nucl. Phys. **B270**, 186 (1986).
41. S. Carlip, "Black hole entropy from conformal field theory in any dimension", hep-th/9812013
42. S. N. Solodukhin, "Conformal description of horizon's states", hep-th/9812056.
43. M. Bañados, "Embeddings of the Virasoro algebra and black hole entropy", hep-th/9811162, to appear in Phys. Rev. Lett.

44. L. Borisov, M.B. Halpern, C. Schweigert, Int. J. Mod. Phys. **A13**, 125 (1998)
45. L. P. Colatto, M. A. De Andrade, F. Toppan, hep-th/9810145.
46. J. Maldacena, Adv. Theor. Math. Phys. **2**, 231 (1998).
47. A. Giveon, D. Kutasov and N. Seiberg, "Comments on string theory on adS(3)", hep-th/9806194
48. M. Henningson, K. Skenderis, J.High Energy Phys 9807, 023 (1998)
49. K. Sfetsos and K. Skenderis Nucl.Phys. **B517**, 179 (1998)
50. S. Hyun, hep-th/9704005
51. K. Skenderis, "Black holes and branes in string theory", hep-th/9901050.

Non Relativistic Limit of a Model of Fermions interacting through a Chern-Simons Field

A.J. da Silva[1]

Instituto de Física, USP
C.P. 66318 - 05389-970, São Paulo - SP, Brazil

Abstract. We study the non relativistic limit of a Model of Fermions interacting through a Chern-Simons Field, from a perspective that resembles the Wilson's Renormalization Group approach, instead of the more usual approach found in most texts of Field Theory. The solution of some difficulties, and a new understanding of non relativistic models is given.

Models of a Chern-Simons [1] field interacting with non relativistic bosons [2] or fermions [3] have being studied in the literature both for its interest in general understanding of field theory by itself as for its application to Condensed Matter Physics [4]. The use of these models face in general, the difficulties of their non renormalizability. This fact is perhaps, the main reason for the interest, on the results of Bergman and Lozano [2], in one loop, later extended to three loops [5]. Their model consists of a non-relativistic boson field ϕ with a quartic self interaction and minimally interacting with a Chern-Simons field A^μ :

$$\mathcal{L} = \phi^* \left(i\frac{d}{dt} + eA^0 \right) \phi - \frac{1}{2m} |(\vec{\nabla} - ie\vec{A})\phi|^2$$
$$-\frac{\lambda_0}{4}(\phi^*\phi)^2 + \frac{\theta}{2}\epsilon^{\mu\nu\rho} A^\mu \partial^\nu A^\rho. \tag{1}$$

The only primitively divergent Green Function is the boson four point function. Up to one loop, the model can be made finite by the choice of a renormalized coupling constant λ through the equation;

$$\lambda_0 = \lambda + \frac{m}{4\pi} \left(\lambda^2 - \frac{4e^2}{m^2\theta^2} \ln\left(\frac{\Lambda}{M}\right) \right) \tag{2}$$

where Λ is an ultraviolet (UV) cut-off and M an arbitrary constant (the renormalization constant) with dimension of mass. Their main observation is that at the critical value, $\lambda^* = |\frac{2e}{m\theta}|$, the one loop contribution vanishes and no renormalization of λ is needed. At this choice of λ the model regains the scale invariance that it has at classical level, and the relative wave function of the two bosons reproduces the Aharonov-Bohm scattering amplitude [8] up two the second Bohr order. The model of non relativistic fermions interacting with the Chern-Simons field was also discussed in [2] and studied in more details in [3]. In this last paper it is shown that the one loop scattering of two fermions with spins of the same sign (in 2+1 dimension the spin is a pseudo-scalar) is finite in one loop, due to the contact interaction represented by the Pauli interaction, that is already present in the minimal interaction of the fermions with the gauge field. As for the scattering of two fermions of opposite spins the Pauli interaction does not have any role and the amplitude is divergent unless a quartic fermionic interaction of the form $c(\Lambda)\,\psi^*\psi\,\phi^*\phi$ where ψ and ϕ represent respectively fermions with spin plus and minus 1/2, and c is a constant that depends logarithmically on the UV cut off Λ. This last fact poses a new problem. If the non relativistic model is thought to be the low energy limit of a more fundamental model of relativistic Dirac fermions interacting with a Chern-Simons field, in the way that this limit is generally taken in most texts [9], it would come from a similar quartic interaction in the relativistic

[1] e-mails: *ajsilva@fma.if.usp.br*

CP484, *Trends in Theoretical Physics II*, edited by H. Falomir, R. E. Gamboa Saraví, and F. A. Schaposnik

fermions. As is well known one such interaction is non renormalizable! We will show that this is, in fact, a false problem. No quartic non renormalizable self interaction is needed in the "parent" relativistic model if a new perspective on the non relativistic limit in field theory is taken. Before going to the description of this new limit, lets us briefly resume, in an example, the "Classical Non Relativistic Limit", and discuss why it is not always correct. Let us consider, in 2+1 dimension, a 2 component Dirac fermion field Ψ, that represents a spin plus fermion and its anti-fermion, interacting with an external electromagnetic field A^μ, as described by the Lagrangian density (the gamma matrices are $\gamma^0 \doteq \sigma^0$,$\gamma^1 \doteq i\sigma^1$ and $\gamma^2 \doteq i\sigma^2$ where σ^μ are the Pauli matrices)

$$\mathcal{L}_{rel} = \bar{\Psi} \left\{ \gamma^\mu \left(i\frac{\partial}{\partial x^\mu} + eA^\mu \right) - m \right\} \Psi. \tag{3}$$

The corresponding equation of motion is:

$$\left(i\frac{d}{dt} + eA^0 \right) \Psi = \left\{ \gamma^0 \vec{\gamma}. \left(-i\vec{\nabla} - e\vec{A} \right) + \gamma^0 m \right\} \Psi. \tag{4}$$

Let us now consider a positive energy solution Ψ of this equation.To make contact with the non relativistic description,in which, the rest energy m of the particles is not included in the solution,let us make in the above equation of motion ,the substitution

$$\Psi = \frac{e^{-imt}}{\sqrt{2m}} \begin{pmatrix} \psi \\ \chi \end{pmatrix}. \tag{5}$$

The result is:

$$\left(i\frac{d}{dt} + eA^0 \right) \psi = i\Pi_- \chi \tag{6}$$

$$\left(i\frac{d}{dt} + eA^0 + 2m \right) \chi = -i\Pi_+ \psi \tag{7}$$

where $\Pi_\pm = \Pi^1 \pm \Pi^2$, and $\Pi^i = -i\frac{d}{dx^i} - eA^i$. If we make the assumptions that: $e|A^0| << m$ and that the momentum space components of ψ and χ are non null only for low momenta and energies, $(|\vec{p}|, E) << m$, then the second equation can be approximately solved for χ , and inserted in the first one, giving:

$$\left(i\frac{d}{dt} + eA^0 \right) \psi = \frac{1}{2m} \left(-i\vec{\nabla} - e\vec{A} \right)^2 + \frac{e}{2m} B\, \psi, \tag{8}$$

where $B = \vec{\nabla} \wedge \vec{A}$ is the magnetic field. The one component spinor ψ represent a fermion with spin plus. The last term is the Pauli magnetic moment–magnetic field interaction term of . The Lagrangian density corresponding to this equation of motion is the so called Pauli Schrödinger (PS) propagator of non relativistic fermions in an electromagnetic field:

$$\mathcal{L}^{class}_{nonrel} = \psi^* \left(i\frac{d}{dt} + eA^0 \right) \psi - \frac{1}{2m} | \left(-i\vec{\nabla} - e\vec{A} \right) \psi|^2 + \frac{e}{2m} B \psi^* \psi \tag{9}$$

The essential facts behind the above non relativistic limit are the assumptions on the strength of A^μ and the momentum space support of the field Ψ (the second assumption is not meaningfull without the first, since a low energy initial state of Ψ can be driven to a relativistic state by the action of a strong A^μ field). Suppose now that A^μ is not an external, controllable field, but is a dynamical field with dynamics given by a Maxwell or Chern-Simons term (that must be thought as added to the Lagrangian (3)). Let us consider in this theory, the scattering of two low energy fermions, their energy and momenta given by $\left(p^0 = m + \frac{|\vec{p}|^2}{2m}, \pm\vec{p} \right)$ with $|\vec{p}| << m$. On the right side of Figure 1 we draw a possible one loop contribution (among others) to this process. The corresponding amplitude is given by a Feynman integral in the loop momentum k^μ:

$$A^{relativ}_{lowenergy} = \cdots \int^\infty d^3k \left(\cdots \frac{i}{(k^0 + p^0)^2 - (\vec{k} + \vec{p})^2 - m^2} \right) \cdots \tag{10}$$

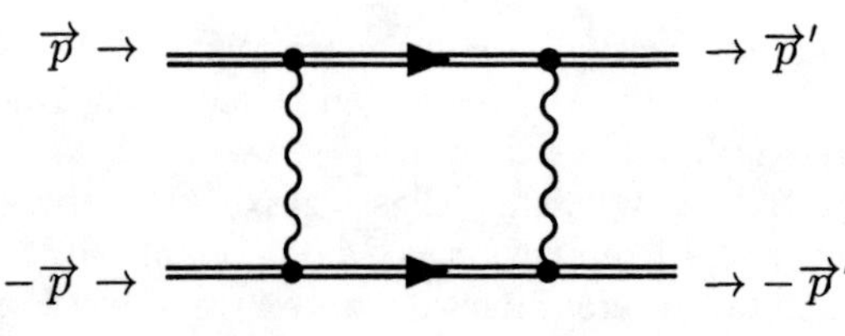

FIGURE 1. Example of a one loop graph contributing to the scattering of two fermions.

The main observation on this equation is that, even if the process we are treating is a low energy one, the amplitude receives contributions of high energy intermediate states, represented by propagators whose dynamics is essentially relativistic, and so, not coming from the Feynman rules of the non relativistic Lagrangian (9). The bigger or lower suppression of the contribution of these high energy states depend on the dynamics of the exchanged A^{μ} field. As we will explicitly see in one example below, for the Chern-Simons field, they effectively give a contribution that can not be understood as coming from the non relativistic Lagrangian (9). What about the description of this same process starting from the non relativistic theory given by (9)? The amplitude for the same process of figure 1 is of the form

$$\mathcal{A}^{class}_{nonrel} = \cdots \int^{\infty} d^3k \left(\cdots \frac{i}{(k^0 + p^0) - (\vec{k} + \vec{p})^2/2m} \right) \cdots \tag{11}$$

where now $p^0 = |\vec{p}|^2/2m$. Here also, the integral extends up to infinity momenta. High energy intermediate states also contribute to the amplitude, even with a bigger weight than to $\mathcal{A}^{relativ}_{lowenergy}$, as seen from the worse UV behavior of the Pauli-Schrödinger(PS) propagator. It must yet be observed, that from the view point of the Special Relativity, the PS propagator misses to represent the propagation of high energy intermediate states. Due to these facts, some authors in Field Theory ([10]) take the view of Non Relativistic Field Theory as a cut off theory. This means that instead of considering ,in a wrong way, the contribution of the high energy intermediate states ,they prefer to decouple them from the theory by limiting the integration in the Feynman integrals up to a maximal energy-momentum, compatible with the newtonian description provided by the PS propagator. This is also a view taken by some authors in optics ([11]). There, the typical energy involved in the scattering processes are of the order of the ionization energy of the atoms, $\alpha^2 m$, where α is the fine structure constant and m is the electron mass. The cut off assumed is $\Lambda = \alpha m$, the inverse of the Bohr radius of the atom, much bigger than the typical energies involved in the scattering processes, and much smaller than the rest energy, m of the electron.

We will take a slight variation of these ideas, suited for understanding the results on non relativistic models with a CS field in the Coulomb Gauge, as treated in the literature [2,3]. We will consider a non relativistic cutoff, only in the spatial momentum $\vec{k}$, of the Feynman integrals; that is, we will calculate the Feynman integrals,firstly freely integrating the k^0 momentum component up to infinity, and then restricting the integration in $|\vec{k}|$ to the region $(0\,,\,\Lambda)$ with Λ chosen so that $|\vec{p}| << \Lambda << m$, where $\vec{p}$ is a characteristic momentum of the low energy process in which of interest, and m is the mass of the fermion field. This choice has the additional technical advantage of not breaking the locality in the time direction. The way, we are proposing, that these cut off non relativistic models are related to originally relativistic ones, is akin to the ideas of the Renormalization Group of Wilson [12]. We will first outline the main ideas in a toy model.

Lets us consider a relativistic field theory in 1 space-time dimension, with dynamics given by a Lagrangian $\mathcal{L}^{rel}(\Phi)$. Its functional generator is given by

$$\mathcal{Z}(j) = \int \mathcal{D}\Phi(p) \, \exp\left(i \int dp \left(\mathcal{L}^{rel}(\Phi) + j\Phi \right) \right) \tag{12}$$

where j is an external source, and $\mathcal{D}\Phi(p) \doteq \prod_0^{\infty} d\Phi(p)$. Suppose that we are only interested in describing "non relativistic"processes involving external particles with momenta p smaller than a certain value $\Lambda << m$, where m is the mass of the field Φ. This limitation can be implemented in the functional generator by choosing the external source to be non null only for the momentum region $(0\,,\,\Lambda)$. The Φ field can be separated in $Phi(p) = \phi(p) + \bar{\phi}(p)$ where ϕ has support in $(0, \Lambda)$ and $\bar{\phi}$ in (Λ, ∞). The integration measure goes in $\mathcal{D}\Phi = \mathcal{D}\phi\,\mathcal{D}\bar{\phi}$, the Lagrangian separates in $\mathcal{L}^{rel}(\Phi) = \mathcal{L}^{rel}(\phi + \bar{\phi}) = \mathcal{L}^{rel}(\phi) + \nabla\mathcal{L}(\phi\,,\,\bar{\phi})$ and $j\,\Phi$ gets reduced to $j\,\phi$. The functional generator becomes

$$\mathcal{Z}(j) = \int \mathcal{D}\phi \exp i \int (\mathcal{L}^{rel}(\phi) + j\,\phi) \int \mathcal{D}\bar{\phi} \exp i \int (\nabla \mathcal{L}(\phi, \bar{\phi}) \tag{13}$$

and can be written in the form

$$\mathcal{Z}(j) = \int \mathcal{D}\phi \exp i \int (\mathcal{L}^{effect}(\phi, \Lambda) + j\,\phi) \tag{14}$$

where $\mathcal{L}^{effect}(\phi\,,\,\Lambda) = \mathcal{L}^{rel}(\phi) + \delta\mathcal{L}(\phi\,,\,\Lambda)$ and

$$\int \delta\mathcal{L}(\phi\,,\,\Lambda) = -i \ln \left(\int \mathcal{D}\bar{\phi} \exp i \int \nabla \mathcal{L}(\phi\,,\,\bar{\phi}) \right) \tag{15}$$

The effects of the high momenta modes $\bar{\phi}$ are incorporated in the effective dynamic of the low energy ones, through the additional term $\delta\mathcal{L}(\phi\,,\,\Lambda)$. It is the effective Lagrangian, $\mathcal{L}^{effect}$, in which the only remaining free momemta modes are the non relativistic ones, and not the original $\mathcal{L}^{rel}$, that will give, through the approximations called Classical Non Relativistic Limit (exemplified above), the same results to low energy processes, as if calculated from the original relativistic model.

The integration in $\bar{\phi}$ in (15) can in general be done by expanding the exponential in a series of powers of $\int \nabla \mathcal{L}(\phi\,,\,\bar{\phi})$. The result will be a series of Feynman graphs with the propagator of $\bar{\phi}$ in the internal lines and the field ϕ in the external legs. This means that the integration in the loop momenta is restricted to the interval $(\,\Lambda\,,\,\infty\,)$. The result is in general Λ dependent (as we will see the result of Bergman and Lozano is an exception) resulting in an Effective Lagrangian $\mathcal{L}^{effect}$ that is dependent on Λ.

Let us now return to the models that we want to treat in 2+1 dimensions. We will start with the relativistic Lagrangian

$$\begin{aligned}\mathcal{L}^{relat} &= \bar{\Psi} \left(\gamma^{\mu} \left(i \frac{\partial}{\partial x^{\mu}} + e\,A_{\mu} \right) - m \right) \Psi \\ &+ \bar{\Phi} \left(\gamma^{\mu} \left(i \frac{\partial}{\partial x^{\mu}} + e\,A_{\mu} \right) + m \right) \Phi + \frac{\theta}{2} \epsilon_{\mu\nu\rho}\, A^{\mu}\, \partial^{\nu}\, A^{\rho}\end{aligned} \tag{16}$$

where Ψ (Φ) is a 2 component Dirac field representing a fermion and anti fermion of spin *plus* (*minus*). In the Coulomb Gauge, the CS propagator is (indices $\mu, \nu, \ldots$ runs from 0 to 2 and indices $i, j, \ldots$ runs over 1 and 2)

$$\Delta_{\mu\nu} \doteq < T A_{\mu}(p) A_{\nu}(-p) > = \frac{1}{\theta}\, \epsilon_{\mu\nu i}\, \frac{k^{i}}{\vec{k}^{2}} \tag{17}$$

$$\tag{18}$$

and will be represented by a wavy line. The Dirac propagators of the relativistic fermions will be represented by double straight lines. Through the same steps that led (2) to (7) we get the Classical Non Relativistic limit of this model

$$\begin{aligned}\mathcal{L}^{class}_{nonrel} &= \psi^{*} \left(i \frac{d}{dt} + eA^{0} \right) \psi - \frac{1}{2m} | \left(-i\vec{\nabla} - e\vec{A} \right) \psi|^{2} + \frac{e}{2m} B\psi^{*}\psi \\ &+ \phi^{*} \left(i \frac{d}{dt} + eA^{0} \right) \phi - \frac{1}{2m} | \left(-i\vec{\nabla} - e\vec{A} \right) \phi|^{2} - \frac{e}{2m} B\phi^{*}\phi \\ &+ \frac{\theta}{2} \epsilon_{\mu\nu\rho}\, A^{\mu}\, \partial^{\nu}\, A^{\rho},\end{aligned} \tag{19}$$

where ψ (ϕ) are anti commuting one-component fields representing a spin *plus*(*minus*) fermion. The fermionic PS propagator will be represented by a single straight line. This model has several different vertices : $F^{*}FA^{0}$, $F^{*}F\vec{A}$, $F^{*}FA^{\mu}A_{\mu}$ and $F^{*}FB$, where F stands for ϕ or ψ. The last vertex (Pauli) gives a local interaction between two fermions, mediated by the the propagator

$$\Delta_{B} \doteq < T B(k) A_{0}(-k) > = \frac{i}{\theta} \tag{20}$$

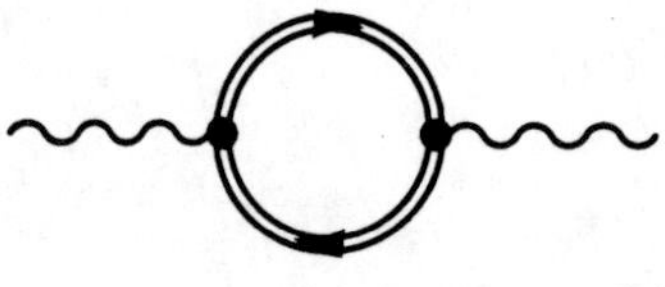

FIGURE 2. Vacuum polarization. The double line represent Dirac fermion propagators, and the wavy line the CS propagator.

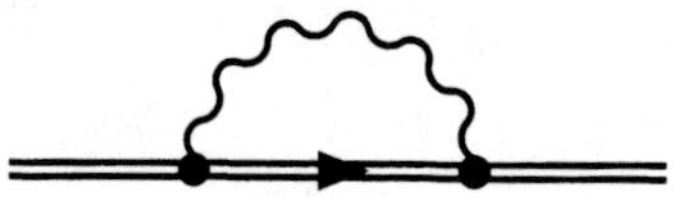

FIGURE 3. Fermion self energy.

that we will represent by an dashed straight line. The fermionic PS Propagator will be represented by a single straight line.

We will leave the result above for future use, and return to the construction of the Effective Non Relativistic Model. This will be done by calculating different low energy processes in the Relativistic Theory and identifying the contributions that come from the low energy intermediate states (and are the same that come from the Classical Non Relativistic Model with a cut off Λ) and the contributions that come from high energy intermediate states, and that must be incorporated in the Effective Non Relativistic Model, through new terms in the Lagrangian. We will restrict the calculations to the one loop order. The sum of Feynman graphs, written as a loop integral can be separated as

$$\mathcal{A}^{relativ}_{lowenergy} = \int^{\infty} d^3kI\left(k^0, \vec{k}, w(p), \vec{p}\right) = \int_0^{\Lambda} d^2k \int_{-\infty}^{\infty} dk^0 I + \int_{\Lambda}^{\infty} d^2k \int_{-\infty}^{\infty} dk^0 I \tag{21}$$

In the low momenta part, both $|\vec{p}|$ and $|\vec{k}|$ are smaller than $\Lambda << m$, and we can safely make the approximation $w(\vec{q}) = m + \frac{\vec{q}^2}{2m}$, for both p and k. The propagators and vertices collapse in the correspondent ones, got from the Lagrangian (16). In the high intermediate energy part this approximation can be taken for w(p) but not for w(k). As $|\vec{p}| << \Lambda$ and the integral is for $|\vec{k}| > \Lambda$, the result, $H(p, \Lambda)$, is analytic in p and can be expanded in a series in p. Every term of this expansion is a contribution to the process, that can be represented by a (new) local term in the Lagrangian of the Effective Non Relativistic Model. The three processes that require renormalization are the Vacuum Polarization Tensor (Figure 2) the Fermion Self energy (Figure 3) and the Vertex Correction (Figure 4). The calculation, of these quantities in covariant gauges are presented in many papers in the literature ([13]). In the Coulomb Gauge it was obtained in [6,7]. The results, separating the contributions of the low (first bracket) and of the high (second bracket) intermediate momenta contributions are respectively

$$\Pi^{lowenergy}_{\mu\nu} = \left[Zero + O(1/m^2) \right] + \left[-i\frac{e^2}{6\pi m}(p^2 g_{\mu\nu} - p_\mu p_\nu) + O(1/m^2) \right] \tag{22}$$

$$\Sigma^{lowenergy}_{\psi or \phi} = \left[Zero + O(1/m^2) \right] + \left[i\,\frac{e^2}{4\pi\theta}\,(\,\pm\vec{\gamma}\cdot\vec{p} - \frac{\vec{p}^2}{m}) + O\left(1/m^2\right) \right] \tag{23}$$

$$e\,A^{\mu}_{external}\,\bar{u}(p')_{\psi or \phi}\Gamma^{lowenergy}_{\mu}(p'-p)\;u(p)_{\psi or\phi}$$
$$= \left[Zero + O(1/m^2) \right] + \frac{e}{2m}\left[\frac{e^2}{2\pi\theta}\epsilon^{ij}\frac{(p'-p)^j}{2m}A^i_{external} + O(1/m^2) \right] \tag{24}$$

As indicated in these formulas, all the contributions to these functions come from the high momenta intermediate states. In fact it is well known that these same functions are zero when calculated in the classical non

FIGURE 4. Contributions (in the Coulomb Gauge) to the scattering of a fermion by an external field A^{μ}_{ext}. The action of the external field is represented by a cross.

relativistic model([2]). As consequence the whole contribution to these functions, come only from the high momenta intermediate states and are independent of the cut off Λ. The effects of these terms in correcting the low energy dynamics of the fermions and the CS field are simulated by adding to the Lagrangian (19) the terms

$$\delta\mathcal{L} = -\frac{1}{4}\frac{e^2}{6\pi m}F_{\mu\nu}F^{\mu\nu} + \frac{e}{2m}\frac{e^2}{2\pi\theta}B\psi^*\psi + \frac{e}{2m}\frac{e^2}{2\pi\theta}B\phi^*\phi. \tag{25}$$

From (25) and (19) we see that the CS field becomes a dynamical propagating field, the so called Maxwell-Chern-Simons field ([1]). We can also see that the magnetic momenta of the spin *plus* and *minus* fermions are corrected to ([7])

$$\mu_{\psi or \phi} = \frac{e}{2m}\left(\pm 1 + \frac{e^2}{2\pi\theta}\right) \tag{26}$$

(these results where obtained previously in the literature in covariant gauges([13])).

Let us now look at the elastic scattering of two low energy fermions. For simplicity we will work in the Center of Momentum Reference Frame in which the incoming fermions have energy and momenta: $(m + \frac{\vec{p}^2}{2m}, \vec{p})$ and $(m + \frac{\vec{p}^2}{2m}, -\vec{p})$ and the outgoing fermions have $(m + \frac{\vec{p}'^2}{2m}, \vec{p}')$ and $(m + \frac{\vec{p}'}{2m}, -\vec{p}')$ with $|\vec{p}| = |\vec{p}'| << \Lambda$. The amplitude is a function of $|\vec{p}|$ and the angle between $\vec{p}$ and $\vec{p}'$. We prefer to write it in terms of $\vec{p}$ and the two vectors $\vec{s} \doteq \vec{p} + \vec{p}'$ and $\vec{q} \doteq \vec{p}' - \vec{p}$. In Figure 5 are shown the non null graphs contributing to this process.

For the scattering of one fermion of spin *plus* and other of spin *minus*, the contributions of these graphs are listed below, separated in two rows. In the first are the contributions of the low intermediate momenta states, $\mathcal{A}(0, \Lambda)$, and in the second row, the local (independent of p) contributions of the high momenta intermediate states, $\mathcal{A}(\Lambda, \infty)$.

$$\mathcal{A}^{++\ relat}_{lowenergy} = \mathcal{A}^{++\ rel}_{lowene}(0,\Lambda) + \mathcal{A}^{++\ rel}_{lowene}(\Lambda,\infty)$$

$$Graph\ 5a = \left[\quad i\frac{e^2}{m\theta}\frac{\vec{s}\wedge\vec{q}}{\vec{q}^2} \quad \right] + \left[\quad 0 \quad \right]$$

$$Graph\ 5b = \left[\quad 0 \quad \right] + \left[\quad \frac{e^4}{6\pi m\theta^2} \quad \right]$$

$$Graphs\ 5c = \left[\quad 0 \quad \right] + \left[\quad \frac{e^4}{2\pi m\theta^2} \quad \right] \tag{27}$$

$$Graph\ 5d = \left[\quad \frac{e^4}{4\pi m\theta^2}\ln\frac{-\vec{q}^2}{\vec{p}^2} \quad \right] + \left[\quad 0 \quad \right]$$

$$Graph\ 5e = \left[\quad \frac{e^4}{4\pi m\theta^2}\ln\frac{\Lambda^2}{\vec{q}^2} \quad \right] + \left[\quad \frac{e^4}{4\pi m\theta^2}\ln\frac{4m^2}{\Lambda^2} \quad \right]$$

Some observations are in order: 1. The $\mathcal{A}(0,\Lambda)$ parts of each graph (of the Relativistic Model) are the same as calculated from the Classical Non Relativistic Model (19) with a cut off Λ, through the graphs drawn on Figure 5, at the right of the corresponding relativistic ones. 2. The $\mathcal{A}(0,\Lambda)$ part of each graph can depend on the non relativistic cut off Λ (see 5e) but the whole graph is independent of Λ ,as can be seen by adding , for each graph, the terms of the first and the second row. 3. Had we interpreted Λ as an UV cut off in the usual way, i.e. $\Lambda \longrightarrow \infty$, and $\mathcal{A}(0,\Lambda)$ would be a divergent amplitude. 4. The $\mathcal{A}(\Lambda,\infty)$ non null contributions of the graphs 5b and 5c could also be get by calculating 5a, starting from the already corrected Effective Lagrangian given by (19) plus (29). 5. The non null $\mathcal{A}(\Lambda,\infty)$ part of diagram 5e instead, is a new term that must be incorporated in the Effective Lagrangian as a local quartic interaction of the form $\psi^*\psi\,\phi^*\phi$. It must be stressed that this term comes from the integration over the high momenta intermediate states of the Renormalizable Relativistic Model; no quartic term of the form $\Psi^*\Psi\,\Phi^*\Phi$ is needed in the "parent"Relativistic Model to generate this quartic term in the Effective Non Relativistic Lagrangian. The Effective Non Relativistic Model incorporating all these terms can be written

$$\begin{aligned}\mathcal{L}^{effect}_{nonrel} &= \psi^*\left(i\frac{d}{dt} + eA^0\right)\psi - \frac{1}{2m}|\left(-i\vec{\nabla} - e\vec{A}\right)\psi|^2 + \frac{e}{2m}(1+\frac{e^2}{2\pi\theta})B\psi^*\psi \\ &+ \phi^*\left(i\frac{d}{dt} + eA^0\right)\phi - \frac{1}{2m}|\left(-i\vec{\nabla} - e\vec{A}\right)\phi|^2 + \frac{e}{2m}(-1+\frac{e^2}{2\pi\theta})B\phi^*\phi \\ &+ \frac{\theta}{2}\epsilon_{\mu\nu\rho}\,A^\mu\,\partial^\nu\,A^\rho - \frac{1}{4}\frac{e^2}{6\pi m}F_{\mu\nu}F^{\mu\nu} \\ &+ \left(\frac{e^4}{4\pi m\theta}\ln\frac{4m^2}{\Lambda^2}\right)\psi^*\psi\,\phi^*\phi.\end{aligned} \tag{28}$$

The calculation of the magnetic moment of the fermions, the propagator of the (Maxwell) Chern-Simons, and of the low energy scattering of two fermions, in this theory, using a cut off Λ (up to one loop), give the same results as the calculation of the same quantities starting from the Relativistic Model (16). For example, the amplitude of scattering of one spin *plus* and one spin *minus* fermion (the sum of the two rows in equation (27)) gives ([14])

$$\mathcal{A}^{++\,eff}_{nonrel} \doteq \mathcal{A}^{++\,rel}_{lowene} = i\frac{e^2}{m\theta}\frac{\vec{s}\wedge\vec{q}}{\vec{q}^2} + \frac{2e^4}{3\pi m\theta^2} + \frac{e^4}{4\pi m\theta^2}\ln\left(\frac{-4m^2}{\vec{p}^2}\right) \tag{29}$$

The calculation starting from the classical Non Relativistic Model, (the sum of terms in the first row in equation (27)) would instead, give the "divergent" result ([3])

$$\mathcal{A}^{++\,class}_{nonrelat} = i\frac{e^2}{m\theta}\frac{\vec{s}\wedge\vec{q}}{\vec{q}^2} + \frac{e^4}{4\pi m\theta^2}\ln\left(\frac{-\Lambda^2}{\vec{p}^2}\right) \tag{30}$$

These results exemplify our main point: taking the non relativistic limit in the Lagrangian and equations of motion (Classical Non Relativistic Limit) and then calculating a process gives in general, a result different than, first calculating the same process in the relativistic theory and later taking the non relativistic limit of the result.

To finish this talk I will turn to the problem that motivated this study: the finite result for $\mathcal{A}^{class}_{nonrel}$ got in ([2]) for the scattering of two bosons and its extension ([3]) to the scattering of *two* spin *plus* fermions (we will think that the two fermions are not identical and we don't need to anti symmetrize the amplitude with respect to the outgoing particles). The non null graphs contributing to this process are the same of figure 5. The result is

$$\begin{array}{lcccc}
\mathcal{A}^{-+\,relat}_{lowenergy} = & \mathcal{A}^{-+\,rel}_{lowene}(0,\Lambda) & + & \mathcal{A}^{-+\,rel}_{lowene}(\Lambda,\infty) \\[2ex]
Graph\ 5a = & \left[\ \frac{e^2}{m\theta}\left(1 + i\frac{\vec{s}\wedge\vec{q}}{\vec{q}^2}\right)\ \right] & + & \left[\ 0\ \right] \\[2ex]
Graph\ 5b = & \left[\ 0\ \right] & + & \left[\ \frac{e^4}{6\pi m\theta^2}\ \right] \\[2ex]
Graphs\ 5c = & \left[\ 0\ \right] & + & \left[\ \frac{e^4}{2\pi m\theta^2}\ \right] \\[2ex]
Graph\ 5d = & \left[\ \frac{e^4}{4\pi m\theta^2}\ln\frac{\vec{q}^2}{\Lambda^2}\ \right] & + & \left[\ \frac{e^4}{4\pi m\theta}\ln\frac{4m^2}{\Lambda^2}\ \right] \\[2ex]
Graph\ 5e = & \left[\ -\frac{e^4}{4\pi m\theta^2}\ln\frac{\Lambda^2}{\vec{q}^2}\ \right] & + & \left[\ \frac{e^4}{4\pi m\theta^2}\left(\ln\frac{4m^2}{\Lambda^2} - 2\right)\ \right]
\end{array} \tag{31}$$

The differences of these results to the ones in (27) come from the Pauli interaction of the magnetic field of each fermion with the magnetic moment of the other fermion. The effects of these interactions cancel in the scattering of a spin *plus* and a spin *minus* fermion and add in the case of *two* spin *plus* fermions. The results for $\mathcal{A}^{effect}$ and $\mathcal{A}^{class}_{nonrel}$ are now

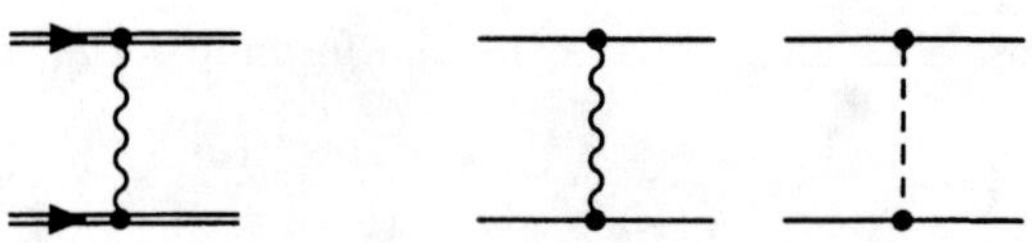

Figure 5a

Figure 5b

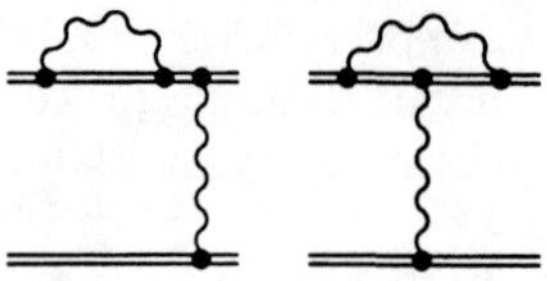

+ 4 diagrams

Figure 5c

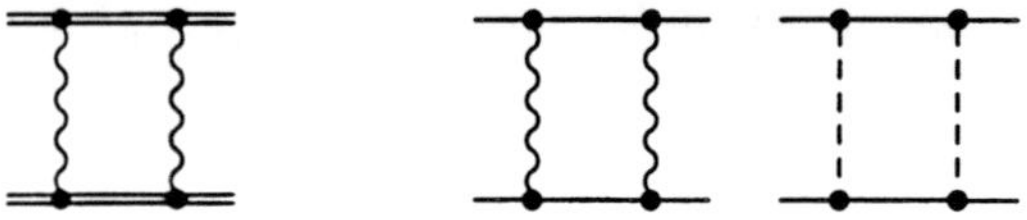

Figure 5d

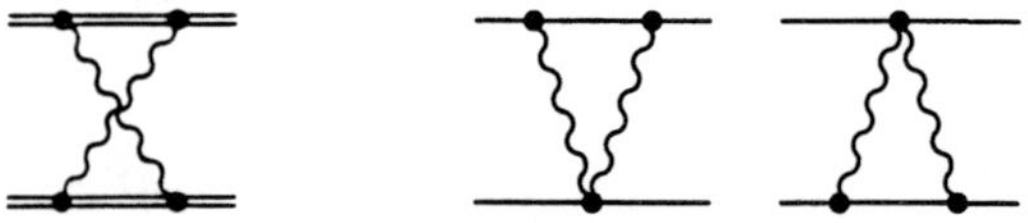

Figure 5e

FIGURE 5. Non null graphs contributing to the scattering of two Dirac fermions. On the right of the diagrams are represented the correspondent graphs in the classical non relativistic model.

$$\mathcal{A}^{-+\,eff}_{nonrel} \doteq \mathcal{A}^{-+\,rel}_{lowene} = \frac{e^2}{m\theta}\left(1 + i\frac{\vec{s}\wedge\vec{q}}{\vec{q}^2}\right) + \frac{e^4}{6\pi m\theta^2} \tag{32}$$

$$\mathcal{A}^{-+\,class}_{nonrelat} = \frac{e^2}{m\theta}\left(1 + i\frac{\vec{s}\wedge\vec{q}}{\vec{q}^2}\right) \tag{33}$$

The unexpected fact that this last result is finite, independent of Λ, is in the literature ([2]) related to the preservation at quantum level, of the scale invariance that the classical non relativistic model presents. In the model of bosons interacting with the CS field this only happens for the special value of the quartic self interaction discussed in the introduction. For fermions the same fact is provided by the Pauli interaction which already appear in the minimal interaction with the CS field; no fine tuning of coupling constants is needed. We here showed another aspect of this indepen dence of Λ. Unusually,not only $\mathcal{A}^{relat}_{lowen}$ is independent of Λ: their high and low momenta intermediate energy contributions are separately independents of Λ. So the difference of the amplitudes got from the Classical or the Effective Non Relativistic Models is only a constant independent of Λ. If the fermions are identical we must anti symmetrize the amplitudes (32) and (29) in the outgoing particles. In this case no difference at all appears in the final result. The amplitude got from both (32) and (33) is : $i\,\frac{2e^2}{m\theta}\frac{\vec{s}\wedge\vec{q}}{\vec{q}^2}$, and gives the Aharonov Bohm scattering amplitude for two identical fermions.

ACKNOWLEDGMENTS

This work was partially supported by CNPq (Conselho Nacional de Desenvolvimento Cientifico e Tecnologico) do Brasil. The author thanks M. Gomes for critically reading the manuscript.

REFERENCES

1. S. Deser, R. Jackiw and S. Templeton, Ann. Phys.(NY) **140**, 372 (1982); S. Deser, R. Jackiw and S. Templeton, Phys.Rev.Lett.**48**, 975 (1982); J. F. Schonfeld, Nucl. Phys. **B185**, 157 (1981).
2. O. Bergman and G. Lozano, Ann. Phys. (NY) **229**, 416 (1994).
3. C. R. Hagen, Phys. Rev. **D56** (1997) 2250.
4. S. C. Zhang, T. H. Hansson and S. Kivelson, Phys. Rev. Lett. **62**, 82 (1989).
5. D. Z. Freedman, G. Lozano and N. Ruiz, Phys. Rev. **D49** 1054 (1994).
6. C. R. Hagen, Ann. Phys. (NY) **157**, 342 (1984).
7. M. Fleck, A. Foerster, H. O. Girotti, M. Gomes, J. R. S. Nascimento and A. J. da Silva, Int. Journ. Mod. Phys. **B12**, 2889 (1997).
8. Y. Aharonov and D. Bohm, Phys. Rev. **115** (1959) 485
9. V. B. Beretetskii, E. M. Lifshitz and L. P. Pitaevskii, Quantum Electrodynamics, Pergamon Press (1980). C. R. Hagen, Phys. Rev. **D31**, O. Bergman, Phys. Rev. **D46**,5474 (1992).
10. W. E. Caswell and G. P. Lepage, Phys. Lett. **167B**, 437 (1986); G. P. Lepage, Proc. of TASI 89, (1989).
11. C. Cohen-Tannoudji, J. Dupont-Roc and G. Grinberg, Introduction to Quantum Electrodynamics, J. Willey and Sons (1989)
12. K. G. Wilson, Phys. Rev.**B4**, 3174 (1971).
13. I. I. Kogan and G. Semenoff, Nucl. Phis. **B368**, 718 (1992), I. I. Kogan, Phys. Lett. **B 262**, 83 (1991), G. Gat and R. Roy, Phys. Lett. **B 340**, 362 (1996).
14. M. Gomes and A. J. da Silva, Phys. Rev. D57 (1998) 3579; H.O. Girotti, M. Gomes, J.R.S. Nascimento and A.J. da Silva, Phys. Rev. **D56** (1997) 3623.(The correspondent problem for scalars interacting with the CS field was studied in M. Gomes, J.M.C. Malbouisson, A.J. da Silva, Phys. Lett. **A236** (1997) 373; **Ibid.**, Int. J. Mod. Phys. **A13** (1998) 3157).

Effective actions at finite temperature

C. D. Fosco [1]

Centro Atómico Bariloche, 8400 Bariloche, Argentina

Abstract. There exist interesting background gauge field configurations such that Γ_{odd}, the parity odd part of the effective action for massive Dirac fermions in $2d+1$ dimensions at finite temperature may be obtained exactly. The method of computation resorts to the chiral anomaly of a dimensionally reduced theory. We present an explicit derivation for the Abelian case in $2+1$ dimensions, providing an alternative 'perturbative' derivation. Then we discuss the general $2d+1$ dimensional case. We finally apply our results to two examples: the 'mixed' CS term in the Dorey-Mavromatos model of planar superconductivity, and a fixed-charge ensemble in the presence of a magnetic field.

I INTRODUCTION

There have recently been interesting results [1–5] in resolving an apparent contradiction concerning the gauge invariance at finite temperature (T) of the induced parity-odd term [2]. General arguments imply that the coefficient of the CS term must be quantized at non-zero temperature [6–8], if the action is to be invariant under topologically non-trivial ('large') gauge transformations. On the other hand, simple perturbative calculations [10]- [19], give the result that the effect of moving to $T \neq 0$ is just to multiply the zero-temperature CS term by a smoothly varying function of T. This is a contradiction, since a (not identically constant) function having only discrete values cannot be smooth.

An important advance was made by Dunne et al [1], who considered a solvable model in $0+1$ dimensions at $T > 0$, in which a CS-like term is generated. They found that the exact effective action at finite temperature was indeed gauge invariant, even though its perturbative expansion produced a result analogous to that found in $2+1$ dimensions, namely, a gauge-non-invariant form.

Subsequent non-perturbative calculations of the effective action in the $2+1$ Abelian case [2] and its explicit temperature-dependent parity-breaking part [3] have shown that the complete effective action is indeed invariant under both large and small gauge transformations, the result of [3], in particular, showing some remarkable similarity to that of Dunne et al [1]. These results have been generalized to $2d+1$ dimensions [5].

Besides its obvious theoretical interest, the resolution of this paradox has relevance for some physical applications. As we shall show in IV A, in the Dorey-Mavromatos (DM) model [20] of two-dimensional superconductivity, the flux quantization phenomenon can only be proved if the (analog of the) exact CS term is used in the effective action for the fermions [4].

As a second example, in IV B we shall discuss a fixed-charge example. The Chern-Simons term links magnetic field and charge, so it will strongly affect the statistical properties of a system in the presence of an external magnetic field, more so when using the fixed-charge ensemble. It is indeed crucial to use the exact induced Chern-Simons term rather than the perturbative one in the derivation of the free energy [21].

The organization of this article is as follows: in II we deal with the Abelian case in $2+1$ dimensions, presenting first a non-perturbative derivation and then a 'perturbative' (all orders) one. Then, in III we discuss the $2d+1$ dimensional case. In IV we present two examples were our results find natural applicability: The Dorey-Mavromatos model mentioned above, and the fixed-charge ensemble in $2+1$ dimensions. Finally, V contains our conclusions.

---

1) fosco@cab.cnea.gov.ar

2) We shall also denote this term the 'Chern-Simons' (CS) term, following the usual (although improper) convention.

CP484, *Trends in Theoretical Physics II*, edited by H. Falomir, R. E. Gamboa Saraví, and F. A. Schaposnik

II THE ABELIAN CASE IN $2+1$ DIMENSIONS

A Non-perturbative derivation

We are interested in evaluating $\Gamma_{odd}(A,M)$, the parity-odd piece of the effective action induced by integrating out massive fermions coupled to an Abelian gauge field A_μ in $2+1$ dimensions at finite temperature. It is defined by

$$\Gamma_{odd}(A,M) = \frac{1}{2}(\Gamma(A,M) - \Gamma(A,-M)) \tag{1}$$

where

$$\exp(-\Gamma(A,M)) = \int \mathcal{D}\psi \mathcal{D}\bar{\psi} \, \exp[-S_F(A,M)], \tag{2}$$

The Euclidean action $S_F(A,M)$ is given by

$$S_F(A,M) = \int_0^\beta d\tau \int d^2x \, \bar{\psi}(\not{\partial} + ie\not{A} + M)\psi . \tag{3}$$

We are using Euclidean Dirac's matrices in the representation

$$\gamma_1 = \sigma_1 \quad \gamma_2 = \sigma_2 \quad \gamma_0 = \sigma_3 \tag{4}$$

where σ_i are the familiar Pauli matrices and $\beta = 1/T$ is the inverse temperature. The label 0 is used to denote the Euclidean time coordinate τ. The fermionic fields in (2) obey antiperiodic boundary conditions in the timelike direction

$$\psi(\beta,x) = -\psi(0,x) \quad , \quad \bar{\psi}(\beta,x) = -\bar{\psi}(0,x) \ , \forall x \, , \tag{5}$$

with x denoting the two space coordinates. The gauge field satisfies periodic boundary conditions instead

$$A_\mu(\beta,x) = A_\mu(0,x) \ , \ \forall x \, . \tag{6}$$

We want to make a calculation which preserves an interesting property of the imaginary time formulation, namely, that there is room for gauge transformations with non-trivial winding around the time coordinate, and any approximation which assumes the smallness of A_0 may put the symmetry under those large transformations in jeopardy.

Let us first discuss the non-trivial gauge transformations at finite temperature. The set of allowed gauge transformations in the imaginary time formalism is defined in the usual way:

$$\psi(\tau,x) \to e^{-ie\Omega(\tau,x)}\psi(\tau,x) \ , \ \bar{\psi}(\tau,x) \to e^{ie\Omega(\tau,x)}\bar{\psi}(\tau,x)$$
$$A_\mu(\tau,x) \to A_\mu(\tau,x) + \partial_\mu\Omega(\tau,x) \tag{7}$$

where $\Omega(\tau,x)$ is a differentiable function vanishing at spatial infinity ($|x| \to \infty$), and whose time boundary conditions are chosen in order not to affect the fields' boundary conditions (5) and (6). It turns out that $\Omega(\tau,x)$ can wind an arbitrary number of times around the cyclic time dimension:

$$\Omega(\beta,x) = \Omega(0,x) + \frac{2\pi}{e}k \tag{8}$$

where k is an integer which labels the homotopy class of the gauge transformation.

We consider now configurations where A_0 is only a function of τ, and A_j is independent of τ. Under these assumptions, the only τ-dependence of the Dirac operator comes from A_0. This dependence can however be erased by a redefinition of the fermionic fields [3], and the timelike component of the gauge is a constant A_0', the mean value of $A_0(\tau)$,

$$\tilde{A}_0 = \frac{1}{\beta}\int_0^\beta d\tau \, A_0(\tau) \, . \tag{9}$$

Note that the spatial components of A_μ remain τ-independent after this redefinition.

After redefining the fermionic fields according to this prescription, the fermionic determinant we should becomes

$$\det(\not{\partial} + ie \not{A} + M) = \int \mathcal{D}\psi \mathcal{D}\bar{\psi} \exp[-S_F(A_j, \tilde{A}_0, M)] , \tag{10}$$

where

$$S_F(A_j, \tilde{A}_0, M) = \int_0^\beta d\tau \int d^2x \, \bar{\psi}(\not{\partial} + ie(\gamma_j A_j + \gamma_0 \tilde{A}_0) + M)\psi , \tag{11}$$

and we removed the primes for the sake of clarity.

Since the Dirac operator in the previous equation is invariant under imaginary time translations it is convenient to perform a Fourier transformation on the time variable for ψ and $\bar{\psi}$

$$\psi(\tau, x) = \frac{1}{\beta} \sum_{n=-\infty}^{+\infty} e^{i\omega_n \tau} \psi_n(x)$$
$$\bar{\psi}(\tau, x) = \frac{1}{\beta} \sum_{n=-\infty}^{+\infty} e^{-i\omega_n \tau} \bar{\psi}_n(x) , \tag{12}$$

where $\omega_n = (2n+1)\frac{\pi}{\beta}$ is the usual Matsubara frequency for fermions. Then the Euclidean action is written as an infinite series of decoupled actions, one for each Matsubara mode

$$S_F(A_j, \tilde{A}_0, M) = \frac{1}{\beta} \sum_{n=-\infty}^{+\infty} \int d^2x \bar{\psi}_n(x) \left[\not{d} + M + i\gamma_0(\omega_n + e\tilde{A}_0) \right] \psi_n(x) \tag{13}$$

where $\not{d}$ is the $1+1$ Euclidean Dirac operator corresponding to the spatial coordinates and the spatial components of the gauge field

$$\not{d} = \gamma_j(\partial_j + ieA_j). \tag{14}$$

As the action splits up into a series and the fermionic measure can be written as

$$\mathcal{D}\psi(\tau, x) \mathcal{D}\bar{\psi}(\tau, x) = \prod_{n=-\infty}^{n=+\infty} \mathcal{D}\psi_n(x) \mathcal{D}\bar{\psi}_n(x) \tag{15}$$

the $2+1$ determinant is an infinite product of the corresponding $1+1$ Euclidean Dirac operators

$$\det(\not{\partial} + ie \not{A} + M) = \prod_{n=-\infty}^{n=+\infty} \det[\not{d} + M + i\gamma_0(\omega_n + e\tilde{A}_0)] . \tag{16}$$

Explicitly, the $1+1$ determinant for a given mode is a functional integral over $1+1$ fermions

$$\det[\not{d} + M + i\gamma_0(\omega_n + e\tilde{A}_0)] =$$
$$\int \mathcal{D}\chi_n \mathcal{D}\bar{\chi}_n \exp\left\{ -\int d^2x \bar{\chi}_n(x)(\not{d} + M + i\gamma_0(\omega_n + e\tilde{A}_0))\chi_n(x) \right\} . \tag{17}$$

In order to compute Γ_{odd} we factorize now these determinants in a piece which is sensitive to the sign of M and a piece which is not. The Euclidean action S_n corresponding to the mode n may be conveniently recasted in the following form

$$S_n = \int d^2x \, \bar{\chi}_n(\not{d} + \rho_n e^{i\gamma_0 \phi_n})\chi_n \tag{18}$$

with

$$\rho_n = \sqrt{M^2 + (\omega_n + e\tilde{A}_0)^2} \; ; \phi_n = \arctan(\frac{\omega_n + e\tilde{A}_0}{M}) \, . \tag{19}$$

We next realize that the change of fermionic variables

$$\chi_n(x) = e^{-i\frac{\phi_n}{2}\gamma_0}\chi'_n(x) \;\; , \;\; \bar{\chi}_n(x) = \bar{\chi}'_n(x)e^{-i\frac{\phi_n}{2}\gamma_0} \tag{20}$$

makes the action S_n independent of ϕ_n. This is not a gauge transformation but a global chiral rotation in the $1+1$ Euclidean fermionic variables. Correspondingly, the fermionic measure picks up an anomalous Fujikawa jacobian [22] so that one ends with

$$\det[\not{d} + M + i\gamma_0(\omega_n + e\tilde{A}_0)] = J_n[A, M] \, \det[\not{d} + \rho_n] \tag{21}$$

where

$$J_n[A, M] = \exp(-i\frac{e\phi_n}{2\pi}\int d^2x \epsilon_{jk}\partial_j A_k) \, , \tag{22}$$

with ϵ_{jk} denoting the $1+1$ Euclidean Levi-Civita symbol.

Recalling the definition of Γ_{odd}, we see that the second factor in expression (21) does not contribute to it, since it is invariant under $M \to -M$. The Jacobian (22), instead, changes to its inverse. As a consequence, the parity odd piece in the effective action is given in terms of the infinite set of n-dependent Jacobians:

$$\exp[-\Gamma_{odd}] = \prod_{n=-\infty}^{n=+\infty} J_n[A, M] \tag{23}$$

or

$$\Gamma_{odd} = -\sum_{n=-\infty}^{n=+\infty} \log J_n[A, M] = i\frac{e}{2\pi}\sum_{n=-\infty}^{n=+\infty} \phi_n \int d^2x \epsilon_{jk}\partial_j A_k \, . \tag{24}$$

There only remains to perform the summation over the ϕ_n's. This can be done by using standard techniques in Finite Temperature Field Theory. We define

$$\mathcal{S} = \sum_{n=-\infty}^{n=+\infty} \arctan(\frac{\omega_n + e\tilde{A}_0}{M}) , \tag{25}$$

whose sign will obviously depend on the sign of M. We make this explicit by rewriting $\mathcal{S}$ as

$$\mathcal{S} = \frac{M}{|M|}\sum_{n=-\infty}^{n=+\infty} \arctan(\frac{\omega_n + e\tilde{A}_0}{|M|}) \tag{26}$$

or, using the expression for ω_n

$$\mathcal{S}(x, y) = \frac{M}{|M|}\sum_{n=-\infty}^{n=+\infty} \arctan(\frac{(2n+1)\pi + x}{y}) \tag{27}$$

where $x = e\beta\tilde{A}_0$, and $y = \beta|M|$ are the two dimensionless parameters built from the original ones. This series must be regularized, and the standard technique consists in subtracting the zero-field value of each term; notice that the sum of these zero-field contributions conditionally converges to 0. Then

$$\mathcal{S}(x, y) = \frac{M}{|M|}\sum_{n=-\infty}^{n=+\infty} \int_0^x du \frac{d}{du}\arctan(\frac{(2n+1)\pi + u}{y}) \tag{28}$$

As the series now converges absolutely we can first perform the summation. The sum to be evaluated is then

$$\sum_{n=-\infty}^{n=+\infty} \frac{y}{y^2+[(2n+1)\pi+u]^2} \tag{29}$$

which is solved by the summation formula

$$\sum_{n=-\infty}^{n=+\infty} \frac{1}{(n-x_1)(n-x_2)} = -\frac{\pi(\cot(\pi x_1)-\cot(\pi x_2))}{x_1-x_2}. \tag{30}$$

After performing the integral we get

$$\mathcal{S} = \frac{M}{|M|}\arctan\left[\tanh(\frac{\beta|M|}{2})\tan(\frac{1}{2}e\beta\tilde{A}_0)\right]. \tag{31}$$

Thus the parity-odd part of Γ finally reads

$$\Gamma_{odd} = i\frac{e}{2\pi}\frac{M}{|M|}\arctan\left[\tanh(\frac{\beta|M|}{2})\tan(\frac{e}{2}\int_0^\beta d\tau A_0(\tau))\right]\int d^2x\epsilon_{jk}\partial_j A_k\,. \tag{32}$$

Some remarks about our result (32): First we observe that this result has the proper zero temperature limit

$$\lim_{T\to 0}\Gamma_{odd} \to \frac{1}{2}\frac{M}{|M|}S_{CS} \tag{33}$$

where S_{CS} is the Chern-Simons action

$$S_{CS} = i\frac{e^2}{4\pi}\int d^3x\epsilon_{\mu\nu\alpha}A_\mu\partial_\nu A_\alpha \tag{34}$$

which shows up in our particular configuration as $\frac{e^2}{2\pi}\int d\tau A_0(\tau)\int d^2x\epsilon_{ij}\partial_i A_j$. So we get the proper induced Chern-Simons term at zero temperature. As it is well known, in the zero temperature case the result is not invariant under large gauge transformations. The quantization of the spatial integral that measures the flux of the magnetic field through a space-like manifold $\tau = constant$ in units of $\frac{2\pi}{e}$ shows that (33) changes by the addition of an odd multiple of $i\pi$ under a large gauge transformation with odd winding number when the magnetic flux is odd. This gauge non-invariance is compensated by the parity anomaly discussed in the Introduction when the complete result is regularized in a gauge invariant scheme.

The same situation occurs in the finite temperature result (32). A large gauge transformation with odd winding number $k = 2p+1$ shifts the argument of the tangent in $(2p+1)\pi$. Although the tangent is not sensitive to such a change, one has to keep track of it by shifting the branch used for arctan definition. This amounts to the same result as in the $T \to 0$ limit: the gauge non-invariance of Γ_{odd} under large gauge transformations is compensated by the parity anomaly $\pm\frac{1}{2}S_{CS}$.

Now we observe that a perturbative expansion in terms of e yields the usual perturbative result

$$\Gamma_{odd} = \frac{1}{2}\frac{M}{|M|}\tanh(\frac{|M|\beta}{2})S_{CS} + O(e^4) \tag{35}$$

where the coefficient of the Chern-Simons term acquires a smooth dependence on the temperature. Were we considering only the first non trivial order in perturbation theory, we would find a clash between temperature dependence and gauge invariance: the gauge non-invariance of the induced CS term is no longer compensated by the parity anomaly.

We observe that the result (32) is not an extensive quantity in Euclidean time. It is however extensive in space, and that is indeed all one expects in Finite Temperature Field Theory. In contrast, the $T = 0$ limit becomes an extensive quantity in space-time, as is expected from zero temperature Field Theory.

B Perturbative derivation

We provide here an alternative derivation of the result obtained in II A, based on straightforward effective action techniques. We then extend it to the case in which the flux is allowed to depend on τ.

Following II A the fermionic determinant is written as an infinite product of the corresponding 1+1 Euclidean Dirac operators

$$\det(\partial\!\!\!/ + ie\,A\!\!\!/ + M) \;=\; \prod_{n=-\infty}^{n=+\infty} \det[d\!\!\!/ + M + i\gamma_0(\omega_n + e\tilde{A}_0)] \;. \tag{36}$$

Then, the effective action $\Gamma(A)$ corresponding to this configuration will be

$$\Gamma(A) \;=\; -\sum_{n=-\infty}^{n=+\infty} \mathrm{Tr}\log\left[d\!\!\!/ + i\gamma_0\tilde{\omega}_n + M\right] \tag{37}$$

where we have defined $\tilde{\omega}_n = \omega_n + e\tilde{A}_0$. This trace cannot, of course, be evaluated explicitly. But if we want to reproduce the result for the induced parity breaking term of [3], it is sufficient to evaluate it up to linear order in A_j, without making any expansion in the time-like component of the gauge field. A naive application of this approach leads, however, to an ambiguous result, as we will now see. Let us call this term $\Gamma^{(1)}(A)$. It is formally given by

$$\Gamma^{(1)}(A) \;=\; -ie\sum_{n=-\infty}^{n=+\infty} \mathrm{Tr}\left[A\!\!\!/(\partial\!\!\!/ + i\gamma_0\tilde{\omega}_n + M)^{-1}\right] \tag{38}$$

where the trace is evaluated over functional and Dirac indices. When written in momentum space, (38) becomes

$$\Gamma^{(1)}(A) \;=\; -ie\sum_{n=-\infty}^{n=+\infty} \int \frac{d^2p}{(2\pi)^2}\,\mathrm{tr}\left[\tilde{A\!\!\!/}(0)(i\,p\!\!\!/ + i\gamma_0\tilde{\omega}_n + M)^{-1}\right] \tag{39}$$

where tr is the Dirac trace, and $\tilde{A}_j(p)$ is the Fourier transform of $A_j(x)$ with respect to the two space variables

$$\tilde{A}_j(p) \;=\; \int d^2x\, e^{-ip\cdot x}\, A_j(x) \;. \tag{40}$$

It is immediate to check that, by rationalizing the denominator and taking the Dirac trace in (39), we obtain 0, which is an unpropitious start to the calculation, and might seem to contradict the results of reference [3]. The way out of this impasse is to realize that, if we are going to deal with an A_j such that the associated magnetic flux is non-zero, then its zero momentum component is necessarily singular, and the result of (39) will be of the ambiguous form $0 \times \infty$. Indeed, from the definition (40) of the Fourier transform, we see that the magnetic flux Φ can be expressed as the following limit

$$\Phi \;\equiv\; \int d^2x\,\epsilon_{jk}\partial_j A_k \;=\; i\lim_{q\to 0}\epsilon_{jk}q_j\tilde{A}_k(q) \;. \tag{41}$$

It follows from (41) that, if we want to have a non-zero flux [3], then the zero-momentum component of $\tilde{A}_j$ necessarily diverges. We somehow need to introduce another momentum ('q') into the problem, and work with the finite quantity Φ.

In order to see how to do this let us consider, in fact, the leading term in the expansion of (38) in powers of $\tilde{A}_0$, which is just the perturbative action

$$\Gamma^{(1,1)}(A) \;=\; (ie)^2\sum_{n=-\infty}^{n=+\infty} \mathrm{Tr}\left[\gamma_0\tilde{A}_0(\partial\!\!\!/ + i\gamma_0\omega_n + M)^{-1}\right.$$

---

3) Zero-flux configurations give $\Gamma = 0$ without any ambiguity. This is consistent with the result of [3].

$$\not{A}(\not\partial + i\gamma_0\omega_n + M)^{-1}] \; . \tag{42}$$

In this expression it is clear that the trace will involve a second momentum integration if $\tilde{A}_0$ is allowed to depend on x as well as on τ:

$$\Gamma^{(1,1)}(\tilde{A}_0(x), \vec{A}(\vec{x})) = (ie)^2 \sum_n \int d^2p d^2q \mathrm{tr}\,[\langle p+q|\gamma_0$$

$$\tilde{A}_0|p\rangle(i\not{p} + i\gamma_0\omega_n + M)^{-1}\langle p|\not{A}|p+q\rangle(i(\not{p}+\not{q}) + i\gamma_0\omega_n + M)^{-1}\Big] \; , \tag{43}$$

where tr means trace over the Dirac indices only. The case of x-independent $\tilde{A}_0$ may then be obtained via the limiting process in which the x dependence of $\tilde{A}_0(x)$ is removed, leading to

$$\Gamma^{(1,1)}(\tilde{A}_0, \vec{A}(x)) = (ie)^2\tilde{A}_0 \sum_{n=-\infty}^{+\infty} \int \frac{d^2p}{(2\pi)^2} \lim_{q\to 0} \mathrm{tr}\left\{\gamma_0(i\not{p} + i\gamma_0\omega_n + M)^{-1}\right.$$

$$\left.\tilde{\not{A}}(q)(i(\not{p}+\not{q}) + i\gamma_0\omega_n + M)^{-1}\right\} \; . \tag{44}$$

Rationalizing the second denominator and taking the Dirac trace reveals the presence of a term proportional to Φ, whose contribution is precisely the odd-parity perturbative action, for our special field configuration. One can proceed in this way, with higher powers of $\tilde{A}_0$. But it turns out that a well-regulated expression is obtained by retaining an extra momentum dependence in only the 'final' propagator as in (44),

$$\Gamma^{(2,1)}(\tilde{A}_0, \vec{A}(\vec{x})) = -(ie)^3\tilde{A}_0^2 \sum_{n=-\infty}^{n=+\infty} \int \frac{d^2q}{(2\pi)^2} \lim_{q\to 0} \mathrm{tr}\left\{\gamma_0(i\not{p} + i\gamma_0\omega_n + M)^{-1}\right.$$

$$\left.\gamma_0(i\not{p} + i\gamma_0\omega_n + M)^{-1}\,\tilde{\not{A}}(q)(i(\not{p}+\not{q}) + i\gamma_0\omega_n + M)^{-1}\right\} \; , \tag{45}$$

and the same is true for higher orders in $\tilde{A}_0$.

But clearly the prospect of evaluating the general term $\Gamma^{(n,1)}$, and then trying to sum up the answer so as to obtain the full $\Gamma^{(1)}$, which is non-perturbative in $\tilde{A}_0$, is unappealing. Fortunately, this is not necessary. Recall that we resorted to perturbation theory in $\tilde{A}_0$ in order to gain insight into the IR behaviour -we were not otherwise forced to expand the denominator in (39) in powers of $\tilde{A}_0$. Indeed, the formulae (44) and (45) indicate that (39) should be interpreted in terms of a limit in which the argument of the trace is replaced by $\not{A}(q)(i(\not{p}+\not{q}) + i\gamma_0\tilde{\omega}_n + M)^{-1}$, and then the limit $q \to 0$ is taken. At this point, however, one realizes that such an expression, though IR regular, is UV divergent. To regulate the UV divergence, we work instead with the derivative of $\Gamma(A)$ with respect to $\tilde{A}_0$, which improves the large momentum dependence of the integrand by adding an extra propagator, and amounts to subtracting the value at $\tilde{A}_0 = 0$. These considerations then lead to the well-behaved expression

$$\frac{\partial}{\partial \tilde{A}_0}\Gamma(A) = -e^2 \sum_{n=-\infty}^{n=+\infty} \lim_{q\to 0} \int d^2p \,\mathrm{tr}\left\{\gamma_0\langle p|(\not\partial + i\gamma_0\tilde{\omega}_n + M)^{-1}\right.$$

$$\left.\tilde{\not{A}}(\not\partial + i\gamma_0\tilde{\omega}_n + M)^{-1}|p+q\rangle\right\} \; . \tag{46}$$

We are now ready to evaluate (46). After taking the trace, one finds that the only non-vanishing contribution to the derivative of $\Gamma(A)$ is

$$\frac{\partial}{\partial \tilde{A}_0}\Gamma(A) = 2Me^2 \sum_{n=-\infty}^{n=+\infty} \lim_{q\to 0}[\epsilon_{jk}q_j\tilde{A}_k(q)] \int \frac{d^2p}{(2\pi)^2} \frac{1}{(p^2 + \tilde{\omega}_n^2 + M^2)^2} \; , \tag{47}$$

which, by using (41), may be put as

$$\frac{\partial}{\partial \tilde{A}_0}\Gamma(A) = -2iMe^2\,\Phi \sum_{n=-\infty}^{n=+\infty} \int \frac{d^2p}{(2\pi)^2}\frac{1}{(p^2+\tilde{\omega}_n^2+M^2)^2}\,. \tag{48}$$

Performing the (convergent) momentum integral in (48) , we obtain

$$\frac{\partial}{\partial \tilde{A}_0}\Gamma(A) = -\frac{iMe^2}{2\pi}\,\Phi \sum_{n=-\infty}^{n=+\infty} \frac{1}{\tilde{\omega}_n^2+M^2}\,. \tag{49}$$

When this expression is integrated term by term over $\tilde{A}_0$, it yields

$$\Gamma(A) = \frac{ie}{2\pi}\,\Phi \sum_{n=-\infty}^{n=+\infty} \arctan[\frac{\omega_n+e\tilde{A}_0}{M}]\,. \tag{50}$$

This series is exactly equal to the one that appears in ref. [3], and indeed, it shows already the identity of this result to the one obtained there by the 'decoupling' change of variables. Thus we conclude that, for one fermionic flavour, and keeping terms linear in A_j, the result for the induced action Γ is

$$\Gamma(A) = \frac{ie}{2\pi}\,\Phi \arctan\left[\tanh\frac{\beta M}{2}\tan(\frac{e}{2}\int_0^\beta d\tau A_0(\tau))\right]\,. \tag{51}$$

There is an extended kind of configuration that can be treated with this method, the case of a gauge field where the constraint of τ-independence of A_j is relaxed, namely,

$$A_0 = A_0(\tau) \quad A_j = A_j(\tau,x)\,. \tag{52}$$

We will, however, calculate this within the approximation of keeping terms linear in A_j. As for the previous configurations, we first go to a gauge where $A_0(\tau)\to\tilde{A}_0$ becomes a constant. After this, A_j will remain τ-dependent. In spite of the fact that there is no τ translation invariance, we perform a Fourier transformation with respect to the imaginary time in the fermionic action, obtaining

$$S_F(A) = \frac{1}{\beta}\sum_{m,n}\int d^2x\bar{\psi}_m(x)\left\{\delta_{m,n}[\not{\partial}+i\gamma_0\tilde{\omega}_n+M]+ie\,\tilde{\not{A}}^{(m-n)}(x)\right\}\psi_n(x) \tag{53}$$

where

$$\tilde{A}_j^{(k)} = \frac{1}{\beta}\int_0^\beta d\tau\, e^{-i\frac{2k\tau}{\beta}}\,A_j(x,\tau)\,. \tag{54}$$

The only change that we have to make in the calculation corresponding to the static A_j case is that now the matrix whose determinant we are evaluating is not diagonal in the space of Matsubara frequencies. Then

$$\Gamma(A) = -\mathrm{Tr}\log\left\{\delta_{m,n}[\not{\partial}+i\gamma_0\tilde{\omega}_n+M]+ie\,\tilde{\not{A}}^{(m-n)}(x)\right\} \tag{55}$$

where now, of course, the trace also affects the discrete frequencies. There is an important simplification which occurs because we are actually dealing with the first order term in A_j. When considering this first order term in the derivative of Γ with respect to $\tilde{A}_0$, we obtain

$$\begin{aligned}\frac{\partial\Gamma(A)}{\partial\tilde{A}_0} &= -e^2\mathrm{Tr}\left\{\gamma_0(\not{\partial}+i\gamma_0\tilde{\omega}_m+M)^{-1}\,\tilde{\not{A}}^{(m-n)}(\not{\partial}+i\gamma_0\tilde{\omega}_n+M)^{-1}\right\}\\ &= -e^2\sum_{n=-\infty}^{n=+\infty}\mathrm{Tr}\left\{\gamma_0(\not{\partial}+i\gamma_0\tilde{\omega}_n+M)^{-1}\,\tilde{\not{A}}^{(0)}(\not{\partial}+i\gamma_0\tilde{\omega}_n+M)^{-1}\right\}\,.\end{aligned} \tag{56}$$

Note that only the zero-frequency component of A_j appears in this expression, which, on the other hand, can now be evaluated analogously to the static-A_j case, by replacing A_j by its zero-frequency component. The final result is then:

$$\Gamma(A) = \frac{ie}{2\pi}\frac{1}{\beta}\int_0^\beta d\tau\Phi(\tau)\arctan\left[\tanh(\frac{\beta M}{2})\tan(\frac{e}{2}\int_0^\beta d\tau' A_0(\tau'))\right]\,, \tag{57}$$

where $\Phi(\tau)\equiv\int d^2x\epsilon_{jk}\partial_j A_k(x,\tau)$.

III THE $2D+1$ DIMENSIONAL CASE

We shall consider here $\Gamma^{2d+1}_{odd}(A,M)$. It is defined formally as in the $2+1$ case (1). The Euclidean γ matrices in $2d+1$ dimensions are denoted as $\gamma_o, \gamma_1, \cdots, \gamma_{2d}$. From the point of view of the $2d$ dimensional theories (with coordinates x) that will arise below, γ_0 will act as a γ_5 chirality matrix.

In order to get an exact result we choose a particular gauge field background which corresponds to a vanishing electric field and a time-independent magnetic field,

$$A_0 = A_0(\tau), \tag{58}$$

$$A_j = A_j(x) \;\; (j = 1, 2, \cdots, 2d) \tag{59}$$

or any equivalent configuration obtained from this by a gauge transformation.

Using the same arguments as in II we can always perform a (non anomalous) gauge transformation of the fermionic fields in the functional integral defining the fermionic determinant in eq.(2), so that the new zero component of the gauge field, $\tilde{A}_0$, is a constant $\tilde{A}_0 = \frac{1}{\beta}\int_0^\beta d\tau\, A_0(\tau)$. Again, the Euclidean action becomes an infinite series of decoupled $2d$ dimensional actions, one for each Matsubara mode. Remarkably, as in II A, the parity odd piece of Γ defined by eqs.(1),(10) can then be factorized [5], for arbitrary d. The outcome of this procedure is [5]

$$\Gamma^{2d+1}_{odd} \;=\; i\,\Phi \int d^2x \mathcal{A}_{2d}[A]\,, \tag{60}$$

where

$$\Phi = \sum_{n=-\infty}^{n=+\infty} \phi_n. \tag{61}$$

and $\mathcal{A}_{2d}[A]$ denotes the $2d$ dimensional chiral anomaly. The phases ϕ_n contain *all* the dependence on the Matsubara frecuencies ω_n, and they are *independent* of the number of spacetime dimensions. Hence the sum over ϕ_n is the same as the one already calculated in ref. [3]. Thus the parity-odd part of Γ reads

$$\Gamma^{2d+1}_{odd} \;=\; i\,\arctan\left[\tanh(\frac{\beta M}{2})\tan(\frac{e}{2}\int_0^\beta d\tau A_0(\tau))\right]\int d^{2d}x\mathcal{A}_{2d}[A]\,. \tag{62}$$

The explicit expression for $D = 2d+1 = 5$ is

$$\Gamma^5_{odd} \;=\; -i\arctan\left[\tanh(\frac{\beta M}{2})\tan(\frac{e}{2}\int_0^\beta d\tau A_0(\tau))\right]\frac{e^2}{16\pi^2}\int d^4x F_{ij}{}^*F_{ij}\,. \tag{63}$$

The anomaly factor in (63) can be non trivial in the Abelian case depending of the properties of the $2d$ manifold $\mathcal{M}$ on which $\mathcal{A}_4$ is defined. For example if $\mathcal{M} = S^2 \times S^2$, $\int \mathcal{A}_4 d^4x = 2$, the smallest value it can take for spin manifolds (for non-spin manifolds it can take also odd values [23]).

In arbitrary dimension $D = 2d$, we quote the form of the Abelian chiral anomaly,

$$\mathcal{A}_{2d} = -\frac{(-e)^d}{(4\pi)^d}\frac{1}{d!}\epsilon_{\mu_1\mu_2\cdots\mu_{2d}}F_{\mu_1\mu_2}\cdots F_{\mu_{2d-1}\mu_{2d}} \tag{64}$$

(see for instance [24]).

Te previous analysis can be extended to the non-Abelian case for a special class of gauge field configurations, generates a Γ^{2d+1}_{odd} with nice topological properties. The model is defined by its Euclidean action

$$S_F \;=\; \int_0^\beta d\tau \int d^{2d}x\;\bar{\psi}(\not{D} + M)\psi \tag{65}$$

where now

$$\not{D}_\mu = \partial_\mu + gA_\mu \tag{66}$$

and the antihermitean gauge connection A_μ corresponds to the Lie algebra of some group G. For computation simplicity we shall consider that G has an Abelian $U(1)$ factor so that we can decompose A_μ as

$$A_\mu = iA^0_\mu + A^a_\mu \tau_a , \tag{67}$$

where A^0_μ is the component corresponding to the Abelian factor $U(1)$, while A^a_μ denotes the ones for the non-Abelian subgroup that for definiteness we shall take to be $SU(N)$. The matrices τ_a are the generators for $SU(N)$, satisfying the relations

$$[\tau_a, \tau_b] = f_{abc}\,\tau_c \quad \tau_a^\dagger = -\tau_a, \quad tr(\tau_a \tau_b) = -\frac{\delta_{ab}}{2}. \tag{68}$$

We now fix the class of gauge configurations we consider to those verifying the conditions

$$\begin{array}{ll} A^0_0 = A^0_0(\tau), & A^0_j = 0, \\ A^a_0 = 0, & A^a_j = A^a_j(x). \end{array} \tag{69}$$

The τ dependence of A_μ, present only through A^0_0, may be eliminated by an Abelian gauge transformation just as in the Abelian case. Then the fermionic action becomes

$$S_F(A^a_j, \tilde{A}^0_0, M) = \int_0^\beta d\tau \int d^{2d}x\, \bar{\psi}(\not{\partial} + g(\gamma_j A_j \tau_a + i\gamma_0 \tilde{A}^0_0) + M)\psi\,. \tag{70}$$

The same steps leading to the calculation of Γ^{2d+1}_{odd} may be performed here, with trivial modifications, except for the fact that the anomaly $\mathcal{A}$ will be the one corresponding to a $2d$ dimensional Abelian chiral rotation for a Dirac fermion in presence of a non-Abelian connection $A^a_j(x)$. This gauge field is to be regarded as an arbitrary $SU(N)$ gauge field for the $2d$ dimensional sector of the theory. The anomaly is then of course the well known "singlet" anomaly [24]

$$\mathcal{A}_{2d} = -\frac{(ig)^d}{(4\pi)^d}\frac{1}{d!}\epsilon_{j_1 j_2 \cdots j_{2d}} tr[F_{j_1 j_2} \cdots F_{j_{2d-1} j_{2d}}]. \tag{71}$$

Now, as the integral of $\mathcal{A}_{2d}$ is proportional to the Pontryagin index of the configuration

$$\int d^{2d}x \mathcal{A}_{2d}(x) = n \tag{72}$$

we may write Γ^{2d+1}_{odd} as

$$\Gamma^{2d+1}_{odd} = i \arctan\left[\tanh(\frac{\beta M}{2})\tan(\frac{g}{2}\int_0^\beta d\tau A_0(\tau))\right] n. \tag{73}$$

Note that this expression is non-trivial only for $D = 2d + 1 > 3$ dimensions since for $2d = 2$ the singlet anomaly vanishes. Depending on the gauge group and the $2d$ manifold over which the anomaly is integrated the Pontryagin index n can be a non trivial integer. Second, it is an object which is sensitive to large gauge transformations in $2d + 1$ spacetime, putting together the winding associated with the timelike direction τ (reflected in A^0_0), with the usual winding transformations in $2d$. However, the restrictions on the background gauge fields that we have imposed do not allow us to analyse general large gauge transformations although we expect that gauge-invariance is preserved.

IV EXAMPLES

A Dorey-Mavromatos model

This model employs two $U(1)$ gauge fields, one the electromagnetic field A_μ, the other a 'statistical' gauge field a_μ, which is also massless.There are $N_f \geq 2$ flavours of four-component fermions, the mass term is parity conserving, and A_μ and a_μ have opposite parity. At zero temperature a 'mixed Chern-Simons (MCS) term' is generated by a fermion loop with one external A and one external a leg, the leading contribution to the action, in powers of derivatives, being

$$\Gamma_{MCS} = N_f \frac{eg}{2\pi} \frac{M}{|M|} \int d^3x \epsilon^{\mu\nu\rho} a_\mu \partial_\nu A_\rho \ , \tag{74}$$

where e and g are the couplings of A and a, respectively.

A familiar phenomenological feature of superconductivity is flux quantization. To explain this, DM argued as follows. At $T \neq 0$, (74) becomes

$$\Gamma_{MCS}(T \neq 0) = i N_f \frac{M}{|M|} \frac{eg}{2\pi} \int_0^\beta d\tau \int d^2x \ \epsilon^{\mu\nu\rho} a_\mu \partial_\nu A_\rho \ . \tag{75}$$

Consider now a configuration in which $a_\mu = (a_0(\tau), \vec{0})$ and $A_\rho = (0, \vec{A}(\vec{x}))$. Its contribution to the action should be invariant under topologically non-trivial gauge transformations on a_0, of the form

$$a_0 \rightarrow a_0 + \partial_\tau \Omega(\tau) \tag{76}$$

with $\Omega(\beta) - \Omega(0) = \frac{2n\pi}{g}$, where n is an integer. Under (76), the variation of $\Gamma_{MCS}(T \neq 0)$ is

$$\delta\Gamma_{MCS}(T \neq 0) = i N_f \frac{M}{|M|} n e \int d^2x \epsilon_{ij} \partial_i A_j \ . \tag{77}$$

Considering then a superconducting annulus enclosing flux Φ, it follows that Φ has to have the value (restoring $\hbar$ and c)

$$\Phi = \frac{m h c}{N_f e} \tag{78}$$

where m is an integer, if $\delta\Gamma_{MCS}(T \neq 0)$ is to be an integer multiple of $2\pi i$, for any n in (77). (78) gives the required result for $N_f = 2$, the value indicated in the DM model. Thus, in this model, the flux quantization comes by requiring invariance under topologically non-trivial gauge transformations, of a CS-like term. Unfortunately, however, (75) is only correct for T infinitesimally close to 0. Indeed, the standard lowest order derivative expansion calculation would yield (75) multiplied by a smooth function of T, as stated above, and this T-dependent factor would destroy the result (78). The blame could be laid on using the lowest order term in perturbation theory, for the field undergoing the large gauge transformation.

In the DM model, the fermionic action is defined to be

$$S_f = \sum_{a=1}^{N_f} \int_0^\beta d\tau \int d^2x \, \bar{\psi}_a (\not\partial + ie \not A + ig \not a \tau_0 + M) \psi_a \tag{79}$$

where the Dirac matrices are in the reducible 4×4 representation:

$$\gamma_\mu = \begin{pmatrix} \sigma_\mu & 0 \\ 0 & -\sigma_\mu \end{pmatrix} \quad \tau_0 = \begin{pmatrix} I_{2\times2} & 0 \\ 0 & -I_{2\times2} \end{pmatrix} , \tag{80}$$

where $\mu = 1, 2, 3$, $I_{2\times2}$ is the 2×2 identity matrix and a is the flavour index.

For the particular configurations $a_0 = a_0(\tau)$, and $A_j = A_j(x)$ (all the other components = 0), we can directly apply the results of II to each of the two terms in the action, just by replacing $\tilde{A}_0$ by $\tilde{a}_0$. This amounts to using an $\tilde{\omega}_n = \omega_n + e\tilde{a}_0$ in the propagators. For N_f four-component fermions we get

$$\Gamma(A, a) = \frac{ie}{\pi} N_f \, \Phi \arctan \left[\tanh \frac{\beta M}{2} \tan(\frac{g}{2} \int_0^\beta d\tau a_0(\tau)) \right] , \tag{81}$$

which is the result that assures the validity of the flux quantization argument for any non-zero temperature, as explained in I.

B Fixed charge ensemble

The partition function $\mathcal{Z}_q$ corresponding to the ensemble with fixed charge q, at a given temperature $T = \frac{1}{\beta}$, for a system described by a quantum Hamiltonian H, and having a conserved additive charge Q ($[H, Q] = 0$), is

$$\mathcal{Z}_q = \int_{-\pi}^{\pi} \frac{d\theta}{2\pi} e^{-i\theta q} \mathcal{Z}_\theta \tag{82}$$

where

$$\mathcal{Z}_\theta = \text{Tr} e^{-\beta H + i\theta Q} . \tag{83}$$

We are assuming the normalization of Q is such that its eigenvalues are just integer numbers. Note that $\mathcal{Z}_\theta$ is formally equivalent to the grand canonical partition function for a system with an *imaginary* chemical potential θ. If the trace in (83) is evaluated using a complete set of simultaneous eigenstates of H and Q, then it follows immediately that (82) will only pick up contributions from quantum states with eigenvalue q for Q. Also by using this complete set one sees that $\mathcal{Z}_\theta$ is a periodic function of θ, with period 2π. Of course, this is closely related to the assumption that particles in the physical spectrum have integer charge. We shall see how this fact turns out to be important for the application to the $2+1$ dimensional case, where this periodicity is tantamount to gauge invariance under large gauge transformations. Alternatively, definition (82) can be justified by noting that

$$P_q = \int_{-\pi}^{\pi} \frac{d\theta}{2\pi} e^{-i\theta(q-Q)} \tag{84}$$

is a projector onto charge-q states. In a fixed-charge ensemble, the fixed charge does not fluctuate at all, as can be shown explicitly by noting that the averages (denoted $\langle \cdots \rangle_q$) of the powers of Q may be written as

$$\langle Q^n \rangle_q = (-i)^n \mathcal{Z}_q^{-1} \int_{-\pi}^{\pi} \frac{d\theta}{2\pi} e^{-i\theta q} \frac{\partial^n}{(\partial\theta)^n} \mathcal{Z}_\theta = q^n , \tag{85}$$

where the periodicity of $\mathcal{Z}_\theta$ has been used in order to ignore terms in the integration by parts. We want to construct the partition function $\mathcal{Z}_q(A)$ for the case of a fermionic field in $2+1$ dimensions in the presence of an external magnetic field (here A is the vector potential corresponding to the magnetic field). We immediately obtain the path-integral representation

$$\mathcal{Z}_\theta(A) = \int \mathcal{D}\bar{\psi}\mathcal{D}\psi$$

$$\exp\left\{ -\int_0^\beta d\tau \int d^2x \bar{\psi}(\tau, x)[\gamma_j D_j + M + \gamma_0(\partial_\tau - i\frac{\theta}{\beta})]\psi(\tau, x) \right\} . \tag{86}$$

It is now evident that (86) corresponds to exactly the same kind of configuration considered in [3], if one makes the identification

$$\tilde{A}_0 = -\frac{\theta}{e\beta} . \tag{87}$$

Periodicity in θ for (86) is equivalent to invariance under large gauge transformations, after this identification is made. We now separate $\mathcal{Z}_\theta$ into its phase and its modulus, which are given by the exponentials of the parity-breaking and parity-conserving parts of the effective action, respectively

$$\mathcal{Z}_\theta = e^{-\Gamma_{odd}(A)} \times e^{-\Gamma_{even}(A)} . \tag{88}$$

We know from [2,3,3] that, for this kind of configuration, Γ_{odd} can be exactly evaluated, and moreover that its periodicity may be assured if the parity anomaly is properly taken into account. As we have assumed that the ensemble corresponds to an *integer* charge q, periodicity of $\mathcal{Z}_\theta$ is required. We shall later on discuss the non-periodic 'gauge anomalous' $\mathcal{Z}_\theta$. The result for Γ_{odd}, including the parity anomaly piece is [3]:

$$\Gamma_{odd}(\theta, A) \; = \; i\frac{e}{2\pi}\frac{M}{|M|}\Phi\left\{ \arctan[\tanh(\frac{\beta|M|}{2})\tan(\frac{\theta}{2})] \; - \; \frac{1}{2}\theta \right\} \tag{89}$$

where $\Phi = \int d^2x\epsilon_{jk}\partial_j A_k$ is the static magnetic flux, and the branch of the arctan is chosen according to the value of θ. The even part of Γ cannot be found exactly, but fortunately there is a well-defined regime where its dependence on θ can be safely ignored. This is the case when $\beta|M| >> 1$, as can be checked explicitly in the calculation of [25], which yields the leading parity conserving contribution to Γ. For example, in a smooth gauge field configuration (though the same holds true without this assumption),

$$\Gamma_{even}(\theta, A_j) \; \simeq \; \Gamma^{(2)}(0, A_j)$$

$$+ \; \frac{e^2\beta}{48\pi M} \frac{\tanh(\frac{\beta M}{2})}{\cos^2(\frac{e\beta\tilde{A}_0}{2}) + \tanh^2(\frac{\beta M}{2})\sin^2(\frac{e\beta\tilde{A}_0}{2})} \int d^2x F_{jk}F_{jk} \; , \tag{90}$$

where the dependence on $\tilde{A}_0$ (and hence on θ) is exponentially suppressed for large $\beta|M|$. A more complete analysis shows that it is not even necessary to have $\beta|M| >> 1$, but already for $\beta|M|$ of order 1 the dependence on $\tilde{A}_0$ can be ignored. Ignoring thus the θ dependence of Γ_{even},

$$\Gamma_{even}(\theta, A_j) \; \simeq \; \Gamma_{even}(0, A_j) \; = \; \Gamma(0, A_j) \tag{91}$$

where the last equality proceeds from the fact that there is no odd part for $\theta = 0$. We can then take the even contribution out of the integral over θ, obtaining

$$\frac{\mathcal{Z}_q(A)}{\mathcal{Z}(A)} \; \simeq \; \int_{-\pi}^{\pi} \frac{d\theta}{2\pi} \, e^{-i\theta q - \Gamma_{odd}(\theta, A)} \; . \tag{92}$$

Note that in the last expression $\mathcal{Z}(A) \equiv \exp[-\Gamma(0, A_j)]$ is the partition function in the presence of a magnetic field in the *canonical* ensemble. This shows that the specific properties of the fixed charge ensemble when $\beta|M|$ is large are determined by Γ_{odd}. Equivalently, in terms of the respective free energies $F \equiv -\frac{1}{\beta}\log Z$,

$$F_q \; - \; F \; \simeq \; -\frac{1}{\beta}\log\left\{ \int_{-\pi}^{\pi} \frac{d\theta}{2\pi} \, e^{-i\theta q - \Gamma_{odd}(\theta, A)} \right\} \; . \tag{93}$$

Now we can consider the behaviour of (92) for different limits: When $\beta M \to \infty$, as the parity anomaly term cancels the induced term coming from the explicit parity breaking mass M, so that Γ_{odd} tends to zero. This means that, when $\beta M \to \infty$,

$$F_q \; \simeq \; F \; \to \; -\frac{1}{\beta}\log[\delta_{q,0}] \tag{94}$$

The meaning of this equation is clear, ensembles with non-zero charge are separated by an infinite free energy barrier, and only the zero charge one is physically possible. When $\beta|M|$ is large but not necessarily zero, ensembles with $q \neq 0$ are possible, and we shall discuss them now. We first note that, due to Parseval's identity, as Γ_{odd} is purely imaginary, we have the sum rule

$$1 \; = \; \sum_{n=-\infty}^{n=+\infty} |\frac{\mathcal{Z}_q(A)}{\mathcal{Z}(A)}|^2 \tag{95}$$

whose physical meaning in this case is that only a very few number of $q's$ shall be accessible with a finite free energy. We shall now derive a more convenient formula for (92) in terms of the dimensionless parameters of the theory. We define the dimensionless quantity $b \equiv \frac{M}{|M|}\frac{e\Phi}{2\pi}$, which essentially measures the magnetic flux in units of the elementary flux quantum $\frac{e\Phi}{2\pi}$. We then note that after some elementary algebra, (92) may be written as follows:

$$\frac{\mathcal{Z}_q(A)}{\mathcal{Z}(A)} \; = \; \int_{-\pi}^{\pi} \frac{d\theta}{2\pi} \, e^{-i\theta(q - \frac{b}{2})} \left(\frac{1 + e^{-2\beta|M|}e^{-i\theta}}{e^{-2\beta|M|} + e^{-i\theta}} \right)^{\frac{b}{2}} \; . \tag{96}$$

The change of integration variable $z = e^{-i\theta}$ maps the integration path to a unit circle in the complex plane:

$$\frac{\mathcal{Z}_q(A)}{\mathcal{Z}(A)} = \frac{i}{2\pi}\oint_C \frac{dz}{z} z^{q-\frac{b}{2}} \left(\frac{1+e^{2\beta|M|}z}{e^{2\beta|M|}+z}\right)^{\frac{b}{2}} \tag{97}$$

which, if b is even, say $b = 2k$ for an integer k, can be evaluated as the sum of the residues over the two poles inside the unit circle. The result of this procedure may be put as

$$\frac{\mathcal{Z}_q(A)}{\mathcal{Z}(A)} = \frac{\Theta(q \le k)}{(k-q)!} \lim_{z->0} \frac{d^{k-q}}{dz^{k-q}} \left[\frac{1+e^{2\beta|M|}z}{e^{2\beta|M|}+z}\right]^k +$$

$$\frac{\Theta(k<0)}{(k-1)!} \lim_{z\to -e^{-2\beta|M|}} \frac{d^{k-1}}{dz^{k-1}} \left[z^{q-k-1}(1+e^{-2\beta|M|}z)^{-k}\right], \tag{98}$$

where the symbol $\Theta(inequality)$ is defined to be one if the inequality is true, and zero otherwise. This is not a closed form but may be exactly evaluated for any set of values for q, k and βM. Note than when the sign of the magnetic flux is the same as the one the mass, k becomes positive, and so the second term in (98) vanishes:

$$\left[\frac{\mathcal{Z}_q(A)}{\mathcal{Z}(A)}\right]_{k>0} = \frac{\Theta(q \le k)}{(k-q)!} \lim_{z->0} \frac{d^{k-q}}{dz^{k-q}} \left[\frac{1+e^{2\beta|M|}z}{e^{2\beta|M|}+z}\right]^k . \tag{99}$$

From a numerical evaluation of this expression, we see that finite temperature effects strongly affect the properties of the free energy. In particular, for $\beta|M|$ of order 1, the maximum of the ratio $\frac{\mathcal{Z}_q(A)}{\mathcal{Z}(A)}$ is reached when q is equal to k. This means that, when the system is heated, the Chern-Simons term makes states with total charge proportional to the total flux more convenient energetically. The situation is qualitatively similar for an odd number of fluxes, though we could only check that numerically.

V CONCLUSIONS

(i) The exact finite temperature effective action induced by massive fermions in arbitrary odd dimensions has the proper behavior under gauge transformations.
(ii) Using a certain class of gauge field configurations, the temperature dependent parity-odd part of the effective action can be calculated *exactly in arbitrary odd dimensions.* The result is a gauge invariant action which is not just a Chern-Simons term with a temperature-dependent coefficient but which reduces, in the low temperature regime, to this product and confirms, at $T = 0$ that massive fermions induce a C-S action.
(iii) An exact calculation is possible because γ_0 in $2d+1$ dimensions can be always taken as the chiral γ_5 matrix in $2d$ dimensions so that the temperature dependent part of the effective action could be decoupled through a γ_0 rotation with constant phase. As it is well-known, the resulting chiral Fujikawa Jacobian can be exactly computed and yields to the $2d$ dimensional chiral anomaly. This gives another example of the connection between C-S terms in odd-dimensions and even-dimensional topological invariants connected to chiral anomalies [26]- [27].
(iv) The temperature dependence of the parity-odd effective action is the same irrespectively of the number of space-time dimensions. This could be attributed to the particular background we considered but the topological nature of the result suggests that a similar result should hold in general. This is also sustained by the fact that the dependence on the $2d$-dimensional components of the gauge-field background occurs through the axial anomaly, also a quantity of topological nature.
(v) The results we derived in a finite temperature Quantum Field Theory language, could also be interpreted in terms of a compactified Euclidean theory in an odd number of dimensions, where the curled coordinate is not necessarily the Euclidean time, but it may be a compact dimension of length $L = \beta$. If this interpretation is adopted, and one takes only the lowest Kaluza-Klein modes for the parity conserving part of the effective action, one then has a $2d$ reduced theory, where the odd part of the effective action we evaluated plays the role of a θ-vacuum term (we assume, of course, that there is also a Yang-Mills action for the gauge field).

ACKNOWLEDGMENTS

The author wishes to thank the organizers for the kind hospitality and for providing financial support for his visit. This visit was also partially supported by ANPCyT and Fundación Antorchas, Argentina.

REFERENCES

1. G. Dunne, K. Lee and C. Lu, Phys. Rev. Lett. **78** (1997) 3434.
2. S. Deser, L. Griguolo and D. Seminara, Phys. Rev. Lett. **79** (1997) 1976; Phys. Rev. **D57** (1998) 7444.
3. C.D. Fosco, G.L. Rossini and F.A. Schaposnik, Phys. Rev. Lett. **79** (1997) 1980 and *errata ibid* **79** (1997). C.D. Fosco, G.L. Rossini and F.A. Schaposnik, Phys. Rev. **D56** (1997) 6547.
4. I.J.R. Aitchison and C.D. Fosco, Phys. Rev. **D57** (1998) 1171.
5. C.D. Fosco, G.L. Rossini and F.A. Schaposnik, Induced parity breaking term in odd dimensions at finite temperature, hep-th/9810199, to appear in Phys. Rev. D.
6. O.Alvarez, Commun. Math. Phys. **100** (1985) 279.
7. D. Cabra, E. Fradkin, G.L. Rossini and F.A. Schaposnik, Phys. Lett. B 383 (1996) 434.
8. N. Bralić, C.D. Fosco and F.A. Schaposnik, Phys. Lett. **B 383** (1996) 199.
9. S. Deser, R. Jackiw and S. Templeton, Phys. Rev. Lett. **48** (1982) 975; Ann. Phys. (N.Y.) **140** (1982) 372.
10. A.J. Niemi and G.W. Semenoff, Phys. Rev. Lett. **51** (1983) 2077.
11. A.J. Niemi, Nucl. Phys. **B251** (1985) 55.
12. A.J. Niemi and G.W. Semenoff, Phys. Rep. **135** (1986) 99.
13. K. Babu, A. Das and P. Panigrahi, Phys. Rev. **D36**, 3725 (1987).
14. A. Das and S. Panda, J. Phys. A: Math. Gen. **25**, L245 (1992).
15. I.J.R. Aitchison, C.D. Fosco and J. Zuk, Phys. Rev. **D48** (1993) 5895.
16. E.R. Poppitz, Phys. Lett. **B252** (1990) 417.
17. M.Burgess, Phys. Rev. **D44** (1991) 2552.
18. W.T. Kim, Y.J. Park, K.Y. Kim and Y. Kim, Phys. Rev. **D 46** (1993) 3674.
19. K. Ishikawa and T. Matsuyama, Nucl. Phys. **B 280** [F518] (1987) 523.
20. N. Dorey and N.E Mavromatos, Nucl. Phys. **B386** (1992) 614.
21. C.D. Fosco, Phys. Rev. **D57** (1998) 6554.
22. K. Fujikawa, Phys. Rev. Lett. **42** (1979) 1195; Phys. Rev. **D21** (1980) 2848.
23. E. Witten, On S Duality in Abelian Gauge Theory, hep-th/9505186.
24. R.D.Ball, Phys. Rep. **182**, 1 & 2 (1989).
25. I.J.R. Aitchison and C. Fraser, Phys. Rev. **D31** (1985) 2605.
26. R. Jackiw,Chern-Simons Terms and their Descendents in Physical Theory, in Santa Fe 1984 proceedings, The Santa Fe Meeting, p.323.
27. L. Alvarez-Gaumé, S. Della Pietra and G. Moore, Ann. of Phys. (NY) **163** (1985) 288.

$\mathcal{N} = 2$ SUPERSYMMETRIC YANG-MILLS THEORIES AND WHITHAM INTEGRABLE HIERARCHIES

José D. Edelstein and Javier Mas

Departamento de Física de Partículas, Universidade de Santiago de Compostela
E-15706 Santiago de Compostela, Spain
E-mail: edels,jamas@fpaxp1.usc.es

Abstract. We review recent work on the study of $\mathcal{N} = 2$ super Yang-Mills theory with gauge group $SU(N)$ from the point of view of the Whitham hierarchy, mainly focusing on three main results: (i) We develop a new recursive method to compute the whole instanton expansion of the low-energy effective prepotential; (ii) We interpret the slow times of the hierarchy as additional couplings and promote them to spurion superfields that softly break $\mathcal{N} = 2$ supersymmetry down to $\mathcal{N} = 0$ through deformations associated to higher Casimir operators of the gauge group; (iii) We show that the Seiberg–Witten-Whitham equations provide a set of non-trivial constraints on the form of the strong coupling expansion in the vicinity of the maximal singularities. We use them to check a proposal that we make for the value of the off-diagonal couplings at those points of the moduli space.

I INTRODUCTION

The study of non-perturbative phenomena in quantum field theory has experienced drastic advances since, in 1994, Seiberg and Witten gave an ansatz for the exact effective action governing the low-energy excitations of $SU(2)$ $\mathcal{N} = 2$ super Yang-Mills theory [1]. It is given in terms of an auxiliary complex algebraic curve, whose moduli space is identified with the quantum moduli space of the low-energy theory $\mathcal{M}_\Lambda$, and a given meromorphic differential, dS_{SW}, that induces a special geometry on $\mathcal{M}_\Lambda$. Apart form its unquestionable beauty, it proved to contain nontrivial dynamical information about the non perturbative behaviour of the theory. For example, the existence of a confinement mechanism when breaking $\mathcal{N} = 2$ to $\mathcal{N} = 1$ by addition of a mass term. The solution was soon extended to the case of $SU(N)$ [2,3]. Still, the price to pay was the need for $\mathcal{N} = 2$ supersymmetry. Soft supersymmetry breaking was shown to preserve the analytic properties of the solution in such a way that exact results in the $\mathcal{N} = 0$ theory could be obtained [4–6].

Interestingly enough, it was soon realized that the Seiberg–Witten solution could be reformulated in terms of certain integrable systems, dS_{SW} being a solution of their averaged (Whitham) dynamics [7]. For example, the periodic Toda lattice is the proper integrable system whose averaged dynamics corresponds to pure $\mathcal{N} = 2$ super Yang-Mills theory for the whole ADE series [8]. The spectral curve Γ_g of the particular integrable system is identified with the auxiliary Seiberg–Witten algebraic curve. Its moduli, in spite of being *local* invariants, evolve with respect to the so-called *slow times* T_n. The system of non-linear equations that describe this evolution, that amounts to adiabatic *deformations* of an hyperelliptic curve, was developed by Whitham [9]. Surprisingly, this system turns out to be itself integrable and receives the generic name of *Whitham hierarchy* (see [10] and references therein).

CP484, *Trends in Theoretical Physics II*, edited by H. Falomir, R. E. Gamboa Saraví, and F. A. Schaposnik

The Whitham dynamics can be thought of as a generalization of the Renormalization Group (RG) flow [7] (see [11] for a review). The corresponding RG equations were recently derived by Gorsky, Marshakov, Mironov and Morozov [12]: the second derivatives of the prepotential with respect to Whitham slow times T_n result to be given in terms of Riemann Theta-functions. We would like to show, in this talk, that this framework is very fruitful both to study many features of the low-energy dynamics of $\mathcal{N}=2$ super Yang-Mills theory, as well as to implement *natural* generalizations of the Seiberg–Witten solution.

We shall start by giving a telegraphical account of the Seiberg–Witten solution to the low-energy dynamics of $SU(N)$ $\mathcal{N}=2$ super Yang-Mills theory and of the basic ideas involved in the Whitham hierachies. We will mainly focus on the fact that they lead naturally to the concept of a prepotential and establish thereby the concrete link between both formalisms. Within this framework, we first develop a new recursive method to compute the whole instanton expansion of the low-energy effective prepotential. Then, we interpret the slow times of the hierarchy as additional couplings and promote them to spurion superfields that softly break $\mathcal{N}=2$ supersymmetry down to $\mathcal{N}=0$ through deformations associated to higher Casimir operators of the gauge group. We discuss in some detail the case of $SU(3)$. Finally, we show that the Seiberg–Witten–Whitham equations provide a set of non-trivial constraints on the form of the strong coupling expansion in the vicinity of the maximal singularities. We use them to check a proposal that we make for the value of the off-diagonal couplings at those points of the moduli space. Most of the work presented in this talk (the first two applications) was developed by the authors in collaboration with Marcos Mariño [13]. The analysis of the strong coupling expansion near the maximal singularities was done in [14]. Further generalizations, like extensions to other Lie algebras and/or inclusion of matter, remain interesting problems of research [15].

II THE SEIBERG–WITTEN SOLUTION

The classical potential of $\mathcal{N}=2$ super Yang-Mills theory with a vector multiplet in the adjoint representation of $SU(N)$ has flat directions. There is a family of inequivalent ground states that constitutes the *classical moduli space* $\mathcal{M}_0$, parametrized by a constant *vev* of the scalar field ϕ in the Cartan sub-algebra

$$< \phi > = \sum_{i=1}^{N-1} a^i H_i = \mathrm{diag}\ (e_1(a^i), \ldots, e_N(a^i)) \ , \tag{1}$$

where $e_i(a) = \lambda^i \cdot a$, λ^i being the i-th fundamental weight of the Lie algebra A_{N-1}. At a generic point of $\mathcal{M}_0$, the unbroken gauge symmetry is $U(1)^{N-1}$. In fact, for every positive root α_+, a couple of gauge bosons $W_\mu^{\pm\alpha_+}$ gets a mass $M_{\alpha_+}(a) = \sqrt{2}|\alpha_+ \cdot a|$ through the Higgs mechanism. They are BPS states with central charge Z_{α_+}, $Z_\alpha \equiv \alpha \cdot a$ and $a = a^i \alpha_i$ with α_i the simple roots. Gauge invariant coordinates for $\mathcal{M}_0$ can be constructed from the characteristic polynomial

$$W_{A_{N-1}}(\lambda, \bar{u}_k) = \det(\lambda - < \phi >) = \lambda^N - \bar{u}_2(a)\lambda^{N-2} - \bar{u}_3(a)\lambda^{N-3} - \cdots - \bar{u}_N(a) \ , \tag{2}$$

whose coefficients, $\bar{u}_k(a)$, are nothing but the Casimir operators. *Each* microscopic theory, characterized by a value of the $\bar{u}_k$, leads to *one* effective field theory at low energies. We may therefore think of the quantum moduli space $\mathcal{M}_\Lambda$ of effective field theories as being parametrized by $u_k(a) \sim < \mathrm{Tr}\,\phi^k >$. It will look like a deformation of $\mathcal{M}_0$, with the *quantum generated scale* Λ as the deformation parameter. The microscopic relations, thus, should be recovered for $\Lambda \to 0$.

The low-energy effective action can be entirely written in terms of a single *holomorphic* function of the Cartan variables, $\mathcal{F}(a^i)$, known as the effective prepotential. By means of symmetry considerations [1–3], the most general expression for $\mathcal{F}(a^i)$ is

$$\mathcal{F} = \frac{1}{2N}\tau_0 \sum_{\alpha_+} Z_{\alpha_+}^2 + \frac{i}{4\pi} \sum_{\alpha_+} Z_{\alpha_+}^2 \log \frac{Z_{\alpha_+}^2}{\Lambda^2} + \frac{1}{2\pi i} \sum_{k=1}^{\infty} \mathcal{F}_k(a) \Lambda^{2Nk} \ , \tag{3}$$

where τ_0 is the bare coupling constant and $\mathcal{F}_k(a)$ are Weyl invariant combinations of a^i, whereas k is the instanton number. The full prepotential is homogeneous of degree two in a^i and Λ. What remains to be computed are the instanton corrections, $\mathcal{F}_k(a)$. It is their exact determination the whole point of the Seiberg–Witten solution. In the semiclassical limit, the first few terms can be computed explicitely in the microscopic theory [16], and their agreement with the output of the Seiberg–Witten solution provides a non-trivial consistecy check of it.

Since the effective prepotential (3) is holomorphic, the imaginary part of its second order derivatives with respect to the Cartan variables, $\text{Im}\tau_{ij}$, are harmonic functions. Therefore, they cannot have a global minimum. However, $\text{Im}\tau_{ij}$ enters the low-energy Lagrangian as an effective coupling constant: it *must* be positive definite. So, τ_{ij} cannot be globally defined. These conditions are automatically fulfilled it τ_{ij} is the period matrix of some Riemann surface. This observation led Seiberg and Witten to the following ansatz:

First: Over each point on $\mathcal{M}_\Lambda$ labelled by u_k, consider a certain hyperelliptic curve. For $SU(N)$ the relevant curve Γ_g is [2]

$$y^2 = P(\lambda, u_k)^2 - 4\Lambda^{2N} \tag{4}$$

with $P = W_{A_{N-1}}$. The moduli space of the family of Riemann surfaces of genus $g = N-1$ written above is to be identified with $\mathcal{M}_\Lambda$. In the classical limit $\Lambda \to 0$, the rational function $y = W_{A_{N-1}}(\lambda, u_k)$ has roots $e_i(a)$. At the quantum level, as y^2 factors into the product $y^2 = y_+ y_-$ with $y_\pm = (P(\lambda, u_k) \pm 2\Lambda^N)$, the points e_i split into two sets of roots of $y_\pm$,

$$e_i(\bar{u}_k) \to e_i^{\pm}(u_k, \Lambda) \equiv e_i(u_{k<N}, u_N \pm 2\Lambda^N) , \tag{5}$$

that become the $2N$ branch points of the Riemann surface.

Second: At the point u_k on $\mathcal{M}_\Lambda$, the (quantum) relations between a^i, a_{Dj} and u_k is given by the period integrals

$$a^i(u) = \oint_{A^i} dS_{SW}(u) \qquad a_{Dj}(u) = \oint_{B_j} dS_{SW}(u) . \tag{6}$$

with dS_{SW} a meromorphic differential given by

$$dS_{SW} = \frac{\lambda P'(\lambda, u_k)}{\sqrt{P^2(\lambda, u_k) - 4\Lambda^{2N}}} d\lambda , \tag{7}$$

and A^i and B_j constitute a symplectic basis of homology cycles of the hyperelliptic curve, with the canonical intersections $A^i \cap A^j = B_i \cap B_j = 0$ and $A^i \cap B_j = \delta^i{}_j$, $i, j = 1, \ldots, N-1$. The prepotential $\mathcal{F}(a)$ is implicitely defined by the equation

$$a_{Di} = \frac{\partial \mathcal{F}(a)}{\partial a^i} . \tag{8}$$

The exact determination of $\mathcal{F}(a)$ involves, in general, the integration of functions $a_{Di}(a)$ for which there is not a closed form available. Therefore, $\mathcal{F}(a)$ will only be calculable in a series expansion.

Third: The BPS spectrum is obtained by integrating the Seiberg–Witten differential dS_{SW} along all nontrivial cycles of the Riemann surface, $\nu(n^e, n_m) = n^e \cdot A + n_m \cdot B$. In fact, this is immediate from the previous point and the fact that the central charge of a state with $n^e{}_i$ units of electric charge and $n_m{}^i$ units of magnetic charge with respect to the i-th $U(1)$ unbroken subgroup can be written as

$$Z(n^e, n_m) = n^e \cdot a + n_m \cdot a_D . \tag{9}$$

Appart from the mass, what remains invariant is the *intersection number* of two BPS states, given by the intersection product of their respective cycles $\nu(n^e, n_m)$ and $\nu'(n'^e, n'_m)$

$$\nu \cap \nu' = n^e \cdot n'_m - n'^e \cdot n_m \in \mathbf{Z} . \tag{10}$$

Notice that this is nothing but the Dirac-Schwinger-Zwanzinger quantization condition [17]. Two dyons are mutually local if they have zero intersection $\nu \cap \nu' = 0$.

Changes in the symplectic basis of homology cycles are performed by means of a symplectic matrix $\Gamma \in Sp(2r, \mathbf{R})$. Accordingly, $\mathbf{a} = (a^i, a_{D\,j})$ transforms as a vector and τ_{ij} as a modular form. Since the central charge, $Z_{\mathbf{n}} = \mathbf{n}^t \cdot \mathbf{a}$ with $\mathbf{n} = (n^e_i, n^j_m)$, is an observable, the invariance of the *non-perturbative* BPS spectrum breaks the continuous duality group $Sp(2r, \mathbf{R})$ down to the discrete subgroup $Sp(2r, \mathbf{Z})$.

Fourth: There are singularities in $\mathcal{M}_\Lambda$, encoded in the quantum discriminant Δ_Λ,

$$\Delta_\Lambda(u_k, \Lambda) = \prod_{i<j} (e_i^+ - e_j^+)^2 (e_i^- - e_j^-)^2 = c\, \Lambda^{2N^2} \Delta_+ \Delta_- \ , \tag{11}$$

at whose zero locus, Σ_Λ, two branch points $e_i^\pm$, $e_j^\pm$ collide. What is the same, a certain homology cycle on the Riemann surface shrinks to zero at Σ_Λ, signaling the appearance of an extra massless state which is generically a dyon. These singularities lie on curves that intersect at points where many BPS states become simultaneously massless. In particular, at the so-called $N-1$ points, exactly $N-1$ mutually local monopoles become massless. This is the maximal number of mutually local simultaneously massless BPS states. The physics of these points was first investigated in Ref. [18]. They remain the vacua of the $\mathcal{N} = 1$ theory upon perturbation of the $\mathcal{N} = 2$ theory by a mass term. For $SU(N)$, $N > 2$, there are regions in $\mathcal{M}_\Lambda$ where *mutually non-local* dyons become simultaneously massless, and the corresponding effective low-energy dynamics seems to be given by a superconformal field theory [19].

III THE UNIVERSAL WHITHAM HIERARCHY

The name *Whitham hierarchy* stands for a wide class of integrable systems of differential equations that describe modulations of solutions of soliton equations [9,20,21]. Following Krichever [10], we define the moduli space of the Whitham hierarchy by

$$\hat{\mathcal{M}}_{g,p} \equiv \{\Gamma_g, P_a, \xi_a(P),\ a = 1, ..., p\} \tag{12}$$

containing the following set of algebraic–geometrical data:

- Γ_g denotes a smooth algebraic curve of genus g.
- P_a is a set of p points (punctures) on Γ_g in generic positions (we will consider, for simplicity, $p = 1$).
- ξ_a are local coordinates in the neighbourhood of the p points, *i.e.* $\xi_a(P_a) = 0$.

From the general theory of meromorphic differentials over Riemann surfaces we know that there are three basic types of Abelian differentials:

i. *Holomorphic differentials, dw_i.* In any open set $U \in \Gamma_g$, with complex coordinate ξ, they are of the form $dw = f(\xi)d\xi$ with f an holomorphic function. The vector space of holomorphic differentials on a genus g Riemann surface has complex dimension g. If the curve is hyperelliptic (4), a canonical basis $\{dw_j\}$ of this vector space is defined through

$$\oint_{A^i} dw_j = \delta^i{}_j \qquad\qquad \oint_{B_i} dw_j = \tau_{ij} \ , \tag{13}$$

where τ_{ij} is the period matrix of the complex curve.

ii. *Meromorphic differentials of the second kind, $d\Omega_{P,n}$.* They have a single pole of order $n+1$ at point $P \in \Gamma$, and zero residue. In local coordinates ξ, we shall adopt the normalization

$$d\Omega_{P,n} = (\xi^{-n-1} + O(1))\ d\xi \ . \tag{14}$$

This fixes $d\Omega_{P,n}$ up to an arbitrary combination of holomorphic differentials. There are several ways to fix this normalization. In the context of integrable theories, the standard way to do it is to require that $d\Omega_{P,n}$ has vanishing A^i-periods

$$\oint_{A^i} d\Omega_{P,n} = 0 \; . \tag{15}$$

iii. *Meromorphic differentials of the third kind,* $d\Omega_{P,0}$. They have first order poles at P and P_0 (a reference point) with opposite residues taking values $+1$ and -1 respectively. In local coordinates $\xi\,(\xi_0)$ about $P\,(P_0)$,

$$d\Omega_{P,0} = (\xi^{-1} + \mathcal{O}(1))d\xi = -(\xi_0^{-1} + \mathcal{O}(1))d\xi_0 \; . \tag{16}$$

The regular part is normalized by demanding that $d\Omega_{P,0}$ has vanishing A^i-periods. The appearance of simple poles in the Seiberg–Witten solution is related to the inclusion of matter hypermultiplets in the fundamental representation [1]. We will only consider in this talk the case of *pure* $SU(N)$, $\mathcal{N} = 2$ super Yang-Mills theory. Thus, we are going to rule out the meromorphic differentials of the third kind from our discussion.

The standard Whitham equations take the following form [21]

$$\frac{\partial d\Omega_n}{\partial T^m} = \frac{\partial d\Omega_m}{\partial T^n} \tag{17}$$

where $d\Omega_n$ is short for $d\Omega_{P,n}$, and T^n are a set of *slow times* the 1-forms may depend upon. According to our previous remark, n will be considered greater or equal than one, unless the contrary is stated. The Whitham hierarchy can be enhanced to incorporate also holomorphic differentials dw_i, with associated parameters α^i, such that

$$\frac{\partial dw_i}{\partial \alpha^j} = \frac{\partial dw_j}{\partial \alpha^i} \qquad \frac{\partial dw_i}{\partial T^n} = \frac{\partial d\Omega_n}{\partial \alpha^i} \qquad \frac{\partial d\Omega_n}{\partial T^m} = \frac{\partial d\Omega_m}{\partial T^n} \; . \tag{18}$$

Equations (18) are nothing but the integrability conditions implying the existence of a *generating differential* dS satisfying

$$\frac{\partial dS}{\partial \alpha^i} = dw_i \qquad \frac{\partial dS}{\partial T^n} = d\Omega_n \; . \tag{19}$$

The Whitham equations hide a certain holomorphic function named *prepotential* $\mathcal{F}(\alpha^i, T^n)$, that can be defined implicitely through the following set of equations

$$\frac{\partial \mathcal{F}}{\partial \alpha^j} = \oint_{B_j} dS \qquad \frac{\partial \mathcal{F}}{\partial T^n} = \frac{1}{2\pi i n} \oint_P \xi^{-n} dS \; . \tag{20}$$

The local behaviour of the generating differential near the puncture P is then

$$dS \sim \left\{ \sum_{n\geq 1} T^n \xi^{-n-1} + 2\pi i \sum_{n\geq 1} n \frac{\partial \mathcal{F}}{\partial T^n} \xi^{n-1} \right\} d\xi \; . \tag{21}$$

An interesting *class of solutions*, and certainly that which is relevant in connection with $\mathcal{N} = 2$ super Yang-Mills theories, is given by those prepotentials that are homogeneous of degree two:

$$\sum_{i=1}^{N-1} \alpha^i \frac{\partial \mathcal{F}}{\partial \alpha^i} + \sum_{n\geq 1} T^n \frac{\partial \mathcal{F}}{\partial T^n} = 2\mathcal{F} \; . \tag{22}$$

The generating differential dS for homogeneous solutions admits the following form [8,10]:

$$dS = \sum_{i=1}^{N-1} \alpha^i dw_i + \sum_{n\geq 1} T^n d\Omega_n \; , \tag{23}$$

and, after (13)–(15), the parameters α^i, and T^n can be recovered from dS as follows:

$$\alpha^i = \oint_{A^i} dS \qquad\qquad T^n = \text{res}_P\, \xi^n dS\ . \tag{24}$$

Inserting (19) and (24) into (22), a formal expression for $\mathcal{F}$ in terms of dS can be obtained [22],

$$\mathcal{F} = \frac{1}{2}\sum_{i=1}^{N-1} \oint_{A^i} dS \oint_{B_i} dS + \frac{1}{4\pi i}\sum_{n\geq 1}\frac{1}{n}\oint_P \xi^n dS \oint_P \xi^{-n} dS\ . \tag{25}$$

Following [22], let us consider the decomposition of dS in a different basis of Abelian differentials,

$$dS = \sum_{n\geq 1} T^n d\hat{\Omega}_n\ , \tag{26}$$

where $d\hat{\Omega}_n$ are meromorphic differentials of the second kind (with the same local behaviour than $d\Omega_n$), whose regular part is fixed by the condition:

$$\frac{\partial d\hat{\Omega}_n}{\partial \text{moduli}} = \text{holomorphic}\ . \tag{27}$$

Notice that we have not added explicitely holomorphic differentials in dS: they are somehow hidden inside the differentials $d\hat{\Omega}_n$. In more concrete terms, the definition of the α^i parameters as given in (24), now forces them to depend on T^n and u_k. Conversely, provided we impose that $d\alpha^i/dT^n = 0$, an implicit set of homogeneous functions $u_k(T^n, \alpha^i)$ of degree zero is obtained, and they solve the Whitham equations:

$$\frac{\partial u_k}{\partial T^n} = -\left(\frac{\partial \alpha^i}{\partial u_k}\right)^{-1}\frac{\partial \alpha^i}{\partial T^n}\ . \tag{28}$$

Finally, from (24) and (26), it is clear that

$$d\hat{\Omega}_m = d\Omega_m + \sum_{i=1}^{N-1}\frac{\partial \alpha^i}{\partial T^m} dw_i\ . \tag{29}$$

IV THE SEIBERG–WITTEN–WHITHAM FORMALISM

The next task is to look for an embedding of the Seiberg–Witten ansatz within the Whitham hierarchy. We will follow very closely, to this end, the approach of Gorsky, Marshakov, Mironov and Morozov [12]. As already noticed in Ref. [7], the curve (4) is the hyperelliptic representation for the spectral curve of the *periodic Toda chain* of length N. It can be written in terms of a complex parameter w as follows (we set $\Lambda = 1$ for convenience)

$$P = \left(w + \frac{1}{w}\right) \qquad\qquad y = \left(w - \frac{1}{w}\right)\ . \tag{30}$$

This defines a natural coordinate in the vicinity of the two points at infinity $\infty_\pm \sim (\pm y = \infty, \lambda = \infty)$. In fact, from Eq.(30), w can be written as a meromorphic function

$$w = \frac{1}{2}(P + y)\ , \tag{31}$$

that near $\infty_\pm$ goes as $w \sim \lambda^{\pm N}$. Then, $\xi_\pm = w^{\mp 1/N}$ are local coordinates at the punctures $\infty_\pm$, which are the points where the relevant meromorphic differential of the Seiberg–Witten solution, dS_{SW}, has its (second order) poles. The times associated to each puncture will be denoted with positive and negative subindices, *i.e.*

$T_{\infty_\pm,n} = T_{\pm n}$. Also, it is convenient to slightly change the normalization of our second-kind differentials to be $d\Omega_{\pm n} \sim \frac{N}{n}\, dw^{\pm n/N}$.

The Seiberg–Witten differential, dS_{SW}, belongs to the *class of solutions* of the Whitham hierarchy that fulfill

$$\frac{\partial dS}{\partial \text{moduli}} = \text{holomorphic} \; . \tag{32}$$

In fact, since its A^i-periods are a^i, one is tempted to identify dS_{SW} as the generating form of the Whitham hierarchy at $\alpha^i = a^i$ and $T_{\pm 1} = 1,\ T_{|n|>1} = 0$,

$$dS_{SW} = a^i dw_i + d\Omega_{\infty_+,1} + d\Omega_{\infty_-,1} \; . \tag{33}$$

Varying Eq.(30) for a given curve, *i.e.* for fixed u_k and Λ, the Seiberg–Witten differential can be written as

$$dS_{SW} = \frac{\lambda P'}{y} d\lambda = \lambda \frac{dw}{w} \; , \tag{34}$$

this making extremely easy to verify the defining property of dS_{SW},

$$\left.\frac{\partial dS_{SW}}{\partial u_k}\right|_{w=const.} = \frac{\lambda^{N-k}}{P'}\frac{dw}{w} = \frac{\lambda^{N-k} d\lambda}{y} = dv_k \quad , \qquad k = 2, 3, ..., N \; . \tag{35}$$

If we hold λ fixed –instead of ω–, there is an additional total derivative in the previous expression. Notice that, from the point of view of the Whitham equations, it matters which coordinates are held fixed as long as there are residues to be computed. It will always be understood that derivatives w.r.t. the moduli are taken at constant w.

Lemma [12]: The meromorphic differentials $d\hat{\Omega}_n$ have the form

$$d\hat{\Omega}_n = R_n \frac{dw}{w} = P_+^{n/N} \frac{dw}{w} \; , \tag{36}$$

where the projection $(\sum_{k=-\infty}^{\infty} c_k \lambda^k)_+ = \sum_{k=0}^{\infty} c_k \lambda^k$. Notice that, from their defining equation (27), the $d\hat{\Omega}_n$ have poles at both punctures, *i.e.* there are not $d\hat{\Omega}_{\pm n}$ but $d\hat{\Omega}_n = d\hat{\Omega}_n + d\hat{\Omega}_{-n}$.

At this point, as shown in Ref. [12], the first and second order derivatives of the prepotential can be computed from Eq.(20), with the following result

$$\frac{\partial \mathcal{F}}{\partial T_n} = \frac{\beta}{2\pi i n} \sum_m m T_m \mathcal{H}_{m+1,n+1} \qquad \qquad \frac{\partial^2 \mathcal{F}}{\partial \alpha^i \partial T^n} = \frac{\beta}{2\pi i n}\frac{\partial \mathcal{H}_{n+1}}{\partial a^i} \; , \tag{37}$$

$$\frac{\partial^2 \mathcal{F}}{\partial T_m \partial T_n} = -\frac{\beta}{2\pi i}\left(\mathcal{H}_{m+1,n+1} + \frac{\beta}{mn}\frac{\partial \mathcal{H}_{m+1}}{\partial a^i}\frac{\partial \mathcal{H}_{n+1}}{\partial a^j}\frac{1}{i\pi}\partial_{\tau_{ij}} \log \Theta_E(0|\tau)\right) \; , \tag{38}$$

where $\beta = 2N$ is the coefficient of the beta function, whereas $\mathcal{H}_{m+1,n+1}$ and $\mathcal{H}_{n+1}$ are homogeneous combinations of the Casimirs constructed as follows:

$$\mathcal{H}_{m+1,n+1} = \frac{N}{mn} \text{res}_\infty \left(P^{m/N}(\lambda) dP_+^{n/N}(\lambda)\right) = \mathcal{H}_{n+1,m+1} \qquad \qquad \mathcal{H}_{m+1} \equiv \mathcal{H}_{m+1,2} \; . \tag{39}$$

The important ingredient is the Riemann's Theta function $\Theta_E(\vec{z}|\tau)$, E being the following even and half-integer characteristic [13]:

$$\vec{\alpha} = (0, \ldots, 0) \qquad \qquad \vec{\beta} = (1/2, \ldots, 1/2) \; . \tag{40}$$

As they stand, however, the expressions given in (37)–(38) are not yet suitable for application to the Seiberg–Witten solution. It is still necessary to define the rescaled times $\hat{T}_n = T_1^{-n} T_n$ and moduli $\hat{u}_k = T_1^k u_k$ (correspondingly, $\hat{\mathcal{H}}_{m+1,n+1} = T_1^{m+n} \mathcal{H}_{m+1,n+1}$) after which, the prepotential of the Seiberg–Witten solution is obtained by identifying T_1 with Λ in the submanifold $\hat{T}_{n>1} = 0$, provided that the moduli space be parametrized by the $\hat{u}_k$ (notice that $\hat{a}^i \equiv \alpha^i(u_k, T_1, \hat{T}_{n>1} = 0) = T_1 a^i(u_k, 1) = a^i(\hat{u}_k, T_1)$) [13]. The restriction to the submanifold $\hat{T}_{n>1} = 0$, yields formulae which are suited for the Seiberg–Witten solution. In particular, from (37) and (38) we obtain

$$\frac{\partial \mathcal{F}}{\partial \log \Lambda} = \frac{\beta}{2\pi i} \hat{\mathcal{H}}_2 \qquad \frac{\partial \mathcal{F}}{\partial \hat{T}_n} = \frac{\beta}{2\pi i n} \hat{\mathcal{H}}_{n+1} , \tag{41}$$

$$\frac{\partial^2 \mathcal{F}}{\partial \hat{a}^i \partial \log \Lambda} = \frac{\beta}{2\pi i} \frac{\partial \hat{\mathcal{H}}_2}{\partial \hat{a}^i} \qquad \frac{\partial^2 \mathcal{F}}{\partial \hat{a}^i \partial \hat{T}_n} = \frac{\beta}{2\pi i n} \frac{\partial \hat{\mathcal{H}}_{n+1}}{\partial \hat{a}^i} , \tag{42}$$

$$\frac{\partial^2 \mathcal{F}}{\partial (\log \Lambda)^2} = -\frac{\beta^2}{2\pi i} \frac{\partial \hat{\mathcal{H}}_2}{\partial \hat{a}^i} \frac{\partial \hat{\mathcal{H}}_2}{\partial \hat{a}^j} \frac{1}{i\pi} \partial_{\tau_{ij}} \log \Theta_E(0|\tau) , \tag{43}$$

$$\frac{\partial^2 \mathcal{F}}{\partial \log \Lambda \, \partial \hat{T}_n} = -\frac{\beta^2}{2\pi i n} \frac{\partial \hat{\mathcal{H}}_2}{\partial \hat{a}^i} \frac{\partial \hat{\mathcal{H}}_{n+1}}{\partial \hat{a}^j} \frac{1}{i\pi} \partial_{\tau_{ij}} \log \Theta_E(0|\tau) , \tag{44}$$

$$\frac{\partial^2 \mathcal{F}}{\partial \hat{T}_m \partial \hat{T}_n} = -\frac{\beta}{2\pi i} \left(\hat{\mathcal{H}}_{m+1,n+1} + \frac{\beta}{mn} \frac{\partial \hat{\mathcal{H}}_{m+1}}{\partial \hat{a}^i} \frac{\partial \hat{\mathcal{H}}_{n+1}}{\partial \hat{a}^j} \frac{1}{i\pi} \partial_{\tau_{ij}} \log \Theta_E(0|\tau) \right) , \tag{45}$$

with $m, n \geq 2$.

The first equation in (41) is precisely the RG equation derived in Ref. [23]. Combining the second equation in (41) and (44), it is easy to obtain an interesting relation between Casimir operators [12]:

$$\frac{\partial \hat{\mathcal{H}}_m}{\partial \log \Lambda} = -\beta \frac{\partial \hat{\mathcal{H}}_2}{\partial \hat{a}^i} \frac{\partial \hat{\mathcal{H}}_m}{\partial \hat{a}^j} \frac{1}{i\pi} \partial_{\tau_{ij}} \log \Theta_E(0|\tau) . \tag{46}$$

Hereafter, we will always work with the scaled coordinates and hats will be omitted everywhere.

V INSTANTON CORRECTIONS

Instanton calculus provides one of the few non-perturbative links between the Seiberg–Witten solution and the microscopic non-abelian field theory that it is supposed to describe effectively at low energies. From the microscopic theory point of view, instanton contributions to the asymptotic semiclassical expansion of the effective prepotential have been computed, and a remarkable agreement with the Seiberg–Witten solution has been found [16]. We shall see in this section that the connection of $SU(N)$ $\mathcal{N} = 2$ super Yang–Mills theory with Toda–Whitham hierarchies embodies in a natural way a recursive procedure to compute the instanton expansion of the effective prepotential up to arbitrary order.

To begin with, let us fix our conventions. We choose the basis $H_k = E_{k,k} - E_{k+1,k+1}$ for the Cartan subalgebra and $E_{k,j}, k \neq j$ for the raising and lowering operators. Let $\{\alpha_i\}_{i=1,\ldots,N-1}$ stand for the simple roots of $SU(N)$ and (α, β) denote the usual inner product constructed with the Cartan-Killing form. The dot product $\alpha \cdot \beta \equiv 2(\alpha, \beta)/(\beta, \beta) = (\alpha, \beta^\vee)$. We have that $\alpha_i \cdot \alpha_j = C_{ij}$, with C_{ij} the Cartan matrix, while $\lambda^i \cdot \alpha_j = \delta^i{}_j$ define the fundamental weights. In particular this means that $\alpha_i = \sum_j C_{ij} \lambda^j$. The simple roots generate the root lattice $\Delta = \{\alpha = n^i \alpha_i | n^i \in \mathbf{Z}\}$.

The instanton expansion of the prepotential was given in eq.(3). We then have, for the LHS of (43),

$$\frac{\partial^2 \mathcal{F}}{\partial(\log\Lambda)^2} = \frac{1}{2\pi i}\sum_{k=1}^{\infty}(2Nk)^2\mathcal{F}_k(Z)\Lambda^{2Nk} \ . \tag{47}$$

The derivative of the quadratic Casimir also has an expansion that can be obtained from the RG equation (first equation in (41)) and the expansion of the prepotential

$$\frac{\partial \mathcal{H}_2}{\partial a^i} = \frac{2\pi i}{\beta}\frac{\partial^2 \mathcal{F}}{\partial a^i \partial \log\Lambda} = C_{ij}a^j + \sum_{k=1}^{\infty} k\mathcal{F}_{k,i}\,\Lambda^{2Nk} \equiv \sum_{k=0}^{\infty} H_i^{(k)}\Lambda^{2Nk} \ , \tag{48}$$

where $\mathcal{F}_{k,i} = \partial\mathcal{F}_k/\partial a^i$. The term involving the couplings that appear in the Theta function Θ_E is

$$i\pi\, n^i\tau_{ij}n^j = \sum_{\alpha_+}\log\left(\frac{Z_\alpha}{\Lambda}\right)^{-(\alpha\cdot\alpha_+)^2} + \frac{1}{2}\sum_{k=1}^{\infty}\ (\alpha{\cdot}\mathcal{F}_k''{\cdot}\alpha)\ \Lambda^{2Nk} \ , \tag{49}$$

where $\alpha = n^i\alpha_i$ and

$$\alpha{\cdot}\mathcal{F}_k''{\cdot}\alpha \equiv \sum_{i,j} n^i\frac{\partial^2\mathcal{F}_k}{\partial a^i\partial a^j}n^j = \sum_{\beta,\gamma\in\Delta}(\alpha\cdot\beta)\frac{\partial^2\mathcal{F}_k}{\partial Z_\beta\partial Z_\gamma}(\gamma\cdot\alpha) \ . \tag{50}$$

For convenience, we have adjusted the bare coupling to $2\pi i\tau_0 = 3N$. We may shift τ_0 to any value by an appropriate rescaling of Λ. This will be reflected in the normalization of the $\mathcal{F}_k$.

Inserting (49) in the Theta function, we obtain

$$\begin{aligned}\Theta_E(0|\tau) &= \sum_{r=0}^{\infty}\sum_{\alpha\in\Delta_r}(-1)^{\rho\cdot\alpha}\prod_{\alpha_+}Z_{\alpha_+}^{-(\alpha\cdot\alpha_+)^2}\prod_{k=1}^{\infty}\left(\sum_{m=0}^{\infty}\frac{1}{2^m m!}(\alpha{\cdot}\mathcal{F}_k''{\cdot}\alpha)^m\ \Lambda^{2Nkm}\right)\Lambda^{2Nr}\\ &\equiv \sum_{p=0}^{\infty}\Theta^{(p)}\Lambda^{2Np} \ , \qquad\qquad \rho = \sum_{i=1}^{N-1}\lambda^i \ .\end{aligned} \tag{51}$$

In the previous expression, $\Delta_r \subset \Delta$ is a subset of the root lattice composed of those lattice vectors α that fulfill the constraint $\sum_{\alpha_+}(\alpha\cdot\alpha_+)^2 = 2Nr$. In particular Δ_1 is the root system, *i.e.* the simple roots together with their Weyl reflections. On the other hand Δ_r, for $r > 1$, will be in general a union of Weyl orbits, since Weyl reflections are easily seen to be an automorphisms of Δ_r. Therefore, $\Theta^{(p)}$ is Weyl invariant by construction. In the logarithmic derivative, Θ_E appears in the denominator, so we need the expansion of the inverse of the Theta function (see Ref. [13] for details):

$$\Theta(0|\tau)^{-1} = \sum_{l=0}^{\infty}\Xi^{(l)}(\Theta)\,\Lambda^{2Nl} \ . \tag{52}$$

Finally, the derivative of the Theta function with respect to the period matrix is given by

$$\begin{aligned}\frac{1}{i\pi}\partial_{\tau_{ij}}\Theta_E(0,\tau) &= \sum_{r=1}^{\infty}\sum_{\alpha\in\Delta_r}(-1)^{\rho\cdot\alpha}(\alpha\cdot\lambda^i)(\alpha\cdot\lambda^j)\prod_{\alpha_+}Z_{\alpha_+}^{-(\alpha\cdot\alpha_+)^2}\prod_{k=1}^{\infty}\exp\left(\frac{1}{2}(\alpha{\cdot}\mathcal{F}_k''{\cdot}\alpha)\Lambda^{2Nk}\right)\Lambda^{2Nr}\\ &\equiv \sum_{p=1}^{\infty}\Theta_{ij}^{(p)}\Lambda^{2Np} \ .\end{aligned} \tag{53}$$

Collecting all the pieces and inserting them back into (43), we find for $\mathcal{F}_k(Z)$ the following expression:

$$\mathcal{F}_k(Z) = -k^{-2}\sum_{p,q,l=0}^{p+q+l=k-1}\ \sum_{ij} H_i^{(p)}H_j^{(q)}\Theta_{ij}^{(k-p-q-l)}\Xi^{(l)} \ , \tag{54}$$

in terms of the previously defined coefficients. If we look at the coefficients in the r.h.s. of Eq.(54), it is easy to see that the expressions they involve depend on $\mathcal{F}_1, \mathcal{F}_2, \dots$ up to $\mathcal{F}_{k-1}$. In fact, although both $H^{(p)}$ and $\Theta^{(p)}$ depend on $\mathcal{F}_1, \dots \mathcal{F}_p$, the indices within parenthesis reach at most the value $k-1$ as $\Theta_{ij}^{(0)} = 0$. Moreover $\Theta_{ij}^{(k)}$ depends on $\mathcal{F}_1, \dots, \mathcal{F}_{k-1}$ since the vector $\alpha = 0$ is missing from the lattice sum. This fact implies the possibility to build up a recursive procedure to *compute all the instanton coefficients* by starting just from the perturbative contribution to $\mathcal{F}(a)$ in (3). The first few instanton contributions, for example, are simply [13]:

$$\mathcal{F}_1 = - \sum_{\alpha \in \Delta_1} (-1)^{\rho \cdot \alpha} Z_\alpha^2 \prod_{\alpha^+} Z_{\alpha^+}^{-(\alpha \cdot \alpha^+)^2} , \tag{55}$$

$$\mathcal{F}_2 = -\frac{1}{4} \left(\sum_{\alpha \in \Delta_1} (-1)^{\rho \cdot \alpha} \prod_{\alpha_+} Z_{\alpha_+}^{-(\alpha \cdot \alpha_+)^2} \left[\mathcal{F}_1 + 2(\alpha \cdot \mathcal{F}_1') Z_\alpha + \frac{1}{2} (\alpha \cdot \mathcal{F}_1'' \cdot \alpha) Z_\alpha^2 \right] \right.$$
$$\left. + \sum_{\beta \in \Delta_2} Z_\beta^2 (-1)^{\rho \cdot \beta} \prod_{\alpha^+} Z_{\alpha^+}^{-(\beta \cdot \alpha^+)^2} \right) , \tag{56}$$

$$\mathcal{F}_3 = -\frac{1}{9} \left(\sum_{\alpha \in \Delta_1} (-1)^{\rho \cdot \alpha} \prod_{\alpha^+} Z_{\alpha^+}^{-(\alpha \cdot \alpha^+)^2} \left[4\mathcal{F}_2 + 4(\alpha \cdot \mathcal{F}_2') Z_\alpha + (\alpha \cdot \mathcal{F}_1')^2 + \frac{1}{2} (\alpha \cdot \mathcal{F}_1'' \cdot \alpha) \left(\mathcal{F}_1 + 2(\alpha \cdot \mathcal{F}_1') Z_\alpha \right) \right. \right.$$
$$\left. + \frac{1}{8} (\alpha \cdot \mathcal{F}_1'' \cdot \alpha)^2 Z_\alpha^2 + \frac{1}{2} (\alpha \cdot \mathcal{F}_2'' \cdot \alpha) Z_\alpha^2 \right] + \sum_{\beta \in \Delta_2} (-1)^{\rho \cdot \beta} \prod_{\alpha_+} Z_{\alpha_+}^{-(\beta \cdot \alpha_+)^2} \left[\mathcal{F}_1 + 2(\beta \cdot \mathcal{F}_1') Z_\beta \right.$$
$$\left. \left. + \frac{1}{2} (\beta \cdot \mathcal{F}_1'' \cdot \beta) Z_\beta^2 \right] + \sum_{\gamma \in \Delta_3} (-1)^{\rho \cdot \gamma} \prod_{\alpha^+} Z_{\alpha^+}^{-(\gamma \cdot \alpha^+)^2} Z_\gamma^2 \right) , \tag{57}$$

etc. The above expressions make patent the recursive character of the procedure.

VI SOFT SUSY BREAKING WITH HIGHER CASIMIR OPERATORS

Sofly–broken supersymmetric models offer the best phenomenological candidates to solve the hierarchy problem in grand–unified theories. The *spurion formalism* [24] provides a tool to generate soft supersymmetry breaking in a neat and controlled manner. To illustrate the method, start from a supersymmetric lagrangian, $L(\Phi_0, \Phi_1, \dots)$ with some set of chiral superfields, and single out a particular one, say Φ_0. If you let this superfield acquire a constant *vev* along a given direction in superspace like, for example, $< \Phi_0 >= c_0 + \theta^2 F_0$, it will induce soft breaking terms, and a vacuum energy of order $|F_0|^2$. Turning the argument around, you could *promote* any parameter in your lagrangian to a chiral superfield, and then *freeze* it along a supersymmetry breaking direction in superspace giving a *vev* to its highest component.

From embedding the Seiberg–Witten solution within the Toda–Whitham framework, we have obtained the analytic depence of the prepotential on some new parameters T_n. In this section, we will interpret these slow times as parameters of a non-supersymmetric family of theories, by promoting them to spurion superfields. In Refs. [4,5,25] this program was initiated with the scale parameter Λ and the masses of additional hypermultiplets, m_i, as the only sources for spurions. The slow times, as Eq.(41) shows, are dual to the $\mathcal{H}_{m+1}$, which are homogeneous combinations of the Casimir operators of the group. This means that we will be able to parametrize soft supersymmetry breaking terms induced by all the Casimirs of the group, and not just the quadratic one. In this way, we shall extend to the $\mathcal{N} = 0$ case the family of $\mathcal{N} = 1$ supersymmetry breaking terms first considered by Argyres and Douglas [19].

We define the *spurion* variables s_n as follows

$$s_1 = -i \log \Lambda \qquad s_n = -i\hat{T}_n , \qquad n = 2, \dots, r = N-1. \tag{58}$$

Our independent coordinates in the prepotential are a^i, s_n. Using (37) one can find explicit expressions for the dual spurions:

$$s_D^1 = \frac{\beta}{2\pi}\left[\mathcal{H}_2 + i\sum_{m\geq 2} m s_m \mathcal{H}_{m+1} - \sum_{m,n\geq 2} m s_m s_n \mathcal{H}_{m+1,n+1}\right] ,$$

$$s_D^n = \frac{\beta}{2\pi n}\left[\mathcal{H}_{n+1} + i\sum_{m\geq 2} m s_m \mathcal{H}_{m+1,n+1}\right] . \tag{59}$$

Notice that, when the spurions s_m are zero, we recover for the variable s_1 the results of [5]. Under the symplectic group $\mathrm{Sp}(2r, \mathbf{Z})$, the spurions are taken to be scalars, $s_m^\Gamma = s_m$. From the point of view of the Toda–Whitham hierarchy, this is natural in that the slow times parametrize deformations of the curve, and should not be affected by duality transformations (which are transformations among symplectic basis of homology cycles of the curve). From the point of view of physics, this invariance is important because their *vev* is an external unambiguous input. We see from (59) that *the dual times are also invariant* under duality transformations.

To break $\mathcal{N}=2$ supersymmetry down to $\mathcal{N}=0$, as anticipated above, we promote the variables s_n to $\mathcal{N}=2$ vector superfields S_n, and then freeze the scalar and auxiliary components to constant vacuum expectation values. We would like to restrict our framework to non-supersymmetric deformations of the *original* pure $SU(N)$ super Yang–Mills theory. Thus, for all $S_n, n \geq 2$, we only keep the top components F_n as a supersymmetry breaking parameter (by $SU(2)_R$ symmetry, we can always rotate the D_n components away). In terms of $\mathcal{N}=1$ superfields we have,

$$S \equiv S_1 = s_1 + \theta^2 F_1 \qquad\qquad S_n = \theta^2 F_n , \qquad n \geq 2 , \tag{60}$$

where s_1 is related, as seen in (58) to the dynamical scale of the theory. The analysis of the soft breaking induced only by S_1 has been done in Refs. [4,5].

As the prepotential has an analytic dependence on the spurion superfields, the effective Lagrangian up to two derivatives and four fermion terms for the $\mathcal{N}=0$ theory is given by the exact Seiberg–Witten solution once the spurion superfields are taken into account. This gives the exact effective potential at leading order and the vacuum structure can be determined. That is, all over the quantum moduli space, the effective action will be

$$\mathcal{L}_{VM} = \frac{1}{4\pi}\mathrm{Im}\left[\int d^4\theta \frac{\partial F}{\partial \Phi^I}\bar{\Phi}^I + \frac{1}{2}\int d^2\theta \frac{\partial^2 F}{\partial\Phi^I \partial\Phi^J} W_\alpha^I W^{\alpha J}\right] , \tag{61}$$

where the capital indices I, J stand both for $i, j = 1, ..., N-1$ labelling abelian chiral $\mathcal{N}=2$ multiples (Φ^i, V^i), and for $m, n = 1, ..., N-1$ that label spurion multiplets (S^m, V_s^m). If we are near a submanifold of the moduli space of vacua where n_H hypermultiplets become massless, the full Lagrangian also contains the hypermultiplet contribution involving pairs of chiral superfields H_a, $\tilde{H}_a$, $a = 1, \ldots, n_H$,

$$\mathcal{L}_{HM} = \sum_a \int d^4\theta \left(H_a^* e^{2n_i^a V^i} H_a + \tilde{H}_a^* e^{-2n_i^a V^i}\tilde{H}_a\right) + \sum_{a,i}\left(\int d^2 \sqrt{2}\Phi^i n_i^a H_a \tilde{H}_a + \text{h.c.}\right) , \tag{62}$$

where the charge of the a-th hypermultiplet with respect to the i-th $U(1)$ factor has been denoted by n_i^a and, in the previous equation, a particular choice of duality frame, a^i, has been made. Namely the vector multiplets (Φ^i, V^i), are such that near the singular subvariety, the light BPS states in the previous lagrangian are weakly coupled, and perturbation theory is reliable. Of course, this amounts to an appropriate choice of the basis of homology cycles (A_i, B^j). Now, if the BPS states becoming massless are mutually local, we can always fix a basis of cycles such that each $U(1)$ couples to one and only one hypermultiplet. This means that $n_i^a = \delta_i^a$ or vanishes, the later case being possible when $n_H < N-1$.

The full effective lagrangian will be the sum of (61) and (62). The effective potential can be computed explicitely resulting in

$$V = B^{mn}F_m F_n^* + \sqrt{2}\,(n^a, b^m)\left(F_m \tilde{h}_a h_a + \bar{F}_m \bar{h}_a \bar{\tilde{h}}_a\right) + 2(n^a, n^b)(h_a \tilde{h}_a \bar{h}_b \bar{\tilde{h}}_b)$$
$$+ \frac{1}{2}(n^a, n^b)(|h_a|^2 - |\tilde{h}_a|^2)(|h_b|^2 - |\tilde{h}_b|^2) + 2|n^a \cdot a|^2(|h_a|^2 + |\tilde{h}_a|^2) , \tag{63}$$

where $n^a \cdot a = \sum_i n_i^a a^i$, and h_a ($\tilde{h}_a$) is the scalar component of H_a ($\tilde{H}_a$). Also we have used the quantities

$$(n^a, n^b) = n_i^a b^{-1\,ij} n_j^b \qquad (n^a, b^m) = n_i^a b^{-1\,ij} b_j{}^m \qquad B^{mn} = b_a{}^m b^{-1\,ab} b_b{}^n - b^{mn} \ , \tag{64}$$

where b is given in terms of the generalized $2(N-1) \times 2(N-1)$ matrix of couplings τ_{IJ},

$$\tau_{ij} = \frac{\partial^2 \mathcal{F}}{\partial a^i \partial a^j} \qquad \tau^n{}_i = \frac{\partial^2 \mathcal{F}}{\partial a^i \partial s_n} \qquad \tau^{mn} = \frac{\partial^2 \mathcal{F}}{\partial s_m \partial s_n} \ , \tag{65}$$

as

$$b_{IJ} = \frac{1}{4\pi} \mathrm{Im}\, \tau_{IJ} \ , \tag{66}$$

and it can be computed all over the moduli space from Eqs.(42)–(45). This is precisely the point where the Seiberg–Witten solution enters the calculations.

To obtain the values of the condensates, we first minimize V with respect to h_a, $\tilde{h}_a$, resulting in $|h_a| = |\tilde{h}_a|$. It is convenient to fix the gauge in the $U(1)^{N-1}$ factors in such a way that

$$h_a = \rho_a \qquad \tilde{h}_a = \rho_a e^{i\beta_a} \ . \tag{67}$$

If the charge vectors n^a are linearly independent, the non-trivial condensates satisfy the equation

$$|n^a \cdot a|^2 + \sum_b (n^a, n^b) \rho_b^2 e^{i(\beta_b - \beta_a)} + \frac{1}{\sqrt{2}} (n^a, b^m) F_m e^{-i\beta_a} = 0 \ , \tag{68}$$

and the effective potential takes the value

$$V = B^{mn} F_m F_n^* - 2 \sum_{ab} (n^a, n^b) \rho_a^2 \rho_b^2 \cos(\beta_a - \beta_b) \ . \tag{69}$$

We will consider in what follows, as an explicit example, the case of $SU(3)$.

VII ANALYSIS OF $SU(3)$

The $\mathcal{N} = 2$ super Yang-Mills theory with gauge group $SU(3)$ has been analyzed in detail in Refs. [3,19]. There are two sets of distinguished singularities in the moduli space of this theory:

i. The three $\mathbf{Z}_2$ vacua, known as $\mathcal{N} = 1$ or maximal points [18], located at $u^3 = 27\Lambda^6/4$, $v = 0$, that give rise to the $\mathcal{N} = 1$ vacua when the theory is perturbed with a mass term of the form $\mathrm{Tr}\Phi^2$ (we denote $u_2 = u$, $u_3 = v$).

ii. The two $\mathbf{Z}_3$ vacua, known as Argyres–Douglas (AD) points [19], located at $u = 0$ and $v = \pm 2\Lambda^3$, where three mutually nonlocal BPS states become simultaneously massless. The low-energy theory there is an $\mathcal{N} = 2$ superconformal theory.

We will briefly describe the situation near both kind of singularities. We set $\Lambda^6 = 4$ for convenience.

The $\mathbf{Z}_2$ vacua

In this subsection we study the soft breaking of the theory near the $\mathcal{N} = 1$ points where two magnetic monopoles become simultaneously massless. To evaluate the second derivatives of the prepotential, we need the values of the periods of the hyperelliptic curve and the structure of the gauge couplings. We will focus on the $\mathcal{N} = 1$ point ($u = 3$,$v = 0$), whereas the values of the quantities at the other two points can be obtained by using the $\mathbf{Z}_3$ unbroken symmetry. The derivatives of the Casimir operators with respect to the dual variables are given by [13]

$$\frac{\partial u}{\partial a_{D\,j}} = -2i \sin\frac{\pi j}{N} \qquad \frac{\partial v}{\partial a_{D\,j}} = -2i \sin\frac{2\pi j}{N} \ . \tag{70}$$

The gauge couplings near the $\mathcal{N}=1$ point have the structure

$$\tau^D_{ij} = \frac{1}{2\pi i}\log\left(\frac{a_{D\,i}}{2\sqrt{3}}\right)\delta_{ij} + (1-\delta_{ij})\tau^{\rm off}_{ij} + \mathcal{O}(a_{D\,i}) \ , \tag{71}$$

where $\tau^{\rm off}_{ij}$, $i \neq j$ are the off-diagonal entries of the coupling constant at the $\mathcal{N}=1$ point, which will be discussed in some detail in the next section. For $SU(3)$, $\tau_{12} = \frac{i}{\pi}\log 2$ [3]. To compute the Theta function, we have to take into account the change of the *electric* characteristic under the symplectic transformation to the magnetic variables, $a_{D\,i}$. Indeed, the transformation law for the Theta function is given by [26]

$$\Theta[\alpha^\Gamma, \beta^\Gamma](\tau^\Gamma|\xi^\Gamma) = e^{i\phi}(\det(C\tau+D))^{1/2}\exp\left[\pi i \xi^t (C\tau+D)^{-1}C\xi\right]\,\Theta[\alpha,\beta](\tau|\xi) \ , \tag{72}$$

where ϕ is a ξ-independent phase and

$$\alpha^\Gamma = D\alpha - C\beta + \frac{1}{2}{\rm diag}(CD^t) \qquad \beta^\Gamma = -B\alpha + A\beta + \frac{1}{2}{\rm diag}(AB^t) \ . \tag{73}$$

Thus, the *magnetic* or *dual* characteristic, D, is

$$\vec{\alpha} = (1/2, 1/2) \qquad \vec{\beta} = (0,0) \ . \tag{74}$$

From the leading behaviour of the Theta function with dual characteristic, we can compute its derivative at the $\mathcal{N}=1$ point of $SU(3)$,

$$\frac{1}{i\pi}\partial_{\tau_{ij}}\log\Theta_D(0|\tau^D)\Big|_{a_{D\,i}=0} = \frac{1}{4}\delta_{ij} - \frac{1}{12}(1-\delta_{ij}) \ . \tag{75}$$

Using (70) and (75), it is easy to check the relation (46) for the Λ derivatives of the Casimir operators.

At the $\mathcal{N}=1$ point, there is a symplectic basis for the hyperelliptic curve, such that the magnetic charge vectors are given by $n^a_j = \delta^a_j$, $a_{D\,i}=0$, and from eq.(68), the condensates are given by

$$\rho_1^2 = \sqrt{\frac{3}{2}}\frac{3}{4\pi^2}|F_1 + \frac{1}{2}F_2| \qquad \rho_2^2 = \sqrt{\frac{3}{2}}\frac{3}{4\pi^2}|F_1 - \frac{1}{2}F_2| \ . \tag{76}$$

We see that the soft breaking induced by the quadratic and cubic Casimirs gives rise to monopole condensation in both $U(1)$ factors, although the condensates are bigger for the soft breaking coming from u (for equal values of the supersymmetry breaking parameters F_1, F_2). In the same way, the vacuum energy associated to these condensates is

$$V_{\rm eff} = -b^{mn}F_m F^*_n = -\frac{9}{4\pi^2}\left(|F_1|^2 + \frac{1}{2}|F_2|^2\right) \ . \tag{77}$$

As expected, the soft breaking associated to u gives lower energy to this vacuum.

The Z_3 vacua

Next we explore the behaviour near the Argyres–Douglas point at ($u=0$, $v=4$). It is convenient to use the parameters ρ and ϵ already introduced in Ref. [19],

$$u = 3\epsilon^2\rho \qquad v - 4 = 2\epsilon^3 \ . \tag{78}$$

The three submanifolds $\rho^3 = 1$ correspond to three massless BPS states which after an appropriate symplectic transformation can be seen to be charged with respect to only one of the $U(1)$ factors, with variables denoted by a^1, a_{D1}. They can be seen to be an electron, a dyon, and a monopole. These submanifolds come together at the AD point, where a nontrivial superconformal field theory is argued to exist [19]. To leading order, the hyperelliptic curve splits at the AD point into a small torus (corresponding to two mutually nonlocal periods a^1, a_{D1} which go to zero) and a big torus with periods $a^2, a_{D2} \sim \Lambda$. The small torus is given by the elliptic curve

$$w^2 = z^3 - 3\rho z - 2 , \tag{79}$$

and the meromorphic Seiberg–Witten differential degenerates on it to

$$\lambda_{SW} = \frac{1}{2\sqrt{2}\pi} \epsilon^{5/2} w dz . \tag{80}$$

The matrix of couplings near the AD point, at leading order, reads [19,27,28]

$$\tau_{11} = \tau(\rho) + \mathcal{O}(\epsilon) \qquad \tau_{12} = -\frac{i}{c}\frac{\epsilon^{1/2}}{\omega_\rho} + \mathcal{O}(\epsilon^{3/2}) \qquad \tau_{22} = \omega + \mathcal{O}(\epsilon) , \tag{81}$$

where ω_ρ is the period of the small torus (with $\mathrm{Im}(\omega_{\rho D}/\omega_\rho) > 0$), c is a nonzero constant and $\omega = \mathrm{e}^{\pi i/3}$. For the dual variables we have similar expressions with $\omega_{\rho D}$ and c_D.

To analyze the Theta function in these variables, we need the appropriate characteristic which, using (73) and the results in [27], is

$$\vec{\alpha} = \vec{\beta} = (1/2, 1/2) . \tag{82}$$

We can already obtain the behaviour of the Theta function as an expansion in ϵ:

$$\Theta(0|\tau) = -\frac{1}{2\pi c}\frac{\epsilon^{1/2}}{\omega_\rho} \vartheta_1'(0|\tau(\rho))\vartheta_1'(0|\omega) + \mathcal{O}(\epsilon^{3/2}) , \tag{83}$$

where $\vartheta_1(\xi|\tau)$ is the Jacobi theta function with characteristic $[1/2, 1/2]$. Now, using that

$$\frac{\vartheta_1'''(0|\tau)}{\vartheta_1'(0|\tau)} = -\pi^2 E_2(\tau) , \tag{84}$$

we find the leading contribution to the derivative of the Theta function

$$\frac{1}{i\pi}\partial_{\tau_{ij}} \log\Theta = \begin{pmatrix} \frac{1}{4}E_2(\tau(\rho)) & \frac{c}{2\pi}\epsilon^{-1/2}\omega_\rho \\ \frac{c}{2\pi}\epsilon^{-1/2}\omega_\rho & \frac{1}{4}E_2(\omega) \end{pmatrix} . \tag{85}$$

Again, it can be checked that the relation (46) for v holds (for u, it is necessary to know the explicit values of the constants).

The analysis of the condensates near the AD point is difficult because one has to take into account mutually nonlocal degrees of freedom, and there is not a Lagrangian description of this theory. In fact, one expects that, in the softly broken theory, a cusp singularity will appear in the effective potential near the AD point, as it happens in $\mathcal{N} = 2$ QCD with gauge group $SU(2)$ and one massive flavour [25]. But we can analyze the monopole condensates along the divisors $\rho^3 = 1$ and their evolution as we approach the AD point. Near each of the submanifolds $\rho^3 = 1$ there is a massless BPS state, and we expect it to condense after breaking supersymmetry down to $\mathcal{N} = 0$. These condensates correspond to mutually nonlocal states but we can assume, following the discussion in Refs. [5,25], that these states do not interact, the condensates being given by the equation [13]

$$\rho_k^2 = -\frac{1}{(b^{-1})_{11}}|a_k|^2 - \frac{\mathrm{e}^{-i\beta_k}}{\sqrt{2}(b^{-1})_{11}} \sum_{n=1,2} F_n (b^{-1})_{kj} b^n_j , \tag{86}$$

where $k = 1, 2, 3$ and a_k are the appropriate local coordinates for each of the massless states (*i.e.* $a_k = a^1$, a_{D1}, $a^1 - a_{D1}$). The quantities $(b^{-1})_{ij}$, $b^n{}_j$ should be also computed in the duality frame dictated by the a_k. This approximation should be good far enough from the AD point. These condensates give only a magnetic Higgs mechanism in one of the $U(1)$ factors, and correspond to the half-Higgsed vacua of [19]. Notice that one should perform a careful numerical study of the equations for the condensates and for the effective potential to know if these partial condensates give the true vacua of the $\mathcal{N} = 0$ theory. As we approach the AD point, $\epsilon \to 0$, we see that the parameters for condensation go to zero for both the quadratic and the cubic Casimirs:

$$\frac{\partial u}{\partial a^1}, \; \frac{\partial v}{\partial a^1} \sim \mathcal{O}(\epsilon^{1/2}), \tag{87}$$

and the mass gap associated to the condensates vanishes at the AD point, like in the $\mathcal{N} = 1$ breaking considered in Ref. [19].

VIII STRONG COUPLING EXPANSION NEAR THE MAXIMAL POINTS

Let us end by applying the Seiberg–Witten–Whitham equations in the strong coupling regime of $\mathcal{N} = 2$ super Yang-Mills theory near its maximal singularities. The case of $SU(2)$ is special in that, being its Cartan subalgebra one–dimensional, the whole strong coupling expansion of the prepotential can be recursively computed *without* an explicit knowledge of the actual solution $(a(u), a_D(u))$ [14], much in the same way than the previously derived instanton corrections. For generic $SU(N)$, however, the SWW equations do not give a closed procedure to obtain the strong coupling expansion of the effective prepotential. Aside from some technical difficulties, the main problem is that, in spite of the fact that a grading in $a_{D\,i}$ and Λ still exists, higher terms of the expansion appear in the equations corresponding to lowest powers of the dual variables spoiling recursivity (see Ref. [14] for details). The SWW equations do not seem to be instrumental to study the full strong coupling expansion of $SU(N)$ $\mathcal{N} = 2$ super Yang–Mills theory.

Other methods have been derived in the literature to tackle the problem of computing the higher threshold corrections to the effective prepotential. For example, in Ref. [29] this has been accomplished by parametrizing the neigborhood of the maximal singularities with a family of deformations of the corresponding auxiliary (singular) Riemann manifold. However, this formalism is not sensitive to quadratic terms in the prepotential. Thus, in particular, it does not give an answer for the couplings between different magnetic $U(1)$ factors at the maximal singularities of the moduli space, $\tau_{ij}^{\rm off}$, introduced in the previous section. The existence and importance of such terms has been first pointed out in Ref. [18] by using a scaling trajectory that smoothly connects the maximal singularities with the semiclassical region. These terms are also important ingredients in the expression of the Donaldson–Witten functional for gauge group $SU(N)$ [27]. To our knowledge, a closed formula for these off-diagonal couplings has not been obtained so far, except for the gauge group $SU(3)$ [3]. Let us then consider the uses of the SWW equations in the solution of this problem.

We have seen before that physical quantities in the neighborhood of any maximal singularity can be translated to a patch in the vicinity of any other by the action of the unbroken discrete subgroup $\mathbf{Z}_N$. We will consider in what follows the point where u_2 is real and positive. The strong coupling expansion of the prepotential at such singular point can be written in terms of appropriate $a_{D\,i}$ variables as[1]

$$\mathcal{F} = \frac{N^2}{2\pi i}\Lambda^2 + \frac{2N\Lambda}{\pi}\sum_{k=1}^{N-1} a_{D\,k} \sin\hat{\theta}_k + \frac{1}{4\pi i}\sum_{k=1}^{N-1} a_{D\,k}^2 \log\frac{a_{D\,k}}{\tilde{\Lambda}_k} + \frac{1}{2}\sum_{k\neq l=1}^{N-1} \tau_{kl}^{\rm off}\, a_{D\,k} a_{D\,l} + \frac{1}{2\pi i}\sum_{s=1}^{\infty} \mathcal{F}_s(a_D)\Lambda^{-s} \,,$$

where $\hat{\theta}_k = \pi k/N$ and the logarithmic term, coming from the one-loop diagram that involves the light monopole, has the appropriate sign and factor making manifest that the theory is non-asymptotically free and that there is a monopole hypermultiplet weakly coupled to each dual photon for $a_{D\,i} \to 0$. The remaining power series expansion comes from the integration of infinitely many massive BPS states: $\mathcal{F}_s(a_D)$ are polynomials of degree $s+2$ in dual variables and $\tilde{\Lambda}_k = e^{3/2}\Lambda \sin\hat{\theta}_k$.

We have seen that the SWW formalism allows us to relate the strong coupling expansion of homogeneous combinations of higher Casimir operators $\mathcal{H}_m$ with that of the prepotential through Eq.(46). Let us first

1) We follow here the conventions of Ref. [18] to fix the first three terms of the expansion.

remark that this equation is also valid for the higher Casimirs h_n themselves [12]: they, as well as their particular combinations encoded in $\mathcal{H}_n$, are homogeneous functions of a_D and Λ of degree n. Thus, at the $\mathcal{N}=1$ singularities, the LHS of Eq.(46) is simply

$$\frac{\partial h_n}{\partial \log \Lambda} = n h_n = \sum_{k=1}^{N} (2\cos\theta_k)^n \ , \tag{88}$$

where we used the fact that the eigenvalues of ϕ are $\phi_i = 2\cos\theta_i$ with $\theta_i = (i-1/2)\pi/N$ [18]. The derivative of the Casimir operators with respect to the dual variables can be computed at the same point of the moduli space, by using the explicit representation of the curve in terms of the Chebyshev polynomials [13], resulting in

$$\frac{\partial h_n}{\partial a_{D\,j}} = -2i \sum_{l=0}^{[n/2-1]} \binom{n-1}{l} \sin(n-2l-1)\hat{\theta}_j \ , \tag{89}$$

whereas, from the expansion of the effective prepotential, the leading couplings at the maximal singularity are given by

$$\tau^D_{ij} = \frac{1}{2\pi i} \log\left(\frac{a_{D\,i}}{\Lambda_i}\right) \delta_{ij} + \tau^{\text{off}}_{ij} \ . \tag{90}$$

The derivative of the Theta function Θ_D with respect to the period matrix has the following expression when evaluated at the $\mathcal{N}=1$ singularity

$$\frac{1}{i\pi}\partial_{\tau^D_{ij}} \log \Theta_D(0,\tau_D) = \frac{1}{4}\left(\sum_{\xi^k=\pm 1} \exp\left(i\frac{\pi}{4}\xi^l \tau^{\text{off}}_{lm}\xi^m\right)\right)^{-1} \sum_{\xi^k=\pm 1} \xi^i\xi^j \exp\left(i\frac{\pi}{4}\xi^l\tau^{\text{off}}_{lm}\xi^m\right) . \tag{91}$$

Now, we can insert the results (88)–(91) into the SWW equations (46) obtaining

$$\begin{aligned} \frac{1}{2N}\sum_{k=1}^{N}(2\cos\theta_k)^n = \sum_{l=0}^{[n/2-1]} \binom{n-1}{l} \sin\hat{\theta}_i \sin(n-2l-1)\hat{\theta}_j \left(\sum_{\xi^k=\pm 1} \exp\left(i\frac{\pi}{4}\xi^l\tau^{\text{off}}_{lm}\xi^m\right)\right)^{-1} \\ \times \sum_{\xi^k=\pm 1} \xi^i\xi^j \exp\left(i\frac{\pi}{4}\xi^l\tau^{\text{off}}_{lm}\xi^m\right) . \end{aligned} \tag{92}$$

We have $N-1$ equations and $(N-1)(N-2)/2$ unknowns (the components of the symmetric matrix τ^{off}_{ij}). Thus, Eq.(92) has predictive power in its own only for $SU(3)$ and $SU(4)$. Indeed, we obtain for these two cases the following values:

$$SU(3): \quad \tau^{\text{off}}_{12} = i/\pi \log 2 \qquad\qquad SU(4): \quad \begin{cases} \tau^{\text{off}}_{12} = \tau^{\text{off}}_{23} = -i/\pi \log(\sqrt{2}-1) \\ \tau^{\text{off}}_{13} = i/\pi \log\sqrt{2} \end{cases} . \tag{93}$$

Notice that our result for $SU(3)$ coincides with that of Ref. [3] while the ones for $SU(4)$ have not been found previously. For higher $SU(N)$, further ingredients would be necessary in order to obtain the off-diagonal couplings at the $\mathcal{N}=1$ singularity. Instead, we can think of Eq.(92) as a *new* constraint that τ^{off}_{mn} must obey. In fact, inspired by the findings in the last section of Ref. [18], we *propose* the following ansatz for τ^{off}_{mn}:

$$\tau^{\text{off}}_{mn} = \frac{2i}{N^2\pi} \sum_{k=1}^{N-1} \sin k\hat{\theta}_m \sin k\hat{\theta}_n \sum_{i,j=1}^{N} \tau^{(0)}_{ij} \cos k\theta_i \cos k\theta_j \ , \tag{94}$$

with $\tau^{(0)}_{ij}$ being given by

$$\tau_{ij}^{(0)} = \delta_{ij} \sum_{k \neq i} \log(2\cos\theta_i - 2\cos\theta_k)^2 - (1-\delta_{ij}) \log(2\cos\theta_i - 2\cos\theta_j)^2 \,. \tag{95}$$

There is no equivalent expression available in the literature to compare with. Nevertheless, we can use precisely the SWW equations (92) in order to make a non-trivial check of our ansatz for the off-diagonal couplings (94)–(95). We have done it numerically up to SU(11) with remarkable sucess [14]. There is a second check that we can do using results that do not rely on Whitham equations at all. Douglas and Shenker showed that the matrix τ^D_{mn} at any point of the scaling trajectory, diagonalizes in the basis $\{\sin k\hat{\theta}_n\}$ with certain particular eigenvalues (see Eqs.(5.9)–(5.12) of Ref. [18]). The couplings (94)–(95) satisfy this restrictive condition in the limit of the scaling trajectory ending at the maximal singularity. As long as our solution (94)–(95) matches two very stringent and independent conditions, we believe that it provides a faithful answer for τ^{off}_{mn} as well as a highly non-trivial test of the Seiberg–Witten–Whitham formalism.

ACKNOWLEDGMENTS

We are indebted to Marcos Mariño for many insightful discussions. J.D.E. would like to thank the organizers of the Meeting *Trends in Theoretical Physics II*, for giving him the opportunity to present these results. The work of J.D.E. is supported by a fellowship of the Ministry of Education and Culture of Spain. The work of J.M. was partially supported by DGCIYT under contract PB96-0960.

REFERENCES

1. N. Seiberg and E. Witten, Nucl. Phys. **B 426** (1994) 19 [Erratum *ibid.* **B 430** (1994) 485]; Nucl. Phys. **B 431** (1994) 484.
2. A. Klemm, W. Lerche, S. Theisen and S. Yankielowicz, Phys. Lett. **B 344** (1995) 169; P.C. Argyres and A.E. Faraggi, Phys. Rev. Lett. **74** (1995) 3931.
3. A. Klemm, W. Lerche and S. Theisen, Int. J. of Mod. Phys. **A 11** (1996) 1929.
4. L. Álvarez-Gaumé, J. Distler, C. Kounnas and M. Mariño, Int. J. Mod. Phys. **A 11** (1996) 4745.
5. L. Álvarez-Gaumé and M. Mariño, Int. J. Mod. Phys. **A 12** (1997) 975.
6. N. Evans, S.D.H. Hsu and M. Schwetz, Nucl. Phys. **B 484** (1997) 124.
7. A. Gorsky, I.M. Krichever, A. Marshakov, A. Mironov and A. Morozov, Phys. Lett. **B 355** (1995) 466.
8. E. Martinec and N.P. Warner, Nucl. Phys. **B 459** (1996) 97; T. Nakatsu and K. Takasaki, Mod. Phys. Lett. **A 11** (1996) 157.
9. G.B. Whitham, *Linear and Nonlinear waves*, Wiley-Interscience, New York, 1974.
10. I. Krichever, `hep-th/9205110`.
11. R. Carroll, `hep-th/9802130`; `hep-th/9804086`.
12. A. Gorsky, A. Marshakov, A. Mironov and A. Morozov, Nucl. Phys. **B 527** (1998) 690.
13. J.D. Edelstein, M. Mariño and J. Mas, Nucl. Phys. **B 541** (1999) 671.
14. J.D. Edelstein and J. Mas, `hep-th/9901006`.
15. K. Takasaki, `hep-th/9901120`; J. Edelstein, M. Gómez–Reino and J. Mas, work in progress.
16. V.V. Khoze, M.P. Mattis and M.J. Slater, Nucl. Phys. **B 536** (1998) 69 and reference therein.
17. P.A.M. Dirac, Proc. Roy. Soc. **A 33** (1931) 60; J. Schwinger, Phys. Rev. **144** (1966) 1087; *ibid* **173** (1968) 1536; D. Zwanziger, Phys. Rev. **176** (1968) 1480.
18. M.R. Douglas and S.H. Shenker, Nucl. Phys. **B 447** (1995) 271, `hep-th/9503163`.
19. P.C. Argyres and M.R. Douglas, Nucl. Phys. **B 448** (1995) 93.
20. H. Flaschka, M.G. Forest and D.W. McLaughlin, Comm. Pure. Appl. Math. **23** (1980) 739.
21. I. Krichever, Funct. Anal. and Appl. **22** (1988) 200.
22. H. Itoyama and A. Morozov, Nucl. Phys. **B 477** (1996) 855; *ibid* **B 491** (1997) 529.
23. M. Matone, Phys. Lett. **B 357** (1995) 342; T. Eguchi and S-K. Yang, Mod. Phys. Lett. **A 11** (1996) 131; J. Sonnenschein, S. Theisen and S. Yankielowicz, Phys. Lett. **B 367** (1996) 145.
24. L. Girardello and M.T. Grisaru, Nucl. Phys. **B 194** (1982) 65; J.A. Helayël-Neto, Phys. Lett. **B 135** (1984) 78.
25. L. Alvarez-Gaumé, M. Mariño and F. Zamora, Int. J. of Mod. Phys. **A 13** (1998) 403; *ibid* **A 13** (1998) 1847; M. Mariño and F. Zamora, Nucl. Phys. **B 533** (1998) 373.
26. H.E. Rauch and H.M. Farkas, *Theta functions with applications to Riemann surfaces*, Williams and Wilkins, 1974.

27. M. Mariño and G. Moore, Commun. Math. Phys. **199** (1998) 25.
28. T. Kubota and N. Yokoi, Prog. Theor. Phys. **100** (1998) 423.
29. E. D'Hoker and D.H. Phong, Phys. Lett. **B 397** (1997) 94.

Spin network quantum gravity

Jorge Griego

Instituto de Física, Facultad de Ciencias
Montevideo, Uruguay
griego@fisica.edu.uy

Abstract. The nonperturbative canonical quantization of General Relativity is successful in providing a physical interpretation of spacetime at the Planck scale. During the last years substantial advances have been achieved which consolidated the mathematical framework of the theory. Spin networks have played a crucial role in the progress of the field.

In this article we make a brief review of the canonical quantization method of gravity, putting a special emphasis in the use of the spin networks. Also the basis for a new formulation of quantum gravity in terms of the Vassiliev invariants are introduced. This new formulation is motivated by some fundamental problems that appear at the dynamical level of the theory.

I INTRODUCTION

The spin networks were introduced by Roger Penrose [1] in the early 70's in an attempt to develop a combinatorial approach of space-time. In this approach the spin networks are defined as trivalent abstract graphs consisting in a set of edges labelled by numbers (the color of the edges) which intersect in some points, called the vertices of the graph. At the vertices the rules of recoupling theory of angular momentum are satisfied.

In quantum gravity the spin networks are embedded in the three dimensional manifold. It turns out that the spin network wavefunctions code a complete set of independent quantum states of gravity, the spin network basis. The introduction of this basis has allowed to solve important problems at the kinematical and dynamical levels of the theory.

In an historic persective, the main steps of the canonical quantization program of gravity can be summarized as follows:

1. The Hamiltonian formulation of General Relativity (Arnowitt, Deser and Misner; 1962 [2]), where spacetime is splitted into space plus time. The canonical variables are the three metric q_{ab} and its conjugated momentum π^{ab}.

2. The canonical quantization of the gravitational field (à la Dirac). This procedure ends with the Wheeler-DeWitt (WdW) equation (1967) [3]. This equation is extremely complicated and no general solution to it is known in the metric representation.

3. The introduction of new variables by Ashtekar (1987) [4] to describe the phase space of General Relativity. The WdW equation becomes a polynomial function of the new variables.

4. The construction of the loop representation by Rovelli and Smolin (1990) [5]. In this representation the WdW equation can be solved exactly.

5. The introduction of the spin network basis by Rovelli and Smolin (1995) [6]. This fact puts on very solid basis the kinematical framework of the theory.

6. The Thiemann's formulation of the WdW equation (1996) [7].

CP484, *Trends in Theoretical Physics II*, edited by H. Falomir, R. E. Gamboa Saraví, and F. A. Schaposnik

The objective of this article is twofold: for one hand we review the main results obtained in the canonical quantization program of gravity based in the Ashtekar new variables and, on the other, we introduce the basis for the formulation of quantum gravity in a new space, the space of Vassiliev invariants. The Vassiliev invariants are associated with the Chern-Simons form and its introduction as a new scenario for quantum gravity is motivated by some fundamental problems that appear in the Thiemann's formulation of the WdW equation.

The article is organized as follows: in Sect. 2 we briefly present the main ideas of the Ashtekar new variables and of the connection and the loop representations. In Sect. 3 we introduce the spin network basis and we show how some important geometric quantities (the area and the volume operators) are realized in this basis. We also consider the Thiemann's formulation of the WdW equation. In Sect. 4 we introduce the principles for a realization of quantum gravity in the arena of Vassiliev invariants.

II PRELIMINARIES

A Some results from the ADM formalism

In the 3+1 decomposition of space-time the symmetries of the theory are expressed in the form of constraints. There are two constraints: the diffeomorphism and the Hamiltonian constraints. The diffeomorphism constraint C_a is given by the covariant derivative of the momentum:

$$C_a = 2D_b\pi^b_a\,. \qquad (1)$$

The smeared form of the constraint by a field $\vec{N}(x)$ is

$$C(\vec{N}) = \int d^3x N^a(x)C_a(x)\,, \qquad (2)$$

and the condition $C(\vec{N}) = 0$ express in the canonical languaje the invariance under spatial diffeomorphisms of General Relativity. The Hamiltonian has a more complicated expression:

$$H = -detqR^{(3)} + (\pi^{ab}\pi_{ab} - \tfrac{1}{2}\pi^2)\,, \qquad (3)$$

where *detq* is the determinant of the three metric and $R^{(3)}$ is the scalar curvature. The constraint

$$H(N) = \int d^3x N(x)H(x) = 0 \qquad (4)$$

is associated with the invariance under time reparametrizations of the theory. They form a first class system of constraints:

$$\{C(\vec{N}), C(\vec{M})\} = C(\mathcal{L}_{\vec{N}}\vec{M}) \qquad (5)$$

$$\{C(\vec{N}), H(M)\} = H(\mathcal{L}_{\vec{N}}M) \qquad (6)$$

$$\{H(N), H(M)\} = C(\vec{K})\,, \qquad (7)$$

where $\mathcal{L}$ indicates the Lie derivative and $K^a := detqq^{ab}(N\partial_b M - M\partial_b N)$. Notice that $\vec{K}$ depends on the configuration variable (the metric). This means that (7) is not a proper algebra and that we have to take care of possible anomalies in the quantization procedure. At the quantum level the classical variables are promoted to operators in the usual way,

$$\hat{q}_{ab}\,\psi(q) = q_{ab}\psi(q)\,, \qquad (8)$$

$$\hat{\pi}_{ab}\,\psi(q) = -i\frac{\delta}{\delta q_{ab}}\psi(q)\,. \qquad (9)$$

The physical states are defined by imposing the quantum constraints:

$$\hat{C}(N)\,\psi(q) = 0\,, \qquad (10)$$

$$\hat{H}(N)\,\psi(q) = 0\,. \qquad (11)$$

The first equation says that the wavefunctions depend only on the three geometry and the last is the Wheeler-deWitt equation. The above equations define the physical Hilbert space in the metric representation. Now we introduce another representation where the configuration variable is a connection.

B The Ashtekar new variables

The classical phase space of gravity can be described by a new set of variables which have the particularity of casting General Relativity as a Yang-Mills theory. To introduce the new variables we make first a $3+1$ decomposition of spacetime with an internal $SO(3)$ symmetry. The internal $SO(3)$ symmetry allows to express the canonical variables in terms of a set of triads[1] E^a_i and fields K^i_a. The configuration fields $K^i_a := K_{ab}E^{bi}$ are given by the contraction of the triad with the the extrinsic curvature K_{ab}. We now perform the canonical transformation:

$$E^a_i = E^a_i \tag{12}$$
$$A^i_a = \Gamma^i_a - iK^i_a , \tag{13}$$

where Γ^i_a is the spin connection compatible with the triad. Notice the imaginary factor i in the definition of A^i_a. This new variable is a complex $(SU(2))$ connection. The pair (A^i_a, E^a_i) is canonical and the constraints in terms of the new variables are:

$$G^i(A,E) = D_a E^{ai} , \tag{14}$$
$$C_a(A,E) = E^b_i F^i_{ab} , \tag{15}$$
$$H(A,E) = \epsilon^{ij}_k E^a_i E^b_j F^k_{ab} , \tag{16}$$

where F^i_{ab} is the curvature of the Ashtekar connection. The last four equations correspond to the usual diffeomorphism and Hamiltonian constraints. The first three equations are extra constraints that stem from the use of triads as fundamental variables. These equations, which have the same form as a Gauss law of an $SU(2)$ Yang-Mills theory, tell us that the formalism is invariant under rotations of the triads.

Notice the huge simplification of the Hamiltonian constraint, which is now a polynomial function of the canonical variables (of quadratic order in each variable). This dramatic simplification, together with the fact that the phase space is the same of a (complex) $SU(2)$ Yang-Mills theory, are the main features of this formulation. The use of complex fields implies that one has to impose reality conditions to recover real results. In spite that this peculiartity does not introduce significative problems at the classical level, in the quantum regime it poses a nontrivial obstacle.

C The connection representation

The casting of General Relativity as a theory of a connection has important implications at the quantum level. The first obvious step is to quantize the theory picking a polarization where the wavefunctions are functions of the connection. The Gauss constraint immediately says that the wavefunctions are gauge invariant, that is, they belong to the space $\mathcal{A}/\mathcal{G}$ of connections modulo gauge transformations. The Poisson algebra of the classical variables can be achieved promoting the connection as a multiplicative operator and the triad as a functional derivative:

$$\hat{A}^i_a \, \psi(A) = A^i_a \, \psi(A) \tag{17}$$
$$\hat{E}^a_i \, \psi(A) = -i \frac{\delta}{\delta A^i_a} \psi(A) . \tag{18}$$

Two factor orderings were considered: the triads to the right and the triads to the left. In the first case we write

$$\hat{G}^i = D_a \frac{\delta}{\delta A^i_a} , \tag{19}$$
$$\hat{C}_a = F^i_{ab} \frac{\delta}{\delta A^i_b} , \tag{20}$$
$$\hat{H} = \epsilon^{ij}_k F^k_{ab} \frac{\delta}{\delta A^j_a} \frac{\delta}{\delta A^k_b} . \tag{21}$$

---

[1] The $a, b, \cdots$ represent space indices, and $i, j, \cdots$ are internal $SO(3)$ indices

This order was first considered by Jacobson and Smolin [8] because the Gauss law and the diffeomorphism constraint generate (formally) gauge and diffeomorphism transformations over the wavefunctions. Solutions to the Hamiltonian constraint are obtained in terms of the Wilson loops (the trace of the holonomy),

$$W_A(\gamma) = Tr[U_A(\gamma)] := Tr[Pe^{\oint_\gamma dy^a A_a(y)}]. \tag{22}$$

Using the properties of the holonomy under gauge transformations it is easy to see that the Wilson loops are gauge invariant quantities. They are also annihilated by the Hamiltonian operator for smooth nonintersecting loops (due to symmetry considerations) [8,11]. However, they are not in the kernel of the diffeomorphism constraint: when a diffeomorphism acts on a Wilson loop, the result is a Wilson loop with the loop displaced by the diffeomorphism. By this reason the Wilson loops cannot be considered candidates for physical states of quantum gravity. In spite of this, the fact that these quantities are annihilated by the Hamiltonian operator had historical relevance. Noticer that up to this discovery no solution to the WdW equation was known in a general case (without making mini-superspace approximations). This result motivated the interest for loops which finally led to the loop representation.

In the factor ordering that puts the triads to the left it is possible to show that the diffeomorphism operator continues generating diffeomorphisms over the wavefunctions (in this case one has to introduce a suitable regularization [9]). In this factor ordering it is useful to consider the Hamiltonian constraint with cosmological constant $H_\Lambda := H + \Lambda/6detq$. The quantum operator is,

$$\hat{H}_\Lambda = \epsilon_k^{ij} \frac{\delta}{\delta A_a^j} \frac{\delta}{\delta A_b^k} F_{ab}^k - \frac{\Lambda}{6} \epsilon^{ijk} \epsilon_{abc} \frac{\delta}{\delta A_a^i} \frac{\delta}{\delta A_b^j} \frac{\delta}{\delta A_c^k} \,. \tag{23}$$

The reason to consider (23) is that there exists a very rich state which is gauge and diffeomorphism invariant and that is annihilated by $\hat{H}_\Lambda$. This state is given by the exponential of the Chern-Simons form built with the Ashtekar connection:

$$\Psi_\Lambda(A) := \exp(-\frac{6}{\Lambda} \int d^3x \epsilon^{abc} Tr[A_a \partial_b A_c + 2/3 A_a A_b A_c]) \,. \tag{24}$$

It is a well known fact that $\Psi_\Lambda(A)$ is gauge and diffeomorphism invariant, and using the property:

$$\hat{E}_i^a \, \Psi_\Lambda(A) = \frac{3}{\Lambda} \epsilon^{abc} F_{bc}^i \, \Psi_\Lambda(A) \,, \tag{25}$$

it is very easy to verify that $\hat{H}_\Lambda \, \Psi_\Lambda(A) = 0$. This nontrivial solution to all the constraints has played a prominent role in the progress made in finding physical states of gravity, and has opened up new connections between general relativity, topological field theories and knot theory [10].

D The loop representation

Using the properties of the Wilson loops one can construct another representation where the wavefunctions depend only on loops [11]. The crucial observation is that the set of Wilson loops, for all loops, constitutes an overcomplete basis of solutions of the Gauss constraint. This means that any gauge invariant function can be expressed as a combination of products of Wilson loops. Introducing the *loop transform*,

$$\psi(\gamma_1, \ldots, \gamma_n) := \int dA \psi(A) W_A(\gamma_1) \ldots W_A(\gamma_n) \,, \tag{26}$$

one can construct then a new quantum representation of a Yang-Mills theory and, in particular, of gravity[2]. The overcompleteness of the loop basis implies that the loop wavefunctions are not all independent. They are constrained by a set of identities, known as the Mandelstam identities. These identities reflect the structure of the particular group being gauged. For the case of quantum gravity this group is $SU(2)$, and the Mandelstam identities have the general form

---

2) Another possible way to define the loop representation is through a non-canonical algebra of quantities, see [5]

$$\sum_k c_k\,\psi([\gamma]_k) = 0\,, \tag{27}$$

where $[\gamma]_k \equiv \gamma_1, \ldots, \gamma_k$ represents a generic multiloop. These identities introduce very nontrivial relations between the loop wavefunctions.

In the loop representation the constraints can be expressed in terms of the *loop derivative*. The loop derivative is a differential operator in the loop space that measures the change of the wavefunction when its argument is changed by an infinitesimal amount:

$$\Delta_{ab}(\pi_o^x)\,\psi(\gamma) := \lim_{\sigma^{ab}\to 0} \frac{\psi(\pi_o^x \circ \delta\gamma \circ \pi_x^o \circ \gamma) - \psi(\gamma)}{\sigma^{ab}}\,. \tag{28}$$

$\delta\gamma$ is an infinitesimal loop of area element σ^{ab} (which lies in the plane ab). The argument of the loop derivative depends on the open path π_o^x going form o to x, by means of which the deformation is introduced at the point x (the $\circ$ operation is the usual composition law between loops or open paths). A loop wavefunction is said to be loop differentiable if the limit exists. In particular, for the Wilson loops one gets,

$$\Delta_{ab}(\pi_o^x)\,W_A(\gamma) = Tr[F_{ab}(x)U_A(\gamma)]\,. \tag{29}$$

One of the advantages of the loop representation is that the Gauss constraint is automatically satisfied by the wavefunctions defined through (26). For the diffeomorphism and the Hamiltonian constraints we get the following expressions in terms of the loop derivative:

$$\hat{C}(\vec{N})\,\psi(\gamma) = \int d^3x N^b(x) \oint_\gamma dy^a \delta(x-y)\Delta_{ab}(\gamma_o^y)\psi(\gamma)\,, \tag{30}$$

$$\hat{H}(N)\,\psi(\gamma) = \int d^3x N(x) \oint_\gamma dy^{[b} \oint_\gamma dz^{a]}\delta(x-y)\delta(x-z)\Delta_{ab}(\gamma_o^x)\psi(\gamma_y^z \circ \gamma_{z\ o}^{y})\,. \tag{31}$$

In both cases the deformation is introduced along the loop γ (the argument of the loop derivative is the portion γ_o^y of the loop). In the case of the Hamiltonian $\gamma_{z\ o}^{y}$ is the portion of the loop going from z to y which includes the basepoint o.

The diffeomorphism constraint acts on functions of loops by infinitesimally deforming the loop argument along the vector $\vec{N}$. The deformation is the same that the loop would suffer if it existed in a spatial manifold on which a diffeomorphism is performed along $\vec{N}$. Therefore, if a wavefunction $\psi(\gamma)$ is to be annihilated by the diffeomorphism constraint it should be invariant under deformations of the loop argument (that is to say, it depends only on the knot class of the loop). This means that the general solutions of the diffeomorphism constraint are given by knot invariants. This is a characteristic of the loop representation where knot invariance codes the difeomorphism invariance of general relativity.

To compute the action of the Hamiltonian one needs a regularization to take into account the two delta functions that appear in (31). Some general problems arise in this context, which have been the subject of intense investigations during the last years [12]. Here we limit to comment the following point: the regularization of the Hamiltonian operator unavoidably introduces a dependence with a background metric (for example, the metric used to point split the delta functions in (31)). This follows from the fact that the Hamiltonian defined by (21) is a density of weight two. In a diffeomorphism invariant context the only natural density is of weight one (the Dirac delta). This means that, in order to generate the correct density weight, one has necessarily to include some dependence with a background metric in the final result. This residual background dependence affects the regulated algebra of the Hamiltonians, which in general does not close at the quantum level.

New solutions to all the constraints are obtained in the loop representation. These solutions are associated with the loop transform of the Chern-Simons state (24). This loop transform defines a regular isotopy invariant[3] $K_\Lambda(\gamma)$, known as the Kauffman bracket invariant [14]:

$$K_\Lambda(\gamma) =< W_A(\gamma) >:= \int dA e^{\frac{i6}{\Lambda}S_{cs}} W_A(\gamma)\,, \tag{32}$$

[3] A regular isotopy invariant is not a true diffeomorphism invariant, but it depends on a framing [13].

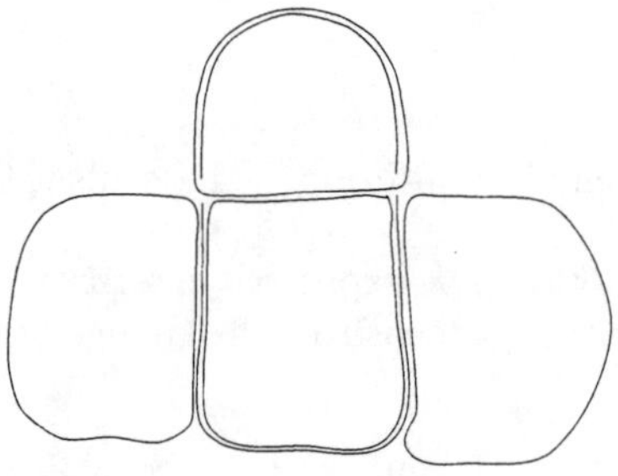

FIGURE 1. A multiloop.

where S_{cs} is the Chern-Simons form. An important property is that the Kauffman bracket admits the following expansion in powers of the the cosmological constant,

$$K_\Lambda(\gamma) = a_o(\gamma) + \Lambda a_1(\gamma) + \Lambda^2(\frac{(a_1(\gamma))^2}{2} + a_2(\gamma)) + \ldots . \tag{33}$$

Here $a_o(\gamma)$ is a constant, $a_1(\gamma)$ is a framing dependent invariant (the Gauss self-linking number of the loop γ) and $a_n(\gamma)$ for $n > 1$ are ambient isotopic invariants. It is possible to demonstrate that these quantities are numerical invariants of finite type, known as the Vassiliev invariants [15]. One can prove that the Kauffman bracket is annihilated by the Hamiltonian constraint with cosmological constant [16] in the loop representation,

$$\hat{H}_\Lambda \, K_\Lambda(\gamma) = (\hat{H} + \Lambda/6 det\hat{q}) \, K_\Lambda(\gamma) = 0 \, . \tag{34}$$

Using this result and the fact that the exponential of the Gauss self-linking number is also annihilated by $\hat{H}_\Lambda$ [17], one gets the relevant result [18],

$$\hat{H} \, a_2(\gamma) = 0 \, . \tag{35}$$

The (true diffeomorphism) invariant $a_2(\gamma)$ nested in the expansion of the Kauffman bracket is a nontrivial physical state of vacuum gravity in the loop representation.

III SPIN NETWORKS IN QUANTUM GRAVITY

A The spin network basis

To explain how the spin networks are introduced in the quantum gravity scenario let us consider the multiloop shown in FIG. 1.

The loops of the multiloop can overlap in some segments and intersect in some points. Following the multiloop we draw a ribbon net in such a way that all the lines of the multiloop run inside the ribbon (FIG. 2). A colored rope designs a set of loop segments that fully overlap, being the color of the rope equals to the number of lines inside the ribbon. The intersection point of several ropes is called a vertex, and the number of ropes emerging from this point gives the valence of the vertex.

Let us now contract each ribbon into a single line. At the end of this process we get a colored graph consisting in a set of edges labelled by numbers and a set of vertices of different valences (see FIG. 3). This graph is embedded in the three dimensional manifold.

The information coded in the graph is not enough to recover the original multiloop. In particular, we do not know how to join the different lines of the ropes that meet at the vertices. This additional information is contained in the ribbon net. This means that to reconstruct the multiloop from the graph we have to blow up each vertex in order to assign a routing to the lines. The graph plus this blow up process are equivalent to a multiloop.

Notice that the connection of the ropes incident to a given vertex would not be in general possible for arbitrary coloring of the edges. Let us consider a trivalent vertex v. The colors of the three edges incident at v can be written in general as: $p_1 = a + b$, $p_2 = a + c$ and $p_3 = b + c$, where a is the number of lines running

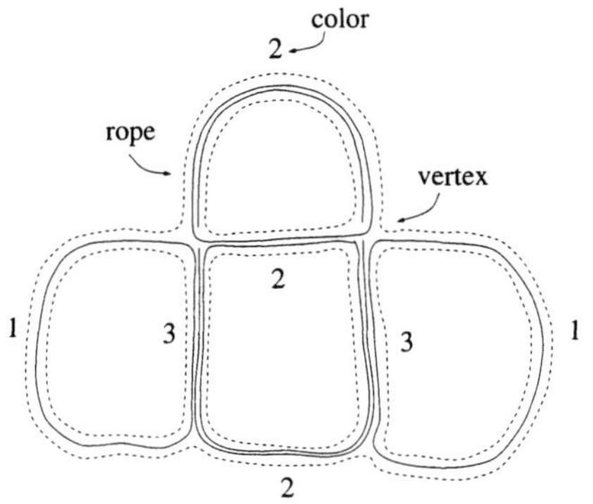

FIGURE 2. The ribbon net.

form the rope 1 to the rope 2, b those running form rope 1 to rope 3 and c from rope 2 to rope 3. a, b and c can be arbitrary positive integers, but not p_1, p_2 and p_3. They have to satisfy the following two compatibility conditions:

$$p_1 + p_2 + p_3 = \text{even}\,, \tag{36}$$

$$p_i \leq p_j + p_k, \quad i, j, k = 1, 2, 3\,. \tag{37}$$

These are the Penrose's rules for the vertices of a trivalent spin network. Vertices of higher valence can be expressed in terms of an inner structure of virtual trivalent vertices [19].

We are now in conditions to define the *spin network state.* To do this we pick up a colored graph Γ_S and construct from it the multiloop $[\gamma]^o_S$ with the condition that at any vertex the lines belonging to the different ropes are connected without any crossing. This procedure is always possible, and it defines a reference multiloop associated with the original graph. The next step is to generate from $[\gamma]^o_S$ others multiloops by antisymmetrizing the lines along each rope. For this we introduce the permutation operator $P_n^{(e)}$ associated with the edge e of color n in the form:

$$P_n^{(e)}\,[\gamma]^o_S := \frac{1}{n!}\sum_{p=1}^{n!}(-1)^{|p|}[\gamma]^p_S\,, \tag{38}$$

where $(-1)^{|p|}$ is the parity of the permutation and $[\gamma]^p_S$ are the $p = 1, \ldots, n!$ multiloops obtained by all possible permutations of the lines running along the rope e. The *spin network state* $\psi(S)$ is then defined as the combination of multiloop wavefunctions obtained by the full antisymmetrization of the graph Γ_S:

$$\psi(S) := \prod_{e \in S} \psi(P_{n_e}^{(e)}\,[\gamma]^o_S)\,. \tag{39}$$

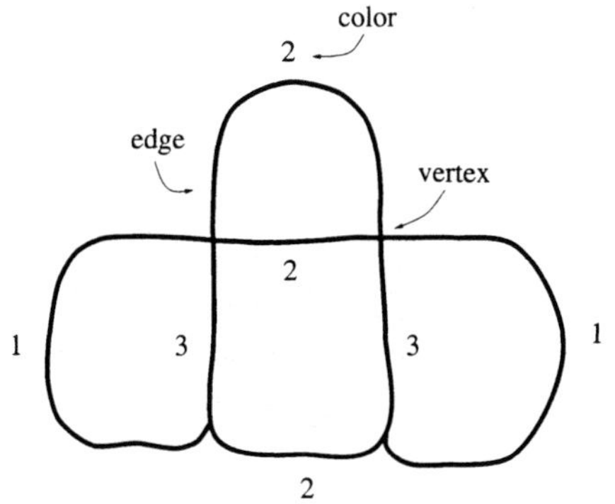

FIGURE 3. The colored graph.

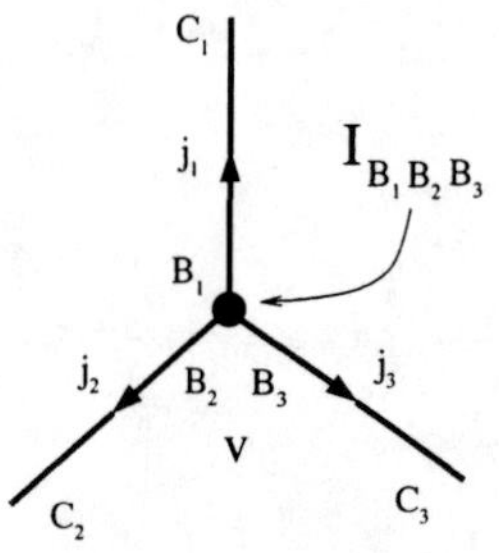

FIGURE 4. The intertwiner at v.

Notice that $\psi(S)$ depends only on the colored graph and not on the rootings through the vertices (because all possible rootings are included in $\psi(S)$). The key fact is that these states are free of the Mandelstam identities. To demonstrate this crucial property in the general case it is more convenient to work in the connection representation [20].

One way to introduce the spin network states in the connection representation is to extend (26) to spin networks:

$$\psi(S) = \int dA \psi(A) W_A(S)\,. \tag{40}$$

The *Wilson net* $W_A(S)$ codes all the information about the spin network in the connection representation. These states can be constructed following a procedure similar to that outlined above. The starting point is now a product of Wilson loops of the form $W_A(\gamma_1)\ldots W_A(\gamma_n)$. We can associate to this product a colored graph in the same way as we did before, with the only difference that an edge e carries now a product of holonomies in the fundamental representation $U_A^{1/2}(e)$. To construct the spin network state we have to symmetrize the product of holonomies associated with the different edges of the graph [6]. By this procedure one gets all the nontrivial irreducible representations of spin $j_e = p_e/2$ for a rope of color p_e. We conclude that, in the connection representation, the edges of the Wilson net will transport in general nontrivial representations $[U_A^{j_e}(e)]^B_C$ of the $SU(2)$ group[4]. In order to distinguish the matrix indices an orientation is assigned to the edges of the graph, with the convention that the upper index corresponds to the tail of the arrow:

$$[U_A^{j_e}(e)]^B_C \to \tag{41}$$

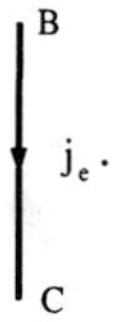

Let e_m, $m = 1, 2, 3$, the three edges of a trivalent vertex v carrying holonomies $[U_A^{j_m}(e_m)]^{B_m}_{C_m}$. We assume that the edges are outgoing from v. If we wish to construct a gauge invariant quantity at v, the upper B_m indices must be contracted with the invariant tensors of the theory, in the following way (see FIG 4):

$$[U_A^{j_1}(e_1)]^{B_1}_{C_1}\,[U_A^{j_2}(e_2)]^{B_2}_{C_2}\,[U_A^{j_3}(e_3)]^{B_3}_{C_3}\,I_{B_1B_2B_3}\,. \tag{42}$$

The quantity $I_{B_1B_2B_3}$ is a gauge invariant matrix associated with the vertex v. These quantities are provided by the recoupling theory of angular momentum (the Clebsh-Gordan intertwiners). If we want to extend the gauge invariance property of (42) to the other indices, the free C_m indices have to be contracted with other invariant tensors. These new intertwiners will be associated with the vertices located at the end of the edges e_m. To get a full gauge invariant expression one has to saturate all the matrix indices of the nontrivial irreducible representations attached to the edges of the graph with intertwiners, one for each vertex of the graph. We conclude that, in the connection representation, a spin network is given by a triple $S = (\Gamma_S, \vec{J}, \vec{I})$ consisting of a embedded graph Γ_S, a coloring $\vec{J} = (j_1, j_2, \ldots)$ of the edges $(e_1, e_2, \ldots)$ by irreducible representations, and an assignement of invariant intertwiners $\vec{I} = (I_1, I_2, \ldots)$ to the vertices $(v_1, v_2, \ldots)$[5]. The spin network state $W_A(S)$ is defined then by the following expression:

---

4) We denote the matrix indices with capital Latin letters from the begining of the alphabet.

5) This assignement is not unique and different choices generate different basis.

$$W_A(S) = W_A(\Gamma_S, \vec{J}, \vec{I}) := Tr[\otimes_{e\in\Gamma_S} U_A^{j_e}(e) \cdot \vec{I}] \,. \tag{43}$$

As we know, these states live in the space $\mathcal{A}/\mathcal{G}$ of connections modulo gauge transformations. This is a nonlinear and infinite dimensional functional space. These two characteristics make the integration on this space a nontrivial problem. This problem was recently solved by Ashtekar et al. [21] using the notion of cylindrical functions. We say that a function $f(A) \in \mathcal{A}/\mathcal{G}$ is cylindrical with respect to a graph Γ if it can be expressed in the form

$$f(A) = f_\Gamma(U_A(e_1), U_A(e_2), \ldots, U_A(e_n)) \,, \tag{44}$$

where f_Γ is a complex valued function of $SU(2)^n$ such that $f(A)$ is gauge invariant. The space of cylindrical functions $Cyl_\Gamma(\mathcal{A}/\mathcal{G})$ is finite dimensional and $Cyl(\mathcal{A}/\mathcal{G}) := \bigcup_\Gamma Cyl_\Gamma(\mathcal{A}/\mathcal{G})$ is dense in $\mathcal{A}/\mathcal{G}$. A diffeomorphism invariant measure $\mu_o(A)$ can be introduced in $\mathcal{A}/\mathcal{G}$ by a projective limit of measures on cylindrical spaces (the Ashtekar-Levandowsky measure). The definition is:

$$\int_{\mathcal{A}/\mathcal{G}} d\mu_o(A)\, f(A) = \int_{\mathcal{A}/\mathcal{G}} d\mu_{o,\Gamma}(A) f_\Gamma(U_A(e_1), \ldots, U_A(e_n)) = \int_{SU(2)^n} d\mu_H(g_1, \ldots, g_n) f_\Gamma(g_1, \ldots, g_n) \,. \tag{45}$$

What we do here is to identify $Cyl_\Gamma(\mathcal{A}/\mathcal{G})$ with n copies of $SU(2)$ and to integrate in the group space using the de Haar measure. It is possible to show that the result does not depend on the particular graph used to represent the cylindrical function. The Ashtekar-Levandowsky measure is gauge invariant and does not depend on any background metric. Using this measure one can define a Hilbert space $\mathcal{L}^2(\mathcal{A}/\mathcal{G})$ with respect to the norm $||\psi|| = (\int_{\mathcal{A}/\mathcal{G}} d\mu_o(A)|\psi|^2)^{1/2}$. The space $\mathcal{L}^2(\mathcal{A}/\mathcal{G})$ is the kinematical arena of quantum gravity.

For a spin network $S = (\Gamma_S, \vec{J}, \vec{I})$, the Wilson net is a cylindrical function and the set $\{W_A(S)\}$ for all S expand the space $\mathcal{L}^2(\mathcal{A}/\mathcal{G})$. This means that $\{W_A(S)\} \in Cyl(\mathcal{A}/\mathcal{G})$ is dense in $\mathcal{L}^2(\mathcal{A}/\mathcal{G})$. The Ashtekar-Levandowsky measure allows to introduce the following inner product in the space:

$$(\psi_1, \psi_2) := \int d\mu_o(A)\, \bar{\psi}_1(A)\psi_2(A) \,, \tag{46}$$

and, using the properties of the de Haar measure and the definition of the spin network states, one gets the result,

$$(W_A(S), W_A(S')) = \delta_{SS'} = \delta_{\Gamma_S\Gamma_{S'}}\delta_{\vec{J}\vec{J}'}\delta_{\vec{I}\vec{I}'} \,. \tag{47}$$

This proves the independence of the spin network states. An important property of the spin network basis is that it diagonalizes relevant geometric quantities, like the area of a surface and the volume of a region.

B The area and volume operators

Classically the area of a surface Σ and the volume of a region R of the three dimensional manifold are given by the following expressions:

$$A(\Sigma) = \int_\Sigma d^2\sigma \sqrt{detq^\Sigma} = \int_\Sigma d^2\sigma \sqrt{n_a n_b E^{ai} E_i^b} \,, \tag{48}$$

$$V(R) = \int_R d^3x \sqrt{detq} = \int_R d^3x \sqrt{\frac{1}{3!}\epsilon_{abc}\epsilon_{ijk} E^{ai} E^{bj} E^{ck}} \,, \tag{49}$$

where $\vec{n}$ is the normal to Σ. These quantities admit quantizations whose action on cylindrical functions is perfectly finite and its spectrum is discrete [22,19]. For the area operator the eigenvalues depend on the number of edges of the spin network that crosses Σ. The explicit result is:

$$\hat{A}(\Sigma)\, W_A(S) = \sum_{i\in S\cup\Sigma} \sqrt{j_i(j_i+1)}\, W_A(S) \,. \tag{50}$$

If the surface Σ does not intersect any edge of the graph the eigenvalue if zero, if only one edge of color j_i crosses Σ the eigenvalue is $\sqrt{j_i(j_i+1)}$, and so on. For the volume operator the result is much more involved to express analytically, so we give here only a schematic description. Let us express $V(R) = \sum_{\Box \in R} V(\Box)$ for a partition of the region R into elementary boxes $\Box$. The operator $\hat{V}_\Box$ has a nontrivial action over the wavefunction $\psi(S)$ only if the box includes a vertex (of valence equal or greater than four). This means that the volume operator has a local action at the vertices of the graph.

The above results provide a very nice picture of quantum spacetime at the Planck scale. We see that the spin network states carry the *quanta of geometry.* This interpretation is supported by the fact that the quantum spacetime can be decomposed in a basis of states that can be visualized as made of quanta of volume (the vertices) separated by quanta of area (the edges). This picture gives a physical meaning to the kinematical arena of quantum gravity in terms of the spin network basis.

C The Thiemann Hamiltonian constraint

The main unsolved problems of the canonical quantization of gravity à la Ashtekar are the reality conditions and the clousure of the regulated constraint algebra. With respect to the reality conditions, they constitute a second class constraint which has to be imposed at the level of probabilities. The hope is that these conditions could be used to select the correct inner product of the theory. Not satisfactory solution to this problem is known at present. On the other hand, the problem in the clousure of the regulated algebra is related to the fact that the Hamiltonian operator is a density of weight two. This introduces a general dependence with a background metric which affects the algebra on general grounds.

Thomas Thiemann [7] has recently proposed a solution for these two problems. The solution is based in the following facts: 1) the use of a kinematical framework similar to that introduced before but defined in terms of *real* connections, and 2) the introduction of a single densitized Hamiltonian constraint whose action over the wavefunctions is finite and anomaly free.

The use of real connections to describe the classical phase space of General Relativity was suggested by Barbero [23]. Barbero considers a generalization of the canonical transformation proposed by Ashtekar (see (13) where the imaginary factor is substituted by a free parameter β:

$$A^i_a = \Gamma^i_a + \beta K^i_a \,. \tag{51}$$

In terms of the new configuration variable the classical constraints are,

$$G^i = D_a E^{ai} \,, \tag{52}$$

$$C_a = E^b_i F^i_{ab} \,, \tag{53}$$

$$H = -\zeta \epsilon^{ij}{}_k E^a_i E^b_j F^k_{ab} + 2\frac{(\beta^2\zeta - 1)}{\beta^2} E^a_{[i} E^b_{j]} (A^i_a - \Gamma^i_a)(A^j_b - \Gamma^j_b) \,, \tag{54}$$

where the factor ζ controls the signature of spacetime ($\zeta = +1$ for a Euclidean signature and $\zeta = -1$ in the Lorenztian case). We see that the Gauss and the diffeomorphism constraints are not affected by the redefinition of A^i_a. This means that all the kinematical properties of the representation will be valid also in this case (as the existence of the spin network basis and the area and volume operators). For the Hamiltonian we get an extra term, which vainshes for Lorentzian signatures only if $\beta^2 + 1 = 0$ (which corresponds to the case of the complex Ashtekar connection). For real values of β the phase space would be described by real fields, but at the price to introduce a more complicated form of the Hamiltonian constraint. For example for $\beta = 1$ we get,

$$H(\zeta = -1, \beta = 1) = \epsilon^{ij}{}_k E^a_i E^b_j (F^k_{ab} - 2R^k_{ab}) \,, \tag{55}$$

where R^k_{ab} is the curvature associated with the spin connection.

Following the ideas of Barbero, Thiemann considers real connections to describe the phase space of Lorentzian gravity. In this way the problem of the reality conditions is solved from the beginning. The second crucial observation is that, in order to quantize the new hamiltonian (55), one *needs* to introduce a factor $1/\sqrt{detq}$. This factor produces a Hamiltonian density of weight one, solving in this way the second general problem mentioned before. The single densitized Hamiltonian considered by Thiemann is:

$$H_{single} = \frac{1}{\sqrt{detq}} H(\zeta = -1, \beta = 1) = \frac{1}{\sqrt{detq}} Tr\{(F_{ab} - 2R_{ab})[E^a, E^b]\} \,. \tag{56}$$

At first sight this expression looks very complicated due to the presence of the nonpolynomical term $1/\sqrt{detq}$ and the curvature R_{ab}. The following two identities allow to write H_{single} in a way appropriate for quantization:

$$\frac{[E^a, E^b]^i}{\sqrt{detq}} = 2\epsilon^{abc}\frac{\delta V}{\delta E^c_i} = 2\epsilon^{abc}\{A^i_c, V\}, \tag{57}$$

$$K^i_a = \frac{\delta K}{\delta E^a_i} = \{A^i_a, V\}, \tag{58}$$

where V is the total volume, $K = \int d^3x\, K^i_a E^a_i$ and the field K^i_a was introduced in Sec. 2.2. Using these identities one can write,

$$\frac{1}{\sqrt{detq}} Tr\{F_{ab}[E^a, E^b]\} = 2\epsilon^{abc} Tr[F_{ab}\{A^i_c, V\}] \equiv H^E, \tag{59}$$

$$\frac{1}{\sqrt{detq}} Tr\{2R_{ab}[E^a, E^b]\} = -8\epsilon^{abc} Tr[\{A_a, K\}\{A_b, K\}\{A_c, V\}] \equiv H'. \tag{60}$$

So we get,

$$H_{single} = H^E + H' = f(A, V, K). \tag{61}$$

The single densitized Hamiltonian is expressed as a function of the connection, the volume and the integrated extrinsic curvature. In what concerns to the quantization of (61), we have at our disposition the following results: : 1) a well defined quantization for the volume in the space of cylindrical functions, and 2) the fact that classically $K = -\{V, \int d^3x H^E(x)\}$ (the integrated extrisic curvature is equal to the time derivative of the total volume with respect to the integrated euclidean Hamiltonian constraint). The following strategy is then appropriate to quantize H_{single}: first one quantizes $H^E(A, V)$ in the space of cylindrical functions using the quantization of V, then one uses the quantizations of H^E and V to quantize K. Finally, we quantize $H'(A, V, K)$. This is the procedure followed by Thiemann. In what follows we consider some details about the quantization of H^E.

To quantize $H^E(A, V)$ in the space of cylindrical functions we need of a regularization. The regularization is achieved by introducing a triangulation $T(M)$ of the three dimensional manifold M by elementary tetrahedra Δ in such a way that integrals over M are expressed as (the limit of) sums over tetrahedra. This means that,

$$H^E(N) = \int d^3x N(x) H^E(x) \overset{regularization}{\longrightarrow} H^E_T(N) = \sum_{\Delta\in T} H^E_\Delta(N). \tag{62}$$

Let $v(\Delta)$ be one of the vertices of Δ and $s_k(\Delta)$ $k = 1, 2, 3$ the three edges of the tetrahedra that meet at $v(\Delta)$. $H^E_\Delta(N)$ has two ingredients: the scalar field $N(x)$ evaluated at some $x \in \Delta$ and the term $Tr[F \wedge \{A, V\}]$. Inside the tetrahedra one can replace the curvature F and the connection A by holonomies evaluated along the edges of Δ through the expressions,

$$\lim_{\Delta\to 0} U_A(\alpha_{ij}(\Delta)) = 1 + \frac{1}{2} F_{ab} s^a_i s^b_j, \tag{63}$$

$$\lim_{\Delta\to 0} U_A(s_k(\Delta)) = 1 + A_c s^c_k, \tag{64}$$

where $\alpha_{ij}(\Delta) = s_i(\Delta) \circ s_{ij}(\Delta) \circ s_j(\Delta)$ is the loop that starting at $v(\Delta)$ closes the triangle formed by $s_i(\Delta)$ and $s_j(\Delta)$ ($s_{ij}(\Delta)$ is the third edge of the triangle). We see that when the tetrahedra shrinks to a point the curvature and the connection are recovered from the finite holonomies $U_A(\alpha_{ij}(\Delta))$ and $U_A(s_k(\Delta))$. Using these results one can write,

$$H^E_\Delta(N) = -\frac{2}{3} N[v(\Delta)]\epsilon^{ijk} Tr[U_A(\alpha_{ij}(\Delta)) U_A(s_k(\Delta))\{U^{-1}_A(s_k(\Delta)), V\}], \tag{65}$$

where we have choosed the vertex $v(\Delta)$ as the reference point to evaluate the field $N(x)$. When $\Delta \to 0$,

$$\lim_{\Delta\to 0} H^E_\Delta(N) = \int_\Delta N Tr[F \wedge \{A, V\}], \tag{66}$$

and then $H^E_T(N) \longrightarrow H^E(N)$ in this limit. The substitutions of the connection and the curvature by finite holonomies living in the tetrahedra prepares the classical expression (59) for quantization in the space of cylindrical functions. The quantization is simply achieved by replacing V for $\hat{V}$ and the Poisson brackets by commutators. From (62) and (65) we get,

$$\hat{H}^E_T(N) = \sum_{\Delta \in T} N_v \hat{H}^E_\Delta \,, \tag{67}$$

with,

$$\hat{H}^E_\Delta = -\frac{2}{3i}\epsilon^{ijk} Tr(U_A(\alpha_{ij}(\Delta) U_A(s_k(\Delta)[U_A^{-1}(s_k(\Delta)), \hat{V}]) \,. \tag{68}$$

The action of the operator $\hat{H}^E_T(N)$ over the cylindrical functions strongly depends on the triangulation T of the three dimensional manifold. For a given cylindrical function different triangulations produce different results. In order to get consistency a further condition is needed. The consistency is fulfilled by adapting the traingulation to the graph Γ of the cylindrical function. Basically, the adaptation of the triangulation to the graph is performed putting the vertices of the tetrahedra $v(\Delta)$ coincident with the vertices $v(\Gamma)$ of the graph and the edges $s_k(\Delta)$ coincident with the initial portions of the edges of the graph meeting at $v(\Gamma)$. This prescription makes the limit $\Delta \to 0$ as the tetrahedras shrink to a point a well defined process.

By this procedure one generates an uncountable family of Hamiltonian operators $\hat{H}^E_{T[\Gamma]}(N)$, one for each graph. It is possible to show that these infinite operators are the projections to cylindrical subspaces of one and the same operator in the Hilbert space. A requirement for this is to restrict the action of the Hamiltonian over diffeomorphism invariant distributions $\psi_D \in Diff\{Cyl(\mathcal{A}/\mathcal{G})\}$ belonging to the dual of the cylindrical functions[6]. One characteristic feature of the regulated operator (67) is that it acts nontrivially only when the adapted tetrahedra includes a vertex of the graph. This is due to the presence of the volume operator in (68). So we can write in general,

$$\hat{H}^E_{T[\Gamma]}(N)\, f_\Gamma = \sum_{v \in V(\Gamma)} N_v \hat{H}^E_v\, f_\Gamma \,, \tag{69}$$

where $V(\Gamma)$ is the set of vertices of Γ. The operator $\hat{H}^E_v$ acts exclusively at the vertices of the graph in the form:

$$\hat{H}^E_v\, f_\Gamma = \frac{8}{E(v)} \sum_{v(\Delta) \in v} \hat{H}^E_\Delta\, f_\Gamma \,, \tag{70}$$

where $\hat{H}^E_\Delta f_\Gamma$ is evaluated acoording to (68). The number $E(v)$ does not depend on the triangulation (it gives the number of non-coplanar triples of edges associated with v). Introducing a diffeomorphism covariant prescription for the rootings of the arcs of the triangulation adapted to the graph it is possible to demonstrate that the result of (69) is finite and independent of the triangulation. The explicit result for the elements of the spin network basis in the case of trivalent vertices is [24],

$\hat{H}^E_v W_A($ 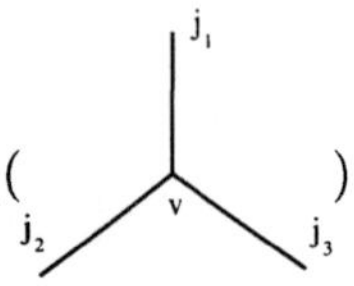

$) = \sum B(j_1, j_2 \pm 1/2, j_3 \pm 1/2) W_A($ 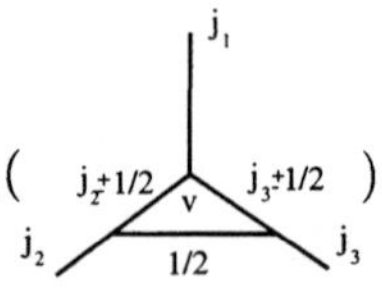

$) +$

$\sum B(j_1 \pm 1/2, j_2, j_3 \pm 1/2) W_A($ 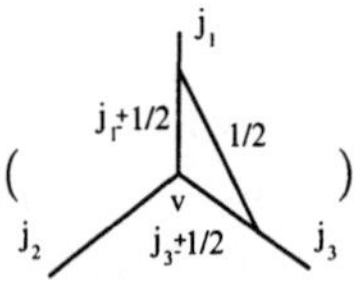

$) + \sum B(j_1 \pm 1/2, j_2 \pm 1/2, j_3) W_A($ 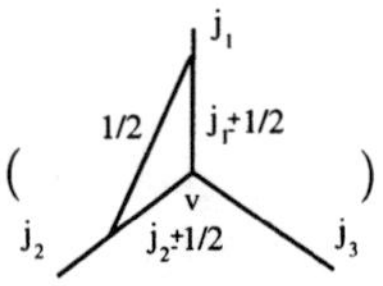

$)$. (71)

The sums are taken over all the values $j_i \pm 1/2$, and $B(\ldots)$ are recoupling coefficients that depend on the coloring of the edges and which includes the action of the volume operator. The inclusion of finite edges of spin 1/2 in the final result is a characteristic of the regularization of H_{single} proposed by Thiemann.

---

6) This means to evaluate $\psi_D[\hat{H}(N)\, f] = \int_{\mathcal{A}/\mathcal{G}} d\mu_o(A) \bar{\psi}_D\, \hat{H}(N)\, f$ for each lapse and any cylindrical function f in the domain of $\hat{H}(N)$.

IV QUANTUM GRAVITY IN THE SPACE OF VASSILIEV INVARIANTS

A The motivations

The Thiemann's formulation of the WdW equation completes in a consistent way the canonical quantization program of gravity. The Hamiltonian constraint equation proposed by Thiemann is well defined, finite and anomaly free. Moreover, a complete set of solutions to all the constraints can be obtained in this approach [25].

In spite of the success of the Thiemann's approach some problems were recently addressed by several authors. The main observed points are: 1) the strong commutativity of the Hamiltonian operator, and 2) the extremely local nature of the physical states.

The strong commutativity means that $[\hat{H}(N), \hat{H}(M)]\psi = 0$ even when ψ is *not* a diffeomorphism invariant wavefunction. It is important to remark that the Thiemann's solutions of the WdW equation are given in terms of diffeomorphism invariant distributions $\psi_D \in Diff\{Cyl(\mathcal{A}/\mathcal{G})\}$ belonging to the dual of the cylindrical space. For these wavefunctions[7],

$$[\hat{H}(N), \hat{H}(M)]\psi_D = \hat{qC}(N, M)\psi_D \equiv 0\,. \tag{72}$$

The Thiemann's proof of the consistency of the constraint algebra is based in the following facts:

1. The regularization of $\hat{H}_{single}(N)$ is such that $[\hat{H}_{single}(N), \hat{H}_{single}(M)]\psi_D = 0$ for $\psi_D \in Diff\{Cyl(\mathcal{A}/\mathcal{G})\}$.
2. An independent regularization of $\hat{qC}(N, M)$[8] can be implemented such that $\hat{qC}(N, M)\psi_D = 0$ in $Diff\{Cyl(\mathcal{A}/\mathcal{G})\}$.
3. We get then $0 = 0$ as a check for the consistency of the algebra. This is the best result we can obtain working in $Diff\{Cyl(\mathcal{A}/\mathcal{G})\}$.

Lewandowski and Marolf [27] have studied the consistency of the algebra in a larger space that includes non-diffeomorphism invariant states $\tilde{\phi}$ (the off shell algebra). In this larger space the Thiemann Hamiltonian operator can be implemented, and they found that

$$[\hat{H}_{Thiemann}(N), \hat{H}_{Thiemann}(M)]\tilde{\phi} = 0\,. \tag{73}$$

The above result does not imply that the regularization proposed by Thiemann is affected by an off shell anomaly. This is because the more general states $\tilde{\phi}$ are also annihilated by the righthand side of (72):

$$\hat{qC}(N, M)_{Thiemann}\tilde{\phi} = 0\,. \tag{74}$$

However, it is possible to propose another quantizations for $qC(N, M)$ such that $\hat{qC}(N, M)_{Others}\tilde{\phi} \neq 0$ [28]. This means that $0 = 0$ does not seem to be a good check for the consistency of the constraint algebra in the larger space, and that the strong commutativity of $\hat{H}_{Thiemann}$ could be a source of anomalies in the theory.

With respect to the physical states, the kernel of $\hat{H}_{Thiemann}$ is computed by an algoritm [25]. The procedure is constructive and a general solution is obtained by *dressing* the vertices with special edges of spin 1/2. We are not going to discuss here the details of this computation. For our analysis the relevant point is that the dressing at different vertices are completely independent processes. This property confers to the solutions a strong local behaviour. As the volume acts locally at the vertices, this means that the physical states allow a division of space into disconnected regions. Using this fact Smolin [29] has argued that the sector of quantum states described by the Thiemann theory is lacked of long ranged correlations and, in consequence, it is not able to include massless particles. This heuristic argument is indicative that the theory fails to reproduce the correct classic limit.

---

7) We denote by $\hat{qC}(N, M)$ the operator corresponding to the righthand side of (7).

8) Notice that it is not possible to realize $\hat{C}$ (the generator of infinitesimal diffeomorphisms) neither $\hat{q}$ (the metric operator) in $Diff\{Cyl(\mathcal{A}/\mathcal{G})\}$ (only the exponentiated form of the diffeomorphism constraint makes sense in the space of diffeomorphism invariant distributions). However, the composite operator $\hat{qC}$ can be realized in this space [26].

The above observations point out the necessity to introduce some modifications into the Thiemann's quantization procedure. One possible way to do this is by looking for a new space where the Hamiltonian (56) could be quantized. This procedure has the advantage to preserve some of the benefits of the Thiemann's approach (like the fact that the Hamiltonian is a density of weight one). The requisites to define this new space are: 1) it belongs to the dual of the cylindrical functions, and 2) its elements must be loop differentiable functions. The first requirement enable us to use a regularization of the type discussed at the end of Sect. 4.4, and the second points out to solve the strong commutativity problem of $\hat{H}_{Thiemann}$. The strong commutativity is essentially a consequence of the trivialization of the limit $\Delta \to 0$ in the Thiemann's regularization scheme. The trivialization of this limit follows form 1) the replacement of the curvature F_{ab} by the finite holonomy $U_A(\alpha_{ij}(\Delta))$ (see (63)), and 2) the fact that the Hamiltonian is evaluated on diffeomorphism invariant distributions $\psi_D \in Diff\{Cyl(\mathcal{A}/\mathcal{G})\}$. Then

$\lim_{\Delta\to 0} \hat{H}^E_\Delta[U_{\alpha_{ij}}]\, f_\Gamma($ 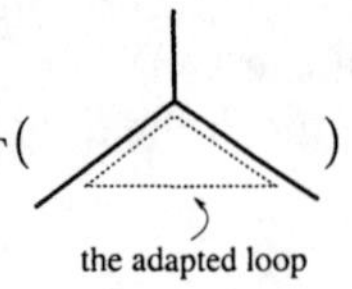

$) = \lim_{\Delta\to 0} f_\Gamma($ 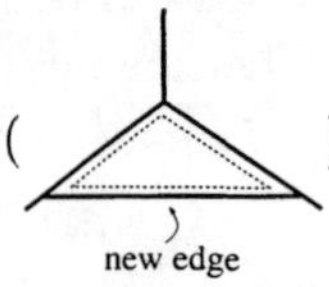

$) \overset{\text{diff. invariance}}{=} f_\Gamma($ 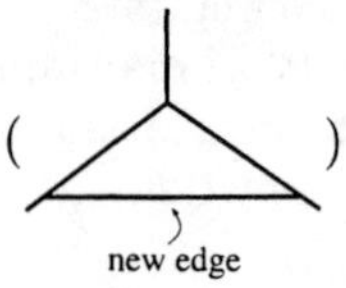

$)\,. \qquad (75)$

By invoking the diffeomorphism invariance the new edge introduced by the finite holonomy can be "disconnected" of the tetrahedra in the limit $\Delta \to 0$. As the result is finite, the limit is trivially taken. The appearance of these new edges in a diffeomorphism invariant context is in the root of the strong commutativity property of $\hat{H}_{Thiemann}$.

B Giving meaning to $\hat{C}$

The first step is to define a suitable space where the generator of infinitesimal diffeomorphisms could be realized. The immediate generalizaton of (30) to the case of spin network wavefunctions is:

$$\hat{C}(\vec{N})\, W_A(S) = \int d^3x N^b(x) \sum_{e\in S} \int_e dy^a \delta(x-y) \Delta^{(e)}_{ab}(e^y_v)\, W_A(S)\,. \qquad (76)$$

The superscript of $\Delta^{(e)}_{ab}(e^y_v)$ indicates that the loop derivative acts on the edge e of the spin network. The loop derivative of the Wilson net is,

$$\Delta^{(e)}_{ab}(e^y_v)\, W_A(S) = F^\alpha_{ab}(y)\, W_A(\ldots U_A(e^y_v)\tau^\alpha U_A(e^{v'}_y)\ldots)\,, \qquad (77)$$

where we have assumed that e is outgoing from v, v' is the end point of e, and τ^α are the $SU(2)$ generators. We see that the loop derivative of a cylindrical function is not a cylindrical function (due to presence of the curvature term). Moreover, if $W_A(\{S\})$ denotes the equivalence class of Wilson nets under diffeomorphisms, then

$$\Delta^{(e)}_{ab}(e^y_v)\, W_A(\{S\}) = \lim_{\sigma^{ab}\to 0} \frac{\text{0 or a finite nonzero term}}{\sigma^{ab}} = \text{ill defined}\,. \qquad (78)$$

This is the same problem that appears when one takes the derivative of a step function. The result is a distribution. Like distributions, the above expression would have sense under integration. We define,

$$\Delta^{(e)}_{ab}\, \psi(S) := \int dA \psi(A)\, \Delta^{(e)}_{ab}\, W_A(S)\,. \qquad (79)$$

A very important loop differentiable wavefunction is given by the generalization of (32) to spin networks:

$$E[S,\kappa] := \int dA e^{-\frac{1}{\kappa}S_{cs}} W_A(S)\,, \qquad (80)$$

with $\kappa \equiv i6/\Lambda$. It is possible to show that $E[S,\kappa]$ is a (framing dependent) invariant of spin networks [30]. To evaluate the loop derivative of $E[S,\kappa]$ we first transform the curvature that appears in (77) to a functional derivative with respect to the connection (using (25)), and then we integrate by parts to apply this derivative over the Wilson net. Using the result,

$$\frac{\delta}{\delta A_c^a(y)} W_A(S) = \sum_{e' \in S} \int_{e'} dz^c \delta(y-z)\, W_A(\ldots U_A(e'^z_v)\tau^a U_A(e'^{v'}_z)\ldots)\,, \tag{81}$$

we get,

$$\Delta_{ab}^{(e)}(\pi_o^y)\, E[S,\kappa] = \kappa\epsilon_{abc} \sum_{e' \in S} (-1)^{2(j_e+j_{e'})} \Lambda_{j_e}\Lambda_{j_{e'}} \int_{e'} dz^c \delta(y-z)\, E \left[\quad , \kappa \right] . \tag{82}$$

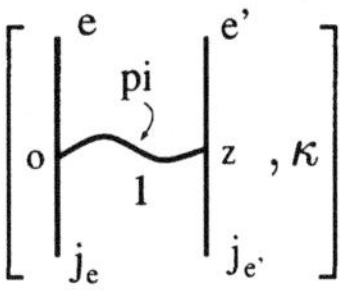

In the above expression the origin o of π_o^y belongs to the edge e of the spin network and $\Lambda_j \equiv \sqrt{j(j+1)}$. We see that the loop derivative of $E[S,\kappa]$ is proportional to the invariant evaluated on a modified spin network which includes a new edge π of spin 1 joining the edges e and e'. As in the case of loops, $E[S,\kappa]$ can be expanded in powers of κ,

$$E[S,\kappa] = E[S,0] + \kappa v_1(S) + \kappa^2 (\frac{(v_1(S))^2}{2} + v_2(S)) + \ldots\,, \tag{83}$$

where $E[S,0]$ is the chromatic evaluation of the invariant (its value for flat connections) and the v's are Vassiliev invariants of spin networks (see [31,32]). An important point is that the loop differentiability of $E[S,\kappa]$ implies the loop differentiability of all the Vassiliev invariants. One gets

$$\Delta_{ab}^{(e)}(\pi_o^y)\, E[S,0] = 0\,, \tag{84}$$

$$\Delta_{ab}^{(e)}(\pi_o^y)\, v_1(S) = \kappa\epsilon_{abc} \sum_{e' \in S} (-1)^{2(j_e+j_{e'})} \Lambda_{j_e}\Lambda_{j_{e'}} \int_{e'} dz^c \delta(y-z)\, E \left[\quad , 0 \right] , \tag{85}$$

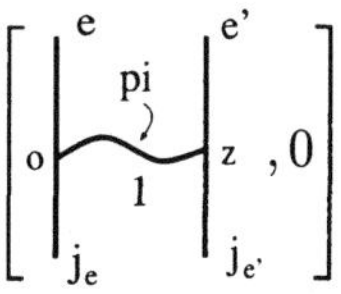

and so on. A (conjectured) property of the Vassiliev invariants is that they classify all knots. In the case of loops we have seen that one of these invariants is a quantum state of gravity (see (35)). Also we have shown that the Vassiliev invariants are loop differentiable wavefunctions of spin networks. In the next section we are going to see that it is possible to introduce appropriate regularizations for the constraints in this scenario [31,32].

C The constraints

The diffeomorphism operator acting over spin network wavefunctions is obtained as a direct extension of (30):

$$\hat{C}(\vec{N})\, E[S,\kappa] = \lim_{\epsilon \to 0} \int d^3z \sum_{e \in \Gamma_S} \int_e dy^b N^a(z,y) \varphi_\epsilon(z-y)\, \Delta_{ab}^{(e)}(e_v^y \pi_y^z)\, E[S,\kappa]\,, \tag{86}$$

where $\varphi_\epsilon(z-y)$ is a regularization of the delta function. The delocalization of the points y and z is needed to take into account the distributional character of (82). Notice that the field $\vec{N}$ can be evaluated at any of the points z or y, and different choices correspond to different prescriptions for the regularization. We can exploit this freedom to fix a prescription where $E[S,\kappa]$ is annihilated by the diffeomorphism operator[9]. It is easy to see that taking the symmetric prescription $N_+^a(z,y) := \frac{1}{2}[N^a(z) + N^a(y)]$ one gets by symmetry considerations,

$$\hat{C}(\vec{N}_+)\, E[S,\kappa] = 0\,. \tag{87}$$

[9] Remember that $E[S,\kappa]$ is a framing dependent invariant. This procedure is equivalent to fix a frame.

In this prescription the algebra of diffeomorphisms trivially closes,

$$[\hat{C}(\vec{N}_+), \hat{C}(\vec{M}_+)]\, E[S,\kappa] = 0\,. \tag{88}$$

The single densitized Hamiltonian (55) can be quantized using a technique similar to that developed in Sect. 4.4. In a first step we realize the operator over the invariant $E[S,\kappa]$, and in a second stage we transport the result over the Vassiliev invariants using the expansion (83). In what follows we limit the discussion to the euclidean part of the single densitized Hamiltonian constraint. The starting point is the definition,

$$\hat{H}^E(N)\, E[S,\kappa] := \int dA e^{-\frac{1}{\kappa} S_{cs}}\, \hat{H}^E_A(N)\, W_A(S)\,, \tag{89}$$

where $\hat{H}^E_A(N) = Tr\{F \wedge [A, \hat{V}]\}$ is the expression for the operator in the space of connections. As this operator is acting over a cylindrical function we can use the same regularization scheme developed in Sect. 4.4. We introduce the triangulation $T(M)$ of space into elementary tetrahedra Δ, but now the edges of the tetrahedra are scaled by a factor ϵ in the form: $s_k(\Delta) = \epsilon \tilde{s}_k(\Delta)$ with $|\tilde{s}_k(\Delta)| = 1$. As before we substitute the connection in $\hat{H}^E_A(N)$ by the holonomy $U_A(s_k)$ and, in what concerns to the curvature F, we use the result (77) to express it in terms of the loop derivative instead of using the result (63) typical of the Thiemann approach. In the case of trivalent vertices we get the following expression for the regulated Hamiltonian operator,

$$\hat{H}^E(N)\, E[S,\kappa] = -\lim_{\epsilon \to 0} \frac{\epsilon^2}{3} \sum_{v \in V(S)} \int d^3y N(y) \varphi_\epsilon(y,v)\, \epsilon^{ijk} \tilde{s}^a_i \tilde{s}^b_j\, a(\vec{J}_v)\, \Delta^{(k)}_{ab}(\pi^y_v)\, E[S,\kappa]\,, \tag{90}$$

where $\varphi_\epsilon(y,v) = \Theta(y,v)/\epsilon^3 Vol(\tilde{\Delta})$, being $\Theta(y,v)$ the step function associated with the tetrahedra (it is equal to one if $y \in \Delta$ and zero otherwise). $Vol(\tilde{\Delta})$ is the volume of the tetrahedra generated by the unitary vectors $\tilde{s_k}$ and $a(\vec{J_v})$ is a coefficient that depends on the coloring of the vertex. This coefficient includes some recoupling factors and the action of the volume operator (which is responsible to particularize the sum over the vertices of S). Evaluating the loop derivative according to (82) we get,

$$\hat{H}^E(N)\, E[S,\kappa] = \sum_{v \in V(S)} N(v)\, a'(\vec{J}_v)\, E[S,\kappa]\,, \tag{91}$$

where $a'(\vec{J}_v)$ is a new recoupling coefficient. The result is finite and proportional to the invariant evaluated on the *same* spin network. This is a characteristic of our regularization scheme. Notice that the path π^y_v included in the argument of the loop derivative tends to the null path in the limit $\epsilon \to 0$. This means that at the end of the process the new edge introduced by the loop derivative is totally absorbed into the recoupling coefficient. Schematically,

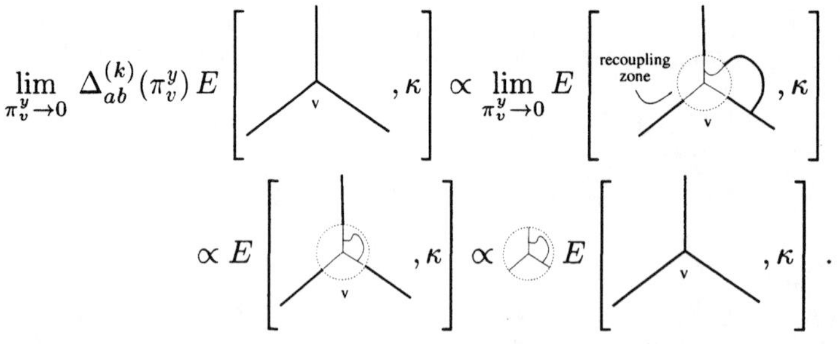

(92)

This fact introduces a substantial difference between our approach and that of Thiemann. We have checked the following results for the commutators:

$$[\hat{C}(\vec{N}_+), \hat{H}^E(M)]\, E[S,\kappa] = \hat{H}^E(\mathcal{L}_{\vec{N}_+} M)\, E[S,\kappa]\,, \tag{93}$$

$$[\hat{H}^E(N), \hat{H}^E(M)]\, E[S,\kappa] = 0\,. \tag{94}$$

These results together with (88) show that the algebra of the constraints closes in the appropriate form.

We have a consistent realization of the constraints over the key state $E[S,\kappa]$. So we have a good chance to construct a new representation of quantum gravity in the space of Vassiliev invariants. The general picture for the action of the Hamiltonian operator that emerges from the expansion (83) is the following:

$$\hat{H}^E(N)\, v_n(S) = \sum_{i<n} v_i(S)\,. \tag{95}$$

For trivalent vertices the above relation is trivially satisfied. Work is in progress to make explicit the above (conjetured) relationship to the general case. We are also considering an extension of the above formulation to include non-diffeomorphism invariant states in order to check the off shell algebra.

V CONCLUSIONS

During more than ten years of intense activity, the program of canonical quantization of gravity based in the Ashtekar new variables has confronted and successfully solved a significative number of fundamental problems. The diffeomorphism invariance poses very special conditions to the quantization of the theory of General Relativity, which has motivated the development of new nonperturbative and background independent techniques. Loop quantum gravity and the Ashtekar-Lewandowski-Baez diffeomorphism invariant integration theory on the space of connections modulo gauge transformations are important by-products of this effort.

More specifically, the three major problems of the theory that have been recently overcome are: 1) the lack of a well defined scalar product, 2) the overcompleteness of the loop basis, and 3) the difficulty of treating the reality conditions. The first problem is related to the definition of a diffeomorphism invariant measure in the space of connections modulo gauge transformations, the second was solved with the introduction of the spin networks basis, and the third with the use of real connections to describe the phase space of General Relativity.

The development of the functional integration calculus on the space of connections modulo gauge transformations and the introduction of the spin network basis have played a crucial role in the definition of the kinematic space of states of quantum gravity. In this scenario a very interesting physical interpretation of the quantum properties of spacetime at the Planck lenght naturally emerges: the spin network states are the conveyors of the quanta of geometry, being the quanta of volume located at the vertices and the quanta of area at the edges of the spin network.

At the dynamical level, a consistent definition of the Hamiltonian constraint has been developed in the space of cylindrical functions. The Hamiltonian has the crucial property of acting only on the vertices of the spin network, which implies that its action is naturally discrete and combinatorial.

While the kinematics of quantum spacetime is well understood, its dynamics is much less clear. The main problem originates from the quantum constraint algebra which, in spite to be anomaly free, it is affected by the strong commutativity problem of the Hamiltonian operator. Moreover, some heuristic arguments point out that the sector of quantum states in the kernel of this Hamiltonian operator does not satisfy the correspondence principle (that is, they do not have the correct classical limit).

One possibility to improve this situation is to consider a realization of the constraints in the space of the Vassiliev invariants. Some of the appealing features of this approach are the fact that the generator of infinitesimal diffeomorphisms can be realized in this space (the Vassiliev invariants are loop differentiable functionals of spin networks) and that the action of the single densitized Hamiltonian constraint is finite and well defined. It is important to remark that solutions in terms of the Vassiliev invariants for the double densitized Hamiltonian constraint exist in the loop representation. It would be very interesting if this kind of solutions are also present in the single densitized version of the constraint. This new sector could bring new insights in the lower energy limit of the theory.

ACKNOWLEDGMENTS

I whish to thank Fidel Schaposnik, Ricardo Gamboa Saravi, Horacio Falomir, and Mariel Santangelo for their hospitality during my stay in Buenos Aires attending the meeting Trends in Theoretical Physics II (Univ. Nal. de La Plata - Univ. Stgo. de Compostela).

REFERENCES

1. R. Penrose, Angular momentum: an approach to combinatorial space time, in *Quantum Theory and Beyond*, ed. T. Bastin, Cambridge Univ. Press, Cambridge, 1971.
2. R. Arnowitt, S. Deser, C.W. Misner, The dynamics of General Relativity, in *Gravitation: An Introduction to Current Research*, ed. L. Witten, Wiley, New York, 1962.
3. B. DeWitt, Phys. Rev. **160**, 1113 (1967).
4. A. Ashtekar, Phys. Rev. Lett. **57**, 2244 (1986); Phys. Rev. **D36**, 1587 (1987).
5. C. Rovelli, L. Smolin, Phys. Rev. Lett. **61**, 1155 (1988); Nucl. Phys. **B331**, 80 (1990).
6. C. Rovelli, L. Smolin, Phys. Rev. **D52**, 5743 (1995).
7. T. Thiemann, Phys. Lett. **380**, 257 (1996); Class. Quan. Grav. **15**, 839 (1988).
8. T. Jacobson, L. Smolin, Nucl. Phys. **B229**, 295 (1988).
9. B. Brugmann, R. Gambini, J. Pullin, Nucl. Phys **B385**, 587 (1992).
10. See for example [11] and references therein.
11. R. Gambini, J. Pullin 1996, *Loops, knots, gauge theories and quantum gravity*, (Cambridge: Cambridge University Press).
12. See for example C. Rovelli, Loop Quantum Gravity, gr-qc/9710008, a review written for "Living Reviews".
13. E. Guadagnini, M. Martellini, M. Mintchev, Nucl. Phys. **B330**, 575 (1990).
14. See [11], chapter 10.
15. V.A. Vassiliev, Cohomology of knot spaces, in *Theory of Singularities and its applications*, ed. V.I. Arnold, Amer. Math. Soc., Providence, RI, 23 (1990); M. Alvarez, J.M.F. Labastida, Nucl. Phys. **b437**, 356 (1995).
16. J. Griego, Nucl. Phys. **B467**, 332 (1996).
17. R. Gambini, J. Pullin, The Gauss Linking Number in Quantum Gravity, in *Knots and Quantum Gravity*, ed. J. Baez, Oxford Univ. Press, 1994.
18. B. Brugmann, R. Gambini, J. Pullin, Phys. Rev. Lett. **68**, 431 (1991).
19. R. Di Pietri, C. Rovelli, Phys. Rev. **D 54**, 2664 (1996).
20. J.Baez, Spin networks in nonperturbative quantum gravity, in *The interface of Knots annd Physics*, ed. L. Kauffman, A.M.S. Providence, 1996; Adv. Math. **117**, 253 (1996).
21. See for example A. Ashtekar, D. Marolf, M. Mourao, Integration on the space of connections modulo gauge transformations, in *The Proceedings of the Lanczos International Centenary Conference*, ed. J.D. Brown et al, SIAM, Philadelphia, 1994.
22. C. Rovelli, L. Smolin, Nucl. Phys. **B442**, 593 (1995); A. Ashtekar, J. Lewandowski, Class. and Quantum Grav. **14**, A55 (1997); Quantum Theory of Geometry II: Volume operators, gr-qc/9711031.
23. F. Barbero, Phys. Rev. **D49**, 6935 (1994).
24. R. Borissov, R. DePietri, C. Rovelli, Class, Quant. Grav. **14**, 2793 (1997).
25. T. Thiemann, Class. Quant. Grav. **15**, 875 (1998).
26. T. Thiemann, Class. Quant. Grav. **15**, 1207 (1998).
27. J. Lewandowski, D. Marolf, Int. J. Mod. Phys. **D7**, 299 (1998).
28. R. Gambini, J. Lewandowski, D. Marolf, J. Pullin, Int. J. Mod. Phys. **D7**, 97 (1998).
29. L. Smolin, The classic limit and the form of the hamiltonian constraint in nonperturbative quantum gravity, gr-qc/9609034.
30. R. Gambini, J. Griego, J. Pullin, Phys. Lett. **B413**, 260 (1997); Phys. Lett. **B425**, 41 (1998).
31. R. Gambini, J. Griego, J. Pullin, Nucl. Phys. **B534**, 675 (1998).
32. C. Di Bartolo, R. Gambini, J. Griego, J. Pullin, Canonical quantum gravity in the Vassiliev invariants arena: I. Kinematical structure, in preparation.

Dynamical content of Chern-Simons Supergravity

Osvaldo Chandía[♮], Ricardo Troncoso and Jorge Zanelli

Centro de Estudios Científicos de Santiago, Casilla 16443, Santiago, Chile
Departamento de Física, Universidad de Santiago de Chile, Casilla 307, Santiago 2, Chile
[♮]*Instituto de Física Teórica, Rua Pamplona 145, Sao Paulo, Brazil*

Abstract. The dynamical content of local *AdS* supergravity in five dimensions is discussed. The bosonic sector of the theory contains the vielbein (e^a), the spin connection (ω^{ab}) and internal $SU(N)$ and $U(1)$ gauge fields. The fermionic fields are complex Dirac spinors (ψ^i) in a vector representation of $SU(N)$. All fields together form a connection 1-form in the superalgebra $SU(2,2|N)$. For $N = 4$, the symplectic matrix has maximal rank in a locally AdS background in which the dynamical degrees of freedom can be identified. The resulting efective theory have different numbers of bosonic and fermionic degrees of freedom.

I INTRODUCTION

Over twenty years ago, Cremmer, Julia and Scherk presented a beautiful $N = 1$ theory of supergravity in 11 dimensions [1], which, apart from the metric and a gravitino (ψ), included a three-form field ($A_{\mu\nu\lambda}$). This theory is quite unique: a larger D or N would give rise to inconsistencies upon compactification to 4 dimensions (i.e., fields with spin greater that 2 or more than one graviton). A "dual" alternative possibility which uses a six form ($A_{\mu_1 \dots \mu_6}$) instead of the three form also leads to inconsistencies [2]. Additionally, it has been also shown that regardless of the compactification arguments, the theory cannot accommodate a cosmological constant in eleven dimensions [3].

One of the puzzling aspects of the theory is the conjecture contained in the original paper by Cremmer, Julia and Scherk in the sense that this theory should be related to another one based on the *osp*(32|1) algebra, a problem that they promised to discuss in a forthcoming article that never appeared. In fact, it can be seen that in a gauge theory for an *osp*(32|1) connection, the anticommutator of the fermionic generators takes the form [4]

$$\{Q, Q\} \approx P_a \Gamma^a + Z_{ab} \Gamma^{ab} + Z_{abcde} \Gamma^{abcde},$$

where one could identify the different components of the connection that accompany the generators in the RHS with the fields in the CJS theory: $A_\mu^a \sim e_\mu^a$, $A_\mu^{ab} \sim A_{\mu\nu\lambda}$, $A_\mu^{abcde} \sim A_{\mu_1 \cdots \mu_6}$. However, no supergravity theory was known to contain all these fields in a simple and natural way. It was therefore a surprise for us to find a family of Lagrangians (one in each odd dimension) which could prove the conjecture of Cremmer, Julia and Scherk to be true [5]. The key ingredients in this new family of supergravity theories is their Chern-Simons form and the fact that the spacetime symmetry is realized in the tangent space and not on the base manifold.

II SUPERGRAVITY IN ODD DIMENSIONS

One of the unique features of gravity in 2+1 dimensions is that it is a genuine gauge theory in the sense of having a fiber bundle structure. This is because the standard Einstein-Hilbert action (both with and without cosmological constant) is a Chern-Simons (**CS**) form $\int < AdA + \frac{2}{3}A^3 >$. As a consequence, the theory is an integrable system, unlike the case of four-dimensional gravity [6]. Furthermore, the simplest $D = 3$, $N = 1$

CP484, *Trends in Theoretical Physics II*, edited by H. Falomir, R. E. Gamboa Saraví, and F. A. Schaposnik

supegravity theory [7] also shares this feature, and as a bonus, the theory is *locally* invariant under the anti-de Sitter group. It is easy to see that these features can be generalized to all odd dimensions: provided one has identified the superalgebra that extends AdS in a given dimension D, it is just a matter of constructing the appropriate Chern-Simons D-form. The resulting theory would be invariant by construction under the right supergroup in which the fields of the theory transform as different pieces of a connection. Thus, the algebra of the supersymmetry transformations is guaranteed to close off shell without requiring auxiliary fields [8].

However, both identifying the algebra and the construction of the CS form involve some subtleties that it is instructive to analyze in detail. The simplest example that captures all the problems –and which yields a theory with propagating local degrees of freedom– occurs in five dimensions. In the following we study the five-dimensional case in detail and indicate where appropriate how the results generalize to other dimensions.

III HIGHER DIMENSIONAL CS THEORIES

In 2+1 dimensions, gravity –or any other Chern-Simons theory for that matter– has no dynamical degrees of freedom. The field equations are

$$F = 0, \tag{1}$$

where F is the (anti-de Sitter or Poincaré) curvature, and it therefore means that all on-shell configurations are locally gauge-equivalent to a flat connection. However, the field equations of a CS theory in dimensions five and higher –for any gauge group– take the form

$$< F_\wedge F_\wedge \cdots {}_\wedge F \, G_M >= 0, \tag{2}$$

where G_M is a gauge generator. This equation in general does not imply a flat connection, allowing for the existence of propagating degrees of freedom.

For dimensions $D > 3$, the dynamical structure of a CS system becomes highly nontrivial. The root of the complexity lies in three independent features inherent of CS theories: **i)** they are gauge systems; **ii)** they have coordinate invariance built in; and **iii)** they are first order systems. Although each of these items are neither exclusive of CS systems, nor particularly difficult to deal with, their conjunction requires special care. As discussed in [9], there are two problems: diffeomorphism and gauge invariances are not completely independent, and because of the first order nature, CS systems have first class constraints inextricably mixed with second class ones. This makes the Dirac matrix hard to express in a simple form and almost impossible to invert.

Moreover, for $D > 3$ the symplectic form (Ω) that multiplies the velocities in (2) is a function of the gauge field $\Omega = \Omega(\mathbf{A})$, and it can degenerate for certain field configurations. In particular, for $D = 5$,

$$\Omega^{ij}_{MN}(\mathbf{A}) = \Delta_{MNP}\epsilon^{ijkl}\mathbf{F}^P_{kl}, \tag{3}$$

where $\Delta_{MNP} \equiv < \mathbf{G}_M\mathbf{G}_N\mathbf{G}_P >$ is an invariant tensor of the gauge group (see Appendix B). At the singular configurations (as $\mathbf{F}^P_{kl} = 0$), the rank of the Dirac matrix is reduced and becomes noninvertible, so that some (or all) second class constraints could actually be viewed as first class, and this further complicates the identification of the propagating degrees of freedom. Outside those regions, however, other dynamical structure is well behaved and the symplectic form has maximal rank.

Here we will not discuss the problems arising from the presence of degenerate surfaces. This seems to be a reasonable point of view as these singularities constitute sets of measure zero in the configuration space of the theory.

The dynamical analysis of a higher-dimensional bosonic CS theory is discussed in [9] and we summarize it here: If the gauge algebra takes the form $G \times U(1)$, where G is a semisimple algebra, the symplectic form can be computed in a background where Ω has maximal rank. The configurations that satisfy this requirement are called generic in the sense that under small deformations the rank of Ω remains maximal. Also, in these configurations, the first and second class constraints can be separated and the degrees of freedom computed. If the algebra has dimension $f > 1$ and the spacetime has dimension $D = 2n + 1 > 3$, it is shown that the number of propagating local degrees of freedom of the theory is

$$g = nf - f - n$$

and, in five dimensions, the symplectic matrix has the form

$$\Omega^{ij}_{MN}(\bar{\mathbf{A}}) = g_{MN}\epsilon^{ijkl}\bar{\mathbf{f}}_{kl}, \tag{4}$$

where the bar stands for background fields, and $\bar{\mathbf{f}}_{kl} = \partial_k \bar{b}_l - \partial_l \bar{b}_k$ is the curvature of the $U(1)$ field.

IV LOCAL ADS$_5$ SUPERGRAVITY

The supersymmetric extension of the anti-de Sitter algebra in five dimensions is $su(2,2|N)$ [10], whose associated connection can be written as,

$$\mathbf{A} = e^a \mathbf{J}_a + \frac{1}{2}\omega^{ab}\mathbf{J}_{ab} + a^\Lambda \mathbf{T}_\Lambda + (\bar{\psi}^r \mathbf{Q}_r - \bar{\mathbf{Q}}^r \psi_r) + b\mathbf{Z},$$

where the generators $\mathbf{J}_a$, $\mathbf{J}_{ab}$, form an AdS algebra ($so(4,2)$), $\mathbf{T}_\Lambda$ ($\Lambda = 1, \cdots N^2-1$) are the generators of $su(N)$, $\mathbf{Z}$ generates a $U(1)$ subgroup and $\mathbf{Q}, \bar{\mathbf{Q}}$ are the supersymmetry generators, which transform in a vector representation of $SU(N)$. The Chern-Simons Lagrangian for this gauge algebra is defined by the relation $dL = i < \mathbf{F}^3 >$, where $\mathbf{F} = \mathbf{dA} + \mathbf{A}^2$ is the (antihermitean) curvature, and $< \cdots >$ stands for the supertrace in the representation described in the Appendix A. Using this definition, one obtains the Lagrangian originally discussed by Chamseddine in [11],

$$L = L_G(\omega^{ab}, e^a) + L_{su(N)}(a^r_s) + L_{u(1)}(\omega^{ab}, e^a, b) + L_F(\omega^{ab}, e^a, a^r_s, b, \psi_r), \tag{5}$$

with

$$\begin{array}{lcl}
L_G & = & \frac{1}{8}\epsilon_{abcde}\left[R^{ab}R^{cd}e^e/l + \frac{2}{3}R^{ab}e^c e^d e^e/l^3 + \frac{1}{5}e^a e^b e^c e^d e^e/l^5\right] \\
L_{su(N)} & = & -Tr\left[a(da)^2 + \frac{3}{2}a^3 da + \frac{3}{5}a^5\right] \\
L_{u(1)} & = & \left(\frac{1}{4} - \frac{1}{N}\right) b(db)^2 + \frac{3}{4l^2}\left[T^a T_a - R^{ab}e_a e_b - l^2 R^{ab}R_{ab}/2\right] b \quad , \\
 & & + \frac{3}{N} f^r_s f^s_r b \\
L_F & = & \frac{3}{2i}\left[\bar{\psi}^r \mathcal{R}\nabla\psi_r + \bar{\psi}^s \mathcal{F}^r_s \nabla\psi_r\right] + c.c.
\end{array} \tag{6}$$

where $a^r_s \equiv a^\Lambda(\tau_\Lambda)^r_s$ is the $su(2,2)$ connection, f^r_s is its curvature, and the bosonic blocks of the supercurvature: $\mathcal{R} = \frac{1}{2}T^a\Gamma_a + \frac{1}{4}(R^{ab} + e^a e^b)\Gamma_{ab} + \frac{i}{4}dbI - \frac{1}{2}\psi_s\bar{\psi}^s$, $\mathcal{F}^r_s = f^r_s + \frac{i}{N}db\delta^r_s - \frac{1}{2}\bar{\psi}^r\psi_s$. The cosmological constant is $-l^{-2}$, and the AdS covariant derivative ∇ acting on ψ_r is

$$\nabla\psi_r = D\psi_r + \frac{1}{2l}e^a\Gamma_a\psi_r - a^s_r\psi_s + i\left(\frac{1}{4} - \frac{1}{N}\right)b\psi_r. \tag{7}$$

where D is the Lorentz covariant derivative.

The above relation implies that the fermions carry a $u(1)$ "electric" charge given by $e = \left(\frac{1}{4} - \frac{1}{N}\right)$. The purely gravitational part, L_G is equal to the standard Einstein-Hilbert action with cosmological constant, plus the dimensionally continued Euler density[1].

The action is by construction invariant –up to a surface term– under the local (gauge generated) supersymmetry transformations $\delta_\lambda A = -(d\lambda + [A, \lambda])$ with $\lambda = \bar{\epsilon}^r \mathbf{Q}_r - \bar{\mathbf{Q}}^r \epsilon_r$, or

$$\begin{array}{lcl}
\delta e^a & = & \frac{1}{2}\left(\bar{\epsilon}^r\Gamma^a\psi_r - \bar{\psi}^r\Gamma^a\epsilon_r\right) \\
\delta\omega^{ab} & = & -\frac{1}{4}\left(\bar{\epsilon}^r\Gamma^{ab}\psi_r - \bar{\psi}^r\Gamma^{ab}\epsilon_r\right) \\
\delta a^r_s & = & -i\left(\bar{\epsilon}^r\psi_s - \bar{\psi}^r\epsilon_s\right) \\
\delta\psi_r & = & -\nabla\epsilon_r \\
\delta\bar{\psi}^r & = & -\nabla\bar{\epsilon}^r \\
\delta b & = & -i\left(\bar{\epsilon}^r\psi_r - \bar{\psi}^r\epsilon_r\right).
\end{array}$$

---

[1] The first term in L_G is the dimensional continuation of the Euler (or Gauss-Bonnet) density from two and four dimensions, exactly as the three-dimensional Einstein-Hilbert Lagrangian is the continuation of the the two dimensional Euler density. This is the leading term in the limit of vanishing cosmological constant ($l \to \infty$), whose local supersymmetric extension yields a nontrivial extension of the Poincaré group [12].

As can be seen from (6) and (7), for $N = 4$ the b field looses its kinetic term and decouples from the fermions (the gravitino becomes uncharged with respect to the $U(1)$ field). The only remnant of the interaction with the b field is a dilaton-like coupling with the Pontryagin four forms for the AdS and $SU(N)$ groups (in the bosonic sector). As it is also shown in the Appendix A, the case $N = 4$ is also special at the level of the algebra, which becomes a superalgebra with a $u(1)$ central extension.

In the bosonic sector, for $N = 4$, the field equation obtained from the variation with respect to b states that the Pontryagin four form of AdS and $SU(N)$ groups are proportional . Consequently, if the curvatures approach zero sufficiently fast at spatial infinity, there is a conserved topological current which states that, for the spatial section, the second Chern characters of AdS and $SU(4)$ are proportional. Consequently, if the spatial section has no boundary, the corresponding Chern numbers are related. Using the fact that $\Pi_4(SU(4)) = 0$, the above implies that the Hirzebruch signature plus the Nieh-Yan number of the spatial section cannot change in time.

V SYMPLECTIC FORM

We will consider a background that is a solution for the field equations, for which Ω has maximum rank. Realizing this is in general a difficult task. However, an amazing simplification happen when $N = 4$, which has its root in the form of the invariant tensor Δ_{MNP}. As stated in the Appendix, considering the spliting: $M = \{M', Z\}$, when $N = 4$ we have "the accident": $\Delta_{ZZZ} = 0$, and $\Delta_{ZM'N'} = -\frac{i}{4} g_{M'N'}$, where $g_{M'N'}$ is the Killing metric of $PSU(2,2|4)$.

This fact will help us find an adequate background. Consider any locally AdS spacetime with pure gauge matter fields (Bosons and fermions), with the exception of the b field. That is, a background $\bar{\mathbf{A}}$ such that $\mathbf{F}^{AB} = \mathbf{F}_\Lambda = \psi_s = \bar{\psi}^r = 0 \neq \mathbf{F}^Z$.

It is easy to see also that for $N = 4$ the conditions that make the separation between first and second class constraints possible in the construction of [9] can also be applied, even if in this case the algebra is bigraded and *not* a direct sum of a semisimple one and $u(1)$, but a central extension.

Indeed, it can be directly checked that the symplectic form takes the form

$$\Omega^{ij}_{MN}\left[\bar{\mathbf{A}}\right] = \begin{bmatrix} 0_{4\times 4} & 0 \\ 0 & \widehat{\Omega}^{ij}_{M'N'} \end{bmatrix} \tag{8}$$

where $\widehat{\Omega}^{ij}_{M'N'}$ is generically an invertible matrix. In fact, for a flat AdS curvature (e.g., a spacetime of constant negative curvature and vanishing torsion) the symplectic matrix takes the form

$$\Omega^{ij}_{MN}\left[\bar{\mathbf{A}}\right] = \begin{bmatrix} 0 & 0 & 0 & 0 & 0 \\ 0 & \eta_{[AB][CD]} & 0 & 0 & 0 \\ 0 & 0 & g_{\Lambda_1\Lambda_2} & 0 & 0 \\ 0 & 0 & 0 & 0 & 2\delta^r_n\delta^\beta_\alpha \\ 0 & 0 & 0 & -2\delta^m_s\delta^\alpha_\beta & 0 \end{bmatrix} \otimes -\frac{i}{4}\epsilon^{ijkl}(\partial_k\bar{b}_l - \partial_l\bar{b}_k). \tag{9}$$

The nonvanishing block in the RHS can be recognized as the Killing metric for $PSU(2,2|4)$, while the factor on the right is the space-dual of the b field, $*db$.

This shows that the requirement for the algebra to be of the form $G \times U(1)$ in order to decouple first and second class constraints is sufficient but not necessary. In general, any semisimple (super) group with a $U(1)$ central extension [as in case of $N = 4$ super AdS$_5$ discussed above] will be sufficient too.

One can now count the degrees of freedom for the effective theory in this background. There are 15 generators of AdS$_5$, 15 for $SU(2,2)$, and 1 for $U(1)$, plus 4×4 Q^i_α's and an equal number of $\bar{Q}^\alpha_i$'s. According to the argument outlined in [9], this makes $f = 63$, $n = 2$, and a total of 61 independently propagating degrees of freedom.

The previous result is puzzling. There can be no matching of fermionic and bosonic degrees of freedom in this case. In fact, there seems to be a hidden subtlety in this counting because, at least in the perturbative sense, the number of degrees of freedom around this background is different. Consider a fluctuation in the connection around a fixed background $\bar{\mathbf{A}}$,

$$\mathbf{A}^M = \bar{\mathbf{A}}^M + \mathbf{u}^M, \tag{10}$$

where the dynamical fields will be the spatial components of $\mathbf{A} = \mathbf{A}^M \mathbf{G}_M$, with M ranging over the whole supergroup indices. Then, for small $\mathbf{u}$, the effective action in the linearized approximation, is given by

$$\begin{aligned} \mathbf{I}_{Eff}(\mathbf{u}_i^M) &\sim \int < \mathbf{u}\bar{\mathbf{F}}\bar{\nabla}\mathbf{u} > \\ &= \int d^5x \, [\mathbf{u}_i^M \Omega^{ij}_{MN}(\bar{\mathbf{A}})\bar{\nabla}_0 \mathbf{u}_j^N + 2\mathbf{u}_0^M \Omega^{ij}_{MN}(\bar{\mathbf{A}})\bar{\nabla}_i \mathbf{u}_j^N - 2\varepsilon^{ijkl}\mathbf{u}_i^A \Delta_{ABC}\bar{\mathbf{F}}^B_{oj}\bar{\nabla}_k \mathbf{u}_l^C]. \end{aligned} \tag{11}$$

If we consider the AdS background where $\Omega^{ij}_{MN}(\bar{\mathbf{A}})$ takes the form (8), then the counting comes to 58.

VI DISCUSSION AND OUTLOOK

1. From the previous discussion it is clear that for $N = 4$ the theory is extremely simple and the b field almost completely decouples from the rest of the fields. In fact, the b field is analogous to a Lagrange multiplier, and this analogy is completely accurate in the effective action for the perturbations, where the perturbation associated with this field doesn't appear at all in the effective action. Actually, around the same AdS background one cannot distinguish between the effective linearized theory described above from that containing only the Kähler-CS like action [13] described by the second term of $L_{u(1)}$. We will discuss this issue in detail elsewhere.

2. Another important problem is to show that the algebra generated by the conserved charges reproduces an algebra which *is not* isomorphic to the gauge algebra $su(2,2|4)$, but is a non-trivial central extension of $psu(2,2|4)$. Here the result is that this is indeed the case and that the algebra sets bounds to the values of the charges and conditions for the existence of Killing spinors that ensure that the background saturates the Bogomolnyí bound. This issue, in turn, raises the question of identifying the nontrivial BPS states. These states must keep a fraction of supersymmetry; in fact a solution of the field equations with these features is some sort of "topological black hole" [14] and it is reassuring to verify that the extreme case saturates the bound. These problems are going to be discussed in a forthcoming article.

3. The five dimensional theory generates a new four-dimensional superconformal theory at the boundary. This theory is constructed on the generalization of the Kač-Moody extension of the superconformal algebra without central charge, that is, the WZW_4 algebra for $PSU(2,2|1)$. This theory could be a rich test ground in the context of AdS/CFT duality conjecture [15].

4. Another interesting problem is try to generalize what is known about $D = 5$ to higher dimensions. We have shown [5] that the $D = 11$, $N = 32$ theory admits a nontrivial extension of the AdS superalgebra with one abelian generator for which anti-de Sitter space without matter fields is a background of maximal rank, and the gauge superalgebra is realized in the Dirac brackets. On a background like the used in the five dimensional theory, the $D = 11$ theory has 2^{12} fermionic and $2^{12} - 1$ bosonic degrees of freedom, and the (super) charges obey a non-trivial central extension of the $OSP(32|32)$ algebra.

ACKNOWLEDGMENTS

The authors are grateful to R. Aros, M. Contreras, J. Gamboa, J. M. F. Labastida, J. Maldacena, C. Martínez, F. Méndez, P. van Nieuwenhuizen, R. Olea, C. Teitelboim for many enlightening discussions and helpful comments. This work was supported in part through grants 1990189, 1970151, 1980788, and 3960009 from FONDECYT, and grant 27-953/ZI-DICYT (USACH). The institutional support of FUERZA AEREA DE CHILE, I. Municipalidad de Las Condes, and a group of Chilean companies (AFP Protección, Business Design Associates, CGE, CODELCO, COPEC, Empresas CMPC, GENER S.A., Minera Collahuasi, Minera Escondida, NOVAGAS and XEROX-Chile) is also recognized. R.T and J.Z. wish to express their gratitude to Marc Henneaux for his kind hospitality in Brussels and for many fruitful discussions and key comments. Last but not least, we wish to thank the organizers for the warm atmosphere at the meeting in Buenos Aires.

APPENDICES

A SUPERSYMMETRIC EXTENSION OF ADS$_5$ ALGEBRA

As discussed in [10], [8], the supersymmetric extension of the anti-de Sitter algebra in five dimensions is $su(2,2|N)$. This is the Lie algebra associated with the invariance group of the quadratic form $q = \theta^{*\alpha} g_{\alpha\beta}\theta^{\beta} + z^{*r}u_{rs}z^{s}$, with $\alpha, \beta = 1, ..., 4$ and $r, s = 1, ..., N$. Here θ^{α} are complex Grassmann numbers and $g_{\alpha\beta}$ and u_{pq} are sesquilinear metrics, which will be chosen as $g_{\alpha\beta} = i(\gamma_0)_{\alpha\beta}$ and $u_{rs} = \delta_{rs}$. The supersymmetric algebra contains $su(2,2) \oplus su(N) \oplus u(1)$ as the bosonic subalgebra. Using the isomorphism: $su(2,2) \simeq so(4,2)$, the generators are composed by the AdS generators ($\mathbf{J}_{AB}$), with $A, B = 0, ...5$, the $2\times 4N$ (complex) supersymmetry generators ($\mathbf{Q}^{\alpha}_{r}$, $\bar{\mathbf{Q}}^{r}_{\alpha}$), and the rest of the algebra is composed by the generators of internal (Lorentz scalar) $su(N) \otimes u(1)$ ($\mathbf{T}_{\Lambda}$, $\mathbf{Z}$), with $\Lambda = 1, ..., N^2 - 1$.

A natural representation acting in the vector superspace $(\theta^{\beta}; z^{q})$ is given by $(4+N)\times(4+N)$ matrices as follows. Defining $\mathbf{I}_{4\times 4}=\begin{bmatrix} \delta^{\alpha}_{\beta} & 0 \\ 0 & 0 \end{bmatrix}$, $\mathbf{I}_{N\times N} = \begin{bmatrix} 0 & 0 \\ 0 & \delta^{r}_{s} \end{bmatrix}$, the generators are:

•AdS generators

$$\mathbf{J}_{AB} = \frac{1}{2}\Gamma_{AB} \otimes \mathbf{I}_{4\times 4}, \tag{A1}$$

$$= \begin{bmatrix} \frac{1}{2}(\Gamma_{AB})^{\alpha}_{\beta} & 0 \\ 0 & 0 \end{bmatrix}, \tag{A2}$$

• $su(N)$ **generators**

$$\mathbf{T}_{\Lambda} = \tau_{\Lambda} \otimes \mathbf{I}_{N\times N} \tag{A3}$$

$$= \begin{bmatrix} 0 & 0 \\ 0 & (\tau_{\Lambda})^{r}_{s} \end{bmatrix} \tag{A4}$$

where $(\tau_{\Lambda})^{r}_{s}$ are the antihermitean generators of $su(n)$.

•Supersymmetry generators

$$\mathbf{Q}^{\alpha}_{s} = \begin{bmatrix} 0 & 0 \\ -\delta^{\alpha}_{\beta}\delta^{r}_{s} & 0 \end{bmatrix},$$

$$\bar{\mathbf{Q}}^{r}_{\alpha} = \begin{bmatrix} 0 & \delta^{r}_{q}\delta^{\beta}_{\alpha} \\ 0 & 0 \end{bmatrix}.$$

• $u(1)$ **charge**

$$\mathbf{Z} = \frac{i}{4}\mathbf{I}_{B} + \frac{i}{N}\mathbf{I}_{F} \tag{A5}$$

$$= \begin{bmatrix} \frac{i}{4}\delta^{\alpha}_{\beta} & 0 \\ 0 & \frac{i}{N}\delta^{r}_{s} \end{bmatrix}. \tag{A6}$$

The commutators of the bosonic generators are those for the algebra $so(4,2) \oplus su(N) \oplus u(1)$: $[\mathbf{J},\mathbf{J}] \sim \mathbf{J}$, $[\mathbf{T},\mathbf{T}] \sim \mathbf{T}$, $[\mathbf{Z},\mathbf{Z}] = 0$, $[\mathbf{J},\mathbf{T}] = 0$, $[\mathbf{Z},\mathbf{J}] = 0$, $[\mathbf{Z},\mathbf{T}] = 0$. The supersymmetry generators transforms as spinors under AdS (then also under Lorentz), as "vectors" under $su(N)$, and carry $u(1)$ charge,

$$\begin{aligned}
[\mathbf{J}_{AB}, \mathbf{Q}^{\alpha}_{s}] &= -\tfrac{1}{2}(\mathbf{\Gamma}_{AB})^{\alpha}_{\beta}\mathbf{Q}^{\beta}_{s} \\
[\mathbf{J}_{AB}, \bar{\mathbf{Q}}^{r}_{\beta}] &= \tfrac{1}{2}(\mathbf{\Gamma}_{AB})^{\alpha}_{\beta}\bar{\mathbf{Q}}_{\alpha} \\
[\mathbf{T}_{\Lambda}, \mathbf{Q}^{\alpha}_{s}] &= (\tau_{\Lambda})^{r}_{s}\mathbf{Q}^{\alpha}_{r} \\
[\mathbf{T}_{\Lambda}, \bar{\mathbf{Q}}^{r}_{\beta}] &= -(\tau_{\Lambda})^{r}_{s}\bar{\mathbf{Q}}^{s}_{\beta} \\
[\mathbf{Z}, \mathbf{Q}^{\alpha}_{s}] &= -i(\tfrac{1}{4} - \tfrac{1}{N})\mathbf{Q}^{\alpha}_{s} \\
[\mathbf{Z}, \bar{\mathbf{Q}}^{r}_{\beta}] &= i(\tfrac{1}{4} - \tfrac{1}{N})\bar{\mathbf{Q}}^{r}_{\beta}.
\end{aligned} \tag{A7}$$

Finally, the anticommutator reads

$$\left\{\mathbf{Q}_s^\alpha, \overline{\mathbf{Q}}_\beta^r\right\} = -\frac{1}{2}\delta_s^r(\Gamma^a)_\beta^\alpha \mathbf{J}_a + \frac{1}{4}\delta_s^r(\Gamma^{ab})_\beta^\alpha \mathbf{J}_{ab} - \delta_\beta^\alpha(\tau^\Lambda)^r{}_s T_\Lambda + i\delta_\beta^\alpha\delta_s^r \mathbf{Z}. \tag{A8}$$

where $\mathbf{J}_a := \mathbf{J}_{a5}$.

It is clear from the algebra that the case $N = 4$ is a special one: the generator $\mathbf{Z}$ commutes with the rest of the algebra and it is just a central extension, as can be read from the right hand of (A8).

It is important to point out that, if $\mathbf{Z}$ were omitted, the new algebra, $psu(2,2|4)$, does not admit the representation described above, but still satisfies the Jacobi identity. Because of this last feature, the full resulting algebra for $N = 4$, **is not** $psu(2,2|4) \oplus u(1)$.

B THIRD RANK INVARIANT TENSOR FOR N EXTENDED SUPER ADS$_5$

¿From the matrix representation described above, it is straighforward to obtain a third rank invariant tensor for the AdS$_5$ supergroup, which is needed to analyze the dynamics of local AdS$_5$ supergravity.

Let $\mathbf{G}_M$ the generators of $su(2,2|N)$, where the index M ranks from the whole superalgebra: $M = \left\{[AB], \Lambda, Z, \binom{r}{\alpha}, \binom{\beta}{s}\right\}$.

The required tensor is defined through $\Delta_{MNP} = < \mathbf{G}_M, \mathbf{G}_N, \mathbf{G}_P >$, where $< ... >$ stands for the supertrace, which ensure the invariance of the tensor under the action of the group. Because the supertrace is the difference between the trace of the upper and lower diagonal bosonic blocks, the invariant tensor has the form:

$$\begin{array}{lcl}
\Delta_{ZZZ} & = & -i\left(\frac{1}{4^2} - \frac{1}{N^2}\right) \\
\Delta_{[AB][CD][EF]} & = & \frac{i}{2}\epsilon_{ABCDEF} \\
\Delta_{(\Lambda_1)(\Lambda_2)(\Lambda_3)} & = & -Tr\left[(\tau_{\Lambda_1})(\tau_{\Lambda_2})(\tau_{\Lambda_3})\right] \\
\Delta_{Z[AB][CD]} & = & -\frac{i}{4}\eta_{[AB][CD]} \\
\Delta_{Z(\Lambda_1)(\Lambda_2)} & = & -\frac{i}{N}g_{\Lambda_1\Lambda_2} \\
\Delta_{Z\binom{\beta}{s}\binom{r}{\alpha}} & = & i\left(\frac{1}{4} + \frac{1}{N}\right)\delta_\alpha^\beta\delta_s^r \\
\Delta_{[AB]\binom{\beta}{s}\binom{r}{\alpha}} & = & \frac{1}{2}(\Gamma_{AB})_\alpha^\beta\delta_s^r \\
\Delta_{(\Lambda)\binom{\beta}{s}\binom{r}{\alpha}} & = & -\delta_\alpha^\beta(\tau_\Lambda)_s^r,
\end{array}$$

where $\eta_{[AB][CD]} := \eta_{AC}\eta_{BD} - \eta_{AD}\eta_{BC})$, and $g_{\Lambda_1\Lambda_2} = Tr\left[(\tau_{\Lambda_1})(\tau_{\Lambda_2})\right]$ is the Killing metric of $su(N)$.

Note that, again in the special case $N = 4$, and considering the splitting $M = \{M', Z\}$, then $\Delta_{ZZZ} = 0$, and $\Delta_{ZM'N'} = -\frac{i}{4}g_{M'N'}$,where $g_{M'N'}$ is the (invertible) Killing metric of $PSU(2,2|4)$.

REFERENCES

1. E. Cremmer, B. Julia and J. Scherk, *Phys. Lett.* **B76** (1978) 409.
2. H. Nicolai, P. K. Townsend and P. van Nieuwenhuizen, *Lett. Nuovo Cim.* **30** (1981) 315.
3. K. Bautier, S. Deser, M. Henneaux and D. Seminara *Phys. Lett.***B406** (1997) 49.
4. J. A. de Azcárraga, J. M. Izquierdo and P. K. Townsend, *Phys.Lett.***B267** (1991)366.
5. R.Troncoso and J. Zanelli, *Phys. Rev.* **D58**, (1998) R101703 ; *Int. Jour. Theor. Phys.***38** (1999) 1193.
6. E. Witten, *Nucl. Phys.* **B 311** (1989) 46.
7. A. Achúcarro and P. K. Townsend, Phys. Lett. **B180** (1986) 89.
8. R.Troncoso and J. Zanelli, *Chern-Simons Supergravities with Off-Shell Local Superalgebras,* in *Black Holes and Structure of the Universe,* C.Teitelboim and J.Zanelli, editors World Scientific, Sigapore, 1999.
9. M. Bañados, M. Henneaux and L.Garay, *Nucl. Phys.* **B 476**, (1996) 611.
10. W. Nahm, Nucl. Phys. **B135** (1978) 149. J. Strathdee, Int. J. Mod. Phys. **A 2** (1987) 273.
11. A. Chamseddine, *Phys. Lett.* **B233** (1988) 291; *Nucl. Phys.* **B346** (1990) 213.
12. M. Bañados, R. Troncoso and J. Zanelli, *Phys. Rev.* **D54**, (1996) 2605.
13. V. P. Nair and J. Schiff, Phys. Lett. **B 246**, 423 (1990); Nucl. Phys. **B 371**, 329 (1992).
14. M. Bañados, A. Gomberoff and C. Martínez, Class.Quant.Grav.**15**, 3575 (1998).
15. J. Maldacena, Adv. Theor. Math. Phys. **2**, 231 (1998).

Effective temperatures out of equilibrium

Leticia F. Cugliandolo

Laboratoire de Physique Théorique de l'École Normale Supérieure
24 rue Lhomond, 75231 Paris Cedex 05, France and
Laboratoire de Physique Théorique et Hautes Energies, Jussieu
5 ème étage, Tour 24, 4 Place Jussieu, 75005 Paris France

Abstract. We describe some interesting effects observed during the evolution of nonequilibrium systems, using domain growth and glassy systems as examples. We briefly discuss the analytical tools that have been recently used to study the dynamics of these systems. We mainly concentrate on one of the results obtained from this study, the violation of the fluctuation-dissipation theorem and we discuss, in particular, its relation to the definition and measurement of effective temperatures out of equilibrium.

One of the major challenges in physics is to understand the behaviour of systems that are far from equilibrium. These systems are ubiquitous in nature. Some examples are phase separation, systems undergoing domain growth, all types of glasses, turbulent flows, systems driven by non-potential forces, etc. All these systems are "large" in the sense that they are composed of many, $N \to \infty$, dynamic degrees of freedom. Apart from succeeding in predicting the time evolution of their macroscopic properties, one would like to know which, if any, of the thermodynamic notions apply to these nonequilibrium cases.

Systems undergoing domain growth, or phase separation, provide the best known example of a nonequilibrium evolution [1]. Take for instance a magnetic system with ferromagnetic interactions in contact with a thermal bath. If the bath temperature is very high the sample is in its paramagnetic phase and the magnetic moments, or spins, point in random directions. If one next cools down the bath, and hence the sample, through a transition temperature T_c, the system enters the low temperature phase and starts forming *domains* or islands of the two ordered phases, say up and down. For definiteness, let us fix the final temperature to be $0 < T < T_c$. At any time t_w after crossing the transition at the initial time, two types of dynamics appear: (i) *fast* fluctuations of some spins, due to thermal fluctuations, inside the otherwise fully ordered domains; (ii) *slow* motion of the domain walls leading to the growth of the averaged domain size $L(t_w)$. If the size of the sample is infinite, in real life very large, the nonequilibrium domain-growth process can take so long that the sample simply does not equilibrate in the time-window that is accessible experimentally. In other words, below the critical temperature T_c one always has $\tau_{\text{OBS}} < \tau_{\text{EQ}}$ with τ_{OBS} the observation time and τ_{EQ} the equilibration time. The two types of dynamics itemized above are clearly seen in Fig. 1 where three two-dimensional slices of a system undergoing domain-growth are displayed. The pictures are obtained at increasing waiting times after the quench. One sees the domains growing as well as the existence, in each of the snapshots, of some reversed spins inside the otherwise ordered domains.

Scaling arguments have been extensively used to describe the dynamics below T_c; they are based on the assumption (sometimes derivation) of the evolution of the averaged domain size $L(t_w)$ and on further proposals for the space and time-dependence of the correlation functions.

Understanding the physics of glassy materials is perhaps a problem of intermediate difficulty [2,3]. Glassy materials can be of very many different types; one has for instance structural glasses [2] , orientational glasses [4], spin-glasses [5,6], plastics [7], gels and clays [8], glycerol [9], etc. Their hallmark is that below some transition range they also fall out of equilibrium.

The easiest way of preparing a glassy system is again through an annealing [2]. This is implemented by decreasing the temperature of the bath with a given cooling-rate. Take for example the case of a molecular

CP484, *Trends in Theoretical Physics II*, edited by H. Falomir, R. E. Gamboa Saraví, and F. A. Schaposnik

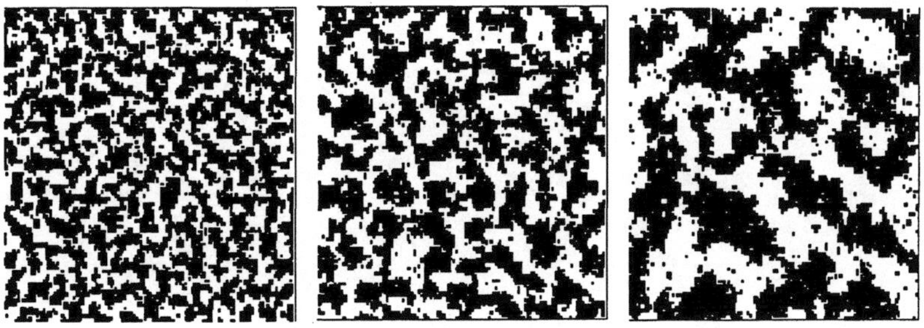

FIGURE 1. Three two-dimensional slices of a domain-growth system at three different waiting times $t_o < t_{w1} < t_{w2} < t_{w3}$. See the text for a discussion.

liquid. For high enough bath temperatures the sample is in its liquid phase and it achieves equilibrium with the bath. At an intermediate bath-temperature range the liquid avoids the crystallization transition, enters a metastable phase and becomes a *super-cooled liquid*, that is to say a liquid with some peculiar properties as, for example, a extremely high viscousity. At an even lower bath-temperature the liquid cannot follow the pace of the annealing and falls out of equilibrium, it becomes a *glass*. If one stops the annealing at any temperature below this range the system stays in its glassy phase for practical purposes forever and is typically an amorphous solid. We talk about a "transition range" since the transition might not be clearcut but depend on the cooling-rate. Actually one can form a glass of probably any substance by choosing a fast enough cooling-rate. Many other routes to the glassy phase are also possible.

There have been proposals to describe the evolution of some glasses, notably spin-glasses, with scaling arguments based on domain growth ideas [10,11]. The assumption is that the glassy dynamics is simply given by the growth of domains of two competing ground states. However, it has been very difficult to prove (or disprove!) either experimentally or numerically that this is indeed the scenario: no "ordered structures" have been identified in general as the growing phases.

Thus, domain growth, phase separation and glassy materials are all "self-sustained"[1] out of equilibrium systems. If one follows their time-evolution, keeping all parameters fixed, in particular the bath-temperature, some of the main features observed during their nonequilibrium evolution are:

Slow dynamics. The evolution is very slow. "One-time quantities", as the energy-density, approach their asymptotic limit with some slow decaying function, say power law, logarithmic or more complicated. It is very important to notice though that even if these one-time quantities can get very close to their asymptotic values, this does not mean that the systems *get frozen* in a metastable state: they are not equilibrated in a restricted region of phase space characterised by these asympotic values. This is most clearly demonstrated by the measurement of "two-time quantities".

Two-time quantities and physical aging. The measurement of these quantities prove that, even if one-time quantities approach a limit, the system is still changing in an important way.

One can distinguish two types of two-time quantities. Those measured during the free evolution of the system, quantifying the spontaneous fluctuations, such as any two-time correlation function, and those measured after applying a small perturbation, such as any response function. These quantities depend on both times involved in the measurement and not only on the time-difference. This shows that the systems neither are equilibrated with the bath nor have approached equilibrium in any metastable state. They are indeed rather far from equilibrium.

The measurement of the spontaneous fluctuations is quite easy to implement in a numerical simulation [12]. One prepares the sample at an initial time t_o and lets it evolve until a waiting time t_w when the system configuration is recorded. One then lets the sample further evolve and computes, at all subsequent times $t \equiv \tau + t_w$, the correlation function between the reference configuration at t_w and the configurations at $\tau + t_w$. These curves depend on both t_w and τ and they are not invariant under time translations showing that the system is out of equilibrium. Furthermore, the decay as a function of τ is slower the longer t_w. This is the phenomenon called *physical aging*: the younger (older) the sample the faster (slower) the decay.

The result of the measurement of a local auto-correlation function is very easy to visualize for a domain

[1] In the sense that no external perturbation is keeping them far from equilibrium.

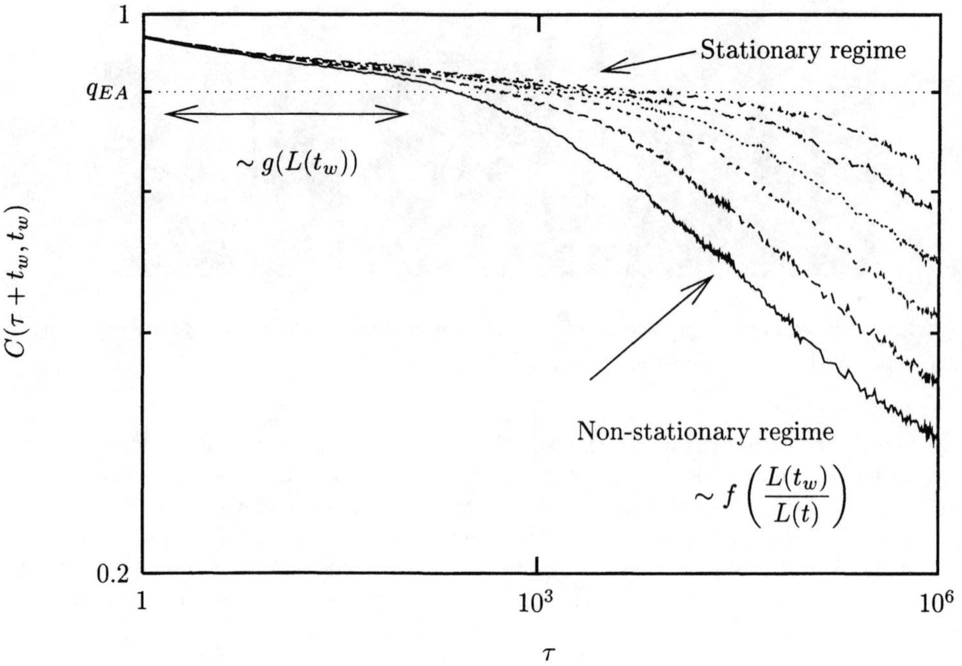

FIGURE 2. The local auto-correlation function for wait-times $t_{w1} < t_{w2} < t_{w3} < t_{w4} < t_{w5} < t_{w6}$. See the text for a discussion of the different regimes.

growth. Take again the case of ferromagnetic domain growth. the dynamic variables are the Ising spins that we encode in a time-dependent N-dimensional vector $\phi(t) = (\phi_1(t), \ldots, \phi_N(t))$ (the index $i = 1, \ldots, N$ labels the spins) and the local auto-correlation function is just the scalar product of two configurations evaluated a different times, $NC(t, t_w) = \langle \phi(t) \cdot \phi(t_w) \rangle$. For Ising spins, the auto-correlation function is normalized to one at equal times. A departure from one measures how different are any two configurations as those shown in Fig. 1. For any fixed t_w the auto-correlation has two distinct regimes depending on the time-difference $\tau \equiv t - t_w$. Let us choose a waiting time t_{w1} and plot $C(\tau + t_{w1}, t_{w1})$ as a function of τ. The curve has a first fast decay from one to a bath-temperature dependent value $q_{EA}(T)$ (the Edwards-Anderson temperature-dependent parameter). This corresponds to the decorrelation associated to the fast flipping of the spins inside the domains. In this regime τ is small compared to an increasing function of the domain size $g(L(t_w))$. When τ increases and becomes of the order of $g(L(t_w))$ one starts seeing the motion of the domain-walls, i.e. the growth of the domains, and the decay slows down. If one repeats this calculation choosing a longer waiting-time $t_{w2} > t_{w1}$, and its associated reference configuration, one observes that the first decay is identical to the one for t_{w1} though it lasts for longer, and that the second regime is notably slower that the one associated to t_{w1}. These features can be easily understood. While τ is smaller than $g(L(t_w))$ the dynamics takes place only inside the domains as thermal fluctuations. The domain walls are ignored and the correlation behaves as if the system were a patchwork of the two equilibrium states. The correlation function decay is then independent of the waiting-time and approaches $q_{EA}(T) = m^2_{\text{EQ}}(T)$. However, after a time-difference of the order of $g(L(t_w))$ the system realizes it has domain walls, the subsequent decay is associated to the motion of the walls and is nonequilibrium in nature. The decay gets slower the longer the t_w simply because the size of the domains reached at t_w is larger.

For glassy systems the correlation functions have exactly the same qualitative behaviour though, as already mentioned, it is not easy to decide if there is any type of order growing. Plots like the one displayed in Fig. 2 have been obtained for an impressive number of glassy models of different nature. Some of them are the 3D Edwards-Anderson model [12], a polymer melt [13], a polymer in a random potential [14], a binary Lennard-Jones mixture [15], etc. Furthemore, the kind of curves were found in several sandpile models and other kind of systems [16].

The measurement of "dc-response" functions or, more precisely, integrated dc-responses is what is usually done experimentally. The starting procedure is similar to the precedent: one prepares the sample at an initial time and lets it freely evolve until t_w. At this waiting time one applies a small, constant, perturbation and then measures the associated integrated response of the system as a function of τ for different t_w. For example, experimenting with spin-glasses one applies a small magnetic field and measures the increase of magnetization [6], manipulating with polymer glasses one applies a stress and measure the tensile creep compliance [7], etc. In all cases, the integrated responses are studied as functions of t_w and τ and they all show aging effects that manifest in a similar way as in the correlation measurement. There is a first increase of the time-integrated susceptibility towards a value $\chi_1(T)$ that does not depend on the waiting-time while there is a second increase of the time-integrated susceptibility towards the equilibrium value $\chi_{\rm EQ}(T)$ that is waiting-time dependent.

This can again be simply visualize in the domain growth problem. The first regime corresponds to the response of the spins inside the domains, i.e. to the response of the full system taken to be roughly a patchwork of independent equilibrium states. The second regime instead corresponds to the response of the domain walls. Since their density decreases as time elapses, one expects this nonequilibrium response to vanish in the long witing-time limit.

"Ac-response" measurements are also usually performed experimentally [4,9]. In these experiments one applies an ac small field of fixed frequency ω at the initial time t_0 and keeps it applied until the measuring time t_w. The relation with the previous results is given by identifying $\omega \sim 1/\tau$. The out of equilibrium character of the evolution is given by an explicit t_w dependence in the relaxation of the in and out of phase susceptibilities.

It is important to notice that these effects are *physical aging* as opposed to *chemical aging.* Physical aging is totally reversible: it suffices to heat the sample above the transition range and cool it back again below it to recover a fully rejuvenated system.

The comparison of spontaneous fluctuation, e.g. a correlation function, to induced fluctuations, measured by its associated response function, is well established for systems evolving in equilibrium. Indeed, this relation involves the temperature of the bath and it is called the *fluctuation-dissipation theorem.* However, for systems that are far from equilibrium this relation does not necessarily hold.

The problem of reaching a theoretical understanding of nonequilibrium physics is important both from a practical and a theoretical point of view. It is obvious that for some applications, one would like to predict the time-evolution of the samples with great precision and avoid undesired changes that depend on the, sometimes unknown, age of the samples. From an analytic point of view, domain-growth and glassy materials are a pattern of out of equilibrium systems whose properties one could try to capture with simple models or simplified approaches to more complex models. The predictions thus obtained can then be experimentally (or numerically) tested in real systems. Importantly enough, one could also try to extend some of these predictions to other nonequilibrium systems such as those externally driven. Some connections between glassy systems and driven systems are discussed in Ref. [17].

How can one modelise domain growth or glassy systems? The "microscopic" constituents and interactions and, consequently, the microscopic models differ from glass to glass. Spin-glasses are composed of magnetic impurities (spins) that occupy fixed random positions in the sample and interact via RKKY interactions; polymer glasses are composed of polymers (strings) that interact via potentials of (oversimplifying) Lennard-Jones type. In the former case, there is quenched (time-independent) disorder in the system that is associated to the random positions of the spins, that give rise to random interactions between them (the RKKY interactions oscillate very rapidly with the distance between the magnetic impurities and change sign in an almost random manner). In the latter case nothing can be interpreted as being quenched disorder. Though one could expect that these two systems (and other type of glasses) behave very differently, the experimental results as well as the recent theoretical developments show that their dynamical behaviour is indeed rather similar. In other words, one can identify certain quantities that have the same qualitative behaviour.

In order to describe the dynamic evolution of a *classical* system in contact with an environment one starts by identifying the relevant variables of the system that evolve in time. One then proposes a Langevin equation, with noise and friction mimicking the coupling of the system to the thermal bath, to determine the time-evolution of the time-dependent variables. These variables are the (soft) spins in the case of a spin-glass, the monomer positions in the case of the polymer glass, etc. This procedure leads to a set of N, the number of dynamic variables in the system, coupled differential equations of second (if inertia is included) or first (if inertia is neglected) order. Obviously, this huge system cannot be solved and one has to resort to some alternative method to further advance in the analysis.

Indeed, one would like to obtain information about "macroscopic" quantities, that can be related to experimental measurements, instead of following the erratic motion of any microcopic variable. The two-point correlation or response functions are macroscopic quantities we would like to monitor. A well-known theoretical method, known under the name of Martin-Siggia-Rose (MSR) formalism [18], allows us to obtain a generating functional, and from it the Schwinger-Dyson integro-differential equations, for these quantities. This method can be applied in complete generality, to any classical model. For models with non-linear interactions of finite range it recquires the calculation of an infinite series of diagrams that one cannot in general resum and express in terms of two-point functions in an explicit form [19]. One then faces the problem of choosing some approximation scheme to simplify this series expansion.

Before discussing how to deal with this problem, let us describe the formalism used to study a *quantum* system in contact with an environment. The Schwinger-Keldysh closed-time path (CTP) formalism was developed to monitor the nonequilibrium time-evolution of a quantum system, and to obtain information about two-time quantities [20]. The environment is usually modelized by a set of harmonic oscillators (infinitely many for each variable in the system) with a spectral distribution of frequencies [21]. The coupling of system and bath is usually chosen to be linear but of course more general situations can be considered. In this way, one obtains the CTP generating functional that, as in the classical case, involves a series expansion that, in general, cannot be obtained explicitly. (The classical limit of the CTP generating functional is the MSR generating functional.)

Typically, two routes are followed to approximate these generating functionals. They are equally applicable in the classical and quantum case and are the following [3]:

- The microscopic models, namely the starting Hamiltonians, are simplified in such a way that the construction can be carried through and that explicit equations can be derived. This is the choice made when one uses, for example, the large N limit of a $O(N)$-model to describe domain growth, fully connected spin models to describe spin-glasses or when one embeds a finite dimensional manifold in an infinite dimensional space to describe an interface motion, the motion of a polymer in a random medium.

- The microscopic models are realistic but some approximation scheme is chosen to select, from the infinite series, a still infinite subset of diagrams that can be resummed to yield an explicit set of dynamic equations for correlations and responses. Many such recipes exist in the literature [22–24], some of them are the mode-coupling approximation, the direct interaction approximation, the self-consistent screening approximation, etc.

These two procedures yield the same "form" of coupled integro-differential causal equations. Actually, in some cases one can show that a simplified microscopic model with infinite range interactions (e.g. the p spin-glass model) yields the same dynamic equations that an approximation scheme (e.g. the mode coupling appoximation) applied to a more realistic model for a glassy material [25,26]. The structure of these equations is always the same: there might be a second time-derivative term if there is inertia, some terms describe the interaction of the system with the bath and some other integral terms describe the interactions in the system (through the self-energy and vertex). It is the explicit form of the self-energy and vertex that is selected by the model or the approximation.

Once one has the equations governing the evolution of the two-time quantities, for all values of the parameters, the question then arises as to which is the phenomenology that they describe.

A combination of analytic and numeric methods are used to study these equations. One can attempt a numerical solution taking advantage of the fact that they are causal. The solution shows that they present a dynamic phase transition at a temperature T_d. Above T_d, the solution reaches, in the long waiting-time limit, a stationary form. All two-time correlations and responses are functions of the time-difference only and are related through the fluctuation-dissipation theorem. The high-temperature dynamic equations for spin-glass were studied in detail by Sompolinsky and Zippelius [27] for spin-glass models, by Götze and collaborators [28] for glass-models and the relation between these two was signalled and investigated by Kirkpatrick, Thirumalai and Wolynes [25] in a series of beautiful papers.

Below T_d, a drawback of the numerical method is that, due to the slowness of the dynamics and the memory of the system, one cannot reach very long time intervals. The numerical solution gives us hints about the structure of the solution but does not give us extremely precise information about more detailed features such as the two-time scaling laws, etc. Nevertheless, the numerical solution sufficed to show that below T_d two-time functions start depending on the waiting-time and that aging is captured by these equations [29].

Below T_d, and in the asymptotic limit of long waiting-time, an analytical solution was developed first for the p spin-glass model [29] and later for other mean-field disordered models such as Sherrington-Kirkpatrick [30] or the motion of manifolds in infinite dimensional random potentials [31,32].

One of the main ingredients of this solution [29,30] concerns the *fluctuation-dissipation theorem* that relates, in equilibrium, the spontaneous to the induced fluctuations. Indeed, if one follows the dynamics of a classical system that is in equilibrium with a bath, one can easily show that

$$R(t,t') \equiv \left. \frac{\delta \langle O(t) \rangle}{\delta h(t')} \right|_{h=0} = \frac{1}{T} \frac{\partial}{\partial t'} \langle O(t) O(t') \rangle \; \theta(t-t') = \frac{1}{T} \frac{\partial}{\partial t'} C(t,t') \; \theta(t-t') \; , \tag{1}$$

with $O(t)$ any observable taken to have zero mean for simplicity and h an infinitesimal field acting a time t' that modifies the energy of the system according to $V \to V - hO$ and that is not correlated with the equilibrium configuration of the system.

In the glassy phase, this relation does not hold. This does not come as a surprise since the equilibrium condition under which it can be proven does not apply. What really comes as a surprise is that the modification of the relation between response and correlation takes a rather simple form for domain-growth and glassy systems.

A way to quantify the modification of FDT in the out of equilibrium phase and to use it to classify different systems is the following. Let us integrate the response function over a time-interval going from a waiting-time t_w to a final time t:

$$\chi(t,t_w) = \int_{t_w}^{t} dt' \, R(t,t') \; . \tag{2}$$

This yields a time-integrated susceptibility that is exactly what is measured experimentally. Next, we compare this integrated-susceptibility to the auto-correlation function. *In equilibrium*, one can use FDT to show that

$$\chi(t,t_w) = \frac{1}{T} \left(C(t,t) - C(t,t_w) \right) \; . \tag{3}$$

Hence, if one draws a plot of χ against C, for increasing t_w, using $\tau = t - t_w$ as a parameter, in the large t_w limit the plot will approach a straight line of slope $-1/T$ joining $(\lim_{t\to\infty} C(t,t), 0)$ and $(0, \chi_{\rm EQ})$. From now on and without loss of generality we take $\lim_{t\to\infty} C(t,t) = 1$. Any departure from this straight line signals a modification of FDT and a departure from equilibrium.

The analytic solution of simplifed models shows that, in the *nonequilibrium* phase, this construction converges to a limiting curve given by

$$\lim_{t_w \to \infty, C(t,t_w)=C} \chi(t,t_w) = \frac{1}{T_{\rm EFF}(C)} (1 - C) \tag{4}$$

where $T_{\rm EFF}(C)$ is a function of the correlation C. We shall discuss the notation and justify the name of this function below. In the large t_w limit two distinct regimes develop in the χ vs C curve. There is a first straight line of slope $-1/T$, joining $(1,0)$ and $(q_{EA}(T), \chi_1(T))$. This characterises what is called the FDT regime. The straight line then breaks and the χ vs C curve goes on in a different manner. The subsequent behaviour depends on the model. Indeed, three families have been identified:

- Models describing domain growth like, for example, the $O(N)$ model in D dimensions in the large N limit. In this case, one follows the local correlation $NC(t,t_w) \equiv \langle \boldsymbol{\phi}(\boldsymbol{x},t) \cdot \boldsymbol{\phi}(\boldsymbol{x},t_w) \rangle$ and its associated local susceptibility. The plot for $C \le q_{EA}(T)$ is flat [34]. The susceptibility gets stuck at its value $\chi_1(T)$ while the correlation continues decreasing towards zero. The same result holds for the Ohta-Jasnow-Kawasaki approximation to the $\lambda\phi^4$ model of phase separation [35].

- Models describing structural glasses like, for example, the so-called F_{p-1} models of the mode-coupling approach or the p spin-glass models. In this case the χ vs C plot, for $C \le q_{EA}(T)$, is a straight line of slope larger than $-1/T$ [29].

- Models describing spin-glasses like, for example, the Sherrington-Kirkpatrick model. In this case the χ vs C plot, for $C \le q_{EA}(T)$, is a non-trivial curve [30].

This "classification" in three families has been checked numerically for more realistic models. Many numerical simulations using either Montecarlo (MC) techniques or molecular dynamics (MD) have shown that several models fall into the expected cathegories. Some models belonging to the first group are the $2D$ Ising model with conserved and non-conserved order parameter [36], the site diluted ferromagnet and the random field Ising model [37] and the $2D$ Ising model with ferromagnetic exchange and antiferromagnetic dipolar interactions [38]. The binary Lennard-Jones mixtures are a standard model for the glass transtion. Both MC and MD simulations show that they belong to the second class [39]. Finally, MC simulations of finite dimensional spin-glass models, the three and four dimensional Edwards-Anderson model, yield the third kind of behaviour [40]. Another particularly interesting problem, relevant for the physics of dirty superconductors [32], is the one of a manifold diffusing in a random potential. The analytic prediction using an infinite dimensional embedding space depends on the nature of the quenched random potential, namely on it being short or long range correlated [33]. This prediction is partially confirmed by the simulations in finite dimensional transverse space with the proviso of a very interesting modification that is not captured by the infinite dimensional approach [41]. Besides, numerical simulations of lattice-gas modes with kinetic constraints [42] and sandpile models [43] also show FDT violations.

Once this modification of FDT in the nonequilibrium situation is identified, several questions arise, all connected with the initial purpose of checking which thermodynamic concepts can be applied, perhaps after some modifications, to the nonequilibrium case. In the following we discuss three interesting issues.

- Why is there always a two-time regime, when C first decays from its equal times value to $q_{EA}(T)$, where FDT holds?

For the domain-growth problem the presence of this piece is easy to justify. In this time-scale one only sees the dynamics and the effect of the perturbation inside the domains. Since one can then ignore the presence of domain walls, the equilibrium relation between correlation and response is expected to hold. Of course, one cannot easily extend this argument to a more general situation. There is however a totally general reason for having FDT when $C \geq q_{EA}(T)$ and it is the following.

For any system in contact with an environment, with bounded correlation functions and without non-potential forces[2] the departure from FDT is bounded by [44]

$$|T\chi(t,t_w) - C(t,t) + C(t,t_w)| \leq K \int_{t_w}^{t} dt' \left(-\frac{1}{\gamma N} \frac{d\mathcal{H}(t')}{dt'} \right)^{1/2} \tag{5}$$

where K is a finite constant and γ the friction coefficient that characterises the coupling to the bath. The Kubo $\mathcal{H}$-function is defined as [46]

$$\mathcal{H}(t') \equiv \int d\phi d\dot{\phi}\, P(\phi,\dot{\phi},t) \left(T \ln P(\phi,\dot{\phi},t) + V(\phi) + \frac{m\dot{\phi}^2}{2} \right) , \tag{6}$$

with $P(\phi,\dot{\phi},t)$ the time-dependent probability distribution, $V(\phi)$ the potential energy and m a mass. The $\mathcal{H}$-function satifies $\dot{\mathcal{H}} \leq 0$ for all times and it vanishes only for the canonical distribution.

From this bound one sees that if $\mathcal{H}(t)$ falls to zero faster than $1/t$ no FDT violations are allowed in the long t_w limit since the right-hand-side in Eq. (5) vanishes. Instead, if $\mathcal{H}(t)$ falls to zero in a slower manner, FDT is imposed by the bound for small time-differences but violations are allowed for longer time-differences. This argument proves that there is always a region of correlations close to $C = 1$ in which FDT holds, even for a system that is not close to equilibrium.

- Can one identify the slope of the plot with an inverse effective temperature and call it $-1/T_{\text{EFF}}(t,t_w) = -1/T_{\text{EFF}}(C)$?

About ten years ago, in the context of weak-turbulence, Hohenberg and Shraiman [47] proposed to define an effective temperature through the departure from FDT. However, a detailed analysis of this quantity and its properties was not given in this reference.

Indeed, one expects that any quantity to be defined as a nonequilibrium effective temperature must fulfill the requirements associated to the intuitive idea of temperature. The first property to check is if this effective temperature is measurable by a thermometer that is weakly coupled to the system, in a statistical manner, at any chosen waiting time [48]. This property can be proven by studying the time-evolution of the thermometer coupled to M identical copies of the system, all of age t_w, and by verifying that this equation becomes a

[2] Other bounds can be found if diffusion and/or non-potential forces are allowed [44,45].

Langevin equation in the presence of a thermal bath characterised by a *coloured noise* with correlation given by the system's correlation and response given by the system's response.

Thus, if the system has several time-scales characterized by different values of the effective temperatures[3]

$$C(t,t_w) = C^{\text{FDT}}(t,t_w) + C^{(1)}(t,t_w) + C^{(2)}(t,t_w) + \ldots \tag{7}$$

$$R(t,t_w) = R^{\text{FDT}}(t,t_w) + R^{(1)}(t,t_w) + R^{(2)}(t,t_w) + \ldots \tag{8}$$

with

$$R^{\text{FDT}}(t,t_w) = \frac{1}{T}\frac{\partial}{\partial t_w} C^{\text{FDT}}(t,t_w)\,\theta(t-t_w) \qquad R^{(i)}(t,t_w) = \frac{1}{T^{(i)}}\frac{\partial}{\partial t_w} C^{(i)}(t,t_w)\,\theta(t-t_w) \tag{9}$$

one can select which value $T^{(i)}$ is measured by choosing the internal time-scale of the thermometer. Say, for example, that the thermometer is a harmonic oscillator of internal frequency ω_o. Then, one chooses the system time-scale to be explored, and hence the value of the effective temperature to be measured, by comparing ω_o to t_w.

Many desirable "thermodynamic" properties of T_{EFF} defined in this way can also be checked, [48] for example:

(i) T_{EFF} controls the direction of heat flows.

(ii) T_{EFF} controls partial equilibrations between observables in a system that evolve in the same time-scales and interact strongly enough.

(iii) Let us take two different glasses, in contact with a single bath of temperature T. These glasses are constructed in such a way that when they are not in contact each of them has a piecewise $T_{\text{EFF}}(C)$ of the form

$$T_{\text{EFF}}^{\text{SYST 1}}(C) = \begin{cases} T & \text{if } C > q_{EA}^{(1)} \\ T^{(1)} & \text{if } C < q_{EA}^{(1)} \end{cases} \qquad T_{\text{EFF}}^{\text{SYST 2}}(C) = \begin{cases} T & \text{if } C > q_{EA}^{(2)} \\ T^{(2)} & \text{if } C < q_{EA}^{(2)} \end{cases} \tag{10}$$

with $T^{(1)} \neq T^{(2)}$. One can then reproduce the experiment of setting two observables in contact by coupling these two systems through a small linear coupling between their microscopic variables. The result is that above a critical (though small) value of the coupling strength the two values the effective temperatures below q_{EA} equal while below the same critical value of the coupling strength the values remain unaltered. One concludes that if the two observables interact strongly the systems arrange their time-scales in such a way to partially thermalise.

The presence of non trivial effective temperatures in glycerol out of equilibrium is presently being checked experimentally by Grigera and Israeloff [49]. Their results show that, at fixed measuring frequency $\omega_o \sim 8$ Hz, this system has an effective temperature $T_{\text{EFF}} > T = 180K$ until measuring times of at least 10^5 sec, that is to say of the order of days! (Note that the bath temperature T is below $T_c = 187$ K.)

Further support to the notion of effective temperatures comes from the study of the effect of quantum fluctuations on the same family of models [50]. Below a critical line, that separates glassy from equilibrium phases, and in the slow dynamic regime, one finds violations of the quantum fluctuation dissipation theorem. These are characterised by the replacement of the bath temperature by an effective temperature $T_{\text{EFF}}(t,t_w)$. The effective temperature is again piecewise. It coincides with the bath-temperature T when C is larger than q_{EA} and it is different when C goes below q_{EA}. This nonequilibrium value has the nice property of being non-zero even at zero bath-temperature. Again, this result can be interpreted within the domain growth example. Whenever one looks at short time-differences with respect to the waiting-time one explores the quantum and thermal fluctuation in the bulk, i.e. one observes a quantum equilibrium dynamics that satisfies the quantum FDT. Instead, when τ is comparable to $g(L(t_w))$ one observes the domain wall motion. These are macroscopic objects for which quantum fluctuation do not have a strong effect. This can be seen, for example, in the form of the FDT violations: they look classical though with an effective temperature that depends on the strength of quantum fluctuations.

- Do effective temperatures in out of equilibrium systems emerge from a symmetry breaking?

3) See Ref. [30] for a precise definition of two-time scales.

In the classical case, one can study the structure of time-scales and effective temperatures with the help of the supersymmetric formulation of stochastic processes [51,52]. Indeed, it is well-known that the effective action in the MSR generating functional is invariant under a supersymmetric group (with a possible symmetry breaking due to the initial condition).

In the kind of glassy systems we deal with, there is a neat separation of time-scales in the long waiting-time limit. This allows us to separate the dynamics in the fast scale from the dynamics in the slow time-scales. The equation governing the slow time-scales, have an enlarged symmetry: they acquire an invariance under super-reparametrizations. The only solution that respects the large symmetry is a trivial, constant one. Hence, in order to have non-trivial dynamics in the long waiting-time limit, the system has to spontaneously break the super-reparametrization invariance. One can prove that the choice of effective temperatures is intimately related to the spontaneous breaking of this invariance [53].

A similar analysis in the quantum case remains to be developed.

In conclusion, we have summarized some interesting features of the slow out of equilibrium dynamics of domain growth and glassy systems. We have explained why these features arise in the domain-growth case. A similar understanding has not been reached for glassy systems yet. With the purpose of developing a "visual" understanding of glassy physics, a careful analysis of the statistics and organisation of the configurations visited by a glass model during its nonequilibrium evolution is in order.

The use of simplified models or, alternatively, self-consistent approximations to more realistic ones have yielded a number of very interesting results and new predicitons. In particular, these models capture much of the aging phenomenology of glassy systems. Surprisingly enough, even puzzling effects of temperature cyclings during aging in spin-glasses, and the absence of these effects in other kind of glasses, can be described by fully-connected models [54]. Some of these new predictions, notably the modification of FDT, have been tested numerically and experiments are now being performed. Obviously, it is desirable to go beyond these approximations and study more realistic models in finite dimensions. This, however, is a very difficult task.

There have been innumerable attempts to define a temperature for an out of equilibrium system. In particular, in the context of glassy materials, a "fictive temperature" is often introduced to describe some of the experimental findings [7]. The effective temperature discussed in this article has the most welcome property of being measurable, hence being open to experimental tests. As far as we have checked the definition, it also has the welcome property of conforming to the common prejudices one has of a temperature. Of course there are still many open questions related to it. Just to mention one, let us say that it would be very interesting to extend the analytical experiment of "coupling a thermometer to a system" to the quantum case.

ACKNOWLEDGMENTS

I wish to especially thank J. Kurchan with whom I have done much of the work on this subject and H. Castillo for suggestions concerning the preparation of this manuscript.

REFERENCES

1. For a review of domain growth see A. J. Bray, *Adv. Phys.* **43**, 357 (1994).
2. C. A. A. Angell, Science **267**, 1924 (1995).
3. For a review of the theoretical approach to glassy dynamics see J-P Bouchaud, L. F. Cugliandolo, J. Kurchan and M. Mézard, *Out of equilibrium dynamics in spin-glasses and other glassy systems*, cond-mat/9702070, in *Spin-glasses and random fields*, A. P. Young ed. (World Scientific, Singapore).
4. F. Alberici-Kious, J-P Bouchaud, L. F. Cugliandolo, P. Doussineau and A. Levelut; Phys. Rev. Lett. **81**, 4987 (1998).
5. K. Binder and A. P. Young; Rev. Mod. Phys. **58**, 801 (1986). M. Mézard, G. Parisi and M. A. Virasoro; *Spin glass theory and beyond* (World Scientific, Singapore, 1987). K. H. Fischer and J. A. Hertz, *Spin Glasses*, (Cambridge Univ. Press, 1991).
6. L. Lundgren, P. Svedlindh, P. Nordblad and O. Beckmann, Phys. Rev. Lett. **51**, 911 (1983). E. Vincent, J. Hammann, M. Ocio, J.P. Bouchaud and L. F. Cugliandolo; *Slow dynamics and aging* in Sitges Conference on Glassy Systems, M. Rubí ed. (Springer-Verlag, 1997), cond-mat/9607224.
7. L. C. E. Struick, *Physical aging in amorphous systems and other materials* (Elsevier, Houston, 1978). I. Hodge, Science **267**, 1945 (1996). G. B. Mc Kenna, J. Res. NIST **99**, 169 (1994) and references therein.

8. D. Bonn, H. Tanaka, H. Kellay, G. Wegdam and J. Meunier; Europhys. Lett. (1998). D. Bonn, H. Tanaka and J. Meunier, Europhys. Lett. **45**, 52 (1999).
9. R. L. Leheny and S. Nagel; Phys. Rev. **B57**, 5154 (1998).
10. D.S. Fisher and D.A. Huse; Phys. Rev. Lett. **56**, 1601 (1986); Phys. Rev. **B38**, 373 (1988).
11. G. Tarjus and D. Kivelson, J. Chem. Phys. **109**, 5481 (1998).
12. H. Rieger, *Ann Rev. of Comp. Phys.* II, ed. D. Stauffer (World Scientific, Singapore, 1995).
13. W. Paul and J. Baschnagel, in *Monte Carlo dynamic simulations in polymer science*, K. Binder ed. (Oxford University Press, 1995).
14. H. Yoshino, J. Phys. **A29**, 1421 (1996). A. Barrat, Phys. Rev. **E55**, 5651 (1997).
15. W. Kob and J-L Barrat, Phys. Rev. Lett. **78**, 4581 (1997).
16. S. Boettcher and M. Paczuski, Phys. Rev. Lett, **79**, 889 (1997). S. Boettcher, Phys. Rev. **E56**, 6466 (1997). O. Sotolongo-Costa, A. Vázquez, J. C. Antoranz, *Aging and Lévy distributions in sandpiles* cond-mat/9901086.
17. J. Kurchan, *Rheology, and how to stop aging*, cond-mat/9812347, Proceedings of *Jamming and Rheology: constrained dynamics on microscopic and macroscopic scales*, ITP, Santa Barbara, 1997.
18. C. P. Martin, E. Siggia and H. A. Rose, Phys. Rev. **A8**, 423 (1973). H. K. Janssen, Z. Phys. **B23**, 377 (1976); *Dynamics of critical phenomena and related Topics*, Lecture Notes in Physics **104**, C. P. Enz ed., (Springer-Verlag, Berlin, 1979).
19. C. de Dominicis and P. C. Martin, J. Math. Phys. **5**, 14 (1964), *ibid* 31 (1964). L. Dolan and R. Jackiw, Phys. Rev. **D9**, 2904 (1974), *ibid* 3320 (1974). J. M. Cornwall, R. Jackiw and E. Tomboulis, Phys. Rev. **D10**, 2428 (1974).
20. J. Schwinger, J. Math. Phys. **2**, 407 (1961). L. V. Keldysh, Zh. Eksp. Teor. Fiz. **47**, 1515 (1964), Sov. Phys JETP **20**, 235 (1965).
21. R. P. Feynman and F. L. Vernon, Ann Phys. **24**, 118 (1963).
22. R. Kraichnan, J. Fluid. Mech. **5**, 497 (1959), *ibid* **7** (1961) 124.
23. K. Kawasaki, Ann. Phys. **61**, 1 (1970).
24. A. J. Bray, Phys. Rev. Lett. **32**, 1413 (1974).
25. T. D. Kirkpatrick and D. Thirumalai; Phys. Rev. **B36**, 5388 (1987). T. R. Kirkpatrick and P. Wolynes, Phys. Rev. **A35**, 3072 (1987). Phys. Rev. **B36**, 8552 (1987). T. R. Kirkpatrick, D. Thirumalai, P. G. Wolynes, Phys. Rev. **A40**, 1045 (1989).
26. S. Franz and J. Hertz, Phys. Rev. Lett. **74**, 2114 (1995).
27. H. Sompolinsky and A. Zippelius, Phys. Rev. Lett. **47**, 359 (1981), Phys. Rev. **B25**, 6860 (1982).
28. W. Götze, in *Liquids, freezing and glass transition*, eds. JP Hansen, D. Levesque, J. Zinn-Justin Editors, Les Houches 1989 (North Holland). W. Götze and L. Sjögren, *Rep. Prog. Phys.* **55**, 241 (1992).
29. L. F. Cugliandolo and J. Kurchan, Phys. Rev. Lett. **71**, 173 (1993); Phil. Mag. **B71**, 501 (1995).
30. L. F. Cugliandolo and J. Kurchan; J. Phys. **A27**, 5749 (1994).
31. S. Franz and M. Mézard; Europhys. Lett. **26**, 209 (1994); Physica **A209**, 1 (1994). L. F. Cugliandolo and P. Le Doussal; Phys. Rev. **E53**, 1525 (1996).
32. T. Giamarchi and P. Le Doussal, *Statics and Dynamics of Disordered Elastic Systems*, cond-mat/9705096, in *Spin Glasses and Random Fields*, ed. A.P. Young, World Scientific (Singapore) 1998.
33. L. F. Cugliandolo, J. Kurchan and P. Le Doussal; Phys. Rev. Lett. **76**, 2390 (1996).
34. L. F. Cugliandolo and D. S. Dean, J. Phys. **A28**, 4213 (1995), *ibid* L453 (1995).
35. L. F. Cugliandolo, J. Kurchan and G. Parisi, *J. Phys.* I (France), **4**, 1641 (1994). L. Berthier, J-L Barrat and J. Kurchan, *Response Function of Coarsening Systems*, cond-mat/9903091.
36. A. Barrat, Phys. Rev. **E57**, 3629 (1998).
37. G. Parisi, F. Ricci-Tersenghi and J. J. Ruiz-Lorenzo, cond-mat/9811374.
38. D. A. Stariolo and S. A. Cannas; *Violation of the fluctuation-dissipation theorem in a two-dimensional Ising model with dipolar interactions*, cond-mat/9903136.
39. G. Parisi, Phys. Rev. Lett. **79**, 3660 (1997). W. Kob and J-L Barrat, Phys. Rev. Lett. **78**, 4581 (1997). J-L Barrat and W. Kob, *Fluctuation dissipation ratio in an aging Lennard-Jones glass*, cond-mat/9806027 and in preparation.
40. S. Franz and H. Rieger, J. Stat. Phys. **79**, 749 (1995). E. Marinari, G. Parisi, F. Ricci-Tersenghi, J. J. Ruiz-Lorenzo, J. Phys. **A31**, 2611 (1998).
41. H. Yoshino, Phys. Rev. Lett. **81**, 1493 (1998).
42. M. Sellitto, *Fluctuation-dissipation ratio in lattice-gas models with kinetic constraints*, cond-mat/9804168.
43. M. Nicodemi, *Dynamical response functions in models of vibrated granular media*, cond-mat/9809346.
44. L. F. Cugliandolo, D. S. Dean and J. Kurchan; Phys. Rev. Lett. **79**, 2168 (1997).
45. L. Laloux and P. Le Doussal, Phys. Rev. **E57**, 6296 (1998).
46. R. Kubo, M. Toda and N. Hashitume, *Statistical Physics II. Nonequilibrium Statistical Mechanics*, Springer-Verlag, 1992.
47. P. C. Hohenberg and B. I. Shraiman, Physica **D37**, 109 (1989).
48. L.F. Cugliandolo, J. Kurchan and L. Peliti, Phys. Rev. **E55**, 3898 (1997).

49. N. E. Israeloff and T. Grigera, Europhys. Lett. **43**, 308 (1998). T. Grigera and N. E. Israeloff, in preparation.
50. L. F. Cugliandolo and G. Lozano, Phys. Rev. Lett. **80**, 4979 (1998); Phys. Rev. **B59**, 915 (1999).
51. J. Zinn-Justin, *Quantum field theory and critical phenomena*, Oxford University Press, 1996.
52. J. Kurchan, J. Phys. (France) I **2**, 1333 (1992). S. Franz and J. Kurchan, Europhys. Lett. **20**, 197 (1992).
53. L. F. Cugliandolo and J. Kurchan, Physica **A263**, 242 (1999), cond-mat/9807226, and in preparation.
54. L. F. Cugliandolo and J. Kurchan, cond-mat/9812229, to appear in Phys. Rev. **B**.

Quantum corrections to the geodesic equation

Diego A. R. Dalvit [1], and Francisco D. Mazzitelli [2]

Departamento de Física "J.J. Giambiagi", FCEN, UBA, Pabellón 1, Ciudad Universitaria, 1428 Buenos Aires, Argentina

Abstract. In this talk we will argue that, when gravitons are taken into account, the solution to the semiclassical Einstein equations (SEE) is not physical. The reason is simple: any classical device used to measure the spacetime geometry will also feel the graviton fluctuations. As the coupling between the classical device and the metric is non linear, the device will not measure the 'background geometry' (i.e. the geometry that solves the SEE). As a particular example we will show that a classical particle does not follow a geodesic of the background metric. Instead its motion is determined by a quantum corrected geodesic equation that takes into account its coupling to the gravitons. This analysis will also lead us to find a solution to the so-called gauge fixing problem: the quantum corrected geodesic equation is explicitly independent of any gauge fixing parameter.

I INTRODUCTION

In quantum field theory there are many physical situations where one is interested in the dynamical evolution of fields rather than in S-matrix elements. The effective action (EA) is a useful tool to obtain the equations that govern such dynamics including the backreaction effects due to quantum fluctuations. However, there are two important problems that should be solved before one can get meaningful equations.

On the one hand, when the usual effective action is used to derive evolution equations, these turn out to be neither real nor causal. The cause is that the EA gives evolution equations for "in-out" matrix elements of the background fields. In order to obtain real and causal equations for expectation values, a different EA ("in-in" EA) has been introduced, which permits a correct approach to initial value problems [1]. On the other hand, both the in-out and the in-in effective actions are not physical quantities off-shell. This is most easily seen in the context of gauge theories, where the EA depends on the gauge fixing condition. The scattering matrix is constructed going on-shell, and therefore it does not suffer from this problem. The equations of motion, on the contrary, are obtained from the off-shell EA, and are thus gauge fixing dependent. The standard approach to tackle this problem is to consider the Vilkovisky-DeWitt effective action [2], which is specifically built to give a reparametrization, gauge invariant action. However, this action suffers from another type of arbitrariness, namely the dependence on the supermetric in the space of fields that is introduced in its definition [3,4].

Backreaction effects on the spacetime metric are relevant in different physical situations like gravitational collapse and black hole evaporation. Any discussion of the backreaction problem should include the effect of gravitons which contribute to the one loop effective stress tensor with terms of the same order as those coming from ordinary matter fields [5]. When graviton loops are included, the metric $g_{\mu\nu}$ that solves the semiclassical Einstein equations depends on the gauge fixing, and as such it is not physical. As an example we can mention calculations of compactification radii in Kaluza-Klein theories [6]. In view of the dependence of the results on the gauge fixing, people turned to the Vilkovisky-DeWitt EA as a way to overcome this setback [7]. However it was eventually shown that this approach was also incomplete because the results depend on the supermetric for the fields manifold [3].

In this talk we put forward a solution different from that advocated in the Vilkovisky-DeWitt EA. Our point is that, due to its interaction with the quantum fluctuations of the gravitational field, a test particle will not

[1] dalvit@df.uba.ar
[2] fmazzi@df.uba.ar

CP484, *Trends in Theoretical Physics II*, edited by H. Falomir, R. E. Gamboa Saraví, and F. A. Schaposnik

follow a $g_{\mu\nu}$-geodesic. Instead its motion is governed by a quantum corrected geodesic equation, which must be gauge fixing independent. *Therefore, the solution of the backreaction problem consists of two steps: to solve the semiclassical Einstein equations and to extract the physical quantities from the solution. It is this second point that, to our knowledge, has never been considered before.*

In order to illustrate these facts we will consider the calculation of the leading quantum corrections to the Newtonian potential. As has been pointed out in [8,9], when General Relativity is looked upon as an effective field theory, low energy quantum effects can be studied without the knowledge of the (unknown) high energy physics. The leading long distance quantum corrections to the gravitational interactions are due to massless particles and only involve their coupling at energies low compared to the Planck mass. Using this idea, many authors have calculated the leading quantum corrections to the Newtonian potential computing different sets of Feynman diagrams [8–11]. Instead of evaluating diagrams and S-matrix elements, we are here concerned with a covariant calculation based on the effective action and effective field equations. This covariant approach is more adequate to study many interesting problems in which one considers fluctuations around non-flat backgrounds, like black hole evaporation, gravitational collapse and backreaction in cosmological settings, among others. We shall first compute the semiclassical Einstein equations for the backreaction problem starting from the standard EA and show how they depend on the gauge fixing. Using a corrected geodesic equation we will deduce a physical quantum corrected Newtonian potential, which does not depend on any gauge fixing parameter. We will also discuss briefly the quantum corrections to the geodesic equation in a cosmological context.

II THE SEMICLASSICAL EINSTEIN EQUATIONS

The action for gravity coupled to a heavy particle (a classical source) has the form [12]

$$S_G + S_M = \frac{2}{\kappa^2}\int d^4x\sqrt{-\bar{g}}\bar{R} - M\int\sqrt{-\bar{g}_{\mu\nu}dx^\mu dx^\nu}, \tag{1}$$

where $\bar{R}$ is the curvature scalar, $\bar{g}_{\mu\nu}$ is the metric tensor, $\bar{g} = \det\bar{g}_{\mu\nu}$, and $\kappa^2 = 32\pi G$, with G being Newton's constant. In the background field method quantum fluctuations of the gravitational field may be expanded around a background metric, $\bar{g}_{\mu\nu} = g_{\mu\nu} + \kappa s_{\mu\nu}$ and a function χ^μ is chosen to fix the gauge, which is implemented through a gauge fixing action

$$S_{\rm gf} = -\frac{1}{2\kappa^2}\int d^4x\sqrt{-g}g_{\mu\nu}\chi^\mu\chi^\nu. \tag{2}$$

We shall consider the one-parameter family of gauge fixing functions, the so-called λ family,

$$\chi^\mu(\lambda) = \frac{1}{\sqrt{1+\lambda}}\left[g^{\mu\gamma}\nabla^\sigma s_{\gamma\sigma} - \frac{1}{2}g^{\gamma\sigma}\nabla^\mu s_{\gamma\sigma}\right]. \tag{3}$$

For gauge fixing functions linear in the metric fluctuations, ghosts decouple from the fluctuations $s_{\mu\nu}$ and only couple to the background fields. The one loop effective action for the background metric is obtained from integrating out quantum fluctuations and implies the evaluation of functional determinants for gravitons and ghosts in the presence of the background fields. For the pure gravitational action S_G, the one loop divergencies in the DeWitt gauge $\lambda = 0$ have been calculated long ago using dimensional regularization and turn out to be local terms quadratic in the curvature tensors [13]. They read

$$\Delta S_G^{\rm div} = \frac{2}{(4-d)96\pi^2}\int d^4x\sqrt{-g}\left[\frac{21}{10}R_{\mu\nu}R^{\mu\nu} + \frac{1}{20}R^2\right], \tag{4}$$

where we have omitted the Gauss-Bonnet term, which is a topological invariant in $d = 4$ spacetime dimensions. Apart from the local parts, the one loop EA also has non-local components. These have been computed up to quadratic order in the curvature tensors through a resummation procedure of the Schwinger DeWitt expansion for the action [14]. In what follows we shall be working to order $\mathcal{O}(R^3)$ at the level of the action and to order $\mathcal{O}(R^2)$ in the equations of motion. The non-local, non-analytic terms proportional to $\ln(-\Box)$ are the relevant ones in order to compute the leading quantum corrections. They can be read off from the divergencies in Eq.(4) in a manner outlined in [8,14]

$$\Delta S_G^{\rm nl} = -\frac{1}{96\pi^2}\int d^4x\sqrt{-g}\Big[\frac{21}{10}R_{\mu\nu}\ln(-\Box)R^{\mu\nu}+ \frac{1}{20}R\ln(-\Box)R\Big]. \quad (5)$$

The second term in Eq.(1) introduces an additional contribution to the EA. Following the method described in [15], one can compute the new divergence, from which it is possible to find the $\ln(-\Box)$ part of the EA arising from the presence of the mass M. After a long calculation we get

$$\Delta S_M^{\rm nl} = -\frac{1}{64\pi^2}\int d^4x\sqrt{-g}\,[M_{\mu\nu\rho\sigma}\ln(-\Box)M^{\rho\sigma\mu\nu}+ 2M_{\mu\nu\rho\sigma}\ln(-\Box)\left(P^{\rho\sigma\mu\nu}-\frac{1}{6}R\delta^{\rho(\mu}\delta^{\sigma\nu)}\right)\Big], \quad (6)$$

where

$$M_{\mu\nu\rho\sigma}(y) = \frac{M\kappa^2}{8}\int d\tau\delta^4(y-x(\tau))\times [g_{\mu\nu}\dot{x}_\rho\dot{x}_\sigma+2\dot{x}_\mu\dot{x}_\nu\dot{x}_\rho\dot{x}_\sigma], \quad (7)$$

and

$$P^{\rho\sigma\mu\nu} = 2R^{\rho(\mu\sigma\nu)}+2\delta^{(\rho[\mu}R^{\sigma)\nu]}-g^{\rho\sigma}R^{\mu\nu}- g^{\mu\nu}R^{\rho\sigma}-R\delta^{\rho(\mu}\delta^{\sigma\nu)}+\frac{1}{2}g^{\rho\sigma}g^{\mu\nu}R. \quad (8)$$

Here indeces in parenthesis or brackets imply symmetrization with a 1/2 factor.

As we will calculate long distance corrections to the Newtonian potential, we can assume that the mass M is a classical static "point mass", although its size should be much larger than its Schwarzschild radius and the Planck length, in order to justify the weak field approximation to be done in what follows. Its contribution to the nonlocal part of the EA is

$$\Delta S_M^{\rm nl} = \frac{7M\kappa^2}{1536\pi^2}\int d^4x\sqrt{-g}R\ln(-\nabla^2)\delta^3(\vec{x}), \quad (9)$$

where the nonlocal operator $\ln(-\nabla^2)$ acts on the delta function as $\ln(-\nabla^2)\delta^3(\vec{x}) = -\frac{1}{2\pi r^3}$ [16]. Adding the classical and quantum contributions of the EA and taking functional derivations with respect to the metric, it is straightforward to compute the semiclassical Einstein equations including backreaction of gravitons. They can be derived from the in-in EA, or by taking twice the real and causal part of the in-out equations of motion. Up to linear order in curvatures they are

$$\frac{1}{8\pi G}(R_{\mu\nu}-\frac{1}{2}Rg_{\mu\nu}) = M\delta^0_\mu\delta^0_\nu\delta^3(\vec{x}) - \frac{1}{96\pi^2}\left[\frac{21}{10}\ln(-\nabla^2)H^{(2)}_{\mu\nu}+\frac{1}{20}\ln(-\nabla^2)H^{(1)}_{\mu\nu}\right]+ \frac{7M\kappa^2}{768\pi^2}(\nabla_\mu\nabla_\nu-g_{\mu\nu}\nabla^2)\ln(-\nabla^2)\delta^3(\vec{x}), \quad (10)$$

where we have introduced the tensors $H^{(1)}_{\mu\nu} = 4\nabla_\mu\nabla_\nu R-4g_{\mu\nu}\nabla^2R$ and $H^{(2)}_{\mu\nu} = 2\nabla_\mu\nabla_\nu R-g_{\mu\nu}\nabla^2R-2\nabla^2R_{\mu\nu}$. Here we have used the fact that the mass M is static to replace $\Box\to\nabla^2$.

In order to solve these equations for the background metric we shall make perturbations around flat spacetime, $g_{\mu\nu} = \eta_{\mu\nu}+h_{\mu\nu}$ with $\eta_{\mu\nu} = \mathrm{diag}(-+++)$. We choose the harmonic gauge for the background perturbation metric. It is worth mentioning that this choice is completely independent of the gauge fixing problem for the quantum fluctuations. The 00 component for the perturbation $h_{\mu\nu}$ turns out to be

$$h_{00}(\lambda=0) = \frac{2GM}{r}\left[1+\frac{43G}{30\pi r^2}-\frac{7G}{12\pi r^2}\right]. \quad (11)$$

The first term is due to the presence of the classical mass M (for simplicity we consider only the Newtonian limit, that is, we do not include classical corrections from general relativity). The second and third terms are quantum corrections. The former stems pure gravitational contributions (vacuum polarization) while the latter arises from the coupling of the mass M to gravitons.

The above result is valid only for the DeWitt $\lambda = 0$ gauge. For any other gauge (not only for the λ family of gauge fixings) one has to add a new contribution to the nonlocal part, which should vanish on-shell. Keeping up to quadratic order in curvatures, the requirement that the effective action be gauge fixing independent on-shell fixes the most general form such a nonlocal term can have

$$\Delta S = \int d^4x\sqrt{-g}\,[aR_{\mu\nu}\ln(-\Box)\mathcal{E}^{\mu\nu} + bRg_{\mu\nu}\ln(-\Box)\mathcal{E}^{\mu\nu} + \mathcal{O}(\mathcal{E}^2_{\mu\nu})]\,. \tag{12}$$

Here $\mathcal{E}^{\mu\nu}$ is the classical extremal $\mathcal{E}^{\mu\nu} = -\frac{2}{\kappa^2}(R^{\mu\nu} - \frac{1}{2}Rg^{\mu\nu}) + \frac{1}{2}T^{\mu\nu}$, where

$$T^{\mu\nu}(y) = M\int d\tau\,\dot{x}^\mu\dot{x}^\nu\delta^4(y - x(\tau)), \tag{13}$$

is the energy-momentum tensor of the classical source and a and b are constants that depend on which particular gauge is used. The reason for omitting terms quadratic in the extremal in Eq. (12) is that, when the equations of motion are perturbatively solved, they vanish identically. This new contribution to the EA modifies the semiclassical equations (10), which will now depend on a and b. Solving the modified equations we obtain the metric in a general gauge

$$h_{00} = \frac{2GM}{r}\left[1 + \frac{43G}{30\pi r^2} - \frac{7G}{12\pi r^2} + \frac{a - 2b}{r^2}\right]. \tag{14}$$

The last term is the extra contribution to the perturbation arising from a gauge fixing different from the DeWitt one. For example, for the λ family we have $a(\lambda) = -\frac{5\lambda\kappa^2}{48\pi^2}$ and $b(\lambda) = \frac{5\lambda\kappa^2}{96\pi^2}$. *It is then clear that the metric that solves the backreaction equations for the one loop quantized gravity depends on which particular function one chooses to fix the gauge.*

III QUANTUM CORRECTED GEODESIC EQUATION: NEWTONIAN LIMIT

The dependence on the gauge fixing of the gravitons is an obstacle to think of a solution to the SEE as the metric of spacetime. This obstacle is not 'technical' (as implicitly assumed in previous works) but physical: since any classical device couples to gravitons, the solution to the SEE will not, in general, have a clear physical interpretation.

To analyze this problem, we will consider the simplest classical device: a test particle of mass m in the presence of the quantized gravitational field $\bar{g}_{\mu\nu}$. A physical observable should be the motion of this particle. We consider that the mass of this particle is much smaller than M, which allows us to neglect all contributions of the test particle to the solution of the one loop corrected equation (14). Now comes the key ingredient: in order to determine how this test particle moves, one also has to take into account the fact that it couples to the quantum metric $\bar{g}_{\mu\nu}$ through a term $-m\int\sqrt{-\bar{g}_{\mu\nu}(x)dx^\mu dx^\nu}$, where x^μ denotes the path of the test particle. Therefore there will be an extra contribution to the one loop EA due to this coupling to gravitons, which in turn will introduce a correction to the geodesic equation. This contribution is, up to linear order in m,

$$\Delta S_m = \int d^4x\sqrt{-g}\left[-\frac{1}{64\pi^2}m_{\mu\nu\rho\sigma}\ln(-\Box)M^{\rho\sigma\mu\nu} - \frac{1}{32\pi^2}m_{\mu\nu\rho\sigma}\ln(-\Box)\left(P^{\rho\sigma\mu\nu} - \frac{1}{6}R\delta^{\rho(\mu}\delta^{\sigma\nu)}\right) + \frac{a}{2}R_{\mu\nu}\ln(-\Box)T^{\mu\nu}_m + \frac{b}{2}Rg_{\mu\nu}\ln(-\Box)T^{\mu\nu}_m\right], \tag{15}$$

where the tensor $m_{\mu\nu\rho\sigma}$ is the one given in Eq.(7) with M replaced by m, and $T_m^{\mu\nu}$ is the energy-momentum tensor for the test particle, given in Eq.(13), with the same replacement. The first two terms correspond to the $\lambda = 0$ gauge fixing, and the last two are extra terms appearing for any other gauge. In the weak, nonrelativistic Newtonian limit, the quantum corrected geodesic equation reads

$$\frac{d^2\vec{x}}{dt^2} - \frac{1}{2}\vec{\nabla}h_{00} = \frac{1}{m}\frac{\delta\Delta S_m}{\delta\vec{x}}. \tag{16}$$

Note that h_{00}, given in Eq. (14), depends on a and b. The term on the rhs can be computed following the same methods we used to solve the backreaction problem. In that way

$$\frac{\delta\Delta S_m}{\delta\vec{x}} = \left[\frac{5G}{12\pi} - a + 2b\right]\vec{\nabla}\left(\frac{GmM}{r^3}\right). \tag{17}$$

Plugging this expression into the corrected geodesic equation we see that those gauge fixing dependent terms arising from the backreaction metric cancel exactly those coming from the coupling of the test particle to gravitons. In this way we obtain a physical, gauge fixing independent Newtonian potential $V(r)$ which we read from $d^2\vec{x}/dt^2 = -\vec{\nabla}V$, namely

$$V(r) = -\frac{GM}{r}\left[1 + \frac{43G\hbar}{30\pi r^2c^3} - \frac{7G\hbar}{12\pi r^2c^3} + \frac{5G\hbar}{12\pi r^2c^3}\right], \tag{18}$$

where we have restored units ($\hbar$ and c). Note that the long distance quantum correction above is extremely small to be measured. *However the specific number is less important than the conceptual fact that the potential and motion of the test particle are gauge fixing independent.*

IV COSMOLOGICAL BACKGROUND GEOMETRIES

Up to here we have considered the quantum corrections to the metric and to the test particle trajectory in the Newtonian approximation. We will now briefly comment about the quantum corrections to the geodesics in a cosmological background.

For simplicity, instead of working with a general gauge fixing term (as we did before), we will fix completely the gauge of the quantum fluctuations and describe gravitons by two massless, minimally coupled scalar fields. Moreover, we will assume that we are able to obtain a cosmological solution to the SEE, including graviton fluctuations, and will focus only in the computation of quantum corrections to the geodesic equation. The solution to the SEE will be described by the line element $ds^2 = -dt^2 + a^2(t)(dx^2 + dy^2 + dz^2)$ for some function $a(t)$.

The corrections to the geodesic equation can be computed following a procedure similar to the one described in the previous section. The coupling between the gravitons and the test particle modifies the classical action for the particle. Kepping only the corrections that are linear in the mass m the corrected geodesic equation reads

$$\begin{aligned}\frac{d^2t}{d\tau^2} + a\dot{a}\frac{d\vec{x}}{d\tau}^2 &= 4\pi G\frac{dx^j}{d\tau}\frac{dx^k}{d\tau}\frac{dx^l}{d\tau}\frac{dx^m}{d\tau}\frac{\partial}{\partial t}G_{jklm}\\ \frac{d}{d\tau}[a^2\frac{d\vec{x}}{d\tau}] &= 16\pi G\frac{d}{d\tau}[G^n_{jkl}\frac{dx^j}{d\tau}\frac{dx^k}{d\tau}\frac{dx^l}{d\tau}]\end{aligned} \tag{19}$$

Here a dot denotes derivative with respect to t. G_{ijkl} is the (renormalized) coincidence limit of the graviton two point funtion $< h_{ij}(x)h_{kl}(x') >$. Note that G_{ijkl} only depends on t.

Taking into account the symmetries of the problem, the above equations can be easily solved. Let us assume that, when the graviton contribution is neglected, the particle moves in the x direction ($x = x^1$). Hence, in this case,

$$\begin{aligned}\frac{dx}{d\tau} &= \frac{\alpha}{a^2}\\ \frac{dt}{d\tau} &= \sqrt{1 + \frac{\alpha^2}{a^2}}\end{aligned} \tag{20}$$

where α is a constant related to the velocity of the test particle. Note that $\alpha \to \infty$ in the null limit. The above equation describes, of course, the classical trajectory of the particle.

After a strightforward calculation we obtain, to first order in the quantum corrections,

$$\frac{dx}{d\tau} = \frac{\alpha}{a^2}[1 + \frac{32}{3}\pi G < \varphi^2 > \frac{\alpha^2}{a^2}]$$
$$\frac{dt}{d\tau} = \sqrt{1 + \frac{\alpha^2}{a^2}[1 + \frac{32}{3}\pi G < \varphi^2 > \frac{\frac{\alpha^4}{a^4}}{1 + \frac{\alpha^2}{a^2}}]} \tag{21}$$

where $< \varphi^2 >$ is the renormalized coincidence limit of the two point function for a massless, minimally coupled scalar field in the background described by $a(t)$.

¿From the above equation we can find the quantum correction to the physical velocity of the test particle, induced by its coupling to the gravitons,

$$\frac{dx}{dt} = \frac{\alpha a^{-2}}{\sqrt{1 + \alpha^2 a^{-2}}}[1 + \frac{32}{3}\pi G < \varphi^2 > \frac{\frac{\alpha^2}{a^2}}{1 + \frac{\alpha^2}{a^2}}] \tag{22}$$

In the null limit $\frac{dx}{dt} \simeq \frac{1}{a}[1 + \frac{32}{3}\pi G < \varphi^2 >]$ describes the graviton correction to the cosmological redshift. We expect this correction to be very small for $t \gg t_{Planck}$

Had we considered a different gauge fixing, we would have obtained a different expression for the quantum corrections to the geodesic equation in a cosmological context. However, the solution to the SEE described by $a(t)$ would also depend on the gauge fixing. As in the case of the Newtonian potential, both dependences should cancel when computing the trajectory of the test particle.

V CONCLUSIONS

We hope to have convinced you that if one is interested in solving the backreaction problem including the graviton contribution, *it is not enough to solve the semiclassical Einstein equations. The solutions are gauge fixing dependent and not physical.* Rather one has to look for physical observables. As an illustration of this point we have shown, in the Newtonian approximation, that a classical test particle does not follow a geodesic in the background metric. Moreover, the trajectory is a physical observable independent of the gauge fixing procedure. We have also computed the quantum corrected geodesic equation for a test particle in a Robertson Walker background.

If the calculations presented here are performed using the Vilkovisky-DeWitt effective action, we expect the dependence on the supermetric to cancel and the resulting quantum corrected geodesic equations to coincide with the ones obtained by means of the conventional effective action.

Similar ideas to the one proposed here can be applied to more general situations and even to the analysis of the mean value equations of any gauge theory, for example when computing gluon backreaction effects on classical solutions to Yang Mills theories.

The results presented in this talk are contained in Refs. [17,18].

ACKNOWLEDGMENTS

We acknowledge the support from Universidad de Buenos Aires, Fundación Antorchas and CONICET (Argentina).

REFERENCES

1. J. Schwinger, J. Math. Phys. **2**, 407 (1961); L. V. Keldysh, Zh. Eksp. Teor. Fiz. **47**, 1515 (1964) [Sov. Phys. JETP **20**, 1018 (1965)].
2. G. A. Vilkovisky, Nucl. Phys. **B234**, 125 (1984); B. S. DeWitt, *Quantum Field Theory and Quantum Statistics* (Adam Hilger, Bristol, 1987).
3. S. D. Odintsov, Phys. Lett. **B262**, 394 (1991)

4. R. Kantowski and C. Marzban, Phys. Rev. D**46**, 5449 (1992).
5. N. D. Birrell and P. C. D. Davies, *Quantum Fields in Curved Space* (Cambridge University Press, London, 1982).
6. G. Kunstatter and H. P. Leivo, Nucl. Phys. **B279**, 641 (1987).
7. S. R. Huggins, G. Kunstatter, H. P. Leivo and D. J. Toms, Phys. Rev. Lett. **58**, 296 (1987).
8. J. F. Donoghue, Phys. Rev. Lett. **72**, 2996 (1994).
9. J. F. Donoghue, Phys. Rev. D**50**, 3874 (1994).
10. I. Muzinich and S. Vokos, Phys. Rev D**52**, 3472 (1995).
11. H. Hamber and S. Liu, Phys. Lett. B**357**, 51 (1995).
12. Our metric has signature $(-+++)$ and the curvature tensor is defined as $\bar{R}^{\mu}{}_{\cdot\nu\alpha\beta} = \partial_\alpha \Gamma^{\mu}_{\nu\beta} - \ldots$, $\bar{R}_{\alpha\beta} = \bar{R}^{\mu}{}_{\cdot\alpha\mu\beta}$ and $\bar{R} = \bar{g}^{\alpha\beta}\bar{R}_{\alpha\beta}$. We use units $\hbar = c = 1$.
13. G. 't Hooft and M. Veltman, Ann. Inst. H. Poincare A **20**, 69 (1974).
14. G. A. Vilkovisky, in *Quantum Theory of Gravity*, ed. S.M. Christiensen, (Adam Hilger, Bristol, 1984).
15. A. Barvinsky and G. Vilkovisky, Phys. Rep. **119**, 1 (1985).
16. D. Dalvit and F.D. Mazzitelli, Phys. Rev. D**50**, 1001 (1994), *ibid* **52**, 2577 (1995).
17. D. Dalvit and F.D. Mazzitelli, Phys. Rev. D**56**, 7779 (1997)
18. D. Dalvit and F.D. Mazzitelli, in preparation.

Semiclassical Series from Path Integrals

C. A. A. de Carvalho[1]

Instituto de Física, Universidade Federal do Rio de Janeiro, Cx. Postal 68528, CEP 21945-970, Rio de Janeiro, RJ, Brasil

R. M. Cavalcanti[2]

Institute for Theoretical Physics, University of California, Santa Barbara, CA 93106-4030, USA

Abstract.
We derive the semiclassical series for the partition function in Quantum Statistical Mechanics (QSM) from its path integral representation. Each term of the series is obtained explicitly from the (real) minima of the classical action. The method yields a simple derivation of the exact result for the harmonic oscillator, and an accurate estimate of ground-state energy and specific heat for a single-well quartic anharmonic oscillator. As QSM can be regarded as finite temperature field theory at a point, we make use of the field-theoretic language of Feynman diagrams to illustrate the non-perturbative character of the series: it contains all powers of $\hbar$ and graphs with any number of loops; the usual perturbative series corresponds to a subset of the diagrams of the semiclassical series. We comment on the application of our results to other potentials, to correlation functions and to field theories in higher dimensions.

I INTRODUCTION

Semiclassical series have a long history in Quantum Mechanics which goes back to the early days of the Schrödinger equation. Their first terms are dictated by classical trajectories, and serve as the initial step of iterative procedures which yield all other terms [1].

Path integral representations for correlation functions have been developed more recently [2–6]. Being sums over trajectories, they provide a natural derivation of semiclassical results through the stationary phase method. Indeed, the saddle points of the path integrals correspond to classical trajectories which dictate the first terms of the series [7–9]. However, despite the many applications in Quantum Mechanics and Field Theory, most discussions which used path integrals never went beyond the first term of a semiclassical series. Notable exceptions were the works of DeWitt-Morette [10] and Mizrahi [11] in Quantum Mechanics.

Semiclassical methods for finite temperature field theories [12–14] also remained restricted to derivations of the first term of a semiclassical series [15], even when the problem was reduced to Quantum Statistical Mechanics [16,17], viewed as field theory at a point (zero spatial dimension). Some references resorted to extensions to the complex plane [18–20] to include complex paths required to describe Fourier transformed quantities but, again, those treatments were not concerned with obtaining the whole series.

In this talk we will present a systematic path integral procedure to generate semiclassical series in Quantum Statistical Mechanics. It leads to the construction of each term of the series from the solution(s) of the classical equations of motion. We will focus our attention on the partition function, and use the method of steepest descent which, in this case, only requires *real* solutions as saddle-points [21]. The restriction of our analysis to

---

[1] E-mail: aragao@if.ufrj.br

[2] Present address: Instituto de Física, Universidade de São Paulo, Cx. Postal 66318, CEP 05315-970, São Paulo, SP, Brasil. E-mail: rmoritz@fma.if.usp.br

CP484, *Trends in Theoretical Physics II*, edited by H. Falomir, R. E. Gamboa Saraví, and F. A. Schaposnik

quantum-mechanical systems (i.e., field theories at a point and at finite temperature) will allow us to construct the semiclassical propagator needed to generate the terms of the series.

In our contribution to last year's Workshop [22], we had already outlined the procedure mentioned in the previous paragraph. However, the approach we used then was quantum-mechanical, as opposed to the field-theoretic language we shall adopt in the present article. Reference [1] describes both approaches and gives a detailed account of the results which will be quoted here.

This article is organized as follows. Section II presents the derivation of the semiclassical series for a generic one-dimensional potential of the single-well type in field-theoretic language, which allows for a simple connection with the works of references [10,11]; the presentation is a natural extension of textbook material [23], and profits from the clear account of reference [20]. Section III applies our results to the harmonic oscillator and to the single-well quartic anharmonic oscillator; for the latter, we compute the ground-state energy and the specific heat. Section IV presents our conclusions, comments on extensions to double-well type potentials, to QSM in higher dimensions and to field theories.

II QUANTUM STATISTICAL MECHANICS

The partition function for a one-dimensional quantum-mechanical system consisting of a particle of mass m in the presence of a potential $V(x)$ in equilibrium with a thermal reservoir at temperature β^{-1} can be written as a path integral:

$$Z(\beta) = \int_{-\infty}^{\infty} dx_0 \int_{x(0)=x_0}^{x(\beta\hbar)=x_0} [\mathcal{D}x(\tau)]\, e^{-S/\hbar}, \tag{1}$$

$$S[x] = \int_0^{\beta\hbar} d\tau \left[\frac{1}{2} m \left(\frac{dx}{d\tau}\right)^2 + V(x)\right]. \tag{2}$$

For convenience we define the dimensionless quantities $q \equiv x/x_N$, $\theta \equiv \omega_N \tau$, $\Theta \equiv \beta\hbar\omega_N$, $U(q) \equiv V(x_N q)/m\omega_N^2 x_N^2$ and $g \equiv \hbar/m\omega_N x_N^2$, where ω_N^{-1} and x_N are natural time and length scales of the problem, respectively. In terms of these quantities we rewrite the partition function as

$$Z(\Theta) = \int_{-\infty}^{\infty} dq_0 \int_{q(0)=q_0}^{q(\Theta)=q_0} [\mathcal{D}q(\theta)]\, e^{-I/g}, \tag{3}$$

$$I[q] = \int_0^{\Theta} d\theta \left[\frac{1}{2}\dot{q}^2 + U(q)\right], \tag{4}$$

where the dot denotes differentiation in θ.

We generate a semiclassical series for $Z(\Theta)$ by: (i) finding the minima $q_c(\theta)$ of the Euclidean action I, i.e., the stable classical paths that solve the Euler-Lagrange equation of motion, subject to the boundary conditions; (ii) expanding the Euclidean action around these classical paths; (iii) deriving a quadratic semiclassical propagator by neglecting terms higher than second order in the expansion; (iv) using that propagator to compute higher (than quadratic) order contributions perturbatively.

For the sake of simplicity, we shall restrict our analysis to potentials of the single-well type, twice differentiable, and such that $U'(q) = 0$ only at the minimum of U, which we shall assume to be at the origin (see Fig. 1). This guarantees that, given q_0 and Θ, there will be a *unique* classical path satisfying the boundary conditions. Multiple-well potentials force us to consider more than one classical path for certain choices of q_0 and Θ. This phenomenon has been analyzed, for a double-well type potential, using the language of catastrophes and bifurcations [24]. Semiclassical series for the double-well quartic oscillator will be presented elsewhere [21].

The Euler-Lagrange equation ($U' \equiv dU/dq$)

$$\ddot{q} - U'(q) = 0, \tag{5}$$

subject to the boundary conditions $q(0) = q(\Theta) = q_0$, describes the motion of a particle in the potential *minus* U. Its first integral is

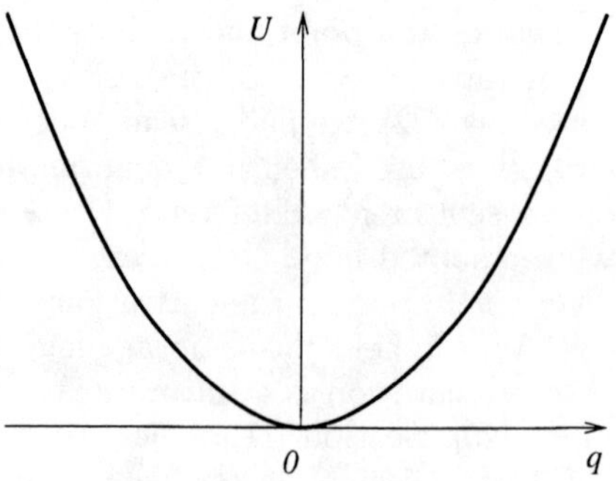

FIGURE 1. $U(q)$.

$$\frac{1}{2}\dot{q}^2 = U(q) - U(q_t), \tag{6}$$

where q_t denotes the single turning point (since we have an inverted single well) of the motion, defined implicitly by

$$\Theta = 2\int_{q_0}^{q_t} \frac{dq}{v(q, q_t)}, \tag{7}$$

where $v(q,q') \equiv \text{sign}(q'-q)\sqrt{2[U(q)-U(q')]}$; equation (7) is a consequence of integrating (6). Thus, for a single well, given q_0 and Θ, the classical path will go from q_0, at $\theta = 0$, to $q_t = q_t(q_0, \Theta)$, at $\theta = \Theta/2$, and return to q_0 at $\theta = \Theta$. (Note that $\text{sign}(q_t) = \text{sign}(q_0)$.)

The action for this classical path has a simple expression in terms of its turning point:

$$I[q_c] = \Theta\, U(q_t) + 2\int_{q_0}^{q_t} dq\, v(q, q_t), \tag{8}$$

where we have used (6). The first term in (8) corresponds to the high-temperature limit of $Z(\Theta)$, where classical paths collapse to a point ($q_t \to q_0$). The last term will be negligible for potentials that vary little over a thermal wavelength $\lambda = \hbar\sqrt{\beta/m}$. However, by decreasing the temperature it will become important and bring in quantum effects.

We now expand the action around the classical path. Letting $q(\theta) = q_c(\theta) + \eta(\theta)$, with $\eta(0) = \eta(\Theta) = 0$, we obtain

$$I[q] = I[q_c] + I_2[\eta] + \delta I[\eta], \tag{9}$$

where

$$I_2[\eta] \equiv \frac{1}{2}\int_0^{\Theta} d\theta\, \left\{\dot{\eta}^2(\theta) + U''[q_c(\theta)]\,\eta^2(\theta)\right\}, \tag{10}$$

$$\delta I[\eta] \equiv \int_0^{\Theta} d\theta\, \delta U(\theta, \eta) = \sum_{n=3}^{\infty} \frac{1}{n!}\int_0^{\Theta} d\theta\, U^{(n)}[q_c(\theta)]\,\eta^n(\theta). \tag{11}$$

Inserting (9) into (3) and expanding $e^{-\delta I/g}$ in a power series yields

$$Z(\Theta) = \int_{-\infty}^{\infty} dq_0\, e^{-I[q_c]/g} \int_{\eta(0)=0}^{\eta(\Theta)=0} [\mathcal{D}\eta(\theta)]\, e^{-I_2[\eta]/g} \sum_{m=0}^{\infty} \frac{1}{m!}\left(-\frac{\delta I[\eta]}{g}\right)^m. \tag{12}$$

The summation in (12) can be written more explicitly as

$$\sum_{m=0}^{\infty} \frac{1}{m!}\left(-\frac{\delta I[\eta]}{g}\right)^m = 1 + \sum_{m=1}^{\infty} \frac{(-1)^m}{g^m m!} \prod_{j=1}^{m} \left[\sum_{n_j=3}^{\infty} \frac{1}{n_j!}\int_0^{\Theta} d\theta_j\, U^{(n_j)}[q_c(\theta_j)]\,\eta^{n_j}(\theta_j)\right]. \tag{13}$$

As a consequence, one is led to compute integrals of the following type:

$$\langle \eta(\theta_1)\cdots\eta(\theta_k)\rangle \equiv \int_{\eta(0)=0}^{\eta(\Theta)=0} [\mathcal{D}\eta(\theta)]\, e^{-I_2[0,\Theta;\eta]/g}\, \eta(\theta_1)\cdots\eta(\theta_k). \tag{14}$$

Such integrals emerge naturally as functional derivatives of the following generating functional:

$$\mathcal{Z}[J] = \int_{\eta(0)=0}^{\eta(\Theta)=0} [\mathcal{D}\eta(\theta)]\, e^{-\frac{1}{g}\left\{I_2[0,\Theta;\eta]-\int_0^\Theta d\theta\, J(\theta)\,\eta(\theta)\right\}}. \tag{15}$$

Indeed,

$$\langle \eta(\theta_1)\cdots\eta(\theta_k)\rangle = g^k \left.\frac{\delta^k\, \mathcal{Z}[J]}{\delta J(\theta_1)\cdots\delta J(\theta_k)}\right|_{J=0}. \tag{16}$$

In order to compute $\mathcal{Z}[J]$, we define

$$\eta(\theta) = \tilde{\eta}(\theta) + \int_0^\Theta d\theta'\, \mathcal{G}(\theta,\theta')\, J(\theta'), \tag{17}$$

where $\tilde{\eta}(0) = \tilde{\eta}(\Theta) = 0$, and $\mathcal{G}(\theta,\theta')$ satisfies

$$\left\{-\frac{\partial^2}{\partial\theta^2} + U''[q_c(\theta)]\right\} \mathcal{G}(\theta,\theta') = \delta(\theta-\theta'), \qquad \mathcal{G}(0,\theta') = \mathcal{G}(\Theta,\theta') = 0. \tag{18}$$

Inserting (17) in (15), and noting that $[\mathcal{D}\eta(\theta)] = [\mathcal{D}\tilde{\eta}(\theta)]$, we obtain

$$\mathcal{Z}[J] = e^{\frac{1}{2g}\int_0^\Theta d\theta \int_0^\Theta d\theta'\, J(\theta)\,\mathcal{G}(\theta,\theta')\,J(\theta')} \int_{\tilde{\eta}(0)=0}^{\tilde{\eta}(\Theta)=0} [\mathcal{D}\tilde{\eta}(\theta)]\, e^{-I_2[0,\Theta;\tilde{\eta}]/g} \tag{19}$$

If we define

$$G_c(\theta_1,\eta_1;\theta_2,\eta_2) = \int_{\eta(\theta_1)=\eta_1}^{\eta(\theta_2)=\eta_2} [\mathcal{D}\eta(\theta)]\, e^{-I_2[\theta_1,\theta_2;\eta]/g}, \tag{20}$$

$$I_2[\theta_1,\theta_2;\eta] = \frac{1}{2}\int_{\theta_1}^{\theta_2} d\theta\, \left\{\dot{\eta}^2 + U''[q_c(\theta)]\,\eta^2\right\}, \tag{21}$$

we finally arrive at

$$\mathcal{Z}[J] = G_c(0,0;\Theta,0)\, \exp\left[\frac{1}{2g}\int_0^\Theta d\theta \int_0^\Theta d\theta'\, J(\theta)\,\mathcal{G}(\theta,\theta')\,J(\theta')\right]. \tag{22}$$

Using this result, we can now calculate (16). The result is simply

$$\langle \eta(\theta_1)\cdots\eta(\theta_k)\rangle = g^{k/2}\, G_c(0,0;\Theta,0) \sum_P \mathcal{G}(\theta_{i_1},\theta_{i_2})\cdots\mathcal{G}(\theta_{i_{k-1}},\theta_{i_k}), \tag{23}$$

if k is even, and zero otherwise. $\sum_P$ denotes sum over all possible pairings of the θ_{i_j}. Inserting this into (12) and (13) yields the semiclassical series for $Z(\Theta)$.

We still have to solve Eq. (18). This can be easily done if one notes that, for $\theta \neq \theta'$, it is a homogeneous second-order differential equation. Therefore, $\mathcal{G}(\theta,\theta')$ can be constructed from a linear combination of two linearly independent solutions $\eta_a(\theta)$ and $\eta_b(\theta)$ of the equation

$$\ddot{\eta} - U''[q_c(\theta)]\,\eta = 0. \tag{24}$$

Indeed

$$\mathcal{G}(\theta,\theta') = \begin{cases} a_-\eta_a(\theta) + b_-\eta_b(\theta), & \theta < \theta' \\ a_+\eta_a(\theta) + b_+\eta_b(\theta), & \theta > \theta'. \end{cases} \tag{25}$$

Continuity imposes

$$\mathcal{G}(\theta'+\epsilon,\theta') = \mathcal{G}(\theta'-\epsilon,\theta'), \tag{26}$$

whereas (18) leads to

$$\frac{\partial}{\partial\theta}\mathcal{G}(\theta,\theta')\Big|_{\theta=\theta'+\epsilon} - \frac{\partial}{\partial\theta}\mathcal{G}(\theta,\theta')\Big|_{\theta=\theta'-\epsilon} = -1, \tag{27}$$

with $\epsilon \to 0^+$. (26), (27) and the boundary conditions completely determine the coefficients in (25). The final result is

$$\mathcal{G}(\theta,\theta') = \frac{\Omega(0,\theta_<)\,\Omega(\theta_>,\Theta)}{\Omega(0,\Theta)}, \tag{28}$$

where $\theta_<(\theta_>) \equiv \min(\max)\{\theta,\theta'\}$, and $\Omega(\theta_1,\theta_2)$ is the function

$$\Omega(\theta,\theta') \equiv \eta_a(\theta)\,\eta_b(\theta') - \eta_a(\theta')\,\eta_b(\theta). \tag{29}$$

In the Appendix we show that $G_c(\theta_1,\eta_1;\theta_2,\eta_2)$ can also be obtained from the two linearly independent solutions of Eq. (24), $\eta_a(\theta)$ and $\eta_b(\theta)$; furthermore, we show how to construct those two functions from the solution $q_c(\theta)$ of the classical equation of motion. This completes the steps needed to write down any term of the series: all that is required is $q_c(\theta)$!

III QUANTUM OSCILLATORS

In this section we will apply our construction to the harmonic oscillator and to the single-well quartic oscillator. The harmonic case is designed to illustrate the compactness of our general formulae, which immediately yield the exact answer — there is no need to compute functional determinants from eigenvalue problems! The anharmonic case is designed to illustrate their power — the first term of the semiclassical series for the partition function allows us to extract a very good estimate of the ground-state energy and of the specific heat.

A The Harmonic Oscillator

In this subsection, we study the potential

$$V(x) = \frac{1}{2}m\omega^2x^2. \tag{30}$$

Choosing $\omega_N = \omega$ and $x_N = \sqrt{\hbar/m\omega}$, and introducing the dimensionless quantities of section II, we have $g = 1$ and

$$U(q) = \frac{1}{2}q^2. \tag{31}$$

Integrating (6) leads to

$$q_c(\theta) = q_t\,\cosh(\theta - \frac{\Theta}{2}), \tag{32}$$

The relation between q_0 and q_t is obtained by taking $\theta = \Theta$ in (32):

$$q_0 = q_c(\Theta) = q_t\,\cosh(\Theta/2). \tag{33}$$

The action for the classical path is

$$I[q_c] = q_0^2 \tanh(\Theta/2). \tag{34}$$

Following the Appendix, the functions η_a and η_b are given by

$$\eta_a(\theta) = \dot{q}_c(\theta) = q_t \sinh(\theta - \frac{\Theta}{2}), \tag{35}$$

and

$$\eta_b(\theta) = \dot{q}_c(\theta)\, Q(\theta) = -q_t^{-1} \cosh(\theta - \frac{\Theta}{2}), \tag{36}$$

It follows that $\Omega_{ij} = \sinh(\theta_j - \theta_i)$ and $W_{ij} = \cosh(\theta_j - \theta_i)$, leading to

$$\begin{aligned} G_c(\theta_1, \eta_1; \theta_2, \eta_2) = & \frac{1}{\sqrt{2\pi \sinh(\theta_2 - \theta_1)}} \\ & \times \exp\left\{ -\frac{1}{2\sinh(\theta_2 - \theta_1)} \left[\cosh(\theta_2 - \theta_1)(\eta_2^2 + \eta_1^2) - 2\,\eta_1\eta_2\right] \right\}. \end{aligned} \tag{37}$$

Using (80) we obtain $\Delta = 2\pi \sinh\Theta$. Since the problem is quadratic, its exact solution is then given by

$$Z(\Theta) \equiv \int_{-\infty}^{\infty} dq_0\, e^{-I[q_c]/g} \Delta^{-1/2} \tag{38}$$

Inserting (34) and the value of Δ yields the well-known result

$$Z(\Theta) = \int_{-\infty}^{\infty} \frac{e^{-q_0^2 \tanh(\Theta/2)}}{\sqrt{2\pi \sinh\Theta}}\, dq_0 = \frac{1}{2\sinh(\Theta/2)}. \tag{39}$$

B The Single-well Quartic Oscillator

In this subsection, we study the potential

$$V(x) = \frac{1}{2} m\omega^2 x^2 + \frac{1}{4}\lambda x^4. \tag{40}$$

Choosing $\omega_N = \omega$ and $x_N = \sqrt{m\omega^2/\lambda}$, and introducing the dimensionless quantities of section II, we have $g = \lambda\hbar/m^2\omega^3$ and

$$U(q) = \frac{1}{2} q^2 + \frac{1}{4} q^4. \tag{41}$$

Integrating (6) leads to [25,26]

$$q_c(\theta) = q_t \,\mathrm{nc}(u_\theta, k), \tag{42}$$

where $\mathrm{nc}(u,k) \equiv 1/\mathrm{cn}(u,k)$ is one of the Jacobian Elliptic functions [25–27], and

$$u_\theta = \sqrt{1+q_t^2}\left(\theta - \frac{\Theta}{2}\right), \qquad k = \sqrt{\frac{2+q_t^2}{2\,(1+q_t^2)}}. \tag{43}$$

For future use, we note that (43) can be rewritten as

$$u_\theta = \frac{2\theta - \Theta}{2\sqrt{2k^2-1}}, \qquad |q_t| = \sqrt{\frac{2\,(1-k^2)}{2k^2-1}}. \tag{44}$$

The relation between q_0 and q_t is obtained by taking $\theta = \Theta$ in (42):

$$q_0 = q_c(\Theta) = q_t \, \mathrm{nc}\, u_\Theta. \tag{45}$$

(We shall often omit the k-dependence in the Jacobian Elliptic functions.)

The action for the classical path (42) is

$$I[q_c] = \Theta\, U(q_t) + \sqrt{2} \int_{|q_t|}^{|q_0|} dq \sqrt{(q^2 + q_t^2 + 2)(q^2 - q_t^2)}. \tag{46}$$

Performing the integral (Ref. [25], formula 3.155.6) and replacing q_0 by the r.h.s. of (45), we obtain

$$I[q_c] = \Theta\left(\frac{1}{2} q_t^2 + \frac{1}{4} q_t^4\right) + \frac{4}{3}\left\{ -\sqrt{1+q_t^2}\left[\mathrm{E}(\varphi_\Theta, k) + \frac{1}{2} q_t^2 \, u_\Theta\right] \right.$$
$$\left. + \, \mathrm{sn}\, u_\Theta \left(1 + \frac{1}{2} q_t^2 \, \mathrm{nc}^2 u_\Theta\right) \sqrt{1 + \frac{1}{2} q_t^2 \left(1 + \mathrm{nc}^2 u_\Theta\right)} \right\}, \tag{47}$$

where $\mathrm{E}(\varphi, k)$ denotes the Elliptic Integral of the Second Kind and $\varphi_\theta \equiv \arccos[q_c(\theta)/q_0] = \arccos(\mathrm{cn}\, u_\theta)$.

For the construction of the quadratic semiclassical propagators $G_c(\theta, \eta; \theta', \eta')$ and $\mathcal{G}(\theta, \theta')$ we shall need

$$\eta_a(\theta) = \dot{q}_c(\theta) = q_t \sqrt{1 + q_t^2}\, \mathrm{sn}\, u_\theta \, \mathrm{dn}\, u_\theta \, \mathrm{nc}^2 u_\theta \tag{48}$$

and

$$Q(\theta) = q_t^{-2}(1+q_t^2)^{-3/2}\left[\left(1 - \frac{1}{k^2}\right) u_\theta + \left(\frac{1}{k^2} - 2\right) \mathrm{E}(\varphi_\theta, k) \right.$$
$$\left. - \frac{\mathrm{cn}\, u_\theta \, \mathrm{dn}\, u_\theta}{\mathrm{sn}\, u_\theta} + (k^2 - 1)\frac{\mathrm{cn}\, u_\theta \, \mathrm{sn}\, u_\theta}{\mathrm{dn}\, u_\theta}\right]. \tag{49}$$

We may then obtain $\eta_b(\theta) = \dot{q}_c(\theta)\, Q(\theta)$ and, thus, Ω_{12} and W_{12} from (76) and (77). Finally, use of (79) and (28) will yield the desired propagators.

For the series expansion of the partition function, we shall need

$$\delta U(\theta, \eta) = q_c(\theta)\, \eta^3 + \frac{1}{4}\eta^4, \tag{50}$$

obtained from (11). Therefore, we have to consider not only the usual quartic vertex, but an additional time(θ)-dependent cubic term. This completes the set of ingredients needed to write down a semiclassical series for any correlation function. In the next subsection, we shall concentrate on the first term of the series for $Z(\Theta)$, which yields the quadratic approximation.

1 *The quadratic approximation for $Z(\Theta)$*

From the knowledge of the classical action and of the Van Vleck determinant, we define

$$Z_2(\Theta) \equiv \int_{-\infty}^{\infty} dq_0 \, e^{-I[q_c]/g} \Delta^{-1/2} \tag{51}$$

as the quadratic approximation to $Z(\Theta)$. To perform the integral over q_0 one must write $I[q_c]$ and Δ solely in terms of q_0 (and Θ), but except in rare cases this is not an easy task. Usually, it is much simpler to write these quantities in terms of q_t [see Eq. (45)], and so it is natural to trade q_0 for q_t as the integration variable in (51). This is much simplified by the fact that the Jacobian of the map $q_0 \to q_t$ is simply related to the van Vleck determinant. In fact, Eqs. (7) and (81) imply

$$\left(\frac{\partial q_0}{\partial q_t}\right)_\Theta = -\frac{(\partial\Theta/\partial q_t)_{q_0}}{(\partial\Theta/\partial q_0)_{q_t}} = \frac{1}{2} v(q_0, q_t)\left(\frac{\partial\Theta}{\partial q_t}\right)_{q_0} = -\frac{U'(q_t)\,\Delta}{4\pi g\, v(q_0, q_t)}. \tag{52}$$

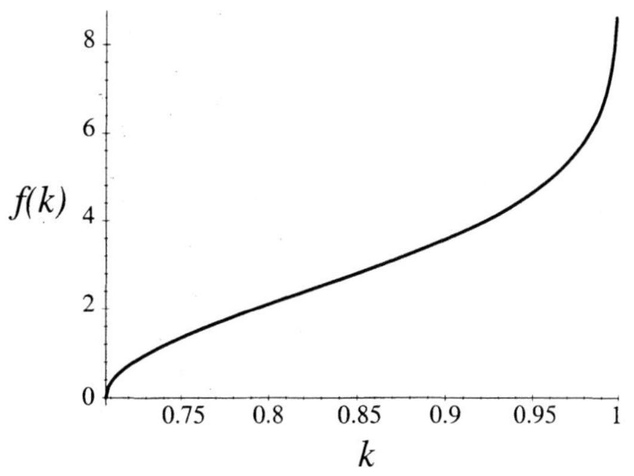

FIGURE 2. Graph of $f(k)$.

Eq. (51) then becomes

$$Z_2(\Theta) = -\frac{1}{4\pi g}\int_{q_\Theta^-}^{q_\Theta^+} dq_t\,\frac{U'(q_t)\,\Delta^{1/2}}{v(q_0,q_t)}\,e^{-I[q_c]/g} \equiv \int_{q_\Theta^-}^{q_\Theta^+} dq_t\, D(q_t,\Theta)\, e^{-I[q_c]/g}, \tag{53}$$

where $q_\Theta^\pm \equiv \lim_{q_0\to\pm\infty} q_t(q_0,\Theta)$.

The expression above is valid for single-well potentials in general. Now, let us especialize to the potential (41). $I[q_c]$ is given by (47), and using (80) and (49) one can write $D(q_t,\Theta)$ as

$$D(q_t,\Theta) = \frac{(1+q_t^2)^{1/4}}{\sqrt{4\pi g}}\left[\frac{1-k^2}{k^2}\,u_\Theta + \frac{2k^2-1}{k^2}\,E(\varphi_\Theta,k) + \frac{\operatorname{cn}u_\Theta\operatorname{dn}u_\Theta}{\operatorname{sn}u_\Theta} + (1-k^2)\,\frac{\operatorname{cn}u_\Theta\operatorname{sn}u_\Theta}{\operatorname{dn}u_\Theta}\right]^{1/2}. \tag{54}$$

From (45) it follows that $q_0 \to \infty$ when $\operatorname{cn}(u_\Theta,k) = 0$, which occurs when $u_\Theta = \mathrm{K}(k)$, where $\mathrm{K}(k)$ is the Complete Elliptic Integral of the First Kind. Using (44), this condition can be written as an equation in k:

$$\frac{\Theta}{2\sqrt{2k^2-1}} = \mathrm{K}(k). \tag{55}$$

The graph of $f(k) \equiv 2\sqrt{2k^2-1}\,\mathrm{K}(k)$ is plotted in Fig. 2. It increases monotonically from zero (at $k = 1/\sqrt{2}$) to infinity (as $k \to 1$), and so for each nonnegative value of Θ Eq. (55) has a unique solution, which we denote by k_Θ. Eq. (44) then gives the corresponding value of q_Θ^+ ($q_\Theta^- = -q_\Theta^+$, since $U(-q) = U(q)$).

2 Limiting cases of the quadratic approximation

Expression (53) may be used to compute $Z_2(\Theta)$ numerically for any value of Θ. However, certain limiting cases may be dealt with analytically. These limits are: the harmonic oscillator ($g \to 0$), high temperatures ($\Theta \to 0$), and low temperatures ($\Theta \to \infty$).

The limit $g \to 0$ of (53) does yield the partition function of the harmonic oscillator, as required, since $V(x) = \frac{1}{2}m\omega^2x^2$ when $g = 0$. In order to arrive at this result, we note that in this limit one can perform the integral (53) using the steepest descent method. Details of the derivation can be found in [1].

At high temperatures, $\Theta \to 0$ and (55) is solved for $k_\Theta \to 1/\sqrt{2}$, and so $q_\Theta^+ \to \infty$. It follows that

$$Z_2(\Theta) \overset{\Theta\to 0}{\sim} \sqrt{\frac{1}{2\pi g\Theta}}\int_{-\infty}^{\infty} dq\, e^{-\Theta\, U(q)/g}, \tag{56}$$

or, equivalently,

$$Z_2(\beta) \overset{\beta\to 0}{\sim} \sqrt{\frac{m}{2\pi\hbar^2\beta}} \int_{-\infty}^{\infty} dx\, e^{-\beta V(x)}, \tag{57}$$

with $V(x)$ and $U(q)$ defined in (40) and (41). This is, clearly, the "classical" limit for the partition function with a pre-factor that incorporates quantum fluctuations.

At low temperatures, $\Theta \to \infty$ and (55) is solved for $k_\Theta \to 1$. A careful derivation [1] leads to

$$Z_2(\Theta) \overset{\Theta\to\infty}{\sim} \int_{-q_\Theta^+}^{q_\Theta^+} \frac{dq_t}{\sqrt{4\pi g}}\, e^{-I[q_c]/g}, \tag{58}$$

with q_Θ^+ given by

$$q_\Theta^+ = \sqrt{\frac{2k'^2_\Theta}{1-2k'^2_\Theta}} \approx 4\sqrt{2}\, e^{-\Theta/2}, \tag{59}$$

and $I[q_c]$ given by

$$I[q_c] = \frac{4}{3}\left[\left(1+\frac{1}{2}q_t^2\,\mathrm{nc}^2 u_\Theta\right)^{3/2} - 1\right] + \mathcal{O}\left(\Theta\, e^{-\Theta}\right). \tag{60}$$

3 Applications

We shall now apply the quadratic semiclassical approximation to obtain the ground-state energy and the curve for the specific heat as a function of temperature. These two applications will teach us about the power of the approximation.

In order to compare (58) with the expected low-temperature limit of the partition function, $Z(\Theta) \sim e^{-\Theta\,\varepsilon_0(g)}$ (where $\varepsilon_0(g) \equiv E_0(g)/\hbar\omega$ is the dimensionless ground state energy), it is convenient to rewrite it in a form in which the Θ-dependence can be analyzed more easily. This can be done by changing the integration variable back to q_0. Since $q_t\, \mathrm{nc}\, u_\Theta = q_0$ and q_Θ^+ is the value of q_t corresponding to $q_0 \to \infty$, one has

$$Z_2(\Theta) \overset{\Theta\to\infty}{\sim} \int_{-\infty}^{\infty} \frac{dq_0}{\sqrt{4\pi g}} \left(\frac{\partial q_t}{\partial q_0}\right)_\Theta \exp\left\{-\frac{4}{3g}\left[\left(1+\frac{1}{2}q_0^2\right)^{3/2} - 1\right]\right\}. \tag{61}$$

When $\Theta \gg 1$ it is possible to write an approximate expression for $q_t(q_0, \Theta)$ [1], thus allowing to write the integrand in (61) solely in terms of q_0 and Θ. The final result is

$$Z_2(\Theta) \overset{\Theta\to\infty}{\sim} \frac{2\, e^{-\Theta/2}}{\sqrt{\pi g}} \int_{-\infty}^{\infty} dq_0 \frac{\exp\left\{-\frac{4}{3g}\left[\left(1+\frac{1}{2}q_0^2\right)^{3/2} - 1\right]\right\}}{\sqrt{1+\frac{1}{2}q_0^2}\left(1+\sqrt{1+\frac{1}{2}q_0^2}\right)}. \tag{62}$$

This gives $\varepsilon_0(g) = 1/2$, indicating that the quadratic approximation is insufficient to yield corrections to the ground state energy of the harmonic oscillator. On the other hand, if one recalls that the partition function can be written as

$$Z(\Theta) = \int_{-\infty}^{\infty} \rho(\Theta; q, q)\, dq, \tag{63}$$

where

$$\rho(\Theta; q, q) = \sum_n e^{-\Theta\varepsilon_n} |\psi_n(q)|^2 \overset{\Theta\to\infty}{\sim} e^{-\Theta\varepsilon_0} |\psi_0(q)|^2 \tag{64}$$

is the diagonal element of the density matrix, one may take the square root of the integrand in (62) as an approximation to the (unnormalized) wave function of the ground state. To test the accuracy of this

TABLE 1. Ground state energies for different values of g ($\hbar = m = \omega = 1$).

g	E_0(semiclassical)[a]	E_0(exact)[b]	error(%)
0.4	0.559258	0.559146	0.02
1.2	0.639765	0.637992	0.28
2.0	0.701429	0.696176	0.75
4.0	0.823078	0.803771	2.40
8.0	1.011928	0.951568	6.34

[a] $\langle\phi_0|H|\phi_0\rangle/\langle\phi_0|\phi_0\rangle$, where $\phi_0(q_0)$ is the square root of the integrand in Eq. (62).

[b] Values quoted from Ref. [28].

approximation, we have evaluated the expectation values of the energy for some values of g and compared them with high precision results found in the literature. As Table 1 shows, the ground state energy computed with this "semiclassical" wave function differs from the exact one by less than 1% even for g as large as 2.

Another concrete problem that can be treated is the calculation of the specific heat of the quantum anharmonic oscillator. It can be written in terms of $Z(\Theta)$ as

$$C = \Theta^2 \left[\frac{1}{Z}\frac{\partial^2 Z}{\partial \Theta^2} - \left(\frac{1}{Z}\frac{\partial Z}{\partial \Theta} \right)^2 \right]. \tag{65}$$

This expression was computed using MAPLE for a few values of Θ and the coupling constant value $g = 0.3$. The result is depicted in Fig. 3, which also exhibits the curve of specific heat of the *classical* anharmonic oscillator (solid line). As expected, the results agree when the temperature is sufficiently high, but, in contrast to the classical result, the semiclassical approximation is qualitatively correct at low temperatures too, dropping to zero as $T \to 0$.

This result, together with the estimate for the ground-state obtained previously, shows that the quadratic approximation works very well, being quite accurate at high temperatures, and still reliable at lower temperatures. In the next subsection, we will comment on why this is so.

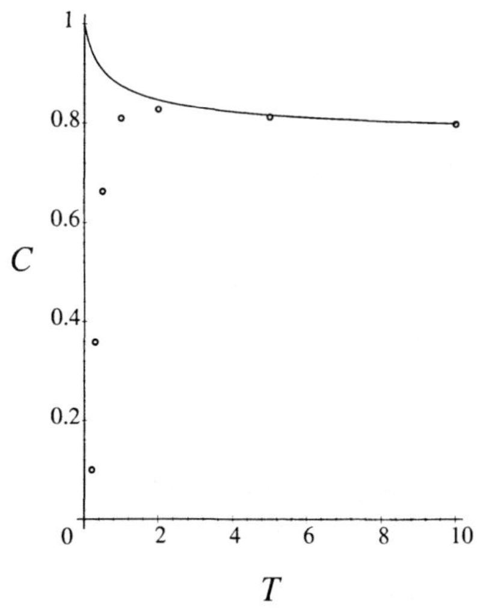

FIGURE 3. Specific heat vs. temperature ($T = 1/\Theta$) for the quantum (diamonds) and classical (solid line) anharmonic oscillator. $g = 0.3$.

4 Beyond quadratic

In this subsection, we shall compute a first correction G_1 to the quadratic approximation, which corresponds to the $m = 1$ term in (12). Using (50), we obtain

$$Z(\Theta) = Z_2(\Theta) - \frac{1}{g}\int_{-\infty}^{\infty} dq_0\, e^{-I[q_c]/g} \int_0^{\Theta} d\theta \left[q_c(\theta)\langle\eta^3(\theta)\rangle + \frac{1}{4}\langle\eta^4(\theta)\rangle \right] + \ldots. \tag{66}$$

Eq. (23) yields $\langle\eta^3\rangle = 0$ and $\langle\eta^4\rangle = 3g^2\, G_c(0,0;\Theta,0)\, \mathcal{G}^2(\theta,\theta)$. Inserting these results in (66) and changing the integration variable from q_0 to q_t gives

$$Z(\Theta) = \int_{q_\Theta^-}^{q_\Theta^+} dq_t\, D(q_t,\Theta)\, e^{-I[q_c]/g}\, [1 - g a_1(q_t,\Theta) + \ldots], \tag{67}$$

where

$$a_1(q_t,\Theta) = \frac{3}{4}\int_0^{\Theta} d\theta\, \mathcal{G}^2(\theta,\theta). \tag{68}$$

Because of the complicated form of $\mathcal{G}(\theta,\theta)$, it is not a simple task to compute $a_1(q_t,\Theta)$. However, we can estimate the magnitude of this term without much effort. Indeed, as shown in [1], $\mathcal{G}(\theta,\theta)$ obeys the following inequality:

$$\mathcal{G}(\theta,\theta) \leq \frac{\theta(\Theta-\theta)}{\Theta} \qquad (0 \leq \theta \leq \Theta). \tag{69}$$

Therefore,

$$a_1(q_t,\Theta) \leq \frac{\Theta^3}{40}. \tag{70}$$

This shows that this correction to the quadratic approximation, Eq. (51) or (53), can be neglected whenever the condition $g\Theta^3/40 \ll 1$ is satisfied; this is compatible with the numerical agreements obtained in the applications of the quadratic approximation.

The next term in the expansion for $Z(\Theta)$, which corresponds to the $m = 2$ term in (12), has a piece with a factor g and one with a factor g^2. The former comes from the product of $\langle\eta^6\rangle \sim g^3$ with the overall g^{-2}, whereas the latter involves $\langle\eta^8\rangle \sim g^4$. The Feynman diagrams which correspond to the $m = 1$ and $m = 2$ contributions are depicted in Fig. 4. Note that the last two of the $m = 2$ diagrams involve the three-leg vertex in (50), which depends explicitly on $q_c(\theta)$. Ordinary perturbation theory corresponds to the subset of graphs which do not contain the three-leg vertex with the replacement of $\mathcal{G}(\theta,\theta')$ by the corresponding (free) expression for the harmonic oscillator. In fact, $\mathcal{G}(\theta,\theta')$ can be expanded in terms of its (free) harmonic oscillator expression and insertions of $(U''[q_c] - 1)$, already an indication of its non-perturbative nature. Alternatively, we may obtain the perturbation theory diagrams by letting $q_c(\theta) \to 0$.

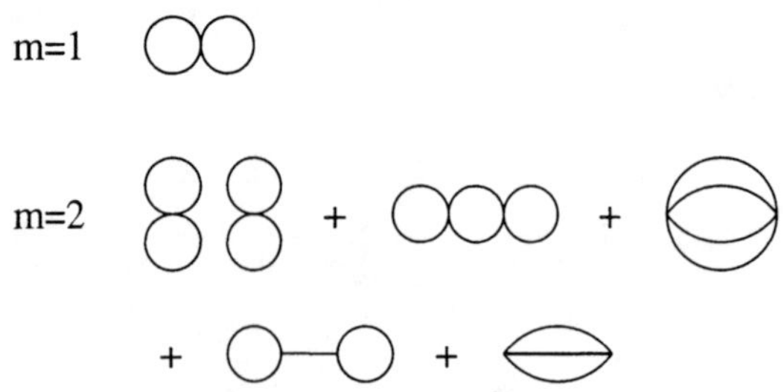

FIGURE 4. Feynman graphs for $m = 1$ and $m = 2$.

IV CONCLUSIONS

The results of section II can be generalized to higher-dimensional Quantum Statistical Mechanics, just as in Quantum Mechanics, where this was accomplished in [10,11]. The generalization to potentials which allow for more than one classical solution, such as the double-well quartic anharmonic oscillator, requires a subtle matching of the series around each appropriate saddle-point (i.e., the minima). This is presently under investigation [21].

An extension of our results to field theories is hampered by the fact that we do not know how to construct semiclassical propagators in general. The technical simplifications which appear in Quantum Mechanics cease to exist. However, our methods may still be of use in problems where classical solutions have a lot of symmetry (e.g., spherical symmetry) so that we can reduce them to effective one-dimensional problems. There are many such examples in Physics: instantons, monopoles, vortices and solitons are a few of the backgrounds that fall into that category. We are currently pursuing this line of investigation.

Finally, we should remark that the field-theoretic treatment can be used to compute any correlation function of interest, in the usual manner. Therefore, a semiclassical series can be written down for any physical quantity once it is expressed in terms of the relevant correlation functions.

ACKNOWLEDGMENTS

The authors acknowledge support from CNPq, FAPERJ, FAPESP and FUJB/UFRJ. RMC was also supported in part by the NSF under Grant No. PHY94-07194. CAAC thanks the organizers of the Workshop for their kind hospitality.

APPENDIX A

It remains to show how one can obtain $G_c(\theta_1,\eta_1;\theta_2,\eta_2)$ from the classical path. For this, we use the fact that the action I_2 is quadratic in η, and so the path integral in (20) is completely determined by the extremum $\eta_e(\theta)$ of $I_2[\theta_1,\theta_2;\eta]$, which satisfies Eq. (24), subject to the boundary conditions $\eta(\theta_1)=\eta_1$ and $\eta(\theta_2)=\eta_2$. Thus,

$$G_c(\theta_1,\eta_1;\theta_2,\eta_2) = G_c(\theta_1,0;\theta_2,0)\, e^{-I_2[\theta_1,\theta_2;\eta_e]/g}, \tag{71}$$

where, after an integration by parts,

$$I_2[\theta_1,\theta_2;\eta_e] = \frac{1}{2}\left[\eta_2\,\dot{\eta}_e(\theta_2) - \eta_1\,\dot{\eta}_e(\theta_1)\right]. \tag{72}$$

We can obtain $\eta_e(\theta)$ by finding the linear combination of any two linearly independent solutions, $\eta_a(\theta)$ and $\eta_b(\theta)$, of (24) which satisfies $\eta_e(\theta_1)=\eta_1$ and $\eta_e(\theta_2)=\eta_2$. The result is

$$\eta_e(\theta) = \frac{\eta_1\,\Omega(\theta,\theta_2) + \eta_2\,\Omega(\theta_1,\theta)}{\Omega(\theta_1,\theta_2)}, \tag{73}$$

with $\Omega(\theta,\theta')$ as defined in (29). We may then write

$$I_2[\theta_1,\theta_2;\eta_e] = \frac{1}{2\,\Omega_{12}}\left[W_{12}\,\eta_2^2 + W_{21}\,\eta_1^2 - (W_{11}+W_{22})\,\eta_1\eta_2\right], \tag{74}$$

where $\Omega_{ij} \equiv \Omega(\theta_i,\theta_j)$ and $W_{ij} \equiv \partial\Omega_{ij}/\partial\theta_j$. (Note that W_{ii} is the Wronskian of η_a and η_b computed at θ_i.)

Explicit expressions for $\eta_a(\theta)$ and $\eta_b(\theta)$ can be obtained as follows. By differentiating (5) with respect to θ, one can verify that $\eta_a(\theta)=\dot{q}_c(\theta)$ satisfies (24). For the second solution, we take $\eta_b(\theta)=\dot{q}_c(\theta)\,Q(\theta)$, where $Q(\theta)$ is defined as

$$Q(\theta) = Q(0) + \int_0^\theta \frac{d\theta'}{\dot{q}_c^2(\theta')} \tag{75}$$

for $\theta < \Theta/2$, $Q(\theta) = -Q(\Theta - \theta)$ for $\theta > \Theta/2$, and $Q(0)$ is chosen so as to make $\dot{\eta}_b(\theta)$ continuous at $\theta = \Theta/2$. One can easily check, using (5), that $\eta_b(\theta)$ indeed satisfies (24). (Alternatively, one could use a procedure introduced by Cauchy [10,11], and differentiate the classical solution $q_c(\theta)$ with respect to any two parameters related to its two constants of integration.) We can now write explicit expressions for Ω_{12} and W_{ij}:

$$\Omega_{12} = \dot{q}_c(\theta_1)\,\dot{q}_c(\theta_2)\,[Q(\theta_2) - Q(\theta_1)], \tag{76}$$

$$W_{ij} = \dot{q}_c(\theta_i)\,U'[q_c(\theta_j)]\,[Q(\theta_j) - Q(\theta_i)] + \frac{\dot{q}_c(\theta_i)}{\dot{q}_c(\theta_j)}. \tag{77}$$

As a final step, the pre-factor in (71) can be derived [1] using the methods of Refs. [23,20]. The result is

$$G_c(\theta_1, 0; \theta_2, 0) = \left[\frac{W_{11}}{2\pi g\,\Omega_{12}}\right]^{1/2}. \tag{78}$$

From (77), one easily finds $W_{ii} = 1$. Therefore, our quadratic semiclassical propagator is given by

$$G_c(\theta_1, \eta_1; \theta_2, \eta_2) = \frac{1}{\sqrt{2\pi g\,\Omega_{12}}}\exp\left[-\frac{1}{2g\,\Omega_{12}}\left(W_{12}\,\eta_2^2 + W_{21}\,\eta_1^2 - 2\,\eta_1\eta_2\right)\right]. \tag{79}$$

As promised, it is completely determined by the classical solution.

Finally, we note that the van Vleck determinant Δ is a by-product of (79):

$$\Delta(q_0, \Theta) = G_c^{-2}(0, 0; \Theta, 0) = 2\pi g\,\Omega(0, \Theta) = 4\pi g\,\dot{q}_c^2(0)\,Q(0). \tag{80}$$

As shown explicitly in [1], one can express Δ as

$$\Delta = \frac{4\pi g\,[U(q_t) - U(q_0)]}{U'(q_t)}\left(\frac{\partial\Theta}{\partial q_t}\right)_{q_0}. \tag{81}$$

Together with (8), this shows that one does not need to know $q_c(\theta)$ in order to write the first term in the semiclassical series; it is enough to know $q_t(q_0, \Theta)$.

REFERENCES

1. C. A. A. de Carvalho, R. M. Cavalcanti, E. S. Fraga and S. E. Jorás, hep-th/9810045, Ann. Phys. (N.Y.), to appear (1999).
2. R. P. Feynman and A. R. Hibbs, *Quantum Mechanics and Path Integrals* (McGraw-Hill, New York, 1965); R. P. Feynman, *Statistical Mechanics* (Addison-Wesley, New York, 1972).
3. L. S. Schulman, *Techniques and Applications of Path Integration* (John Wiley, New York, 1981).
4. R. J. Rivers, *Path Integral Methods in Quantum Field Theory* (Cambridge University Press, Cambridge, 1987).
5. U. Weiss, *Quantum Dissipative Systems* (World Scientific, Singapore, 1993).
6. H. Kleinert, *Path Integrals in Quantum Mechanics, Statistics and Polymer Physics* (World Scientific, Singapore, 1995).
7. M. C. Gutzwiller, J. Math. Phys. **8**, 1979 (1967); **12**, 343 (1971).
8. R. F. Dashen, B. Hasslacher and A. Neveu, Phys. Rev. D **10**, 4114, 4130, 4138 (1974); **11**, 3424 (1975); **12**, 2443 (1975).
9. R. Rajaraman, Phys. Rep. **21**, 227 (1975).
10. C. DeWitt-Morette, Commun. Math. Phys. **28**, 47 (1972); **37**, 63 (1974); Ann. Phys. (N.Y.) **97**, 367 (1976).
11. Maurice M. Mizrahi, J. Math. Phys. **17**, 566 (1976); **19**, 298 (1978); **20**, 844 (1979).
12. C. W. Bernard, Phys. Rev. D **9**, 3312 (1974).
13. L. Dolan and R. Jackiw, Phys. Rev. D **9**, 3320 (1974).
14. S. Weinberg, Phys. Rev. D **9**, 3357 (1974).
15. R. F. Dashen, Shang-keng Ma and R. Rajaraman, Phys. Rev. D **11**, 1499 (1975).
16. B. J. Harrington, Phys. Rev. D **18**, 2982 (1978).
17. L. Dolan and J. Kiskis, Phys. Rev. D **20**, 505 (1979).
18. A. Lapedes and E. Mottola, Nucl. Phys. **B203**, 58 (1982).

19. R. D. Carlitz and D. A. Nicole, Ann. Phys. (N.Y.) **164**, 411 (1985).
20. D. Boyanovsky, R. Willey and R. Holman, Nucl. Phys. **B376**, 599 (1992).
21. S. E. Jorás, Ph. D. thesis, Universidade Federal do Rio de Janeiro (1998); S. E. Jorás and C. A. A. de Carvalho, in preparation.
22. C. A. A. de Carvalho and R. M. Cavalcanti, in *Trends in Theoretical Physics*, edited by H. Falomir, R. E. Gamboa-Saravi and F. A. Schaposnik, AIP Conference Proceedings **419**, 94 (1997).
23. J. Zinn-Justin, *Quantum Field Theory and Critical Phenomena* (Oxford University Press, Oxford, 1993).
24. C. A. A. de Carvalho and R. M. Cavalcanti, Braz. J. Phys. **27**, 373 (1997).
25. I. S. Gradshteyn and I. M. Ryzhik, *Table of Integrals, Series, and Products* (Academic Press, New York, 1965).
26. P. F. Byrd and M. D. Friedman, *Handbook of Elliptic Integrals for Engineers and Physicists* (Springer-Verlag, Berlin, 1954).
27. M. Abramowitz and I. A. Stegun (eds.), *Handbook of Mathematical Functions* (Dover, New York, 1965).
28. F. Vinette and J. Čížek, J. Math. Phys. **32**, 3392 (1991); W. Janke and H. Kleinert, Phys. Rev. Lett. **75**, 2787 (1995).

Features of Type I compactifications and their heterotic duals.

G. Aldazabal

CNEA, Centro Atómico Bariloche,
8400 S.C. de Bariloche, and CONICET, Argentina.

Abstract. In this note we indicate the basic ingredients needed for the construction of Type IIB $N = 1, D = 4$ orientifolds. In particular we introduce a Cartan-Weyl basis description, which appears to be specially useful for establishing the connection with dual candidate heterotic pairs. Hints for finding such candidate pairs are presented.

I INTRODUCTION

Today it is generally accepted (although far for been proved) that the five known consistent perturbative string theories are nothing but manifestations of a more fundamental underlying theory, called M-theory.

Thus, the two Type II closed string theories with two supersymmetries, the two closed heterotic theories with gauge groups $E_8 \times E_8$ and $SO(32)$ and $N = 1$ supersymmetries and Type I theory containing open and closed strings in ten dimensions, so different from each other at a perturbative level, would correspond to different expansions of M-theory. Moreover, "duality transformations" should exist relating all of these theories among themselves.

Last five years have provided us with exciting theoretical developments and different types of evidence sustaining such a picture (see for instance [1] and references therein).

A key ingredient in this picture is S-duality [2], a duality transformation containing the inversion of the coupling constant, and therefore relating a weakly coupled theory with another theory at strong coupling. The existence of such a transformation opens the way to understand a strong coupling regime by using ordinary perturbative techniques.

In particular, Type I string theory and heterotic string theory with gauge group $SO(32)$ are conjectured to be strong-weak duals [3,4] in $D = 10$ dimensions. This duality manifests in a more entangled way when such theories are compared in lower compactified dimensions. We will refer to some aspects of this Type I-heterotic duality in $D = 6, 4$ dimensions in what follows. Nevertheless, it is worth recalling that, unlike other string theories, compactifications of Type I perturbative theory to less than ten dimensions deserved very little attention (see however [5–7]) until very recently. Thus comparisons with heterotic models is a research object of study at present.

The reasons for such poor information about Type I compactifications may be probably due to the technical complexity of such constructions, in part originated in the fact that such theories are not constrained by modular invariance, as closed oriented strings instead are. Consistency may be interpreted here as tadpole cancellation requirements.

From the phenomenological point of view, perturbative $SO(32)$ Type I theory appears as acceptable as heterotic theories, in the sense that Standard Model gauge group can be accommodated (unlike Type II perturbative theories). Nevertheless, the above mentioned difficulties added to the (related) fact that we must deal with quite awful objects like Möbius strips or Klein bottles (compared with the "simplicity" of heterotic compactifications) delayed the development of deeper studies from this point of view.

It was the understanding of *Dp*-branes as carriers of RR charges, which opened the possibility for a more systematic study of such compactifications. Thus, in last years many new models in $D = 8, 6$ and $D = 4$

CP484, *Trends in Theoretical Physics II*, edited by H. Falomir, R. E. Gamboa Saraví, and F. A. Schaposnik

dimensions have been constructed (see [8] for references). Such models are interesting by themselves and in particular, in $D = 4$ dimensions a sort of "Type I phenomenology" is becoming possible today [9].

In section 1 we offer a general description of Type I $N = 1$ orbifold compactifications as Type IIB orientifolds in $D = 4$ dimensions. We introduce a Cartan-Weyl basis description which highly simplifies computation of the spectra and that appears particularly suitable for comparisons with heterotic orbifold models. We present a list of consistent models for different orbifold twists.

In section 2 we discuss different features of heterotic-Type I duality. Most of the present note is based on work done in [8,10] where the reader is referred for details.

II $D = 4$, $N = 1$, TYPE IIB ORIENTIFOLDS

In this section we summarize the basic ingredients [5,7,11,8] and notation needed in the construction of $D = 4, N = 1$ orientifold. The reader is referred to [8] for further details.

Type II theory is invariant under a world sheet parity operation Ω, or orientation reversal exchanging left and right movers. Dividing out by this symmetry leads to Type I theory which possesses half the number of supersymmetries. In a Type IIB orientifold, the toroidally compactified theory is divided out by the joint action of a discrete symmetry group G_1, (like Z_N or $Z_N \times Z_M$) together with Ω. Ω action can be accompanied by extra operations, thus leading to generic orientifold group $G_1 + \Omega G_2$ with $\Omega h \Omega h' \in G_1$ for $h, h' \in G_2$.

In this article we refer to the cases $G_1 = G_2 = Z_N$ (some $G_1 = Z_N \times Z_M$ models are also presented). $D = 4$ orientifolds are obtained by acting with the twist Ω on Type IIB string compactified on T^6/G_1. The allowed orbifold groups, acting crystallographically on T^6 and leading to $N = 1$ unbroken supersymmetry were classified in [13]. The list, with corresponding twist vector eigenvalues $v = (v_1, v_2, v_3)$ associated to the Z_N orbifold twist θ is given in Table 1.

Orientifolding closed Type IIB string introduces a Klein-bottle unoriented surface. Amplitudes on such a surface contain tadpole divergences. In order to eliminate such unphysical divergences Dp-branes must be, generically, introduced. In this way, divergences occurring in the open string sector cancel up the closed sector ones and produce a consistent theory. Tadpole cancellation is interpreted as cancellation of the charge carried by RR form potentials. Namely, dividing Type II B theory by Ω and the orbifold twists introduces fixed "orientifold planes" which act as sources of RR fields. Dp-branes are also charged with respect to such fields and are introduced in such a way that the sum of all charges vanishes.

For Z_N, with N odd, only D9-branes are required. They fill the full space-time and six dimensional compact space. For N even, whenever an element $\Omega R_i R_j$ is present in the orientifold group, where R_i (R_j) is an order two twist of complex compact plane $i(j)$, D5_k-branes with world volume filling space-time and $k \neq i, j$ compact plane are required.

In what follows we consider cases with only one set of five branes. Z_N twists in Table 1 were organized in such a way that, for even N, the order two element $R = \theta^{N/2}$ reflects Y_1 and Y_2 complex plane variables and thus the corresponding orientifolds have 5_3-branes, filling space-time and compact dimension Y_3.

Thus, orientifold models spectra have contributions both from a closed sector and an open string sector. The former is obtained by performing the orientifold projection on the original Type IIB string states. It contains the $N = 1$ supergravity multiplet and a series of moduli (uncharged with respect to the brane gauge groups) states. For completeness we present examples of such states in Table 2

Open string states are denoted by $|\Psi, ab\rangle$, where Ψ refers to world-sheet degrees of freedom while the a, b Chan-Paton indices are associated to the open string endpoints lying on Dp-branes and Dq-branes respectively.

These Chan-Paton labels must be contracted with a hermitian matrix λ^{pq}_{ab}. The action of an element of the orientifold group on Chan-Paton factors is achieved by a unitary matrix $\gamma_{g,p}$ such that $g : \lambda^{pg} \to \gamma_{g,p} \lambda^{pq} \gamma^{-1}_{g,q}$. We denote by $\gamma_{k,p}$ the matrix associated to Z_N orbifold twist θ^k acting on a Dp-brane.

TABLE 1. Z_N actions in D=4.

Z_3	$\frac{1}{3}(1,1,-2)$	Z'_6	$\frac{1}{6}(1,-3,2)$	Z'_8	$\frac{1}{8}(1,-3,2)$
Z_4	$\frac{1}{4}(1,1,-2)$	Z_7	$\frac{1}{7}(1,2,-3)$	Z_{12}	$\frac{1}{12}(1,-5,4)$
Z_6	$\frac{1}{6}(1,1,-2)$	Z_8	$\frac{1}{8}(1,3,-4)$	Z'_{12}	$\frac{1}{12}(1,5,-6)$

TABLE 2. Number of chiral multiplets in closed string sectors for some Z_N and $Z_N \times Z_M$, D=4, N=1 type IIB orientifolds. A dilaton multiplet must be added in the untwisted sector.

Twist Group	Untwisted moduli	Twisted moduli
Z_3	9	27
$Z_3 \times Z_3$	3	81
Z_7	3	21
Z_6	5	29
Z_6'	4	42
$Z_3 \times Z_6$	3	71
Z_{12}	3	27

Consistency under group transformations imposes restrictions on the representations γ_g. For instance, from $\Omega^2 = 1$ and $\theta^N = 1$ it follows that

$$\gamma_{\Omega,p} = \pm\gamma^T_{\Omega,p} \quad \text{(II.1)}$$

$$\gamma_{N,p} = \pm 1 \quad \text{(II.2)}$$

etc.

Tadpole cancellation imposes further constraints on twist matrices γ_g. For instance, when considering D9-branes, the plus sign is forced upon us by untwisted sector tadpole cancellation. Also a total number of 32 D9-branes is called for. In agreement with Gimon and Polchinski action, analyzed in [11] we will also chose $\gamma_{\Omega,5} = -\gamma^T_{\Omega,5}$ for five branes. Again, tadpole cancellation requires 32 five-branes. These constraints lead to $SO(32)$ and $USp(32)$ groups in the 99 and 55 open string sectors respectively. Also notice that consistency $(\Omega\theta^k)^2 = \theta^{2k}$ and the above convention for signs in first row of (II.2) lead to

$$\gamma^*_{k,p} = \gamma_{\Omega,p}\,\gamma_{k,p}\gamma_{\Omega,p} \quad \text{(II.3)}$$

for $p = 9, 5$.

Therefore, for a Z_N orbifold twist action, with $N = 2P$ ($N = 2P + 1$) for N even (odd), a generic matrix $\gamma_{\theta,p}$ can be written as

$$\gamma_{1,p} = (\tilde{\gamma}_{1,p}, \tilde{\gamma}^*_{1,p}) \quad \text{(II.4)}$$

where $*$ denotes complex conjugation. $\tilde{\gamma}$ is a $N_p \times N_p$ diagonal matrix given by

$$\tilde{\gamma}_{1,p} = \mathrm{diag}\,(\cdots, \alpha^{V_j} I_{n^p_j}, \cdots, \alpha^{V_P} I_{n^p_P}) \quad \text{(II.5)}$$

with $\alpha = e^{2i\pi/N}$. $V_j = \frac{j}{N}$ with $j = 0, \ldots, P$ corresponds to a "with vector structure" action ($\gamma^N = 1$) while $V_j = \frac{2j-1}{2N}$ with $j = 1, \ldots, P$ describes "without vector structure" action ($\gamma^N_{1,p} = -1$) [1].

If we chose matrices $\gamma_{\Omega,9}$ and $\gamma_{\Omega,5}$

$$\gamma_{\Omega,9} = \begin{pmatrix} 0 & \mathrm{I}_{N_9} \\ \mathrm{I}_{N_9} & 0 \end{pmatrix} \quad ; \quad \gamma_{\Omega,5} = \begin{pmatrix} 0 & -i\mathrm{I}_{N_5} \\ i\mathrm{I}_{N_5} & 0 \end{pmatrix} \quad \text{(II.6)}$$

then (II.2) and (II.3) are satisfied.

In what follows we will be mainly concerned with models without vector structure whenever 5D-branes are present. They actually break the simplectic groups down to unitary subgroups.

Apart from the mentioned "untwisted tadpole" conditions there are "twisted tadpole" equations that further constrain above matrices.

---

1) Following the classification introduced in [14] for six-simensional models.

Tadpole conditions are usually given in terms of traces of twist matrices. Consider the Z_N twist eigenvalues $\frac{1}{N}(t_1, t_2, t_3)$ with t_a given in Table 1. Then for N odd, where only D9-branes are present such conditions read [8,10]

$$[2Tr\gamma_{2k,9}\prod_{a=1}^{3} cos(\frac{\pi k t_a}{M}) - 1] = 0 \tag{II.7}$$

for $k = 1, \ldots, P$.

The even order case is somewhat more involved. An interesting and unexpected feature is that orientifolds containing the order four twist eigenvalues $\frac{1}{4}(1,1,-2)$ (like Z_4, Z_8, Z_8ánd Z_{12}' orientifolds) are inconsistent (in the "without vector structure case"). In fact, it can be shown that tadpole divergences remain, even in the presence of 5-branes [8,10]. In the other cases, absence of tadpole divergences in the even orientifold cases in Table 1 requires

$$sin(\frac{t_3\pi k}{N})A_k = -4(1+(-1)^k)[sin(\frac{t_1\pi k}{N}) + sin(\frac{t_2\pi k}{N})] \tag{II.8}$$

with $k = 1, \ldots P$. Also, $A_k = 0$ for the cases $t_3\pi k = 0$ mod N. We have defined

$$A_k = 4sin(\frac{t_1\pi k}{N})sin(\frac{t_2\pi k}{N})Tr\gamma_{k,5,0} + Tr\gamma_{k,9} \tag{II.9}$$

Symmetric equations with 5 and 9 indices interchanged must also be considered.

Once we have determined the allowed twist matrices, i.e. solutions of above equations, we can write down the spectra of their corresponding consistent models.

The spectrum associated to the 9 and 5-brane orientifold configuration is easily obtained by working in a Cartan-Weyl basis, (see [8]).

The gauge fields living on the world volumes of a pD-brane have associated Chan-Paton factors λ^p corresponding to the gauge group G_p with $G_9 = SO(32)$ and $G_5 = Sp(32)$ (here $Sp(2) = SU(2)$). In Cartan-Weyl basis such generators are organized into charged generators $\lambda_a = E_a$, $a = 1, \cdots, 480$, and Cartan algebra generators $\lambda_I = H_I$, $I = 1, \cdots, 16$, such that

$$[H_I, E_a] = \rho_I^a E_a \tag{II.10}$$

where the $rankG_p$ dimensional vector with components ρ_I^a is the root associated to the generator E_a.

The matrices $\gamma_{1,p}$ and its powers represent the action of the Z_N group on Chan Paton factors, and they correspond to elements of a discrete subgroup of the Abelian group spanned by the Cartan generators. Hence, we can write

$$\gamma_{1,p} = e^{-2i\pi V^p \cdot H} \tag{II.11}$$

Thus, this equation defines a sixteen dimensional vector V^p with coordinates corresponding to the V_j's defined in (II.5) above. Cartan generators are represented as tensor products of σ_3 Pauli matrices. In such a description the massless states are easily found. In fact, in the pp sector the gauge group is obtained by selecting the root vectors satisfying

$$\rho^a \cdot V^p = 0 \,\mathrm{mod}\, \mathbf{Z} \tag{II.12}$$

Recall that root vectors for orthogonal groups are of the form $(\underline{\pm 1, \pm 1, 0, \cdots, 0})$, where underlining indicates that all possible permutations must be considered. In the simplectic case we have to include, in addition, the long roots $(\underline{\pm 2, 0, \cdots, 0})$.

Matter states correspond to charged generators with

$$\rho^a \cdot V^p = v_i \,\mathrm{mod}\, \mathbf{Z} \tag{II.13}$$

Here we are considering the case in which 5-branes sit at fixed points. The reader is referred ref. [8,10] for other situations.

In the 95 sector the subset of roots of $G_9 \times G_5$ of the form

$$P_{(95)} = W_{(9)} \otimes W_{(5)} = (\underline{\pm 1, 0, \cdots, 0}; \underline{\pm 1, 0, \cdots, 0}) \tag{II.14}$$

must be taken into account. Matter states are obtained from

$$P_{(95)} \cdot V^{(95)} = (s_j v_j + s_k v_k) \bmod \mathbf{Z} \tag{II.15}$$

with $s_j = s_k = \pm\frac{1}{2}$, plus (minus) sign corresponding to particles (antiparticles) and $V^{95} = V^9 \otimes V^5$.

These are all the ingredients needed to construct $N = 1$, $D = 4$ consistent orientifold models.

Let us summarize the general structure of the massless spectrum. From closed sector gravity multiplets and moduli type fields are obtained. Open strings stretching between ninebranes 9-branes give rise to a G^9 gauge subgroup of $SO(32)$ (see eq.(II.12) with chiral matter charged under it. Similarly, strings with both ends at 5-branes produce a group G^5 with charged chiral matter. In addition there is the 59 sector with chiral matter charged under both G^9 and G^5.

It is interesting to notice that tadpole cancellation equations ensure cancellation of all gauge and gravitational anomalies. More explicitly, in the even case, equations above ensure cancellation of 55 sector contributions while the symmetric ones ($5 \leftrightarrow 9$) do it for the 99 sector. However, as it is discussed in reference [10], tadpole cancellation is generically a stronger requirement than anomaly cancellation in $D = 4$ dimensions.

In Table 3 we present a list of models obtained from [8].

TABLE 3. Gauge group and charged chiral multiplets in some Z_N and $Z_N \times Z_M$, D=4, N=1 type IIB orientifolds with GP action. Only models with at most one set of 5-branes are shown. All 5-branes sit on the fixed point at the origin so that in models with 5-branes the spectrum is explicitly T-dual.

Twist Group Gauge Group	(99)/(55) matter	(95) matter
Z_3 $U(12) \times SO(8)$	$3(12, 8) + 3(\overline{66}, 1)$	-
$Z_3 \times Z_3$ $U(4)^3 \times SO(8)$	$(\underline{4, 1, 1}, 8_v) + (\underline{\overline{4}, \overline{4}, 1}, 1) + (\underline{6, 1, 1}, 1)$	-
Z_7 $U(4)^3 \times SO(8)$	$(\underline{4, 1, 1}, 8_v) + (\underline{\overline{4}, \overline{4}, 1}, 1) + (\underline{6, 1, 1}, 1)$ $+(\overline{4}, 4, 1, 1) + (1, \overline{4}, 4, 1) + (4, 1, \overline{4}, 1)$	-
Z_6 $(U(6)^2 \times U(4))^2$	$2(15, 1, 1) + 2(1, \overline{15}, 1)$ $+2(\overline{6}, 1, 4) + 2(1, 6, \overline{4})$ $+(\overline{6}, 1, \overline{4}) + (1, 6, 4) + (6, \overline{6}, 1)$	$(6, 1, 1; 6, 1, 1) + (1, \overline{6}, 1; 1, \overline{6}, 1)+$ $(1, 6, 1; 1, 1, \overline{4}) + (1, 1, \overline{4}; 1, 6, 1)+$ $(\overline{6}, 1, 1; 1, 1, 4) + (1, 1, 4; \overline{6}, 1, 1)$
Z_6' $(U(4)^2 \times U(8))^2$	$(\overline{4}, 1, 8) + (1, 4, \overline{8}) + (\underline{6, 1}, 1)+$ $(4, 1, 8) + (1, \overline{4}, \overline{8}) + (\overline{4}, 4, 1) + (1, 1, 28)$ $+(1, 1, \overline{28}) + (4, 4, 1) + (\overline{4}, \overline{4}, 1)$	$(\overline{4}, 1, 1; \overline{4}, 1, 1) + (1, 4, 1; 1, 4, 1)+$ $(1, \overline{4}, 1; 1, 1, 8) + (1, 1, 8; 1, \overline{4}, 1)+$ $(4, 1, 1; 1, 1, \overline{8}) + (1, 1, \overline{8}; 4, 1, 1)$
$Z_3 \times Z_6$ $(U(2)^6 \times U(4))^2$	$(2, 2, 1^5) + (1^2, 2, 2, 1^3) + (1^4, 2, 2, 1)+$ $(1^4, 2, 1, 4) + (1^5, 2, \overline{4}) + (1, 2, 1^2, 2, 1^2)$ $+(1^3, 2, 1, 2, 1) + (2, 1^5, 4) + (2, 1^4, 2, 1)$ $+(1^2, 2, 1^3, \overline{4}) + (1, 2, 1^4, \overline{4})+$ $(1^2, 2, 1, 2, 1^2) + (1^3, 2, 1^2, 4) + 4(1^7)$	$(2, 1^6; 1, 2, 1^5) + (1^2, 2, 1^4; 1^3, 2, 1^3)+$ $(1^4, 2, 1^2; 1^5, 2, 1) + (1^5, 2, 1; 1^6, \overline{4})$ $+(1^4, 2, 1^2; 1^6, 4)$ + same with groups reversed
Z_{12} $(U(3)^4 \times U(2)^2)^2$	$(\underline{\overline{3}, 1}, \underline{\overline{3}, 1}, 1, 1) + (\underline{3, 1, 1, 1}, 2, 1)+$ $2(1, 3, 1, 1, 2, 1) + 2(3, 1, 1, 1, 1, 2)+$ $2(1, 1, 1, 3, 2, 1) + 2(1, 1, 3, 1, 1, 2)+$ $(\underline{3, 1, 1}, 1, 1, 1)$	$(\overline{3}, 1^5; 1, \overline{3}, 1^4) + (1, 3, 1^4; 1^5, 2)+$ $(3, 1^5; 1^4, 2, 1) + (1^2, 3, 1^3; 1^4, 2, 1)+$ $(1^2, \overline{3}, 1^3; 1^3, \overline{3}, 1^2) + (1^3, 3, 1^2; 1^5, 2)$ + same with groups reversed

III COMPARISON WITH HETEROTIC STRING

As we mentioned in the Introduction, Type I theory in ten dimensions is expected to be strong/weak dual to $SO(32)$ heterotic string. Some clues for the idea that the two theories are connected are that they both have $N = 1$ supersymmtry in $D = 10$ and the gauge group is $SO(32)$. Moreover, both theories present the "same type" of massless fields i.e. the dilaton, the graviton and an antisymmetric tensor field. However, perturbative expansions in each of these theories differ and thus, if both theories are related, they should be describing different regimes.

An analysis of Type I and Heterotic effective actions suggests that one is mapped into the other if corresponding dilaton fields are exchanged in the following way,

$$\Phi^H \to -\Phi^I \tag{III.1}$$

Since $\lambda =< e^{\phi} >$ is the coupling constant, this equivalence suggests that the weak coupling limit of one theory is describing the strong coupling limit of the other and viceversa. We refer to [3,4] for a detailed analysis of this conjecture.

As it stands, checking this relation appears out of reach of perturbation theory in ten dimensions, since at least one of the theories should be known at strong coupling.

The situation looks different when we compactify down to lower dimensions D. In fact, a simple dimensional reduction (fields do not depend on internal $10 - D$ coordinates) leads to [3,14–16]

$$\Phi^H_D = \frac{(6-D)}{4}\Phi^I_D - \frac{(D-2)}{8} ln V_{10-D} \tag{III.2}$$

which, in terms of coupling constants becomes

$$\lambda^H_D \to \lambda^{I\,\frac{(6-D)}{4}} / V_{10-D}^{\frac{(D-2)}{8}} \tag{III.3}$$

In ten dimensions, the strong coupling result (III.1) is recovered. Interestingly enough, in fewer dimensions a weak-weak coupling duality for regions of moduli space seems indeed possible. For instance for a moderate Type I compactification volume a weak-weak coupling duality is expected in $D = 4$ while a large volume will lead us to a strong-weak coupling, producing complementary information.

The Cartan-Weyl basis construction presented above is very suggestive for establishing connections with heterotic orbifolds. In a heterotic orbifold the action of the orbifold twist on the internal compactified coordinates (characterized by the eigenvalues $v = \frac{1}{N}(t_1, t_2, t_3)$, as given in Table 1), must be accompanied by a corresponding action into the gauge degrees of freedom. Such action can be realized by a vector shift V_{het} such that NV_{het} belongs to a sixteen dimensional lattice $\Gamma_8 \times \Gamma_8$ of $E_8 \times E_8$ or Γ_{16} of $Spin32/Z_2$. It is the latter situation which is relevant for establishing connections with Type I theory.

The spectrum for each model is subdivided in sectors. For a Z_N heterotic orbifold there are N sectors twisted by $\theta^j, j = 0, 1, \cdots, N-1$. Each particle state is created by a product of left and right vertex operators $L \otimes R$. The mass of the states follow from

$$\begin{aligned} m_R^2 &= N_R + \frac{1}{2}(r + j\,v)^2 + E_j - \frac{1}{2} \quad ; \\ m_L^2 &= N_L + \frac{1}{2}(P + j\,V_{het})^2 + E_j - 1 \end{aligned} \tag{III.4}$$

Here r is an $SO(8)$ weight with $\sum_{i=1}^{4} r_i =$ odd and P is a gauge lattice vector with $\sum_{I=1}^{16} P^I =$ even. E_j is the twisted oscillator contribution to the zero point energy, $E_j = \sum \frac{1}{2}|v_a|(1 - v_a|)$. The multiplicity of states satisfying eq. (III.4) in a θ^j sector is given by the appropriate generalized GSO projections [17]. The gravity multiplet, charged and neutral chiral fields appear in the untwisted sector. Twisted sectors contain charged chiral multiplets.

Massless states in the untwisted sector require that lattice vectors verify $P^2 = 2$ and thus only $SO(32)$ vectors are selected. Spinorial weights do not contribute to this sector. Moreover, gauge bosons verify $P.V_{het} = int$, while matter states are selected by $P.V_{het} = v_i$ modulo integers. Interestingly enough, these are the same kind of conditions we had found in eqs.(II.12) and (II.13) for the 99 sector of Type I string if the replacement $V_{het} \to V^9$ is performed. Thus, a natural mapping between (99) sector of Type I string and untwisted sector of $SO(32)$ heterotic string is suggested. Namely,

$$\text{Type I} \longleftrightarrow SO(32) \text{ Heterotic}$$

$$\psi^{\mu}_{-\frac{1}{2}}|0,ab\rangle\,\lambda^{(0)}_{9,ab} \longleftrightarrow \psi^{\mu}_{-\frac{1}{2}}|0\rangle_R \otimes |P^I\rangle_L\;,|\partial X^I\rangle_L \quad I=1,\cdots,16$$

$$\psi^{i}_{-\frac{1}{2}}|0,ab\rangle\,\lambda^{(i)}_{9,ab} \quad i=1,2,3 \longleftrightarrow \psi^{i}_{-\frac{1}{2}}|0\rangle_R \otimes |P^I\rangle_L\;,|\partial X^I\rangle_L \quad i=1,2,3$$

$$\gamma_{1,9} = \exp(-2i\pi V_{(99)}\cdot H) \longleftrightarrow V_{het} = V_{(99)},\; NV_{het} \in \Gamma \tag{III.5}$$

where Γ is the $Spin(32)/Z_2$ lattice and $P \in \Gamma$ are the gauge quantized momenta of the heterotic string. We see that the identification

$$V_{het} = V^9 \tag{III.6}$$

allows us to distinguish potentially dual models.

However, more ingredients must be taken into account. On the one hand we notice that, for even order orientifolds with D5-branes, the rank of the gauge group (see Table 3) is generically greater than 16 (or in general 22), the maximum rank allowed for a consistent perturbative heterotic model. Thus, if such models are effectively dual to each other, Type I cases would be giving us a nonperturbative sector of heterotic. Recall that non-perturbative solitonic solutions of heterotic string are NS fivebranes, with the correct properties to be duals to D5-branes. On the other hand odd order orientifolds have rank sixteen and have a chance to lead to the same model as in the heterotic side. Notice that even in these cases, when untwisted heterotic sector and 99 Type I sector match, the closed sector moduli fields shown in Table 2 and twisted sectors heterotic states (see eq.(III.4)) still appear.

Another fact to be considered is that perturbative heterotic models must obey the modular invariance constraint

$$N(V_{het}^2 - v^2) = even \tag{III.7}$$

ensuring level matching condition $m_R = m_L$.

One can check that only the shifts V^9 corresponding to odd N ($Z_3, Z_3 \times Z_3, Z_7$) obey such constraints, while none of the even order twists do. We see that, even if eq.(III.6) is a step forward in the identification of dual models, a deeper analysis is called for. Nevertheless, some progress has been achieved. Indeed, perturbative heterotic duals for the three odd orientifolds were proposed in refs. [15,18,19]. A study of the effective potentials on both sides shows that some fields become adequately massive so that massless spectra completely match.

Models with D5-branes are somewhat more complex since solitonic five branes contributions on the heterotic side should be taken into account. The situation is not so bad in $D = 6$ dimensions where such contributions, identified as small instantons either in the bulk or located at fixed points, have been studied in several cases [20–22].

Let us first address the problem of modular invariance. Notice that it is easy to obtain new shifts that produce the same untwisted spectrum and *are* modular invariant also for even N. It is enough to consider any of the even order shifts in section 3 and to do the replacement

$$V_{het} = V_{(99)} \longrightarrow V_{het} = V_{(99)} - (0,0,0,\cdots,0,1) \tag{III.8}$$

What does this redefinition imply? A $D = 6$ analysis performed in ref. [22] can give us a hint. Indeed, there the simplest Z_2, D=6 BPGP [11] orientifold is considered. The Chan-Paton shifts correspond to $V^9 = V^5 = \frac{1}{4}(1,1,...,1,1)$. The above rules for $D = 6$ case lead to an open string spectrum which has a $U(16)_9 \times U(16)_5$ gauge group with hypermultiplets in $2(\mathbf{120},\mathbf{1}) + 2(\mathbf{1},\mathbf{120}) + (\mathbf{16},\mathbf{16})$ representations. This would be the case if all eight dynamical D-five-branes coincided at the same fixed point. This would be non perturbative from the heterotic point of view. However, if half a five-brane is located at each of the 16 fixed points, the gauge group is $U(16)_9 \times U(1)_5^{16}$ with hypermultiplets transforming as $2(\mathbf{120}) + 16(\mathbf{16}) + 20(\mathbf{1})$. Moreover, the $U(1)_5^{16}$ group is broken and swallows sixteen of the singlet hypermultiplets in a variation of the Green-Schwarz mechanism [14]. This particular BSGP model massless spectrum coincides with the $SO(32)$ heterotic perturbative [14] model described the shift $V_{het} = V^9 - (0,0,0,\cdots,0,1) = \frac{1}{4}(1,\cdots,1,-3)$. We should notice that while the $16(\mathbf{16})$ hypermultiplets of this model originate in a standard (perturbative) twisted sector, in the orientifold they originate in the (59) open string sector. The closed string spectrum produces twenty $K3$ moduli.

Thus, we have found a perturbative heterotic model which has a dual perturbative Type I candidate.

Interestingly enough, a non perturbative heterotic candidate for the Type I case with all D5-branes at the origin can also be found. From above considerations this would require to consider a non modular invariant shift on the heteretic side. Since modular invariance is actually a perturbative concept this should not be a problem. Moreover, it is expected to be explicitly violated in the presence of fivebranes equation. In fact, to see this, notice that (III.7) is nothing but the orbifold version of the conservation of the magnetic charge associated to the antisymmetric tensor field form H. Namely

$$Q_{TOT} = 0 = \int_X dH = \int_X (F^2 - R^2) = \sum_f Q_f \tag{III.9}$$

where X is the compact space and the last term emphasizes that contributions to the charge come from integrals around fixed points f in the orbifold case. Since five branes are magnetic sources for H, we expect this equation to be modified in their presence to give

$$Q_{TOT} = \sum_f Q_f + n_5 = 0 \tag{III.10}$$

with n_5 the number of solitonic fivebranes. In terms of level matching conditions we thus expect a modification of the form

$$N\,(V^2 - v^2) + 2ME_5(f) = \text{even} \tag{III.11}$$

where E_5 is an adequate shift in the mass formula induced by the presence of five branes, which explicitly violates modular invariance. Now, if modular invariance is violated then, as a consequence, anomaly cancellation is no longer assured. Spectra computed with such shifts will be generically anomalous. This is just as well since small instantons have gauge and matter degrees of freedom which will also be anomalous. It is the sum of all contributions which must be anomaly free This is , in effect, what actually happens with the heterotic dual of the particular BSGP model with gauge group $U(16)_9 \times U(16)_5$ with all D5- branes at the origin [22]. Several examples of this kind in which standard modular invariant constraints are violated have been constructed and provide non trivial evidence of Type I - heterotic duality in $D = 6$.

The D=4, N=1 even orientifolds discussed above appear to behave in a similar way concerning type I-heterotic duality. It seems that the heterotic duals of orientifolds with all 5-branes sitting at the origin are non-perturbative D=4, N=1 heterotic orbifolds in which the usual modular invariance constraints are violated and non-perturbative gauge groups and fields arise due to small instanton effects. Although, as we mentioned above, for each of the orientifolds of even order a perturbative heterotic candidate dual obeying modular invariance constraints can be found, the full massless spectra cannot be matched with orientifolds. In effect, on the one hand these perturbative vacua are lacking the extra degrees of freedom associated to the 5-branes. Furthermore, these perturbative heterotic vacua have extra charged matter fields in their twisted sectors which are also absent in their orientifold counterparts. From this point of view the D=6, Z_2 GP model is exceptional, since there is one configuration of the 5-branes (two in each of the 16 fixed points) which precisely matches the *perturbative* Z_2 heterotic orbifold that we mentioned above. With that configuration there is no gauge group left from the 5-brane sector. In the case of generic D=4, N=1 orientifolds of even order such privileged 5-brane configurations seem difficult to find, in the case they exist at all.

To exemplify the above discussion let us consider a perturbative heterotic candidate dual of the Z_6' orientifold presented in Table (3). The corresponding shift is $V^5 = V^9 = \frac{1}{12}(1111555553333333)$ which does not satisfy the modular invariance constraint. The mapping in eq. (III.5) suggests constructing a perturbative heterotic orbifold with a shift $V_{het} = V_{(99)}$ appropriately shifted as in (III.8). As we said, the untwisted sector fully matches the (99) sector of the orientifold. On the other hand, the twisted sectors have the following content

$$\begin{aligned} \theta,\theta^5 &: 12(4,1,1) + 12(1,\bar{4},1) \\ \theta^2,\theta^4 &: 9(6,1,1) + 9(1,6,1) + 6(\bar{4},4,1) + 3(4,\bar{4},1) + 18(1,1,1) \\ \theta^3 &: 4[(1,1,8) + (1,1,\bar{8}) + (\bar{4},1,1) + (1,4,1)] + 8[(4,1,1) + (1,\bar{4},1)] \end{aligned} \tag{III.12}$$

One can check that the contribution to non-Abelian anomalies coming from θ, θ^5 sectors is cancelled by that coming from the θ^2, θ^4 sectors. The contribution of the θ^3 particles exactly cancels against the untwisted sector. Notice that the content of the θ^3 twisted sector is identical to that of the (59) sector of the Z_6' orientifold, except for the obvious fact that multiplicities coming from the number of fixed points in the heterotic orbifolds are representations with respect to the (55) gauge group on the orientifold side. Although this coincidence would suggest that this perturbative heterotic model could be dual to the Z_6' orientifold, there is nothing in the orientifold resembling the spectrum of $\theta^n, n = 1,2,4,5$ sectors. Rather, one would expect this model to be dual to some *non-perturbative type I* vacuum which has solitonic states reproducing those sectors.

IV SUMMARY AND FINAL COMMENTS

We have presented a set of requirements needed for the construction of consistent compactifications of Type I string down to four dimensions and presented some explicit examples. An unexpected result, mentioned above only laterally, is that consistent models containing Z_4 twist action $1/4(1,1,-2)$ [8,10] do not seem to exist. They contain tadpoles which cannot be cancelled by 5-branes. An indication of this fact is that models of this type, with generic twist matrices as defined above are always anomalous [10]. Leaving technicalities aside this an interesting behaviour which would deserve further investigation on the light of dualities.

We have also indicated how heterotic-Type I dual candidates can be identified. The knowledge of small instantons effects in $D = 6$ allows to explicitly construct non heterotic perturbative models with massless spectra exactly matching Type I ones. Non perturbative effects in $D = 4$ are less known. Nevertheless, the study of gauge theories on the world volume of D5-branes in four dimensions, as computed above, may give us some hints about non perturbative contributions on the heterotic side.

A last comment regarding the presence of spinor representations is in order. Since the perturbative gauge group of Type I string is $SO(32)$, we do not find spinors of $SO(32)$ subgroups compactifications in the massless spectrum. This is not the case in heterotic string where we actually have a $Spin(32)/Z_2$ structure and where spinors frequently appear in twisted sectors. From heterotic/Type I duality we expect them to appear as non perturbative massive states on Type I side. It has been recently proposed that they correspond to solitonic $D0$-branes of Type I string. Moreover, the presence of a D(-1)-brane was also conjectured [23,24]. Establishing the actual correspondence of spinor states on heterotic models and massive states of Type I in six and four dimensional compactifications is an open and interesting problem.

ACKNOWLEDGMENTS

My warm thanks to A. Font, L. Ibañez and G. Violero, with whom the work in which this note is based was done, for specially enjoyable collaboration.

REFERENCES

1. J. Schwarz, Nucl. Phys. B55 (1997) 1, Proc. Suppl., hep-th/9607201; hep-th/9711029;
   S. Forste and J. Louis, hep-th/9612198.
   A. Sen, hep-th/9802051.
2. A. Font, L. Ibañez, D. Luest and F. Quevedo, Phys. Lett. B249 (1990) 35.
3. E. Witten, Nucl. Phys. B443 (1995) 85, hep-th/9503124.
4. J. Polchinski and E. Witten, Nucl. Phys. B460 (1995) 535, hep-th/9510169
5. A. Sagnotti, in Cargese 87, *Strings on Orbifolds*, ed. G. Mack et al. (Pergamon Press, 1988) p. 521.
6. P. Horava, Nucl. Phys. B327 (1989) 461; Phys. Lett. B231 (1989) 251;
   J. Dai, R. Leigh and J. Polchinski, Mod.Phys.Lett. A4 (1989) 2073;
   R. Leigh, Mod.Phys.Lett. A4 (1989) 2767.
7. G. Pradisi and A. Sagnotti, Phys. Lett. B216 (1989) 59;
   M. Bianchi and A. Sagnotti, Phys. Lett. B247 (1990) 517.
8. G. Aldazabal, A. Font, L.E. Ibáñez and G. Violero, Nucl. Phys. B536 (1999) 29, hep-th/9804026.
9. L.E. Ibáñez, hep-th/9802103.
   L. Ibañez, .Burgess, F. Quevedo, hep-th/9810535.
10. G. Aldazabal, D. Badagnani, L.E. Ibáñez and A. Uranga ,*Tadpole versus anomaly cancellation in* $D = 6, 4$ *compact Type IIB orientifolds*, to appear.
11. E. Gimon and J. Polchinski, Phys.Rev. D54 (1996) 1667, hep-th/9601038.
12. A. Dabholkar and J. Park, Nucl. Phys. B472 (1996) 207, hep-th/9602030.
13. L. Dixon, J.A. Harvey, C. Vafa and E. Witten, Nucl. Phys. B274 (1986) 285.
14. M. Berkooz, R. G. Leigh, J. Polchinski, J. H. Schwarz, N. Seiberg and E. Witten, Nucl. Phys. B475 (1996) 115, hep-th/9605184.
15. C. Angelantonj, M. Bianchi, G. Pradisi, A. Sagnotti and Ya.S. Stanev, Phys. Lett. B385 (1996) 96, hep-th/9606169.
16. I. Antoniadis, H. Partouche and T.R. Taylor, Nucl. Phys. BCCC (1997) 160, hep-th/9706211.
17. L.E. Ibáñez, J. Mas, H.P. Nilles and F. Quevedo, Nucl. Phys. B301 (1988) 157;
    A. Font, L.E. Ibáñez, F. Quevedo and A. Sierra, Nucl. Phys. B331 (1990) 421.

18. Z. Kakushadze, Nucl. Phys. B512 (1998) 221, hep-th/9704059.
19. Z. Kakushadze and G. Shiu, Phys. Rev. D56 (1997) 3686, hep-th/9705163.
20. E. Witten, Nucl. Phys. B460 (1996) 541, hep-th/9511030.
21. K. Intrilligator, Nucl. Phys. B496 (1997) 177, hep-th/9702038.
22. G. Aldazabal, A. Font, L.E. Ibáñez, A.M. Uranga and G. Violero, hep-th/9706158.
23. A. Sen, hep-th/9808141;hep-th/ 9809111
24. E. Witten, hep-th/9810188

QUANTIZATION OF $2D$ DILATON GRAVITY THEORY

Victor O. Rivelles

Instituto de Física, Universidade de São Paulo
Caixa Postal 66318, 05315–970, São Paulo, SP, Brazil
E-mail: rivelles@fma.if.usp.br

Abstract. We discuss the relation between canonical and Schrödinger quantization of the CGHS model.

The CGHS model [1] is described by the action

$$S = \int d^2x\, \sqrt{-g}\, e^{-2\phi}\, (R + 4g^{\mu\nu}\partial_\mu\phi\partial_\nu\phi - \Lambda)\,, \tag{1}$$

where R is the scalar curvature, ϕ is the dilaton and Λ the cosmological constant. The interest in the CGHS model stem from the fact that the model allows black hole formation and Hawking radiation at the semi–classical level [2]. It can be reformulated as a topological field theory of the BF type with the gauge group being the extended Poincaré group [3]. Its supersymmetric version is also known [4]. Many properties of the CGHS model were derived at the semi–classical level. However it is possible to rewrite the action Eq.(1) in a quadratic form [5,6] after performing a canonical transformation. The resulting action describes two scalar fields $r^a(x), a = 1, 2$ corresponding to combinations of the gravity and dilaton fields and the reparametrization symmetry of the original geometric action is retained as a Hamiltonian constraint. Without loss of generality we can take the cosmological constant $\Lambda = 1$ and the action reduces to

$$L = \pi_a\dot{r}^a + \lambda_0 H_0 + \lambda_1 H_1, \tag{2}$$

where $\pi_a(x)$ is the canonical momentum of $r^a(x)$ and $\lambda_a(x)$ are the Lagrange multipliers which implement the reparametrization constraints which are the components of the energy–momentum tensor of the two massless scalar fields $r^a(x)$

$$\begin{aligned} H_0 &= \frac{1}{2}(\pi^a\pi_a + r'^a r'_a), \\ H_1 &= \pi_a r'^a. \end{aligned} \tag{3}$$

Here a dot (dash) indicates differentiation with respect to time (space). The indices $a, b, \ldots$ are raised and lowered with a Minkowskian type metric $\eta_{ab} = diag(1, -1)$. This means that the canonical variables π_a, r^a appear in an indefinite quadratic form in Eqs.(2, 3). We will then say that the field $r^0(x)$ has positive signature while $r^1(x)$ has negative signature.

The theory described by Eqs.(2, 3) looks very simple since there are no interaction terms. In the gauge $\lambda_0 = 0, \lambda_1 = 1$ it describes two massless scalar fields with opposite signature and with a vanishing energy–momentum tensor. We would expect that the physical states should be the direct product of states for each degree of freedom separately. However, there are subtle correlations due to the constraints Eqs.(3) and the Hilbert space has not a direct product structure.

The canonical quantization of the theory is upset with anomalies. Due to the normal ordering in the constraints Eqs.(3) there appears the well known Virasoro anomaly in the algebra of the energy–momentum tensor. One possible way to remove the anomaly is to add background charges to the scalar fields [7]. By

CP484, *Trends in Theoretical Physics II*, edited by H. Falomir, R. E. Gamboa Saraví, and F. A. Schaposnik

choosing appropriately the value of the background charges the anomaly can be made to cancel. We can also understand the situation from the Schrödinger point of view [8] and we will show how the constraints have to be modified in order to take into account ordering problems.

The usual way to incorporate background charges is to consider an improved energy–momentum tensor derived from an action with a surface term linear in the fields. In the present case we find

$$T_{\mu\nu} = \frac{1}{2}\partial_\mu r^a \partial_\nu r_a - \frac{1}{4}\eta_{\mu\nu}\partial^\rho r^a \partial_\rho r_a + \frac{1}{2}Q_a \partial_\mu \partial_\nu r^a - \frac{1}{4}\eta_{\mu\nu} Q_a \Box r^a. \tag{4}$$

The constraints $H_0 = (T_{++} + T_{--})/2$ and $H_1 = (T_{++} - T_{--})/2$ are now

$$H_0 = \frac{1}{2}(\pi^a \pi_a + r'^a r'_a) + Q_a r''^a, \tag{5}$$

$$H_1 = \pi_a r'^a + Q_a \pi'^a. \tag{6}$$

The Poisson bracket algebra of the constraints acquires a classical central charge due to the surface term

$$\begin{aligned}
\{H_0(x), H_0(y)\} &= (H_1(x) + H_1(y))\,\delta'(x-y),\\
\{H_0(x), H_1(y)\} &= (H_0(x) + H_0(y))\,\delta'(x-y) - Q^a Q_a \delta'''(x-y),\\
\{H_1(x), H_1(y)\} &= (H_1(x) + H_1(y))\,\delta'(x-y).
\end{aligned} \tag{7}$$

Therefore the new constraints have a first class algebra only if $Q_a Q^a = 0$. A careful analysis of the canonical transformation which brings the original CGHS model Eq.(1) (written in terms of the dilaton and gravity fields) to the quadratic form Eq.(2) (written in terms of r^a) shows that the two Lagrangians differ by a surface term [5]. This surface term can be written in the form $Q_a \Box r^a$ with $Q_a Q^a = 0$. This is necessary to retain the reparametrization symmetry of the original model. However, since the quantum theory is afflicted with anomalies we are allowed to modify it by adding a Wess–Zumino field to cancel the anomaly. Since in two–dimensions a scalar field can serve as its own Wess–Zumino field we can add an improvement term to the constraints with an appropriate value of $Q_a Q^a$ to cancel the anomaly. So, in general, $Q_a Q^a$ will no longer vanish and the classical theory will loose reparametrization invariance. However, it will be recovered at the quantum level.

Before performing the canonical quantization with the new constraints let us first consider a single massless scalar field. In the canonical approach it has an expansion in terms of Fock space operators $a^\dagger(k)$ and $a(k)$ associated with particles of positive and negative energy respectively. The conventional vaccum is defined as $a(k)|0>= 0$ so that the Hilbert space is positive definite and the energy of the states is also positive. This gives rise to a central charge $c = 1$ in the energy–momentum tensor algebra when normal ordering is taken into account. An alternative choice for the vaccum is to take $a^\dagger(k)|0>= 0$. In this case the Hilbert space is no longer positive definite, the energy of the states is negative and the central charge is $c = -1$. For conventional theories this choice of the vaccum is not allowed.

Let us now consider a single scalar field with negative signature. In the canonical approach there is a crucial change of sign in the canonical momentum which leads to a change of sign in the algebra of creation and annihilation operators. Now if the vaccum is defined as $a(k)|0>= 0$ then the Hilbert space is not positive definite but the energy of the states is positive and the central charge is $c = 1$. For the other choice of the vaccum $a^\dagger(k)|0>= 0$ the Hilbert space is positive definite, the energy is negative and $c = -1$. Then the quantum theory of a scalar field with negative signature has troubles for any choice of the vaccum.

When a background charge Q is added its effect is just to shift the value of the central charge. A short calculation shows that for the conventional scalar field we have for the usual choice of the vaccum $a(k)|0>= 0$ the value $c = 1 + 12\pi Q^2$ while for the vaccum $a^\dagger(k)|0>= 0$, $c = -1 + 12\pi Q^2$. For the scalar field with negative signature and vaccum $a(k)|0>= 0$ we have $c = 1 - 12\pi Q^2$, while for the vaccum $a^\dagger(k)|0>= 0$ we find $c = -1 - 12\pi Q^2$. This is summarized in Table .

Then the canonical quantization of the CGHS model allows several possibilities for the vanishing of the total central charge. If no background charges are present we can achieve $c = 0$ by choosing the vaccum $a_0(k)|0>= a_1^\dagger(k)|0>= 0$. Note that since our Hamiltonian is zero we have no troubles with the positivity of the energy. If background charges with $Q_a Q^a = 0$ are present we must do the same vaccum choice. If the background charges have $Q_a Q^a \neq 0$ then the vanishing of the central charge requires $Q_a Q^a = \pm 1/(6\pi)$ depending on which

TABLE 1. Hilbert space norm, energy sign and central charge for all possible vaccum choices of a scalar field

Signature	Vaccum	Norm	Energy	Central Charge
positive	$a\|0>=0$	positive definite	positive	$1+12\pi Q^2$
positive	$a^\dagger\|0>=0$	not positive definite	negative	$-1+12\pi Q^2$
negative	$a\|0>=0$	not positive definite	positive	$1-12\pi Q^2$
negative	$a^\dagger\|0>=0$	positive definite	negative	$-1-12\pi Q^2$

vaccum is choosen. There are two possiblities: $a_0(k)|0>=a_1(k)|0>=0$ or $a_0^\dagger(k)|0>=a_1^\dagger(k)|0>=0$. Either possibility is troublesome since positivity of the Hilbert space is compromised. We will also meet difficulties for the case $Q_aQ^a\neq 0$ in the Schrödinger representation. These results are presented in Table 2.

We now consider the Schrödinger representation. The Scrödinger functional Ψ is a functional of r^a, $\Psi(r^a)$, and π_a is realized as a functional derivative $\pi_a(x)=-i\delta/\delta r^a(x)$. In the Schrödinger representation there is no normal products to be taken into account. The only source of ambiguity is in the operator ordering. So the questions about the value of the central charge are difficult to be posed in this formalism. The relevant point here is whether there is a first class algebra of quantum constraints so that physical states can be properly defined.

When the Poisson bracket algebra of the constraints Eqs.(7) is replaced by the respective commutator algebra we obtain the same central charge proportional to Q_aQ^a. The algebra of the constraints is not first class and physical states can not be defined unless $Q_aQ^a=0$. Alternatively we could try to modify the constraints to take normal ordering into account in order to recover a first class algebra. So let us consider the effect of normal ordering in each term of the constraints. Let us assume again that we have a single scalar field ϕ. Assuming that $\phi(x)$ and $\pi(x)$ have canonical commutation relations we find that

$$:\phi'(x)\pi(y):=\phi'(x)\pi(y)-\frac{i}{2}\delta'(x-y), \tag{8}$$

for any choice of the vaccum and for any signature of the field. This means that

$$:H_1(x):=r'^a\pi_a+Q_a\pi'^a-i\lim_{y\to x}\delta'(x-y), \tag{9}$$

which does not depend on which vaccum is choosen. This is the same ambiguity that we find if we consider the operator ordering in H_1. Since π_a and r^a have canonical commutation relations there is an ambiguity in the term $\pi_a r'^a$ in Eq.(6) with the same form as in Eq.(9). Then the coefficient of the $\delta'(x-y)$ term is not fixed. For each choice of this coefficient we have an operator ordering prescription. This is also consistent with the commutator algebra of the constraints. It is independent of the value for this coefficient as it is easily verified.

Let us now consider the effect of normal ordering in H_0. If the field ϕ has positive signature

$$:\pi(x)\pi(y):=\pi(x)\pi(y)\mp\frac{1}{2}\omega(x-y), \tag{10}$$

TABLE 2. Choices of the backgound charge and vaccum for vanishing central charge in the CGHS model

Background Charges	Vaccum	Norm
$Q_a=0$	$a_0\|0>=a_1^\dagger\|0>=0$	positive definite
$Q_a=0$	$a_0^\dagger\|0>=a_1\|0>=0$	not positive definite
$Q_aQ^a=0$	$a_0\|0>=a_1^\dagger\|0>=0$	positive definite
$Q_aQ^a=0$	$a_0^\dagger\|0>=a_1\|0>=0$	not positive definite
$Q_aQ^a=-\frac{1}{6\pi}$	$a_0\|0>=a_1\|0>=0$	not positive definite
$Q_aQ^a=\frac{1}{6\pi}$	$a_0^\dagger\|0>=a_1^\dagger\|0>=0$	not positive definite

where

$$\omega(x-y)=\frac{1}{2\pi}\int dk\,|k|e^{ik(x-y)}. \tag{11}$$

The upper sign in Eq.(10) is for the usual vaccum $a|0>=0$ while the lower sign is for the unusual vaccum $a^{\dagger}|0>=0$. If the field ϕ has negative signature then

$$:\pi(x)\pi(y):=\pi(x)\pi(y)\pm\frac{1}{2}\omega(x-y), \tag{12}$$

with the upper (lower) sign for the usual (unusual) vaccum. The same structure holds for $\phi'(x)\phi'(y)$. Therefore we find that

$$:H_0:=\frac{1}{2}(\pi^a\pi_a+r'^a r'_a)+Q_a r''^a+\frac{c}{2}\lim_{y\to x}\omega(x-y), \tag{13}$$

where $c=0,\pm2$ is the sum of the central charges of r^0 and r^1. This takes into account all possible choices of the vaccum.

The equations for the physical states are then

$$r^a(x)\frac{\delta\Psi}{\delta r^a(x)}+Q^a\left(\frac{\delta\Psi}{\delta r^a(x)}\right)'-\alpha\lim_{y\to x}\delta'(x-y)\Psi=0, \tag{14}$$

$$\frac{1}{2}\left(-\frac{\delta^2\Psi}{\delta r^a(x)\delta r_a(x)}+r'^a(x)r'_a(x)\Psi\right)+Q_a r''^a(x)\Psi+\frac{c}{2}\lim_{y\to x}\omega(x-y)\Psi=0, \tag{15}$$

where α is a constant which will select the operator ordering prescription. The simplest choice is to take $\alpha=0$ and, as we will see, we can find solutions with this prescription. So we will adopt it from now on.

We will now look for the vaccum state in the Schrödinger representation. This is most easily done going to the canonical formalism and expressing the creation and annihilation operators in terms of r^a and π_a. We find

$$a_a(k)=\frac{1}{\sqrt{4\pi|k|}}\int dx\,\left(|k|r^a(x)\pm i\pi_a(x)\right), \tag{16}$$

where the upper sign holds for $a=0$ and the lower sign for $a=1$. As we have seen before the vaccum is the same for $Q_a=0$ and $Q_aQ^a=0$ cases. It is defined by

$$a_0(k)\Psi_{vaccum}=a_1^{\dagger}(k)\Psi_{vaccum}=0, \tag{17}$$

and the solution is known [7]

$$\Psi_{vaccum}=\det{}^{\frac{1}{2}}\left(\frac{\omega}{\pi}\right)\exp\left[-\frac{1}{2}\int dx\,dy\,\left(r^0(x)\omega(x-y)r^0(y)+r^1(x)\omega(x-y)r^1(y)\right)\right]. \tag{18}$$

Since $\omega(x-y)$ Eq.(11) has a positive kernel this vaccum is normalizable. For the case $Q_aQ^a\neq0$ the vaccum would be defined by

$$a_0(k)\Psi_{vaccum}=a_1(k)\Psi_{vaccum}=0 \tag{19}$$

or

$$a_0^{\dagger}(k)\Psi_{vaccum}=a_1^{\dagger}(k)\Psi_{vaccum}=0. \tag{20}$$

These equations do not have normalizable solutions. The solution of Eq.(19), for example, is given by [7]

$$\Psi=\exp\left[-\frac{1}{2}\int dx\,dy\,\left(r^0(x)\omega(x-y)r^0(y)-r^1(x)\omega(x-y)r^1(y)\right)\right]. \tag{21}$$

The minus sign in front of the second term makes the wave functional non normalizable since the kernel is positive. This shows that there is no vaccum state for $Q_aQ^a \neq 0$ in the Schrödinger representation. In the canonical analisys we found that in this case the Hilbert space is not positive definite.

We now look for physical states in Eqs.(14,15). For the case $Q_a = 0$ they are already known [7]. Taking $Q_a = c = 0$ in Eq.(14,15) we obtain

$$\Psi_{Q=0} = \exp\left(\pm\frac{i}{2}\int dx\, \epsilon_{ab} r^a(x) r'^b(x)\right). \tag{22}$$

For the case $Q_aQ^a = 0$ we still have $c = 0$ in Eq.(15) and we find [8]

$$\Psi_{Q_aQ^a=0} = \exp\left[\pm\frac{i}{2}\int dx\, \epsilon_{ab} r^a(x) r'^b(x) \pm i\int dx\, \epsilon_{ab} Q^a r'^b(x) \ln\left(\epsilon_{cd} Q^c r'^d(x)\right)\right]. \tag{23}$$

This solution reduces to the former solution when $Q_a = 0$.

Having found new physical states still leaves open the main difficulty of this approach: how to extract the space–time geometric properties from the Hilbert space. It is also necessary to compare the non–perturbative results obtained in this approach with the semi–classical results. We must solve these issues in the two–dimensional models, where the problems are tractable, before embarking in realistic four or higher dimensional gravity theories with propagating gravitons.

ACKNOWLEDGMENTS

This work is partially supported by FAPESP. The work of V. O. Rivelles is partially supported by CNPq.

REFERENCES

1. C. G. Callan, S. B. Giddings, J. A. Harvey and A. Strominger, *Phys. Rev.* D**45** (1992) R1005.
2. J. G. Russo, L. Susskind and L. Thorlacius, *Phys. Lett.* **B292** (1992) 13; T. Banks, A. Dabholkar, M. Douglas and M. O'Loughlin, *Phys. Rev.* **D45** (1992) 3607; L. Susskind and L. Thorlacius, *Nucl. Phys.* **B382** (1992) 123.
3. D. Cangemi and R. Jackiw, *Phys. Rev. Lett.* **69** (1992) 233.
4. V. O. Rivelles, *Phys. Lett.* **B321** (1994) 189; D. Cangemi and M. Leblanc *Nucl. Phys.* **B420** (1994) 363; M. M. Leite and V. O. Rivelles, *Class. Quantum Grav.* **12** (1995) 627.
5. A. Miković, *Black Holes and Nonperturbative Canonical 2D Dilaton Gravity*, hep-th/9402095 (1994)
6. D. Cangemi and R. Jackiw, *Phys. Lett* **B337** (1994) 271.
7. D. Cangemi, R. Jackiw and B. Zwiebach, *Ann. Phys.* **245** (1996) 408.
8. F. F. Cassemiro and V. O. Rivelles, "Canonical and functional Schrodinger quantization of two-dimensional dilaton gravity,"hep-th/9812096 (to be published in Phys. Lett. B)

INSTANTON-LIKE EXCITATIONS IN 2D FERMIONIC FIELD THEORY

J.L. Cortés[†1], J. Gamboa[‡2], I. Schmidt[§3] and J. Zanelli[‡¶ 4]

[†] *Departamento de Física Teórica, Universidad de Zaragoza, Zaragoza 50009, Spain*
[‡] *Departamento de Física, Universidad de Santiago de Chile, Casilla 307, Santiago 2, Chile*
[§] *Departamento de Física, Universidad Técnica F. Santa Maria, Valparaíso, Chile*
[¶] *Centro de Estudios Científicos de Santiago, Casilla 16443, Santiago, Chile*

In this article we review some recent results previously reported in [1,2], where a new type of nonperturbative excitations was identified in the Thirring and Schwinger models. This effect is explicitly observed in the light-cone quantization using functional methods. The effect arises from the need to compactify the x^- coordinate in the light-cone, which in effect changes the spacetime topology.

Integrating out the right-handed fermions (ψ_R) yields an effective action for left-handed fermions (ψ_L) which contains excitations similar to abelian instantons produced by a composite of ψ_L. Right-handed fermions don't have a similar effective action and therefore, quantum mechanically, the symmetry $\psi_L \leftrightarrow \psi_R$ must be broken as a result of the topology. The conserved charge associated to the topological states is quantized. Different cases with only fermionic excitations or bosonic excitations or both can occur depending on the boundary conditions and the value of the coupling.

Alternatively, one can start from an action including an auxiliary vector field which, upon integration of ψ_R, carries instanton excitations [2]. In what follows, we discuss the first approach and mention in the conclusions the differences with the second alternative.

In recent years quantization in the light cone frame has been extensively studied in connection with the discovery of new non-perturbative effects that would be unobservable in the standard spacetime quantization [3–5].

There are several reasons that make the light cone quantization a method radically different compared to the standard quantization. The dispersion relation $k_\mu k^\mu = m^2$becomes $k^+k^- = m^2$, in the light cone. This implies that the particles and antiparticles occupy disconnected sectors of momentum space. Furthermore, if the momentum k^+ is discretized by compactifying x^-, the energy k^- is also quantized and nonzero. Thus, the light cone momentum is never singular if the spacetime topology is $S^1 \times \Re$ [6].

The above comments show that the quantization in the light cone is naturally defined over a manifold with non-trivial topology. As a consequence, one could expect new physical effects originated in the implicit difference in topology from standard spacetime quantization.

From this point of view, the massless Thirring model is an example where one could investigate the new effects that emerge in the light-cone frame. We will show below that, contrary to the standard quantization, the non-trivial topology of the light-cone spacetime induces abelian instanton-like excitations, *i.e.* a purely quantum mechanical effect that appears in the calculation of the fermionic determinant.

Let us consider the massless Thirring model in the light cone frame

$$\mathcal{L} = i\psi_L^\dagger \partial_+ \psi_L + i\psi_R^\dagger \partial_- \psi_R - 2g^2(\psi_L^\dagger \psi_L)(\psi_R^\dagger \psi_R), \tag{1}$$

1) E-mail: cortes@leo.unizar.es
2) E-mail: jgamboa@lauca.usach.cl
3) E-mail: ischmidt@fis.utfsm.cl
4) E-mail: jz@cecs.cl

CP484, *Trends in Theoretical Physics II*, edited by H. Falomir, R. E. Gamboa Saraví, and F. A. Schaposnik

where $\partial_\pm = \frac{\partial}{\partial x^\pm}$ and x^+ and x^- play the role of time and space respectively.

One should note that (1) is invariant under the charge conjugation symmetry $\psi_{L,R} \leftrightarrow \psi^\dagger_{L,R}$, that is expected to be conserved at the quantum level. However in order to quantize the system following the path integral methods it is more convenient to write (1) as follows

$$\begin{aligned}\mathcal{L} &= \psi^\dagger_R(i\partial_- - 2g^2\psi^\dagger_L\psi_L)\psi_R + \psi^\dagger_L(i\partial_+)\psi_L,\\ &= \psi^\dagger_R(i\partial_- + 2A_-)\psi_R + \psi^\dagger_L(i\partial_+)\psi_L.\end{aligned} \tag{2}$$

where $A_- = -g^2\psi^\dagger_L\psi_L$.

The symmetry under charge conjugation including the auxiliary field A_- is now

$$\psi^\dagger_L \leftrightarrow \psi_L, \quad \psi^\dagger_R \leftrightarrow \psi_R, \quad A_- \to -A_-. \tag{3}$$

Thus, the generating functional,

$$Z = \int \mathcal{D}\psi^\dagger_R \mathcal{D}\psi_R \mathcal{D}\psi^\dagger_L \mathcal{D}\psi_L \; e^{iS}, \tag{4}$$

after integrating over the right handed fields is

$$Z = \int \mathcal{D}\psi^\dagger_L \mathcal{D}\psi_L \det(i\partial_- + 2A_-) \; e^{i\int dx^+dx^- \psi^\dagger_L i\partial_+\psi_L}, \tag{5}$$

where $i\partial_- + 2A_-$ is a one-dimensional Dirac operator. It should be noted that $A_- = A_-(x^-, x^+)$ is a function of two variables. The eigenvalues λ_n of the equation

$$[i\partial_- + 2A_-]\varphi_n = \lambda_n\varphi_n, \tag{6}$$

are parametric functions of x^+, which can be determined integrating (6) with x^+ fixed.

On the other hand, as $i\partial_- + 2A_-$ describes a fermionic system, usual practice would be to solve (5) using antiperiodic boundary conditions. However, the periodic topology of x^- also allows for twisted boundary conditions

$$\psi(x^+, \sigma) = e^{2\pi i\gamma(x^+)}\psi(x^+, 0), \tag{7}$$

where the real parameter γ should be fixed by quantum consistency.

The solution of (6) is

$$\varphi_n(x^+, x^-) = e^{i\int_0^{x^-} dy[2A_-(x^+,y) - \lambda_n]}\chi_0, \tag{8}$$

where χ_0 is a Grassmann spinor independent of x^- . Using (7) the eigenvalues are found to be

$$\lambda_n = -\frac{2\pi(\gamma + n)}{\sigma} + \frac{2}{\sigma}\int_0^\sigma dx^- A_-. \tag{9}$$

Using ζ-function regularization, the determinant is found to be

$$\Gamma_\gamma(A) = \det(i\partial_- + 2A_-) = \mathcal{N}\sin(\Phi - \pi\gamma), \tag{10}$$

where $\Phi = \int_0^\sigma dx^- A_-$ is the "magnetic flux" produced along the surface $x^+ = const.$ and $\mathcal{N}$ is a normalization constant.

Thus, the effective quantum theory for the left handed fields reads

$$Z = \int \prod_{x^+} \mathcal{D}\psi^\dagger_L \mathcal{D}\psi_L \; \Gamma_\gamma(A) \; e^{i\int dx^+dx^- \psi^\dagger_L(i\partial_+)\psi_L}, \tag{11}$$

with $\Gamma_\gamma(A)$ given by (10).

In order to preserve quantum mechanically the charge conjugation symmetry, the boundary conditions have to be consistent with the transformation $\psi^\dagger \leftrightarrow \psi$. This requires that

$$e^{2\pi i\gamma} = e^{-2\pi i\gamma} \tag{12}$$

and then γ must be an integer (periodic boundary conditions) or a half-integer (antiperiodic boundary conditions). The next step is to observe that the boundary condition (7) is invariant under the shift $\gamma \to \gamma + 1$. This implies

$$\Gamma_\gamma(A) = \Gamma_{\gamma+1}(A), \tag{13}$$

which requires $\sin(\Phi - \pi\gamma) = \sin(\Phi - \pi(\gamma + 1))$, or equivalently

$$\sin(\Phi - \pi\gamma) = 0. \tag{14}$$

In view of the assertion (12), the only consistent solutions of (14) are:

$$\Phi = (m + \frac{1}{2})\pi, \quad \gamma = m' + \frac{1}{2}, \quad \text{with} \quad m, m' \in \mathbf{Z}, \tag{15}$$

or,

$$\Phi = m\pi, \quad \gamma = m', \quad \text{with} \quad m, m' \in \mathbf{Z}. \tag{16}$$

In summary, the Thirring fields obey either antiperiodic boundary conditions with half-integer flux, or periodic boundary conditions with integer flux.

As a consequence of (14) the fermionic determinant satisfies the condition

$$\Gamma_\gamma(A) = \Gamma_{-\gamma}(-A). \tag{17}$$

and the theory is invariant under charge conjugation. In fact, condition (14) means that the generating function for the effective theory vanishes. This should come as no surprise because, as noted by 't Hooft, the path integral for a massless fermion vanishes when the Fermi field couples to a gauge field with nontrivial topology [7] (see also [8]).

This result shows that there are two points of view to analyze this problem. Before integrating out ψ_R each Fermi field interacts with a background made out of fermions of the opposite chirality. After integrating out one of the two species, the remaining field self interacts with its own condensate $\psi^\dagger\psi$. In our discussion $\psi_L^\dagger\psi_L$ plays the role of an external background gauge field A_- with non-trivial topology interacting with the $\psi_L^\dagger$. As a consequence of the light-cone quantization, the left-right symmetry is broken because x^- is compactified but not x^+. Thus, although the starting classical action (1) is symmetric under $\psi_L \leftrightarrow \psi_R$, quantum mechanically only left-handed charge is conserved.

In fact, the conserved Noether charge associated with a rigid phase rotation of the effective action (11) is

$$Q_- =: \int_0^\sigma dx^- \psi_L^\dagger \psi_L, \tag{18}$$

while a similar conserved charge for the right-handed fermions is not defined in the effective theory. Furthermore, this charge is quantized by virtue of (15) or (16). In the previous form of the path integral (4), the right-handed charge $Q_+ =: \int_0^\sigma dx^- \psi_R^\dagger \psi_R$, is classically conserved but obeys an anomalous quantum algebra with Q_-.

At this point one could note that the expression for the determinant of the one-dimensional Dirac operator obtained by using ζ-function is not the most general expression for the regularized determinant. The general form of the regularized one-dimensional determinant is

$$\Gamma_\gamma(A) = e^{a+b\xi} \sin\xi, \tag{19}$$

where $\xi = \Phi - \pi\gamma$ and a and b are arbitrary coefficients that depend on the regularization procedure. This general form is obtained starting from the formal expression of $\Gamma_\gamma(A)$ as an infinite product of eigenvalues depending on ξ. In order to give meaning to this product one considers the logarithm as a (divergent) sum and takes enough derivatives with respect to ξ (two) until a convergent sum, which is the series representation of $\sin^{-2}\xi$, is obtained. The linear expression $a + b\xi$ gives the most general ξ-dependence due to the divergence of the one-dimensional determinant. The result of ζ-function regularization corresponds to the particular choice of the regularization parameter $b = 0$ and e^a is the normalization constant $\mathcal{N}$ in (10).

If one takes the general form (19) for the regularized one-dimensional determinant, the consistency condition (13) leads now to

$$\left[1 + e^{-b\pi}\right] \sin(\Phi - \pi\gamma) = 0 \tag{20}$$

instead of (14), and this implies once more a quantization of the flux, (15) or (16) , for $b \neq i(2n+1)$. The possibility to have a charge conjugation invariant theory without a quantized flux by choosing appropiately the regularization, $b = i(2n+1)$, deserves further investigation.

We end up this discussion by pointing out that an important information on the non-perturbative spectrum of the model can be read from the relation, $\Phi = -g^2 Q_-$, between the quantized flux Φ and the conserved charge Q_- which counts the number n_L of left-handed fermions in a fixed-x^+ surface.

In the case of periodic boundary conditions $m\pi = -g^2 n_L$; then one has $m = n_L = 0$, i.e. only neutral bosonic excitations, unless g^2/π is a rational number. If $g^2/\pi = p/q$ then the quiral charge has to be a multiple of q and the magnetic flux will be $\Phi = -\pi n_L p/q$. For q even one has bosonic excitations while for q odd the spectrum contains both bosonic and fermionic excitations.

In the case of antiperiodic boundary conditions one has $(m + 1/2)\pi = -g^2 n_L$ and g^2/π has to be a rational number with even denominator. Only states, characterized by an integer n, with a chiral charge $n_L = (2n+1)q/2$ and a flux $\Phi = -(n + 1/2)p\pi$ appear in the spectrum; for $q/2$ even one has a purely bosonic spectrum but in the case $q/2$ odd only fermionic excitations are present.

The results presented here can be summarized as follows: (i) A fermionic topological instanton-like excitation arising from the compactification of the x^-; (ii) Quantum mechanically, the massless Thirring model is analogous to the $\lambda\phi^4$ theory in 1+1 dimensions, where an abelian instanton is also possible (see *e.g.* [9]; (iii) different situations with only fermionic excitations, bosonic excitations or both can be identified ; (iv) the quantum system breaks the classical left-right symmetry.

The lesson we've learned is that a simple self-interacting fermionic system can have many more excitations than those expected perturbatively. This last statement is particularly interesting for QCD_4 at low energies where this kind of theory - v.i.z. the Nambu-Jona-Lasinio model- is expected. It remains to be proven whether a similar construction can be carried out in higher dimensions.

To end up let us point out that a quantization of the Thirring model in the light-cone compatible with the conservation of the vector current is possible if an auxiliary vector field is introduced before integrating out the fermion fields. In this formulation the non-trivial topology of space-time, due to the compatification of the x^- coordinate in the light-cone, leads to a quantization condition due to the boundary conditions. New excitations are also identified in this case as a consequence of the contribution of the zero modes. This formulation can be applied beyond the Thirring model; for an analysis of the excitations and vacuum structure of the Schwinger model and chiral Schwinger model see [2].

ACKNOWLEDGMENTS

We thank the organizers of the meeting for their kind hospitality in Buenos Aires. This work was partially supported by CICYT (Spain) project AEN-97-1680 and grants 1980788, 1960229, 1960536, 7960001, and 7980045 from FONDECYT-Chile, and DICYT-USACH. I.S. is a recipient of a Cátedra Presidencial en Ciencias-Chile. Institutional support to CECS from Fuerza Aerea de Chile and a group of private companies (Business Design Associates, CGE, Codelco, Copec, Empresas CMPC, Minera Colahuasi, Minera Escondida, Novagas, and Xerox-Chile), is also acknowledged.

REFERENCES

1. J. L. Cortés, J. Gamboa, I. Schmidt and J. Zanelli, *Phys. Lett.* **B 444** (1998) 451.
2. J. L. Cortés and J. Gamboa, hep-th/990141, *Phys. Rev.* **D** (to be published).
3. S. Brodsky, H.-C. Pauli and S. Pinsky, hep-th/9705477, *Phys. Rep.* (in press).

4. M. M. Brisudova, R. Perry and K.G. Wilson, *Phys. Rev. Lett.* **78**, 1227 (1997); S. D. Glazek and K. G. Wilson, *Phys. Rev.* **D57**, 3558 (1998).
5. D. Bigatti and L. Susskind, hep-th/9711063.
6. T. Maskawa and K. Yamawaki, *Prog. Theor. Phys.* **56** 270 (1976); S. Brodsky and H. -C. Pauli, *Phys. Rev.***32**, 1993 (1985).
7. G. 't Hooft, *Phys. Rev. Lett.***37**, 8 (1976); *i.b.i.d. Phys. Rev.***D14**, 3432 (1976).
8. J. Kiskis, *Phys. Rev.***15D**, 2329 (1977).
9. S. Coleman, *Aspects of Symmetry*, Cambridge University Press; A. M. Polyakov, *Gauge Fields and Strings*, Harwood (1988).

AUTHOR INDEX